Tectonic Uplift and Climate Change

Tectonic Uplift and Climate Change

Edited by
William F. Ruddiman
University of Virginia
Charlottesville, Virginia

Plenum Press • New York and London

Library of Congress Cataloging-in-Publication Data

On file

ISBN 0-306-45642-7

© 1997 Plenum Press, New York
A Division of Plenum Publishing Corporation
233 Spring Street, New York, N.Y. 10013

http://www.plenum.com

10 9 8 7 6 5 4 3 2 1

Printed in the United States of America

Contributors

Michael A. Arthur • Department of Geosciences and Earth System Science Center, Pennsylvania State University, University Park, Pennsylvania 16802.

Elizabeth K. Berner • Department of Geology and Geophysics, Yale University, New Haven, Connecticut 06520-8109.

Robert A. Berner • Department of Geology and Geophysics, Yale University, New Haven, Connecticut 06520-8109.

Joel D. Blum • Earth, Ecosystem and Ecological Sciences Program, Earth Sciences Department, Dartmouth College, Hanover, New Hampshire 03755.

Anthony J. Broccoli • Geophysical Fluid Dynamics Laboratory/NOAA, Princeton University, Princeton, New Jersey 08542.

Thure E. Cerling • Department of Geology and Geophysics, University of Utah, Salt Lake City, Utah 84112.

Peter Copeland • Department of Geosciences, University of Houston, Houston, Texas 77204-5503.

Donald J. DePaolo • Center for Isotope Geochemistry, Department of Geology and Geophysics, University of California at Berkeley, Berkeley, California 94720-4767

Louis A. Derry • Department of Geological Sciences, Cornell University, Ithaca, New York 14853-1504.

John M. Edmond • Department of Earth, Atmospheric, and Planetary Sciences, Massachusetts Institute of Technology, Cambridge, Massachusetts 02139.

Johan P. Erikson • Department of Geology, University of California at Davis, Davis, California 95616.

Christian France-Lanord • Centre de Recherches Petrographiques et Geochimiques-CNRS, BP20 54501, Vandoeuvre-les-Nancy, France.

Youngsook Huh • Department of Earth, Atmospheric, and Planetary Sciences, Massachusetts Institute of Technology, Cambridge, Massachusets 02139.

Teresa E. Jordan • Department of Geological Sciences, Cornell University, Ithaca, New York 14853-1504.

Lee R. Kump • Department of Geosciences and Earth System Science Center, Pennsylvania State University, University Park, Pennsylvania 16802.

John E. Kutzbach • Center for Climatic Research, University of Wisconsin-Madison, Madison, Wisconsin 53706.

Syukuro Manabe • Geophysical Fluid Dynamics Laboratory/NOAA, Princeton University, Princeton, New Jersey 08542.

Sean E. McCauley • Center for Isotope Geochemistry, Department of Geology and Geophysics, University of California at Berkeley, Berkeley, California 94720-4767.

John D. Milliman • School of Marine Science, College of William and Mary, Gloucester Point, Virginia 23062.

Timothy C. Partridge • Climatology Research Group, University of the Witwatersrand, WITS 2050, Johannesburg, South Africa.

William J. Pegram • Department of Geology and Geophysics, Yale University, New Haven, Connecticut 06520-8109.

Warren L. Prell • Department of Geological Sciences, Brown University, Providence, Rhode Island 02912.

I. Colin Prentice • School of Ecology, Lund University, 223 62 Lund, Sweden.

Maureen E. Raymo • Department of Earth, Atmospheric, and Planetary Sciences, Massachusetts Institute of Technology, Cambridge, Massachusetts 02139.

James H. Reynolds III • Department of Geosciences and Anthropology, Western Carolina University, Cullowhee, North Carolina 28723.

David Rind • Goddard Space Flight Center, Institute for Space Studies, New York, New York 10025.

Gary Russell • Goddard Space Flight Center, Institute for Space Studies, New York, New York 10025.

William F. Ruddiman • Department of Environmental Sciences, University of Virginia, Charlottesville, Virginia 22903.

Karl K. Turekian • Department of Geology and Geophysics, Yale University, New Haven, Connecticut 06520-8109.

Preface

Plenum's initial inquiry two years ago about a volume on tectonics and climate came at just the right time. Although Plenum had in mind a broad overview treatment of this subject, I felt there was good reason to put together a volume focused entirely on one such connection, the link between plateau–mountain uplift and global climate over the last 50 million years. A series of papers published in the middle–late 1980s had proposed several linkages between uplift and climate, and these hypotheses had been subjected for almost a decade to the usual critical scrutiny by the scientific community. It was my not unbiased sense that the original hypotheses had emerged from this scrutiny largely intact (though, of course, not entirely), and that their scope had in fact been expanded by the subsequent addition of new concepts. If these earlier phases represented the "thesis" and "antithesis" phases of the dialectical method of science, the time now seemed ripe for an attempt at a "synthesis." This seemed particularly desirable because the many publications on this topic lay scattered among disciplinary journals, and there was no central source to provide a complete overview. This volume attempts to fill that need.

Although focused on a single set of hypotheses, this volume is far from narrow in scope. Indeed, it encompasses most of the massive and dramatic transformations of the Earth's surface in recent geologic history. These include: the collision of continents, the uplift of massive plateaus and mountain belts, changes in the position of the jet stream and westerly winds, the creation of monsoon circulations that focus heavy rainfall on uplifted terrain, rapid and intense physical weathering of rocks in plateaus and mountain belts, runoff of sediment-laden rivers to the ocean, chemical weathering of rock and slow removal of CO_2 from the atmosphere, gradual cooling of global climate, formation of permanent ice sheets over Antarctica and Greenland, development of sea-ice cover in the Arctic Ocean, expansion of tundra and boreal forest southward from the Arctic margins of Asia and North America, replacement of tree and shrub vegetation by grassland in the subtropics, and, finally, the onset of periodic fluctuations of massive ice sheets over North America and Eurasia. A story that begins with tectonic uplift in the tropics and subtropics thus ends with glaciation of the polar regions.

For teaching-related purposes, this volume is unique in combining a broad interdisciplinary scope with a close focus on a single central issue (the uplift–

climate connection). It will probably prove most useful in graduate or advanced undergraduate seminars or in combined lecture–seminar courses. Many of the individual chapters could be the basis for a detailed investigation of the methods used and results obtained from specific Earth Sciences disciplines (such as geochemistry, atmosphere and ocean modeling, tectonophysics, and paleobotany). Yet the results from each chapter fit into a larger picture that will expand the breadth of vision of students who are too often focused only on one method or disciplinary area. This approach is well matched to an ongoing trend evident in most research universities, which are creating or encouraging new alignments among component departments in the Earth and Environmental Sciences in an effort to stimulate interdisciplinary research. I believe this volume represents a highly successful example of the kind of research that could emerge from such efforts.

The value of this volume as a reference source for researchers is obvious. All the chapters are current to the very end of 1996, and all the authors are recognized experts in their fields. It was my choice to pick the best people and give them the freedom of a nonrefereed volume, both to encourage cross-disciplinary thinking (and even speculation) and to speed the volume toward timely publication. The most obvious omission in the book is the absence of chapters on the timing of uplift in North America; these were solicited early, but fell through too late in the process for me to be able to obtain others.

I thank the following people for help and encouragement in seeing this project through to rapid completion: the authors, who all did their job in a timely way; Ken Howell, who nudged and nagged at about the right level of frequency and intensity; long-term colleagues Maureen Raymo, John Kutzbach, and Warren Prell, for past and present scholarly collaboration that has been both educational and enjoyable; and both Ginger and Debra Angelo, for literally making this volume possible.

W. F. Ruddiman
Charlottesville

Contents

Chapter 3
Variability in Age of Initial Shortening and Uplift in the Central Andes, 16–33°30′S
Teresa E. Jordan, James H. Reynolds III, and Johan P. Erikson

Chapter 4
Late Neogene Uplift in Eastern and Southern Africa and its Paleoclimatic Implications
Timothy C. Partridge

Part III. General Circulation Model Studies of Uplift Effects on Climate

Chapter 5
Mountains and Midlatitude Aridity
Anthony J. Broccoli and Syukuro Manabe

Chapter 6
The Effects of Uplift on Ocean–Atmosphere Circulation
David Rind, Gary Russell, and William F. Ruddiman

Chapter 7
Possible Effects of Cenozoic Uplift and CO_2 Lowering on Global and Regional Hydrology
John E. Kutzbach, William F. Ruddiman, and Warren L. Prell

Chapter 8
**The Impact of Tibet–Himalayan Elevation on the Sensitivity of the
Monsoon Climate System to Changes in Solar Radiation**
Warren L. Prell and John E. Kutzbach

Chapter 9
**Testing the Climatic Effects of Orography and CO_2 with
General Circulation and Biome Models**
William F. Ruddiman, John E. Kutzbach, and I. Colin Prentice

Part IV. Geological and Geochemical Evidence of Uplift Effects on Weathering and CO$_2$

Chapter 10
Fluvial Sediment Discharge to the Sea and the Importance of Regional Tectonics
John D. Milliman

Chapter 11
The Effect of Late Cenozoic Glaciation and Tectonic Uplift on Silicate Weathering Rates and the Marine ^{87}Sr/^{86}Sr Record
Joel D. Blum

Chapter 12
Himalayan Weathering and Erosion Fluxes: Climate and Tectonic Controls
Louis A. Derry and Christian France-Lanord

Chapter 13
Late Cenozoic Vegetation Change, Atmospheric CO_2, and Tectonics
Thure E. Cerling

Chapter 14
Chemical Weathering Yields from Basement and Orogenic Terrains in Hot and Cold Climates
John M. Edmond and Youngsook Huh

Chapter 15
Silicate Weathering and Climate
Robert A. Berner and Elizabeth K. Berner

Chapter 19
The Marine $^{87}Sr/^{86}Sr$ and $\delta^{18}O$ Records, Himalayan Alkalinity Fluxes and Cenozoic Climate Models
Sean E. McCauley and Donald J. DePaolo

Part V. Synthesis

Chapter 20
The Uplift–Climate Connection: A Synthesis
William F. Ruddiman, Maureen E. Raymo, Warren L. Prell,
and John E. Kutzbach

Tectonic Uplift and Climate Change

Introduction

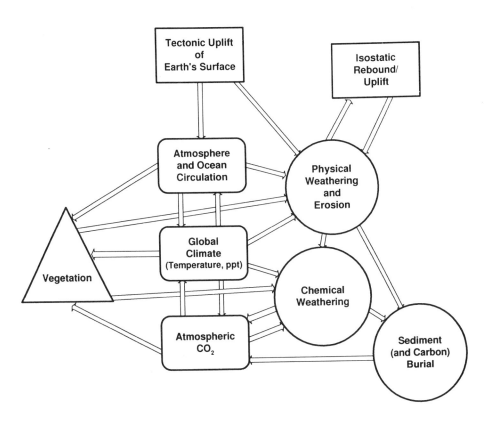

Introduction to the Uplift–Climate Connection

William F. Ruddiman and Warren L. Prell

1. INTRODUCTION

Studying the history of the Earth is inherently a highly interdisciplinary task requiring knowledge in many fields of science. This book addresses a provocative interdisciplinary Earth-history problem: the nature of the linkage between uplift of the Earth's surface and its climatic history. This elusive relationship has interested those studying the history of the Earth for over a century and has in recent years emerged as a focus of wide interest among a large number of scientists.

Uplift and climate change are intricately linked, as this volume will show. (Unless otherwise specified, we use the term "uplift" in this chapter to refer to the raising of the Earth's surface across broad plateau or mountain regions). Uplift of the Earth's surface to higher elevations can alter regional and global climate through a number of physical and chemical mechanisms that involve changes in the Earth's atmospheric and oceanic circulation.

Conversely, climate change, whether driven by uplift or other factors, can alter atmospheric and oceanic circulation in such a way as to change rates and styles of erosion in high mountain terrain. Increased erosion can strip rock layers from high-elevation regions, causing a rebound of the underlying layers in response to removal of the weights of the overburden. This rebound of the Earth's crust is referred to as "isostatic" uplift. Although the remaining layers of rock are in this way uplifted with respect to their previous positions, their rebound only partially compensates for erosional removal of the overlying layers and thus leaves the regional mean surface of the Earth lower than it had been.

This is but one example of the complexities inherent in the intricate relationship between uplift and climate. Addressing these complexities thus

William F. Ruddiman • Department of Environmental Sciences, University of Virginia, Charlottesville, Virginia 22903. *Warren L. Prell* • Department of Geological Sciences, Brown University, Providence, Rhode Island 02912.

Tectonic Uplift and Climate Change, edited by William F. Ruddiman. Plenum Press, New York, 1997.

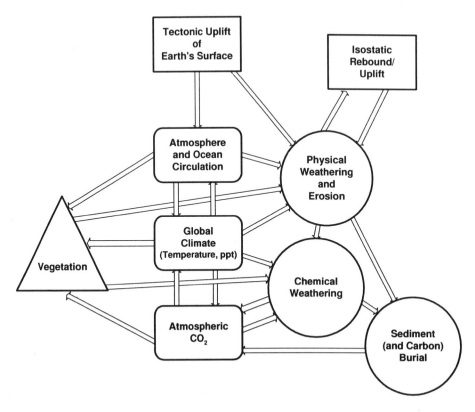

FIGURE 1.   Potential interactions and feedbacks between tectonic uplift and the Earth's climatic and environmental system. Rectangles indicate parts of the Earth's system related to tectonics (uplift); rounded rectangles indicate parts of the Earth's system related to atmospheric and oceanic circulation and climate; circles indicate parts of the Earth's system related to weathering and erosion; and the triangle indicates the vegetational component of Earth's system.

requires adequate knowledge of all the major components of the outer envelope of the Earth: changes in its "solid-rock" configuration (the lithosphere and uppermost asthenosphere); variations in temperature, wind, sensible and latent heat, and CO_2 levels (the atmosphere); fluctuations of water vapor, precipitation, runoff and ocean circulation (the hydrosphere); changes in the erosional and/or climatic impacts of mountain glaciers, continental ice sheets, and sea ice (the cryosphere); and variations in the amount and type of vegetation and carbon on Earth (the biosphere).

Resolving the complexities of the uplift–climate connection also requires insight into the operation of a large array of linkages among several parts of the Earth's system, the most critical of which are summarized schematically in Fig. 1. This volume examines a wide array of evidence pertinent to these linkages in an effort to evaluate the importance of the uplift–climate connection to Earth history.

This chapter consists of three main sections. Section 2 reviews key components of a series of related papers that initially posed several uplift–climate hypotheses; Section 3 summarizes several criticisms and controversies subsequently provoked by these hypotheses; and Section 4 outlines the way in which the individual chapters of this book explore various aspects of the uplift–climate problem.

Following this introduction are 18 chapters that present evidence pertinent to critical connections between uplift and climate. Each focuses on a subset of the array of climate–tectonic linkages identified in Fig. 1. In order to highlight the specific subject areas and linkages under investigation, Fig. 1 is reproduced at the start of each chapter with stippling added to point out the subject area discussed.

The synthesis chapter that ends the book brings together the new evidence put forward in this volume into a critical overview of the current state of the uplift–climate problem. It also targets key issues that need further investigation.

2. INITIAL UPLIFT–CLIMATE HYPOTHESES

Uplift of plateaus and mountains has been proposed as a cause of large-scale changes in climate. Two basic categories of uplift effects on climate are recognized: (1) direct physical impacts on climate by means of changes in the circulation of the atmosphere and ocean; and (2) indirect biochemical effects on climate via changes in atmospheric CO_2 and global temperature caused by chemical weathering of silicate rocks.

2.1. Physical Effects of Uplift on Climate

The uplift–climate hypotheses invoke several physical effects that result from plateau–mountain uplift and that have potentially important impacts on Earth's climate (Fig. 2).

2.1.1. Lapse-Rate Cooling of High-Elevation Surfaces

One effect of plateau uplift is the creation of an elevated surface at cooler upper levels in the atmosphere (Fig. 2a). The planetary lapse rate (6.5°C cooling for each kilometer increase in elevation) causes elevated plateaus to be much colder than nearby low-lying terrain, especially in winter when snow cover may further lower temperatures by albedo-temperature feedback. Older geologic hypotheses that invoked uplift effects on atmospheric circulation[1–3] mainly focused on smaller mountainous regions such as the European Alps or Apenines, now known to have had at most localized cooling effects and no larger-scale impact. Subsequent publications during the mid-1900s[4–6] focused mainly on smaller-scale uplift in high-latitude regions as a potential mechanism for nucleating continental ice sheets. Birchfield and Weertman[7] noted, however, that geologically youthful uplift of the massive Tibetan Plateau must have provided a

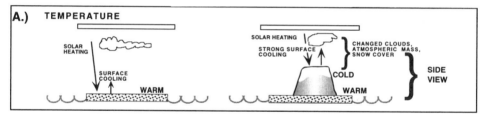

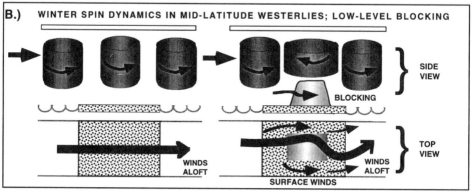

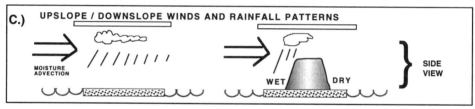

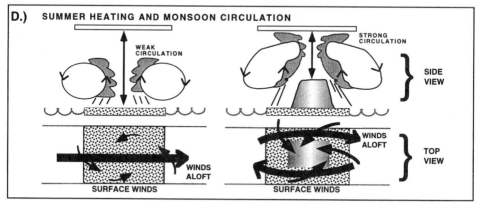

FIGURE 2. Schematic representation (after Kutzbach *et al.*[22]) of the effects of plateau and mountain uplift on atmospheric circulation: (a) Cooling of high-elevation surface. (b) Diversion of winter westerly winds (both at surface and jet stream levels) by high-plateau–mountain barrier. (c) Alteration of precipitation patterns by high orographic barrier intercepting moisture-bearing winds. (d) Monsoonal impact of summer heating of high plateau on large-scale atmospheric circulation and precipitation.

really extensive surface on which snow cover would form in winter, thus cooling the climate on a much larger scale.

2.1.2. Diversion–Amplification of Jet Stream Meanders

A second large-scale physical effect of uplift on the Earth's atmosphere (and ocean) circulation is an intensification of meandering of westerly atmospheric flow at lower middle latitudes, where much of the high-plateau terrain occurs (Fig. 2b). This effect encompasses both the lower-level westerly flow near the Earth's surface and the Rossby waves in the upper-level jet stream flow. This increase in amplitude of the meanders occurs because the high topography forms an obstacle to direct zonal (west-to-east) circulation in the lower troposphere, and because conservation of potential vorticity propagates this low-level diversion to higher (jet stream) altitudes. As a result, westerly circulation is diverted in a broad poleward meander around the uplifted rock obstacles, and broad meanders form upstream and downstream from the high terrain. In addition, orographically enhanced rainfall occurs on the windward side of the mountains, and rainshadow drying on their lee sides (Fig. 2c).

This effect of uplift on meander intensity was initially demonstrated numerically in the 1960s and 1970s by experiments with general circulation models developed to reproduce key features of the Earth's atmospherc circulation.[8–10] Ruddiman et al.[11] pointed out that this theoretical knowledge gained from the atmospheric sciences is directly relevant to geologically recent Earth history, because the huge area of high terrain in Tibet, as well as several other elevated regions on Earth, have been uplifted to their present elevations within the last 40 million years. A comprehensive series of related papers[12–14] then showed that this immense scale of tectonic uplift is capable of altering basic atmospheric circulation patterns across the entire Northern Hemisphere and parts of the Southern Hemisphere as well. Subsequent general circulation model papers investigated other aspects of this problem.[15–17]

2.1.3. Creation–Intensification of Monsoon Circulations

The third direct effect of uplift on the atmosphere (Fig. 2d) is the creation or intensification of seasonally reversing monson circulations, particularly the powerful South Asian monsoon. Summer monsoons are caused by the heating of the land surface by the strong summer sun, rising of the heated air, and compensating inflow of moist air from the ocean, with explosive release of latent heat during the heavy rains that fuel the most intense monsoonal circulation. This summer monsoon circulation is hemispheric in scale: upward motion in low-pressure cells over the plateaus in summer is compensated by broad subsidence of dry air in high-pressure cells over the oceans. The South Asian monsoon is also linked to Southern Hemisphere circulation in the Indian Ocean sector by cross-equatorial wind flow, which contributes most of the latent heat released in the South Asian monsoon.[18]

The monsoonal circulation in winter is directly opposite in sense to that in summer, with radiative cooling of the elevated land mass, sinking of cold dry air over the continent, and outflow of this air over the ocean. The winter monsoon flow is further enhanced by the increased snow cover that develops at high altitudes owing to cooling associated with the prevailing lapse rate. The winter monsoon circulation is also hemispheric in scale: downward motion in the high-pressure cells over the plateaus and nearby areas in winter is compensated by upward motion in low-pressure cells over the oceans.

This large-scale effect of high terrain on the seasonal monsoon circulations was first demonstrated by experiments with general circulation models by Hahn and Manabe.[19] The potential relevance of such experiments to the late Cenozoic history of uplift in southern Asia was briefly alluded to in Prell[20] and was fully investigated for the first time in the trilogy of uplift papers published in 1989.[12-14] Aspects of the impact of Tibetan uplift on the Southeast Asian climate have been more thoroughly examined and quantified in subsequent studies.[21-22]

2.2. Biogeochemical Effects of Uplift on Climate via CO_2

Uplift may also have an indirect but potentially more extensive effect on global climate through the action of the carbon cycle (Fig. 3). Chemical weathering of silicate rocks on land is the primary long-term sink for atmospheric CO_2. CO_2 from the atmosphere combines with groundwater supplied by rainfall to form carbonic acid (H_2CO_3), which slowly attacks silicate rocks by hydrolysis. With the complex compositional array of silicate rocks on the continents simplified to $CaSiO_3$ (a pyroxene mineral), the net reaction that extracts CO_2 from the atmosphere is:

$$CaSiO_3 + H_2CO_3 \rightarrow CaCO_3 + SiO_2$$

| silicate | ground- | ocean | ocean |
| rock | water | plankton | plankton |

Raymo and colleagues[23-24] proposed that tectonic uplift could increase rates of removal of CO_2 from the atmosphere owing to the combined effects of several factors inherent to regions of uplifted terrain: (1) active faulting that exposes fresh unweathered rock; (2) vigorous high-altitude mechanical weathering owing to the steep slopes, lack of vegetation, and vigorous periglacial–glacial weathering processes; and (3) strong summer precipitation from monsoon rains focused on the windward margins of high plateaus or from orographic rainfall in summer or winter on the windward slopes of mountain ranges. These factors generate a large volume of unweathered bedrock in a highly pulverized form that promotes rapid chemical weathering in the wet environment. Chemical weathering continues as the debris is carried downslope by rivers and spread across floodplains and deltas at lower elevations in warmer and wetter climatic regimes.

The basic outline of this concept was proposed almost a century ago by Chamberlin,[25] who had also noted the apparent correlation between mountain

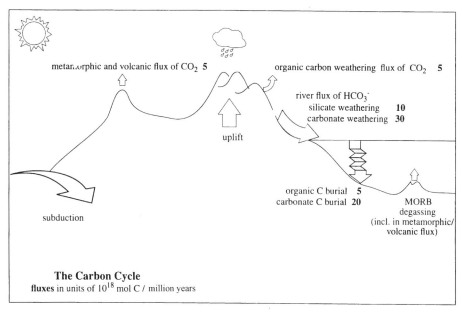

metamorphic and volcanic flux of CO_2 5   organic carbon weathering flux of CO_2 5

river flux of HCO_3^-
silicate weathering   **10**
carbonate weathering   **30**

uplift

organic C burial   **5**
carbonate C burial **20**   MORB
degassing
(incl. in metamorphic/
volcanic flux)

subduction

The Carbon Cycle
fluxes in units of 10^{18} mol C / million years

FIGURE 3. Schematic representation (from Chapter 18) of processes hypothesized to control CO_2 levels in the atmosphere over long tectonic (million-year or longer) time scales: (1) rates of seafloor spreading (controlling rates of CO_2 input from subduction-zone volcanoes and along the midocean ridge system; and (2) rates of chemical weathering of silicate rock on land (controlling CO_2 removal in recently uplifted mountain belts). Estimated fluxes between carbon reservoirs are indicated.

building and glacial climates over geologic history. Chamberlin's complete climatic hypothesis was, however, complex and multifaceted, and it included several other speculative components now known to be incorrect. As a result, the central core of his hypothesis, the uplift–weathering connection, was ignored for most of this century, until Raymo[23] independently devised a similar hypothesis that called on lowered CO_2 to explain an array of newly acquired geochemical data.

The uplift–weathering hypothesis challenges the leading explanation of long-term CO_2 changes: the BLAG hypothesis, an acronym based on the initials of the authors.[26] This hypothesis is based on the fact that the main input of CO_2 to the atmosphere is associated with metamorphism and volcanic degassing along subduction zones and at midocean ridges (Fig. 3). The BLAG hypothesis therefore assumes that global mean rates of seafloor spreading and subduction are the main controls on delivery of CO_2 to the atmosphere, and thus the primary driver of changes in atmospheric CO_2 levels through time.

The BLAG hypothesis calls on chemical weathering for negative feedback to keep the climate system in near-equilibrium. It is hypothesized that this feedback affects broad areas of the Earth's surface in the following way: faster spreading delivers more CO_2 to the atmosphere; this warms the Earth, increases water vapor and vegetation levels, and speeds up chemical reaction rates; these changes increase the rate of global chemical weathering of silicate rock and thus offset most of the spreading-generated CO_2 increase. Conversely, when spreading rates are slow, less CO_2 is delivered to the atmosphere, chemical weathering slows and removes less CO_2, and this keeps the system close to equilibrium. This "thermostat" role for chemical weathering was first conceived by Walker and colleagues.[27]

In contrast, the uplift–weathering hypothesis elevates chemical weathering from a weak negative feedback to the major factor in reducing long-term CO_2 levels and thus cooling global climate. It assumes that massively increased chemical weathering localized to regions of uplifted terrain is the critical factor that controls CO_2 and climate.

3. CHALLENGES TO THE UPLIFT–CLIMATE HYPOTHESES

Several challenges to the uplift–climate hypotheses have arisen in the years since their publication, as summarized below.

3.1. Misinterpretations of the Extent of True (Surface) Uplift

A major challenge to the original uplift–climate hypotheses came from Molnar and England,[28-29] who suggested that many of the geological and paleobotanical studies that have postulated youthful (late Cenozoic) uplift of the Earth's surface are mistaken. They argued that geomorphologists have often mistaken late Cenozoic increases in erosional denudation of high terrain as being caused by concurrent uplift of the Earth's surface, when the erosion may in fact have been caused by climatic changes imposed on terrain that had already been at high elevations for millions of years. In the Molnar–England view, these studies mistakenly inferred surface uplift in places actually undergoing isostatic rebound owing to erosional removal of overburden.

Molnar and England also criticized paleobotanical studies postulating that late Cenozoic shifts toward cold-tolerant vegetation in high-plateau and mountain regions must be due entirely to uplift. Instead, they suggested that the observed vegetation changes must at least in part reflect the superimposed effects of late Cenozoic climatic cooling.

In the end, however, Molnar and England[28] acknowledged that true surface uplift may have been an important factor driving climate change during the Cenozoic. They note that uplift of the immense Tibetan Plateau has certainly occurred within the last 50 million years in association with the well-constrained plate tectonic collision of India and Asia. They further speculate that a powerful positive feedback loop could exist between uplift and climate: true surface uplift in areas like Tibet could alter hemispheric or global climate; this climatic change could in turn cause enhanced erosion of high terrain in areas not at that point tectonically active; and this increase in erosion could then cause isostatic rebound in older mountain belts that had been tectonically inactive. These additional increases in erosion could then further increase global mean rates of chemical weathering, which could further lower CO_2 levels and thus further alter climate, and so on. In short, true surface uplift occurs initially in tectonically active plateau and mountain regions, but then erosion progressively penetrates into, and "reactivates," tectonically inactive mountain belts, as part of an ongoing process that drives the planet progressively deeper into glacial conditions.

In any case, the criticisms in the Molnar–England papers raise valid questions about the extent and timing of Cenozoic uplift in many regions that stand at high elevations today, and about modeling studies based on assumed elevational histories.[12–14] These issues are further investigated in this volume.

3.2. Interpretation of Chemical Weathering Proxies

A second challenge to the uplift–climate hypotheses is tied to the global rate of chemical weathering. Raymo and colleagues[23] invoked the oceanwide $^{87}Sr/^{86}Sr$ ratio in calcareous marine sediments as an index of global chemical weathering rates through the Cenozoic. They argued that the abrupt increase in this ratio over the last 40 million years reflects increased chemical weathering of continental rocks with high $^{87}Sr/^{86}Sr$ values. They inferred that massively increased chemical weathering in the rapidly uplifting Himalayan–Tibetan region has driven a global-scale increase in chemical weathering that has caused CO_2 levels to fall.

The other factor thought capable of driving rapid changes in the $^{87}Sr/^{86}Sr$ ratio is seafloor spreading, which delivers to the ocean hydrothermal fluids with a low $^{87}Sr/^{86}Sr$ ratio along the midocean ridge system. This input is thought to have been nearly constant for the last few tens of millions of years and thus is not regarded as a key factor in the rapid rise of the $^{87}Sr/^{86}Sr$ ratio over the last 40 million years.[30]

Edmond,[31] however, challenged the interpretation that increased global chemical weathering explains the rapid late Cenozoic rise in the $^{87}Sr/^{86}Sr$ ratio. Instead, he argued that the rise reflects erosion of Himalayan source rocks that had become anomalously enriched in ^{87}Sr. In Edmond's view, the crucial explanatory factor is thus changes in the $^{87}Sr/^{86}Sr$ ratio of the material being weathered, rather than in the rates of weathering.

Richter *et al.*[30] found that the most plausible explanation of the abrupt rise in the oceanwide $^{87}Sr/^{86}Sr$ ratio over the last 40 million years is a roughly equal mixture of increased chemical weathering fluxes in the Himalayan–Tibetan region and higher Sr ratios in the runoff from this region. This interpretation means that the $^{87}Sr/^{86}Sr$ ratio cannot be taken as a direct index of global (or even South Asian) chemical weathering. This important issue is also investigated further in this volume.

3.3. Lack of Negative Feedback in the Uplift–Weathering Hypothesis

Another challenge to the uplift–weathering hypothesis is that it lacks a natural feedback process to keep climate within moderate bounds.[32–34] For the late Cenozoic, this criticism reduces to a specific question: Why doesn't the proposed increase in chemical weathering of silicate rocks pull all of the CO_2 out of the small atmospheric reservoir and freeze the Earth?

In the BLAG model,[26] changes in volcanic input of CO_2 drive the basic direction of the CO_2 and climatic responses, but then the climate-sensitive weathering feedback[27] kicks in as a "thermostat" to moderate changes in CO_2 and climate. The uplift–weathering model, in contrast, calls on weathering to drive the system in the first place. This would seem to leave volcanic CO_2 input as the strongest remaining source of a negative CO_2 feedback, and yet there is no plausible mechanism known by which volcanic input would respond to uplift–driven CO_2 changes as a negative feedback. In addition, based on spreading-rate evidence, no net change in CO_2 input from volcanic sources is estimated over the last 30 million years, and so this factor cannot have offset the proposed CO_2 decreases owing to uplift-induced chemical weathering.

Thus the central question remains: What has moderated climate during this interval? An increased flux of CO_2 from the sedimentary carbon reservoir has been invoked to counteract much of the CO_2 loss via silicate weathering during the last 20 million years,[24] but even the sign of carbon transfer to/from this reservoir is debated.[33] Sundquist[34] suggested that the CO_2-driven global Cenozoic cooling would slow the rates of chemical weathering (and CO_2 extraction) in regions outside the activity uplifting orogenic belts, and that this might provide the needed negative feedback. In this view, chemical weathering is thus both the driver of climate change (via increased extraction of CO_2 in areas of active uplift) and the "thermostat" mechanism that moderates the degree of change (via decreased CO_2 extraction in regions outside areas of active uplift). These and other possible feedbacks to the uplift–weathering hypothesis will be considered in this book.

4. OUTLINE OF THE VOLUME

The hypotheses and criticisms reviewed above provide a background context for understanding much of the focus of this book. In this section, we briefly note how individual chapters in the book address these issues.

4.1. Uplift Histories: True Surface Uplift versus Isostatic Rebound

As previously noted, it is critical to distinguish between true surface uplift driven by tectonic factors and the partial restorative isostatic uplift that occurs in response to erosional removal of overlying material. The first section of this book (Chapters 2–4) addresses the issue of the amount of Cenozoic surface uplift in some of the highest and broadest plateau or mountain areas on Earth.

Southeast Asia (Chapter 2) and the central Andes (Chapter 3) are regions of predominant compressional tectonics, and so have been areas of true surface uplift during the Cenozoic. In southern Asia, a recent hypothesis suggests that uplift may not have continued to the present, and that the highest parts of Tibet may even have lost some elevation subsequent to a spurt of rapid uplift about 8 million years ago.[35] In the extensional tectonic regime of East Africa (Chapter 4), Cenozoic surface uplift has been associated with volcanic construction and deep-seated thermal doming.

Unfortunately, two invited chapters addressing the highly contentious issue of the timing of uplift in the American West failed to materialize. In one view, a Cordilleran-style mountain chain in the Far West was created during the Mesozoic by subduction, and the region of high terrain then expanded eastward during the Laramide orogeny 75–45 million years ago.[36] In this view, increased erosion in the America West during the last 20 million years is due to climate change, not to tectonic uplift.[28] In the other view, a broad-scale updoming of much of western North America has occurred mainly in the last 20 million years owing to processes originating in the upper mantle.[37]

Underlying these chapters dealing with uplift histories are fundamental questions about the relationship between plate tectonics and high-plateau–mountain orography. Is the collision of continents indeed so rare in Earth's history that the existence of a huge orographic mass like Tibet is the exception rather than the rule? Conversely, are seafloor spreading and subduction of ocean lithosphere under continents sufficiently persistent through time that they have created Cordilleran-like mountain chains along continental margins at roughly steady-state rates throughout the plate tectonic history of the Earth? And how common (or rare) in geologic history are volcanic constructional episodes related to continental hot spots and mantle plumes that build edifices comparable to the youthful East African Rift Valley and the older South African Plateau? These questions are all related to the larger issue of whether modern orography is unusual compared to the norm across geologic history.

4.2. New General Circulation Modeling Opportunities

The second section of this book (Chapters 5–9) covers new modeling studies of the climatic impacts of uplift. General circulation models (GCMs) are the most complete models of the Earth's circulation available and thus offer a useful means of testing the regional and global impacts of uplift and CO_2 on climate. GCMs have continually evolved toward more complete representations of Earth's climate system, including: (1) a trend toward smaller gridbox sizes, which reproduce more detail of boundary-condition input parameters such as topographic texture, and (2) addition of interactive features such as snow cover, sea ice, and hydrology, which had been set at fixed (modern) values in earlier models.

Chapter 5 examines a number of mechanisms through which higher orography can cause pervasive drying at middle latitudes. Chapter 6 uses a newly developed coupled ocean–atmosphere model to assess the effect of Tibetan uplift on atmospheric and oceanic circulation, particularly in the high-latitude North Atlantic sector. Chapter 7 looks at the impact of uplift on the hydrologic cycle, including the amount of runoff from each continent to each ocean basin. It also examines the combined effect of plausible changes in orography and CO_2 on global hydrology and on extreme rainfall–runoff events in South Asia. Chapter 8 explores the interaction between tectonic uplift and orbital-scale forcing of the monsoon to see whether uplift alters the sensitivity of the Asian monsoon system to orbital-scale changes in summer insolation. Chapter 9 surveys the separate climatic effects of rising orography and decreasing CO_2 and evaluates the combined effects of plausible changes in both factors during the last 20 million years. The Biome model,[38] which converts climate data output from GCMs into plant-functional types, is then used to assess the extent to which uplift and CO_2 can explain late Cenozoic changes in vegetation in the Northern Hemisphere.

4.3. Physical and Chemical Weathering

The third and largest section of the book (Chapters 10–19) addresses issues relevant to the uplift–weathering hypothesis, including the extent to which Cenozoic uplift has increased physical and chemical weathering, and whether or not it has caused a reduction in atmospheric CO_2.

Field evidence of physical and chemical weathering is considered first. Chapter 10 examines the most direct kind of evidence of rates of physical and chemical weathering: suspended and dissolved loads in rivers, particularly those draining high, mountainous terrain. Chapter 11 investigates chemical weathering rates of young glacial moraines in the mountains of Wyoming to assess how rapidly weathering rates decrease after initial exposure of fresh silicate bedrock. It also assesses the possible impact of continental ice-sheet erosion on the ocean $^{87}Sr/^{86}Sr$ ratio and on global mean rates of chemical weathering. Chapter 12 examines sediments deposited in the Indian Ocean Bengal Fan for evidence of Cenozoic denudation of the Himalaya and to assess changes in the relative amount of physical versus chemical weathering in that region. Chapter 13 summarizes carbon-isotopic evidence of simultaneous late Miocene changes in vegetation in South Asia and on three other continents resulting from reductions in atmospheric CO_2. Chapter 14 surveys dissolved loads in major river systems of the world across a range of tectonic and climatic settings to assess the factors that control chemical weathering, and to assess the capacity of these watersheds to extract CO_2 from the atmosphere by weathering of silicate rocks. Chapter 15 examines a recently developed geochemical index ($^{187}Os/^{186}Os$ ratio) that is in part a monitor of chemical weathering on the continents.

Modeling studies of chemical weathering are covered next. Chapter 16 summarizes evidence that rates of chemical weathering are affected by the so-called "greenhouse" climatic factors: temperature, precipitation, and vegetation. Chapter 17 investigates the extent to which carbon-cycle models can be used to evaluate the net rate of carbon burial in ocean sediments. Chapter 18 summarizes the strengths and weaknesses of both the BLAG and uplift-weathering models and then attempts to extract the Cenozoic history of silicate weathering in southern Asia from a mass-balance analysis of the $^{87}Sr/^{86}Sr$ record.

Chapter 18 also summarizes possible negative feedbacks to the uplift–weathering effect. Chapter 19 uses a geochemical mass-balance model to investigate global chemical weathering rates during the Cenozoic and explores a possible source of negative feedback to the uplift–weathering hypothesis.

REFERENCES

1. Dana, J. D. (1856). *Am. J. Sci.* **22**, p. 305.
2. Lyell, C. (1875). *Principles of Geology.* Murray, London.
3. Ramsay, W. (1924). *Geol. Mag.* **61**, p. 152.
4. Flint, R. F. (1957). *Glacial and Pleistocene Geology.* John Wiley, New York.
5. Emiliani, C., and Geiss, J. (1959). *Geol. Rundsch.* **46**, p. 576.
6. Hamilton, W. (1968). *Meteorol. Monogr.* **8**, p. 128.
7. Birchfield, G. E., and Weertman, J. (1983). *Science* **219**, p. 284.
8. Mintz, Y. (1965). Very long-term global integration of the primitive equations of atmospheric motion. WMO Tech. Note No. 66, p. 141.
9. Kasahara, A., and Washington, W. M. (1971). *J. Atmos. Sci.* **28**, p. 657.
10. Manabe, S., and Terpstra, T. B. (1974). *J. Atmos. Sci.* **31**, p. 3.
11. Ruddiman, W. F., Raymo, M. E., and McIntyre, A. (1986). *Earth Planet. Sci. Lett.* **80**, p. 117.
12. Ruddiman, W. F., Prell, W. L., and Raymo, M. E. (1989). *J. Geophys. Res.* **94**, p. 18379.
13. Kutzbach, J. E., Guetter, P. J., Ruddiman, W. F., and Prell, W. L. (1989). *J. Geophys. Res.* **94**, p. 18393.
14. Ruddiman, W. F., and Kutzbach, J. E. (1989). *J. Geophys. Res.* **94**, p. 18409.
15. Ruddiman, W. F., and Kutzbach, J. E. (1991). *Sci. Amer.* **264**, p. 66.
16. Broccoli, A. J., and Manabe, S. (1992). *J. Clim.* **5**, p. 1181.
17. Rind, D., and Chandler, M. (1991). *J. Geophys. Res.* **96**, p. 7437.
18. Hastenrath, S., and Greischar, L. (1983). *J. Geophys. Res.* **98**, p. 6869.
19. Hahn, D. G., and Manabe, S. (1975). *J. Atmos. Sci.* **32**, p. 1515.
20. Prell, W. L. (1984). In: *Climate Processes and Climate Sensitivity*, Geophysics Monograph Series Vol. 29 (J. E. Hansen and T. Takahashi, eds.), pp. 48–57. American Geophysical Union, Washington, D.C.
21. Prell, W. L., and Kutzbach, J. E. (1992). *Nature* **360**, p. 647.
22. Kutzbach, J. E., Prell, W. L., and Ruddiman, W. F. (1993). *J. Geol.* **101**, p. 177.
23. Raymo, M. E., Ruddiman, W. F., and Froelich, P. N. (1988). *Geology* **16**, p. 649.
24. Raymo, M. E., and Ruddiman, W. F. (1992). *Nature* **359**, p. 117.
25. Chamberlin, T. C. (1899). *J. Geol.* **7**, p. 545.
26. Berner, R. A., Lasaga, A. C., and Garrels, R. M. (1983). *Am. J. Sci.* **283**, p. 641.
27. Walker, J. C. G., Hays, P. B., and Kastings, J. F. (1981). *J. Geophys. Res.* **86**, p. 9776.
28. Molnar, P., and England, P. (1990). *Nature* **346**, p. 29.
29. England, P., and Molnar, P. (1990). *Geology* **18**, p. 1173.
30. Richter, F. M., Rowley, D. B., and DePaolo, D. J. (1992). *Earth Planet. Sci. Lett.* **109**, p. 11.
31. Edmond, J. M. (1992). *Science* **258**, p. 1594.
32. Volk, T. (1993). *Nature* **361**, p. 123.
33. Derry, L. A., and France-Lanord, C. (1996). *Paleoceanography* **11**, p. 267.
34. Sundquist, E. T. (1991). *Quat. Sci. Rev.* **10**, p. 283.
35. Molnar, P., England, P., and Martinod, J. (1993). *Rev. Geophys.* **31**, p. 357.
36. Coney, P. J., and Harms, T. A. (1984). *Geology* **12**, p. 550.
37. Morgan, P., and Swanberg, C. A. (1985). *J. Geodyn.* **3**, p. 39.
38. Prentice, I. C., Cramer, W., Harrison, S. P., Leeman, R., Monserud, R. A., and Solomon, A. M. (1992). *J. Biogeogr.* **19**, p. 117.

Evidence of Cenozoic Uplift

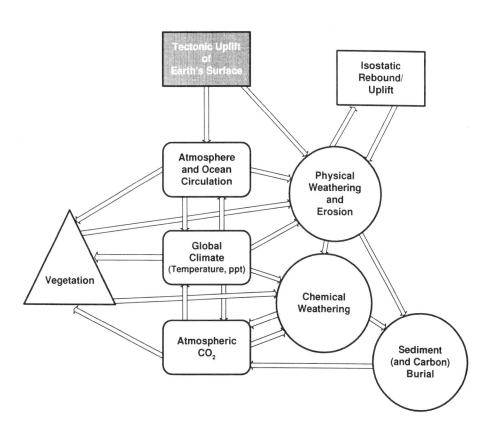

2

The When and Where of the Growth of the Himalaya and the Tibetan Plateau

Peter Copeland

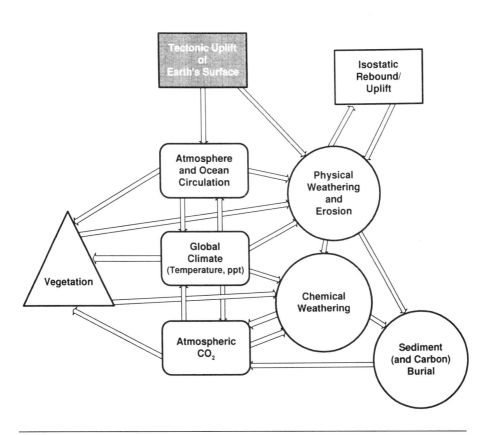

Peter Copeland • Department of Geosciences, University of Houston, Houston, Texas 77204-5503.

Tectonic Uplift and Climate Change, edited by William F. Ruddiman. Plenum Press, New York, 1997.

1. INTRODUCTION

Over the past 50 million years, collision between India and Asia has produced the highest and most extensive plateau on Earth today and the greatest present-day topographic relief (Fig. 1). The Tibetan Plateau and Himalaya may be the largest such feature seen on Earth in the Phanerozoic (i.e., since approximately 542 million years ago). Although this tremendous geographic feature has been ascribed to the effects of the collision between India and Asia since the early part of this century,[1] there are still many important questions regarding the tectonic and geographic history of the region that remain unanswered. As plate tectonic theory has been applied to the region, a detailed understanding has begun to emerge regarding certain small areas, mostly on the edge of the plateau. The ability to integrate information from sources such as seismology and geodesy makes study of this ongoing collision particularly valuable in understanding more ancient orogens, where such information is generally not available. The relative

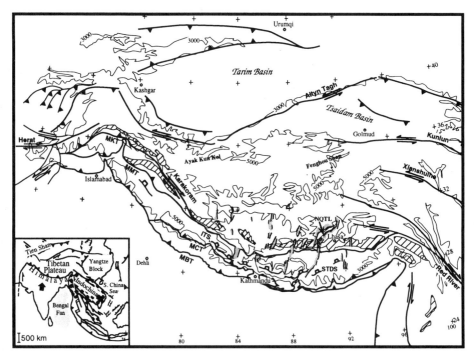

FIGURE 1.   Geographic location of the Himalaya and the Tibetan Plateau showing major geologic features. Heavy lines are faults; thin lines are 3000 and 5000 m topographic contours. Diagonal ruled areas are the Gangdese and related batholiths formed by the subduction of the Tethys oceanic crust in the Cretaceous and Paleogene. Stippled areas are regions of east–west extension of the southern Tibetan Plateau since the late Miocene. MKT = Main Karakoram Thrust, MMT = Main Mantle Thrust, MCT = Main Central Thrust, MBT = Main Boundary Thrust, ITS = Indus–Tsangpo Suture, STDS = Southern Tibetan Detachment System, NQTL = Nyainqen-tanghla range. After Harrison et al.[2]

youth of this orogen allows a precise temporal resolution of events by a variety of methods, which allows better evaluation of the possible mechanisms responsible for the growth of the plateau.[2] The Tibetan Plateau has had an unquestioned and significant effect on the climate of Asia as well as on global circulation patterns,[3–5] and it is our growing understanding of the tectonics of this region that allows more detailed investigations into the variation of Cenozoic climate. We can in many cases date the time of initiation or cessation of important tectonic structures in this orogen, but the evaluation of the relationships between these features and the elevation of the surface of the Earth is a generally less straightforward process.

In this paper I will review the major structures associated with the Indo–Asian collision and the constraints available on the timing of their formation, as well as the extent to which we can use this information to place constraints on the elevation of the surface, which is essential for modeling of past climate.

2. MAJOR STRUCTURES IN THE HIMALAYA AND THE SOUTHERN TIBETAN PLATEAU

The Mesozoic precollisional history of the area now occupied by the Himalaya and Tibetan Plateau is marked by a variety of manifestations of the Wilson cycle,[6–8] which will not be considered here. During the Cenozoic this tectonic activity has continued to modify the geography of Asia, with effects on other global systems including atmospheric circulation, ocean water circulation, and ocean chemistry.

The velocity and position of India relative to Asia can be accurately calculated using ocean floor magnetic anomalies. From 70 to 40 million years ago, India was moving to the north with a speed of more than 100 mm/year.[9] About 40 million years ago the speed was cut in half to about 50 mm/year.[10] The time of initiation of the collision varies along strike of the Himalayan orogen, ranging from as old as early Paleocene in the west[11] (based largely on sedimentologial evidence) to as young as late Eocene in the east.[12] This is also the approximate age of the youngest plutons of the Gangdese batholith in southern Tibet, which represents the roots of the Andean-type continental margin arc that had been built on the southern edge of Asia as the Tethyan oceanic crust was being subducted.[13–15] The youngest marine sedimentary rocks in the eastern Himalaya are of Eocene age.[16]

While the end of the Eocene marked a significant shift in the tectonic environment of southern Tibet, the change did not immediately lead to the raising of the Tibetan Plateau or to any other geographic change that would have affected global circulation or climate. For approximately the first half of the Indo–Asian collision, very little of the convergence between India and Siberia was taken up by lifting the surface of the Earth. Rather, the process of tectonic escape seems to have been the primary mechanism for accommodating the continued northward progression of India earlier than about 25 million years ago. The

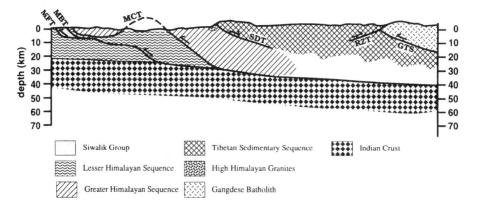

FIGURE 2. Diagrammatic view of the Himalaya and the southern Tibetan Plateau at approximately 86°E. MFT = Main Frontal Thrust, MCT = Main Central Thrust, MBT = Main Boundary Thrust, STD = Southern Tibetan Detachment System, RZT = Renbu-Zedong Thrust, GTS = Gangdese Thrust System (after Harrison et al.[132] and Windley[133]).

continental escape model, as envisaged by Tapponnier et al.,[17-24] shows large portions of what is now Southeast Asia moving along strike-slip faults to make room for India as it advances toward the north. This motion took place along strike-slip faults in eastern China and Indochina, most notably the Red River Fault in Yunnan and Vietnam. Movement along these faults led to the opening of the South China Sea, and dating of the youngest magnetic anomalies in this basin is one of the ways by which we can date the cessation of large-scale tectonic escape as a consequence of the Indo–Asian collision. Brias et al.[23] showed that the history of the opening of the South China Sea is marked by two distinct phases. During the first, from 32 to 27 million years ago, spreading was at an average full rate of approximately 50 mm/year. At the end of this phase there was a ridge jump followed by slower spreading (~35 mm/year), which had ceased altogether by about 16 million years ago. This chronology is consistent with the timing of movement on the Red River Fault in Yunnan and Vietnam.[20,24] Thus it seems that the importance of tectonic escape as a significant mechanism for accommodating the convergence between India and Asia was diminishing in the late Oligocene and was inconsequential by the middle of the Miocene.

While parts of Asia stopped sliding out of the way by the middle Miocene, India's northward progression did not abate at this time nor has it since.[10] Since the collision of India and Asia began, over 2000 km of continued convergence has been accommodated within Asia.[17,18] A substantial fraction of this convergence has been accommodated by one north-dipping fault system in southern Tibet and two major north-dipping faults within the Himalaya (Fig. 2). The northernmost and oldest of the major identified structures is the Gangdese Thrust System (GTS). This north-dipping thrust cuts through the Gangdese batholith for several hundred kilometers along strike in the eastern and central portions of southern

Tibet. Near Zedong, Yin et al.[25] observed a north-dipping shear zone more than 200 m thick that juxtaposes the southern margin of the Gangdese batholith against deformed Tethyan sediments. Kinematic indicators and geological relationships clearly indicate that this fault is a south-directed thrust. Seismic reflection data from Project INDEPTH[26-27] are consistent with the down-dip projection of the GTS in the Lhasa area.

Thermochronometry of samples in the shear zone and undeformed hanging wall constrain initiation of motion up the thrust ramp to 27 ± 1 million years ago. Assuming that thrusting proceeded at a rate similar to the present convergence in the Himalaya (~ 18 mm/year), the tip of the footwall bend would have reached the northern margin of the Gangdese batholith by about 22 million years ago. Total movement on the GTS in this location is estimated to be 46 ± 9 km. While this amount of displacement is not as impressive as the better-known thrusts in the Himalaya, the recognition of the GTS is important for at least two reasons.

First, the timing of the beginning of the GTS coincides with the start of the ebbing of spreading in the South China Sea, suggesting a continuum of accommodation mechanisms, which is required by the seafloor spreading data in the Indian Ocean. The relatively small amount of displacement on this thrust system compared to the Himalayan thrusts may be explained by the fact that some of the convergence between India and Asia was still being accommodated by tectonic escape.

The second important aspect of the recognition of the GTS is that it provides a mechanism to explain the pattern of cooling ages in the Gangdese batholith. Previous work had revealed several places that had experienced brief but significant episodes of rapid cooling at various times since the beginning of the Indo–Asian collision.[15,28] There are six clear examples of rapid cooling later than 40 million years ago: the Quxu, Pachu, Gu Rong, Dagze, and Nyemo plutons, and the area around Yangbajian. The last two of these experienced rapid periods of cooling ca. 9 and 29 million years ago, respectively, whereas all the others experienced a brief period of extremely rapid erosion sometime in the interval from 26 to 15 million years ago. These results are consistent with cooling in response to the formation of a single large-scale feature such as the GTS, which caused enhanced erosion and subsequent cooling of basement rocks diachronously both across and along strike.

The Main Central Thrust (MCT) and the Main Boundary Thrust (MBT) are the two best-known of the major crustal discontinuities in the Indo–Asian orogen. They can be seen in various forms at outcrop in the Himalaya from Pakistan to eastern Nepal[29] (Figs. 1 and 2). The MCT juxtaposes the rocks of the Higher Himalayan Crystalline Sequence (HHC), a series of schists, gneisses, and migmatites, against the Lesser Himalayan Series (LHS), a series of low-grade metasedimentary, metavolcanic, and metagranitoid rocks. At least 100 km of movement has occurred on the MCT[30-34] and perhaps as much as 350 km.[29] Seismic investigation has revealed an extension of north-dipping thrusts in Nepal at a depth of about 35 km beneath southern Tibet[35] (Fig. 2), which has been interpreted as a common décollement for all the Himalayan thrusts and termed the Main Himalayan Thrust (MHT).

At the top of the Greater Himalayan Sequence there is a discontinuous group of Miocene leucogranites,[36] which are thought to be anatectic melts of the HHC induced by the influx of fluids from the LHS just after thrusting on the MCT.[37] Dating these granites has proved difficult in many cases because of the particular geochemistry of the rocks, but most recent investigations are showing these rocks to have been formed in a rather brief interval. In southern Tibet and central and eastern Nepal, the Rongbuk pluton, the Manaslu granite and the Makalu granite have all been investigated by a variety of isotopic approaches that provide insight into their time of crystallization. On the north side of Mount Everest the Rongbuk pluton has been dated by a number of workers[38-41] by several methods with some variability, but the large majority of the results indicates a crystallization about 21 million years old. The age of the Manaslu granite in central Nepal has eluded isotope geochronologists for some years owing to the complicated geochemistry of this and other Himalayan leucogranites, but the most recent effort[40] seems to have overcome these problems and suggests a crystallization about 22 million years ago. A study of the Makalu granite to the west of Mount Everest in eastern Nepal[42] suggested an age between 22 and 24 million years ago. Combined with the Le Fort[37] model of the formation of these granites, the coincidence of these results suggests that the MCT started movement in the latest Oligocene. Thermochronology on rocks in the hanging wall of the MCT in Nepal suggest that it was active in the earliest Miocene[43-45] and that ductile deformation was complete by about 18 million years ago.[43]

Although the Indo–Asian orogen is the product of major plate convergence, not all of the important structures are compressional. The Southern Tibet Detachment System (STDS) or North Himalayan Fault (NHF) is a major down-to-the-north normal fault, which has been described from many locations on the north side of the Himalaya[46-49] (see Figs. 1 and 2). This fault separates the GHS from its weakly metamorphosed structural cover, the Tibetan Sedimentary Sequence (TSS), and it has been interpreted to have resulted from gravitational collapse of an already thickened southern Tibetan crust. Burchfiel et al.[49] estimated total slip on the STDS north of Mount Everest to be in excess of 50 km. The Rongbuk granite, mentioned above, cuts across this fault in the Everest region; thus, the 21 million year crystallization age for that rock constrains at least some of the motion in this fault system to be even older. Thus, if the tectonic interpretations regarding the history of this fault are correct, there was a significant crustal thickness in southern Tibet by the beginning of the Miocene. This suggests that the MCT and STDS were active at the same time, and the geochronological constraints and the models discussed above for these faults suggest a causal relationship between these features.

The surface expression of the MBT (Fig. 2) is a steeply north-dipping fault that marks the contact between the Lesser Himalayan Sequence and the underlying Miocene–Pleistocene Siwalik molasse.[29,52,53] Stratigraphic analysis of the Siwaliks in central Nepal suggests that the MCT was active during the time of Lower Siwalik deposition, with the MBT active mainly during the time that the middle and upper parts of the Siwalik Group were deposited.[54] Based on a change in accumulation rate at several locations in the foreland of Pakistan and

India, Burbank and colleagues[55,56] concluded that the MBT was first active at about 11 million years ago. Norlund et al.[57] dated the boundary between the thin-bedded mudrocks and fine sandstones of the Lower Siwaliks and the coarse-grained sandstones of the Middle Siwaliks in central Nepal between 8.5 and 8.3 million years ago and suggested that this closely followed the initiation of movement on the MBT in Nepal. Until recently, most researchers assumed that displacement on the MCT effectively ceased when the structurally lower MBT became active [50,51] but recent work in the Anapurna[134] and Kathmandu[135] regions suggests there may have been significant displacement on the MCT in the late Miocene. Siwalik deposition was most likely arrested after initial motion of the Main Frontal Thrust (MFT), which is presently active and places the Siwalik Group on top of Quaternary deposits.[32]

A final group of structures that can be observed directly is the series of north–south trending graben in southern Tibet (Fig. 1). Perhaps the most well-received hypothesis to explain these features is the one that suggests that these rifts are accommodating east–west extension that results from the crust in this area attaining the maximum thickness sustainable and subsequently undergoing extensional collapse,[17,58–60] but the reason that this thickness was attained is the subject of greater debate.[61] While these graben are quite numerous, only a few have been closely studied and only one on the plateau itself has been the subject of the kind of geochronological investigation that can provide detailed information on the time of initiation of this extension.

Approximately 80 km northwest of Lhasa lies the Yangbajian Graben, a segment of the Yadong–Gulu Rift, which extends for about 600 km from just north of the High Himalaya onto the east-central part of the plateau.[60] The Yangbajian Graben is unusual among this group of rift basins as most are oriented north–south, but in this case the long axis of the valley runs north–northeast to northeast (this variance in orientation may be due to exploitation of preexisting structures[17,60]). However, the extension direction is essentially the same for all other graben in the region for which reliable measurements are available.[60] Northwest of the graben is the Nyainqentanghla Range. The boundary between the basin and the range is a low-angle detachment shear zone, which is made of amphibolite mylonites with consistent down-to-the-southeast sense-of-shear indicators.[62] Thermochronology of several samples from two transects across this shear zone indicates a significant acceleration of cooling in the footwall of the fault beginning about 8 million years ago and ending about 4 million years ago,[63] consistent with initiation of extension at the end of the Miocene. If the orogenic collapse hypothesis for the development of all of the Tibetan Graben is correct, and if the Yangbajian Graben is typical of these rift structures, these data indicate that the crust of the southern Tibetan Plateau had reached something like its present thickness (or perhaps a bit thicker) by this time.

Based on $^{40}Ar/^{39}Ar$ analysis of a muscovite that filled a fracture in a fault zone some 40 km to the southeast of the Thakkola Graben, just to the north of the High Himalaya in central Nepal, Coleman and Hodges[64] suggested that this rift basin has been active for about 14 million years (similar to the estimate of Mercier et al.[65] based on the stratigraphy of the basin fill) and further concluded that this datum provides a constraint on the time that the maximum elevation of the Tibetan Plateau was attained. However, the assignment of the significance of this datum to the attainment of maximum elevation for the Tibetan Plateau in the broad sense is problematic as the fault in question is closer to the STDS in this area than to the Thakkola Graben itself, and lineations associated with the STDS trend east–west in the region[66] suggesting that muscovite growth within this fault zone may be related to large-scale detachment in the Himalaya rather than to wholesale collapse of the Tibetan Plateau. Perhaps more significantly, the Thakkola Graben does not extend north of the Indus–Tsangpo suture zone and therefore most of the extension in the Thakkola Graben may be the result of either the collapse of thickened Himalayan crust or flexure of the Himalayan arc without regard to other crustal thickening further north,[25,58] but this mechanism cannot account for the extension seen in the rift basins north of a line connecting the east and west syntaxes of the Himalaya.[67]

On the northwest side of the Tibetan Plateau, the Tien Shan (Fig. 1) present another impressive topographic barrier to the interior of the plateau. This region has been undergoing shortening since the early Miocene for a total of at least 100 km.[68–70] The topographic differences among the peaks of the Tien Shan is a consequence of recent rapid deformation and associated uplift.[71,72]

The data from the interior of the Tibetan Plateau are substantially less ample than the relatively abundant and growing data from the Himalaya and the region of Tibet near Lhasa. From the data that are available, however, we have only one documentation of crustal shortening and thickening in this region in the last 50 million years. In the Fenghuo Shan (Fig. 1), Eocene sedimentary rocks are deformed in a fold and thrust system with more than 50% shortening for a total of at least 50 km of transport.[73,74] The lack of documented examples of collision-related shortening and the remarkable flatness of large parts of the plateau[75,76] have led some researchers to suggest models for the tectonic evolution of the Tibetan Plateau in which the entire plateau behaves as a single block,[77–79] however, all direct observation of the geology of the Tibetan Plateau is consistent with distributed shortening,[80] variable in both space and time, being responsible for most of the crustal thickening in Tibet.

3. INDIRECT METHODS OF ASSESSING TECTONISM

While direct observation of important tectonic features and constraints on their possible evolution via cross-cutting relationships and isotope geochronology provide a detailed understanding of a small part of the Indo–Asian collision zone, other indirect approaches have been used to give information on different scales. These studies generally provide information that reflects the average behavior of much larger areas. The approaches that have bearing on the history of the topographic evolution of the Tibetan Plateau include sedimentological and paleontological studies, geochemical analysis of detrital grains and pedogenic carbonate, assessment of the evolution of the geochemistry of world oceans (by a variety of methods), and geophysical studies of the oceanic crust of the Indian Ocean.

The study of the molasse deposits shed off the evolving southern edge of the Tibetan Plateau in Tibet, Pakistan, and Nepal and of sediments deposited in the marine environment gives us a significant understanding of the history of rock that no longer exists: the material that has been eroded off the southern edge of the plateau since collision began. The detrital composition of ancient sandstones has long been recognized as an important indicator of tectonic setting and provenance.[81] The record of many orogenies is now present predominantly (if not exclusively) in the sediments shed from once-evolving mountain belts. However, the utility of sandstone petrology is by no means restricted to ancient rocks. Study of sediments shed off an actively deforming mountain belt offers the ability to possibly pinpoint the source of the detritus as well as its detailed exhumation history prior to transport and deposition.

The material shed off the western Himalaya may now be found in various deposits including the Indus Fan in the northern Arabian Sea. The Muree Formation in northern Pakistan began accumulating during the Eocene.[82] Stratigraphic analysis of the Indus Fan indicates that rapid subsidence began at about 25 million years ago.[83] The beginning of the accumulation of the Siwalik Group in Pakistan was somewhat earlier than 18 million years ago.[52,84] Similar estimates have been made for the Siwaliks in Nepal,[85,86] but no Siwalik section has an exposed base and so we do not have a firm understanding of when significant molasse accumulation began in the foreland basin. However, it does not appear to have preceded the Miocene by a substantial amount of time. Sedimentation in the Bengal Fan was underway by the late Eocene and appears to have prograded to the south rapidly during the last part of the Oligocene and earliest part of the Miocene.[87,88] In southern Tibet, the Kailas conglomerate, just north of the Indus–Tsangpo suture, records the unroofing of the Gangdese batholith in sections which range in thickness from a few hundred to over 3000 m.[29] The time of deposition of this conglomerate has been constrained at three sites in southern Tibet to be early Miocene.[89] Stratigraphic analysis in several areas of a variety of material interpreted to be shed from the Indo–Asian collision area thus suggests that the southern margin of what is now the Tibetan Plateau began to be a significant topographic feature by the late Oligocene or early Miocene.

The studies cited above by and large analyzed the age and thickness of sedimentary strata. While this is an essential and well-established method of assessing the geologic history of the provenance (i.e., the rate and time of erosion), additional information can come from analyzing individual fragments of these strata. These approaches generally emphasize characteristic changes in clastic mineral abundances as an indicator of the dominant tectonic style.[90,91] Critelli and Ingersoll[92] noted a change from quartzolithic to quartzofeldspathic sands and sandstones from the Eocene to Recent in the material shed from the Himalaya and southern Tibetan Plateau, suggesting progressive exhumation of midcrustal terranes and a brief exposure to weathering. However, the questions we may ask about individual sand grains are not limited to their mineralogy, as grains can retain an even more informative record of provenance, including information regarding the age of the source and its cooling history. With the advent of high-sensitivity isotopic techniques, it is now possible to determine isotopic variations with high precision from single grains, even in relatively fine-grained sedimentary rocks.[93-96] When the grain size is too small for precise analysis of individual grains, chemical and isotopic analyses of whole rocks[97-99] can help in our understanding of the evolution of the now-eroded provenance.

One approach to applying isotope geochemistry to sedimentary rocks associated with the Indo–Asian collision is to determine the cooling age of individual sand grains and compare the distribution of such ages to the age of sedimentation. Figure 3 shows the results of such analyses of K-feldspars from sandstones from the southern Bengal Fan collected on ODP Leg 116.[96] The minimum $^{40}Ar/^{39}Ar$ age is plotted against stratigraphic age (based on biostratigraphy), where the $^{40}Ar/^{39}Ar$ age is an estimate of the last time the grain was at a temperature of approximately 200°C. Since volcanism has been shown to have been quite rare in the Himalaya during the Neogene, we can rule out the possibility that the sand grains were last at this temperature while being thrown out of a volcano; therefore this age represents the last time the grains were between 4 and 6 km below the surface, depending on the geothermal gradient. In every stratum investigated there is at least one point that plots on the 1:1 line (the locus of points with cooling age and sedimentation age equal). We can conclude that such grains came from a depth greater than 4 km to the surface and out to the southern Bengal Fan in a very short time, probably less than 500,000 years. This indicates that these grains must have experienced exhumation rates greater than 5 mm/year during this interval. It seems quite unlikely that such rates could be sustained in the absence of significant topographic relief. Therefore, we can conclude from Fig. 3 that some portion of the Himalaya has been a significant topographic feature for at least 16 million years. Because of the paucity of granulite facies rocks in the Himalaya, we can say further that there must have been several locations that experienced these brief episodes of rapid exhumation (the presence of deeply eroded rocks would allow long periods of rapid exhumation in a single location). However, because the source area of the southern Bengal Fan is essentially the entire Himalaya (or at least the eastern three-quarters), we cannot say which location experienced a pulse of rapid erosion at any given time.

A way around this problem is to analyze sedimentary sequences closer to the source, such as the Siwalik Group. Two sections of the Siwaliks at Bakiya Khola and Dhansar Khola in central Nepal have been analyzed in a similar fashion[57,89]; the $^{40}Ar/^{39}Ar$ results from these samples are given in Fig. 4. These data are similar to the results from the Bengal Fan in that there is a significant component of young material and a large spread in the ages (up to 800 million years older than the time of deposition). Because these samples are from the foreland basin and not from the far-offshore marine location in the Bengal Fan, we can make more

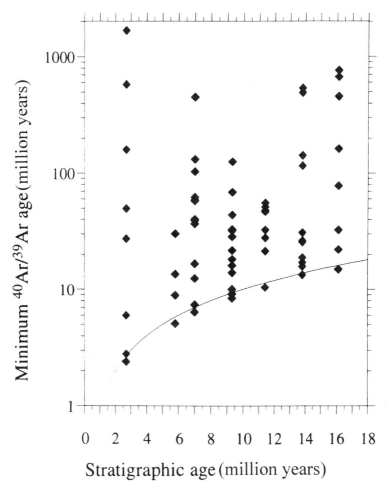

FIGURE 3.   Semilog plot of minimum ^{40}Ar/^{39}Ar ages of detrital feldspars versus stratigraphic age for samples from the southern end of the Bengal Fan. Symbols represent minimum age of age spectrum on individual feldspar crystals and are associated with a closure temperature of about 200°C. Curve represents the 1:1 line.

precise tectonic interpretations about the geographic location of the source area. While we cannot say for sure how big the source area was at any given time, we can say that it was always some small portion of the Himalayas, rather than potentially all of the range. With this idea in mind, the fact that we see a significant range in ages in each sample investigated requires that the source area, however small, be made up of still smaller subregions, which have experienced different uplift–erosion histories. If this were not so, then the rapid erosion characterized by the grains with cooling ages near the sedimentation age would have removed all of the old material within about 2 million years (i.e., there would

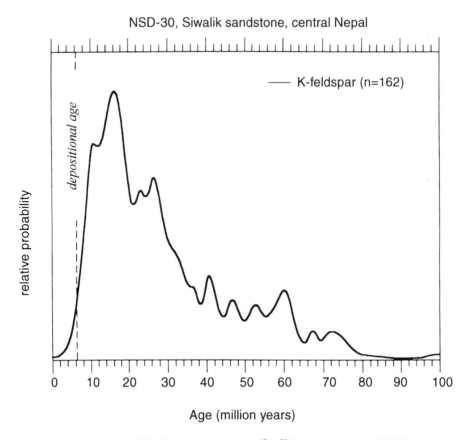

FIGURE 4. Relative probability diagram for minimum ^{40}Ar/^{39}Ar ages from detrital K-feldspars from a sandstone from the Siwalik Group in Dhansar Khola, about 30 km directly south of Kathmandu. Depositional age of this sample based on magnetostratigraphy (from Norlund et al.[57]).

be only young grains present in the population). Since there is still a significant range in age, there must have been significant variation in the source that was feeding this part of the foreland basin during the late Miocene.

An interesting piece of information from this kind of analysis would be the age of the oldest sedimentary rock derived from the Himalaya that does *not* contain any grains with cooling ages equal to the sedimentation age. So far, no such rock has been found, owing largely to the fact that most of the older strata are covered by younger rocks. When such a rock is found, it will give another indication of when the Himalaya first became a major mountain range.

Chemical and isotopic analysis of fine-grained clastic material has also yielded information regarding the erosional and tectonic history of the Himalaya and southern Tibet. Clay minerals comprise evidence of the type of material present in the provenance, as well as past weathering processes. The chemical and mineralogical composition of this material thus provides another record of the types of rocks and the weathering processes they underwent at times in the past. Presently we mostly have such data from the strata of the Bengal Fan.[97-99] These data are of particular use as the chemical and isotopic signature of the major rock types in the Himalaya have been well characterized.[36,100,101] The data from the Indian Ocean indicate that the primary source for the turbidites in the Bengal Fan has been the HHC since at least the early Miocene.[97-99]

Another approach to understanding the unroofing of the Tibetan Plateau and the Himalaya has been to track the variations in ocean chemistry during the Cenozoic.[102-109] The isotopic composition of marine carbonates has been used as a proxy for the isotopic composition of the water in which these types of rocks were formed, and this record has been used to track the integrated chemical fluence into the global ocean over time. The chemical composition of the ocean is determined by the relative contribution of deep-sea hydrothermal activity and terrigenous weathering.

Given that the Himalaya and the Tibetan Plateau have provided a tremendous amount of sediment to the ocean (more than 15 million km^3), several studies have focused on the effect of erosion from this orogen on the chemical and isotopic composition of the oceans. One of the most significant variations in the record of $^{87}Sr/^{86}Sr$ in marine carbonates during the Phanerozoic is the large increase in the last 40 million years, with values more or less constant in the interval from 75 to 40 million years ago at about 0.7078, rising to a value of about 0.7083 about 22 million years ago, and then rising very rapidly to about 0.7089 some 12 million years ago, with a more gradual increase to the present-day value of roughly 0.7092.[110] Richter *et al.*[106] noted that the portion of this curve with the greatest slope, in the interval from 20 to 15 million years ago, corresponds to the period of significant tectonism in the Himalaya and the southern Tibetan Plateau discussed above, and suggested that changes in the Sr flux to the sea from this region alone could account for the changes in the global Sr isotopic values.

Paleoecology is another approach to assess the elevation of the Tibetan Plateau. Paleobotanical information from the Tibetan Plateau has been used to infer mean annual temperature and, from this, elevation.[111-115] Most of these studies have concluded that uplift of the Tibetan Plateau is a young (i.e., post-Miocene) feature brought about by an event common to the entire plateau. However, the composition of floral and faunal assemblages is a function of both elevation and of climatic changes occurring independent of elevation changes. Moreover, the age assignments of nonmarine strata are often rather imprecise and for pre-Pleistocene strata there may not be any close relatives still living with which to compare the fossil species.[116] The geologic data presented above deal primarily with the southern plateau and Himalayan range, whereas the paleoecological samples come largely from the central part of the Tibetan Plateau. Therefore, it may be argued that these two groups of data are reconcilable, but as we will see below there are data that suggest that the plateau was rather well established by the end of the Miocene. Fort[117] has shown a variety of flaws in the arguments for very young uplift of the Tibetan Plateau based on paleoenvironmental data alone.

In the relationship of the Tibetan Plateau to climate, the area of the plateau is as important as the height.[3,5] A high but narrow mountain range will produce a change over an area adjacent to the range but will not have the same sort of effect on global circulation. We can therefore use evidence of climate change as evidence of the establishment of a broad and high Tibetan Plateau. Such evidence has been suggested from two types of data. The first of these recognizes the effect of climate on floral assemblages, the second on the way the Asian monsoon affects oceanic circulation. Plants such as trees and shrubs are in a group called C_3 plants because the fixation of carbon occurs in three-carbon molecules; C_3 plants favor the lighter isotopes of carbon over the heavier in this process. Plants that fix carbon in four-carbon molecules include most grasses; C_4 plants are less discriminating in the isotopic species of carbon used and therefore tend to have "heavier" carbon isotopic values. Carbonate that forms in soils will have an isotopic composition that reflects that of the local biomass, and therefore isotopic analysis of carbonate in paleosol will give an indication of the type of plants that were growing at the time of deposition. Studies of the isotopic composition of pedogenic carbonate in the Siwalik Group in Pakistan and Nepal have yielded similar results.[3,89] In both cases pedogenic carbonate from strata more than 8 million years old show $\delta^{13}C_{PDB}$ values in the range of -9 to -12 per mil, with values from strata less than 6 million years old from -2 to $+3$ per mil. These data indicate that the terrestrial Siwalik Basin was dominated by forests of trees and shrubs prior to 7 million years ago but changed to grassland afterward. Quade et al.[4] suggested that this shift was the result of the intensification of the Asian monsoon, with more concentrated and perhaps more abundant rainfall favoring grasses over forests. However, this isotopic shift has now been observed elsewhere and thus the ultimate cause may be at a greater scale than the development of the Tibetan Plateau.

The interpretation that the Tibetan Plateau reached a threshold size sufficient to induce the Asian monsoon about 8 million years ago is supported by fossil assemblages in the Arabian Sea.[118] Summer upwelling in the Arabian Sea, produced by strong southwesterly winds, produced a distinct fossil assemblage dominated by G. bulloides d'Orbigny. Prior to 8.6 million years ago, this taxon is sparse, but after 7.4 million years ago (save for a brief diminution about 5.5 million years ago), this species accounts for more than 40% of the foraminiferal assemblage. Kroon et al.[118] interpreted this variation to be the result of an intensification of the Asian monsoon in the late Miocene. The monsoonal pattern of precipitation may have been present before this time but not in its present magnitude.

A final piece of indirect evidence for the evolution of the Tibetan Plateau comes from studies of the deformation of the oceanic crust of the Indian plate. It has been known for some time that there is a region of anomalous seismicity south of Sri Lanka, which has been interpreted as a diffuse[119-122] or nascent[123] plate boundary between India and Australia. Analysis of magnetic anomalies suggests a wide plate boundary extending from the Carlsberg Ridge in the west to the Sumatran subduction zone in the east at about 2° to 5°S with north–south shortening in the eastern portion and north–south stretching in the western portion of the deforming zone.[122] Based on a single-channel air gun seismic reflection profile across this zone, Cochran[88] concluded that the deformation in this region began between 8.0 and 7.5 million years ago. However, based on analyses of magnetic anomalies, Royer *et al.*[124] suggested that differential rotation of blocks within the broad plate margin started earlier than 11 million years ago, perhaps as far back as 17 million years ago.

The temporal coincidence of the beginning of extension in the Yangbajian Graben, the change in the foraminiferal assemblage in the Arabian Sea, and the beginning of folding of the oceanic crust near the equator in the Indian Ocean are all data points that can be tied to a common explanation. The Tibetan Plateau may have become large enough to modify the global circulation sufficiently to induce the Asian monsoon in the late Miocene. The fact that we see abrupt changes in the paleontological and geochemical indicators of climate may indicate there is some sort of threshold size for the plateau below which climate is not significantly perturbed in more than a local sense. The fact that this period of apparent climate change corresponds to the time of initiation of two important structural events may indicate that the size of plateau necessary to induce the monsoon is also just about as big as the plateau was ever going to get, as this is the same period during which we see the beginning of new ways of accommodating the continued convergence between Australia and Siberia. Both the extension of the southern Tibetan Plateau and the shortening of the Indian plate are consistent with the plateau achieving a state in which it was now easier to accommodate the convergence in ways other than the distributed shortening that seems to have been dominant during the Miocene.

Based on the indirect and direct evidence summarized here we can evaluate the relative time and importance of the various mechanisms responsible for the accommodation of the convergence between the Indo–Australian plate(s) and Siberia during the Cenozoic. These contributions are schematically described in Fig. 5.

4. RELATIONS BETWEEN TECTONICS AND CLIMATE

While much has been learned about the tectonic evolution of the Himalaya and Tibetan Plateau in the past 15 years, there are still not enough data to unequivocally provide all the answers to the questions relevant to the relationship between the tectonics of the Indo–Asian collision and global climate. What remains at issue includes the uplift history of much of the central part of the plateau and the extent to which we can use information about the deformational history of rocks on the margins of the plateau to establish the elevation of these areas over time. Despite the large gaps in our knowledge, I conclude there are a

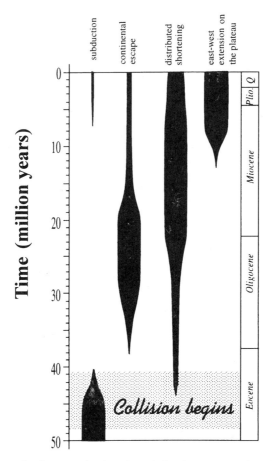

FIGURE 5. Schematic diagram showing the relative importance of various mechanisms of accommodation of the Indo–Australian plate(s) relative to Siberia for the last 50 million years (ordinate is time, abscissa is arbitrary). Relative thickness of symbols represents intensity of activity.

few features of the Himalayan–Tibetan orogen of which we can be reasonably sure:

- The initial collision between India and Asia began in the early Eocene in the northwest and the late Eocene in the east.
- Most of the convergence between India and Asia during the first 20 million years of the collision was accommodated by large-scale movement of blocks of crust to the east and southeast along major strike-slip faults.
- While there was already an Andean-style continental margin arc on the southern margin of the Tibetan Plateau at the beginning of the collision, there is substantial evidence from a variety of sources that suggests an increased topographic gradient in many parts of the region in the interval spanning the Oligo–Miocene boundary.

· The Tibetan Plateau was probably large enough and high enough to have an effect on regional or global atmospheric circulation by the end of the Miocene, launching, or at least intensifying, the Asian monsoon.

What exactly are the dimensions of a plateau sufficient to alter climate? Computer modeling of Earth's atmospheric circulation has something to offer in answer to this question, but it seems that for at least the near term the sense of the output of such calculations will be evaluated by comparison with the geologic record. It may be that the paleobotanical data used to support arguments for very recent uplift of the plateau may be reflecting an increase in elevation above the elevation needed to modify circulation. However, because of the data's high uncertainty, in both an ecological and temporal sense, the models for the evolution of the Tibetan Plateau built solely on paleobotanical data will never, in my opinion, have as good of a foundation as models which rely on geological data or a combination of geology and paleobotany. Some other types of inferences used to argue for a very young and accelerating uplift of Tibet are also questionable. The suggestion that the coarsening-upward character of the Siwalik Group in Nepal and Pakistan indicates an increasing rate of uplift,[103] while not incompatible with the observation, seems quite less likely than the conclusion that the Siwaliks coarsen and thicken upward during the late Miocene and afterward because the source was getting closer to the basin. The southward propagation of the major thrusts in the Himalaya has progressively uplifted material closer to the Siwalik depocenters, but this in no way requires the rate of uplift to have increased; it could have in fact locally decreased, yet still produced coarser sediment to a given location.

Perhaps one of the more obfuscatory terms in the study of the Indo–Asian collision has been the phrase "uplift of the plateau." This usage may suggest to some that the Tibetan Plateau is a single place that can be described with a single (although perhaps complicated) history. One of the fundamental insights to come out of the past two decades of geological investigation in Asia is that in practically every area covered in detail and by practically every method employed, the conclusions are that the continental crust has a variety of complicated histories and that the lengthscale of this tectonic variability can be quite small. Those who model such regions, whether for tectonic or climatological goals, clearly must take this into account.

The exhumation of mountain belts in general, and the Indo–Asian collision in particular, has been investigated by a variety of methods, including studies of the thickness of sedimentary deposits in basins adjacent to orogenic belts, the comparison of the cooling age of detrital minerals with the stratigraphic interval in which they were deposited, and the cooling histories of basement rocks. Each of these approaches provides a different perspective on the scale and rate of erosion and the tectonic processes responsible for exhumation.

Studies of stratigraphic accumulation[11,57,83–87,92] provide insight into the provenance-wide average erosion rate but may not be characteristic of any particular location. Isotopic studies of individual detrital minerals[57,93–96,125] provide two firm points in the evolution of the grain: the time of passing through the closure interval of the mineral in question, and the time of deposition. However, it cannot be determined where the individual mineral grains came from within the source area. The cooling histories of igneous and metamorphic rocks based on isotope thermochronology[15,20,21,25,40,41,43,45,63] provide detailed information about a specific location, but much work is required to characterize a region.

Studies in the Himalaya, Alps, Andes, and Rocky Mountains have shown that: (1) the cooling histories of rocks from orogenic belts are rarely simple (neither linear nor a simple curve), and (2) the exhumation of an orogen cannot be characterized by a simple history. Segments of orogens thousands or even hundreds of kilometers long cannot be summarized by a single description, because such segments are very likely to be made up of several domains that have behaved independently over millions of years. Even in the case where two distant locations can be shown to have been at about the same temperature at the same time, there is only a small likelihood that these sites shared a common (or even similar) exhumation history. The history that has been documented in many locations in the Himalaya (southern Tibet,[15] central Nepal,[40-45,126] and northern Pakistan[127-128]) indicates exhumation variable in both space and time: locations such as Nanga Parbat,[128] Quxu,[15] and Mount Everest[43] are regions a few to tens of kilometers across, which have experienced exhumation histories markedly different from those of adjacent regions. Episodes of rapid exhumation over brief intervals are likely related to the relatively short duration of continuous motion on any single fault or fault zone.[129,130] Such faults cannot move entire mountain ranges at once, but the sum of many small displacements can explain the observed pattern of erosion. In a discussion of the tectonic history of the northwestern Himalaya, Burbank and Reynolds[131] observed

Although the rate of convergence of the Indian subcontinent with Eurasia appears steady... [a] detailed reconstruction of the northwestern Himalayan foredeep indicates that prolonged periods of quiescence and uniform sedimentation were punctuated by brief, intense intervals of deformation as the locus of thrust faulting encroached in a stepwise fashion on the adjacent foredeep. This sporadic tectonism may be regarded as the pulse of an orogenic process which, when integrated over time intervals in excess of 1 million years seems to be continuous.

When all the Indo–Asian collision has been studied in as much detail as northern Pakistan has now been described, it is very likely that such a statement will have been written about each of the subregions of the Himalaya and the Tibetan Plateau.

REFERENCES

1. Argand, E. (1924). *Comp. Rendus.* **1**, p. 171.
2. Harrison, T. M., Copeland, P., Kidd, W. S. F., and Yin, A. (1992). *Science* **255**, p. 1663.
3. Ruddiman, W. F., and Kutzbach, J. E. (1989). *J. Geophys. Res.* **94**, p. 18409.

4. Quade, J., Cerling, T. E., and Bowman, J. R. (1989). *Nature* **342**, p. 163.
5. Kutzbach, J. E., Prell, W. L., and Ruddiman, W. M. (1993). *J. Geol.* **101**, p. 177.
6. Tapponnier, P., Mercier, J. L., Proust, F., Andrieux, J., Armijo, R., Bassoullet, J. P., Brunel, M., Burg, J. P., Colchen, M., Dupré, B., Girardeau, J., Marcoux, J., Mascle, G., Matte, P., Nicolas, A., Li, T., Xiao, X., Chang, C., Lin, P., Li, G., Wang, N., Chen, G., Han, T., Wang, X., Den, W., Zhen, H., Sheng, H., Cao, Y., Zhou, J., and Qiu, H. (1981). *Nature* **294**, p. 405.
7. Chang, C., Chen, N., Coward, M. P., Deng, W., Dewey, J. F., Gansser, A., Harris, N. B. W., Jin, C., Kidd, W. S. F., Leeder, M. R., Li, H., Lin, J., Liu, C., Mei, H. Molnar, P., Pan, Y., Pearce, J. A., Shackleton, R. M., Smith, A. B., Sun, Y., Ward, M., Watts, D. R., Xu, J., Xu, R., Yin, J., and Zhang, Y. (1986). *Nature* **323**, p. 501.
8. Sengör, A. M. C., Cin, A., Rowley, D. B., and Nie, S. (1993). *J. Geol.* **101**, p. 51.
9. Jurdy, D. M., and Gordon, R. G. (1984). *J. Geophys. Res.* **89**, p. 9927.
10. Molnar, P., and Tapponnier, P. (1975). *Science* **189**, p. 419.
11. Beck, R. A., Burbank, D. W., Sercombe, W. J., Riley, G. W., Barndt, J. K., Berry, J. R., Afzal, J., Khan, A. M., Jurgen, H., Metje, J., Cheema, A., Shafique, N. A., Lawrence, R. D., and Khan, M. A. (1995). *Nature* **373**, p. 55.
12. Dewey, J. F., Shackelton, R. M., Chang, C., and Sun, Y. (1988). *Phil. Trans. R. Soc. Lond.* **A327**, p. 379.
13. Xu , R.-H., Schärer, U., and Allègre, C. J. (1985). *J. Geol.* **93**, p. 41.
14. Schärer, U., and Allègre, C. J. (1984). *Earth Planet. Sci. Lett.* **63**, p. 423.
15. Copeland, P., Harrison, T. M., Yun, P., Kidd, W. S. F., Roden, M. K., and Zhang, Y. (1995). *Tectonics* **14**, p. 223.
16. Bordet, P., Colchen, M., Le Fort, P., and Pêcher, A. (1981). *Geodyn. Ser.* **3**, p. 149.
17. Tapponnier, P., Peltzer, G., and Armijo, R. (1986). *Geol Soc. Spec. Publ.* **19**, p. 115.
18. Tapponnier, P., Lacassin, R., Leloup, P. H., Schärer, U., Zhong, D., Wu, H., Liu, X., Ji, S., Zhang, L., and Zhong, J. (1990). *Nature* **343**, p. 431.
19. Leloup, P. H., and Kienast, J.-R. (1993). *Earth Planet. Sci. Lett.* **118**, p.213.
20. Harrison, T. M., Chen Wenji, Leloup, P. H., Ryerson, F. J., and Tapponnier, P. (1992). *J. Geophys. Res.* **97**, p. 7159.
21. Harrison, T. M., Leloup, P. H., Ryerson, F. J., Tapponnier, P., Lacassin, R., and Wenji, C. (1994). In: *The Tectonics of Asia* (A. Yin and T. M. Harrison, eds.), pp. 208–226, Cambridge University Press.
22. Brias, A., Tapponnier, P., and Pautot, G. (1989). *Earth Planet. Sci. Lett.* **95**, p. 307.
23. Brias, A., Patriat, P., and Tapponnier, P. (1993). *J. Geophys. Res.* **98**, p. 6299.
24. Schärer, U., Tapponnier, P., Lacassin. L., Leloup, P. H., Zhong, D., and Zhi., S. (1990). *Earth Planet. Sci. Lett.* **97**, p. 65.
25. Yin, A., Harrison, T. M., Ryerson, F. J., Chen, W., Kidd, W.S.F., and Copeland, P. (1994). *J. Geophys. Res.* **99**, p. 18175.
26. Alsdorf, D., Brown, L., Clark, M., Ross, A., Nelson, D., Cogan, M., Makovsky, Y., Klemperer, S., Zhao, W., and Che, J. (1995). *Geol. Soc. Am. Abstr. with Prog.* **27**, p. A334.
27. Hauck, M. L., Nelson, K. D., and others (1995). *Geol. Soc. Am. Abstr. with Prog.* **27**, p. A336.
28. Copeland, P., Harrison, T. M., Kidd, W. S. F., Xu, R., and Zhang Y. (1987). *Earth Planet. Sci. Lett.* **86**, p. 240.
29. Gansser, A. (1964). *The Geology of the Himalayas*, Wiley Interscience, New York.
30. Brunel , M. (1975). *Acad. Sci. Paris* **280**, p. 551.
31. Pêcher, A. (1989). *J. Metamorphic Petrol.* **7**, p. 31.
32. Schelling, D., and Arita, K. (1991). *Tectonics* **10**, p. 851.
33. Schelling, D. (1992). *Tectonics* **11**, p. 925.
34. Arita, K. (1983). *Tectonophysics* **95**, p. 43.
35. Zhao, W., Nelson, K. D., and Project INDEPTH Team (1993). *Nature* **336**, p. 557.
36. Le Fort, P., Cuney, M., Deniel, C., France-Lanard, C., Sheppard, S. M. F., Upreti, B. N., and Vidal, P. (1987). *Tectonophysics* **134**, p. 39.
37. Le Fort, P. (1981). *J. Geophys. Res.* **86**, p. 10545.
38. Copeland, P., Parrish, R. R., and Harrison, T. M. (1988). *Nature* **333**, p. 760.

39. Hodges, K. V., Parrish, R. R., Housh, T., Lux, D., Burchfield, B. C., Royden, L. H., and Chen, Z. (1992). *Science* **258**, p. 1466.
40. Harrison, T. M., McKeegan, K. D., and Le Fort, P. (1995). *Earth Planet. Sci. Lett.* **133**, p. 271.
41. Hodges, K., Bowring, S., Hawkins, D., and Davidek, K. (1996). *Abstract Volume of the 11th Himalayan–Karakorum–Tibet Workshop, Flagstaff*, p. 63.
42. Schärer, U. (1984). *Earth Planet Sci. Lett.* **67**, p. 191.
43. Hubbard, M. and Harrison, T. M. (1989). *Tectonics* **8**, p. 865.
44. Parrish, R. R., and Hodges, K. V. (1993). *Geol. Soc. Am. Abstr. with Prog.* **25**, p. A174.
45. Coleman, M. E., and Parrish, R. R. (1995). *EOS* **76**, p. F708.
46. Burg, J. P., Brunel, M., Gapais, D., Chen, G. M., and Liu, G. H. (1984). *J. Struct. Geol.* **6**, p. 535.
47. Burchfiel, B. C., and Royden, L. H. (1985). *Geology* **13**, p. 679.
48. Herren, E. (1987). *Geology* **15**, p. 409.
49. Burchfiel, B. C., Chen, Z., Hodges, K. V., Liu, Y., Royden, L. H., Deng, C., and Xu, J. (1992). Geological Society of America Special Paper 269, 41 pp.
50. Le Fort, P. (1975). *Am. J. Sci.* **275A**, p. 1.
51. Molnar, P. (1984). *Ann. Rev. Earth Planet. Sci.* **12**, p. 489.
52. Johnson, N. M., Stix, J., Tauxe, L., Cerveny, P. F., and Tahirkheli, R. A. K. (1985). *J. Geol.* **93**, p. 27.
53. Mugnier, J.-L., Mascle, G., and Faucher, T. (1992). *Bull. Soc. Géol. France* **163**, p. 585.
54. Hérail, G., Mascle, G., and Delacaillau, G. (1986). *Sciences de la Terre* **47**, p. 155.
55. Burbank, D. W., Leland, J., Fielding, E., Anderson, R. S., Brozovic, N., Reid, M. R., and Duncan, C. (1996). *Nature* **379**, p. 505.
56. Meigs, A. J., Burbank, D. W., and Beck, R. A. (1995). *Geology* **23**, p. 423.
57. Norlund, P., Copeland, P., Hall, S. A., Evans, I., ClayPool, P. A., and Ojha, T. P. (1995). *Geol. Soc. Am. Abstr. with Prog.* **26**, p. A334.
58. England, P., and Houseman, G. (1989). *J. Geophys. Res.* **94**, p. 3664.
59. Dewey, J. F. (1988). *Tectonics* **7**, p. 1123.
60. Armijo, R., Tapponnier, P., Mercier, J. L., and Han, T. (1986). *J. Geophys. Res.* **91**, p. 13803.
61. Molnar, P., England, P., and Martinod, J. (1993). *Rev. Geophys.* **31**, p. 357.
62. Pan, Y., and Kidd, W. S. F. (1992). *Geology* **20**, p. 775.
63. Harrison, T. M., Copeland, P., Kidd, W. S. F., and Lovera, O. M. (1995). *Tectonics* **14**, p. 658.
64. Coleman, M., and Hodges, K. V. (1995). *Nature* **374**, p. 49.
65. Mercier, J.-L., Armijo, R., Tapponnier, P., Carey-Gailhardis, E., and Han, T. L. (1987). *Tectonics* **6**, p. 275.
66. Coleman, M. E. (1993). *Geol. Soc. Am. Abstr. with Prog.* **25**, p. A174–175.
67. Dewey, J. F., Cande, S., and Pitman III, W. (1989). *Ecologae Geol. Helv.* **82**, p. 717.
68. Peltzer, G., Tapponnier, P. (1988). *J. Geophys. Res.* **93**, p. 15085.
69. Yin, A., and Nie, S. (1996). In: *Tectonics of Asia* (A. Yin and T. M. Harrison, eds.), pp. 442–485. Cambridge University Press, Cambridge.
70. Burchfiel, B. C., and Royden, L. H. (1991). *Ecologae Geol. Helv.* **84**, p. 599.
71. Molnar, P., Burchfiel, B. C., K'uangyi, L., and Zhao, Z. (1987). *Geology* **15**, p. 249.
72. Burchfiel, B. C., Zhang Peizhen, Wang Yipeng, Zhang Weiqi, Song Fangmin, Deng Qidong, Molnar, P., and Royden, L. (1991). *Tectonics* **10**, p. 1091.
73. Leeder, M. R., Smith, A. B., and Yin, J. (1988). *Phil. Trans. R. Soc. Lond.* **A327**, p. 107.
74. Kidd. W. S. F., Pan, Y. Chang, C., Coward, M. P., Dewey, J. F., Gansser, A., Molnar, P., Shackelton, R. M., and Sun, Y. (1988). *Phil. Trans. R. Soc. Lond.* **A327**, p. 287.
75. Molnar, P. (1989). *Am. Sci.* **77**, p. 350.
76. Feilding , E., Isacks, B., Barazangi, M., and Duncan, C. (1994). *Geology* **22**, p.163.
77. Zhao, W., and Morgan, W. J. (1985). *Tectonics* **4**, p. 359.
78. Zhao, W., and Morgan, W. J. (1987). *Tectonics* **6**, p. 489.
79. Powell, C. McA. (1986). *Earth Planet. Sci. Lett.* **81**, p. 79.
80. Dewey, J. F., and Burke, K. (1973). J. Geol. **81**, p. 683.
81. Dickinson, W. R., and Suczek, C. A. (1979). *AAPG Bull.* **63**, p. 2164.
82. Bosart, P., and Ottiger, R. (1989). *Ecologae Geol. Helv.* **82**, p. 133.

83. Whitting, B. M., and Karner, G. D. (1991). *EOS* **72**, p. 472.

84. Barry, J. C., Lindsay, E. H., and Jacobs, L. L. (1982). *Palaeogeog. Palaeoclim., Palaeoec.* **37**, p. 95.

85. Yoshida, M., Tokuoka, T., Takayasu, K., and Hisatomi, K. (1988). In: *Paleoenvironment of East Asia from the Mid-Tertiary, Vol. 1* (P. Whyte, ed.), pp. 157–169. Center of Asian Studies, University of Hong Kong.

86. Appel, E., Rössler, W., and Corvinus, G. (1991). *Geophys. J. Int.* **105**, p. 191.

87. Curray, J. R. (1994). *Earth Planet. Sci. Lett.* **125**, p. 371.

88. Cochran, J. R. (1990). In: *Proc. ODP, Sci. Res., Vol. 116* (J. R. Cochran, D. A. V. Stow *et al.*, eds.), pp. 397–414. Ocean Drilling Program, College Station, Tex.

89. Harrison, T. M., Copeland, P., Hall, S. A., Quade, J., Burner, A., Ojha, T. P., and Kidd, W. S. F. (1993). *J. Geol.* **101**, p. 159.

90. Dickinson, W. R. (1974), In: *Tectonics and Sedimentation*, pp. 1–27. Soc. Econ. Paleont. Mineral. Spec. Paper 22.

91. Ingersoll, R. V. (1988). *Geol. Soc. Am. Bull.* **100**, p. 1704.

92. Critelli, S., and Ingersoll, R. V. (1994). *J. Sed. Res.* **A64**, p. 815.

93. Froude, D. O., Ireland, T. R., Kinny, P. D., Williams, I. S., Compston, W., Williams, I. R., and Myers, J. S. (1983). *Nature* **304**, p. 616.

94. Cervany, P. F., Naeser, N. D., Zeitler, P. K., Naeser, C. W., and Johnson, N. M. (1988). In: *New Perspectives in Basin Analysis* (K. L. Kleisphehn and C. Paola, eds.), pp. 43–61. Springer, New York.

95. Copeland, P., and Harrison, T. M. (1990). *Geology* **18**, p. 354.

96. Copeland, P., Harrison, T. M., and Heizler, M. T. (1990). In: *Proc. ODP. Sci. Res. Vol. 116* (J. R. Cochran, and D. A. V. Stowe, *et al.*, eds.), pp. 93–114. Ocean Drilling Program, College Station, Tex.

97. Derry, L. A. and France-Lanord, C. (1996). *Earth Planet. Sci. Lett.*, **142**, p. 59.

98. France-Lanord, C., Derry. L. A., and Michard, A. (1993). In: *Himalayan Tectonics* (P. J. Treloar and M. Searle, eds.), pp. 603–621. Geological Society of Lond.

99. Bouquillon, A., France-Lanort, C. Michard, A., and Tiercelin, J.-J. (1990). In: *Proc. ODP, Sci. Res., Vol 116* (J. R. Cochran, D. A. V. Stow, *et al.*, eds.), pp. 43–58. Ocean Drilling Program, College Station, Tex.

100. France-Lanord, C., Sheppard, S. M. F., and Le Fort, P. (1988). *Geochim. Cosmochim. Acta* **52**, p. 513.

101. Denial, C., Vidal, P., Fernandez, A., Le Fort, P., and Peucat, J. J. (1987). *Cont. Min. Petrol.* **96**, p. 78.

102. Jacobsen S. B. (1988). *Earth Planet. Sci. Lett.* **90**, p. 315.

103. Raymo, M. E., Ruddiman, W. F., and Froelich, P. N. (1988). *Geology* **16**, p. 649.

104. Palmer, M. R., and Edmond, J. M. (1989). *Earth Planet. Sci. Lett.* **92**, p. 11.

105. Hodell, D. A., Mead, G. A., and Mueller, P. A. (1990). *Chem. Geo.* **80**, p. 291.

106. Richter, F. M., Rowley, D. B., and DePaolo, D. J. (1992). *Earth Planet. Sci. Lett.* **109**, p. 11.

107. Edmond, J. M. (1992). *Science* **258**, p. 1594.

108. Krishnaswami, S., Trivedi, J. R., Sarin, M. M., Ramesh, R., and Sharma, K. K. (1992). *Earth Planet. Sci. Lett.* **109**, p. 243.

109. France-Lanord, C., and Derry, L. A. (1994). *Geochim. Cosmochim. Acta* **58**, p. 4809.

110. Burke, W. H., Denison, R. E., Hetherington, E. A., Koepnick, R. B., Nelson, H. F., and Otto, J. B. (1982). *Geology*, **10**, p.516.

111. Wu, R. (1981). In: *Proceedings of Symposium of Qinghai-Xizang (Tibet) Plateau, Vol. 1, Geology, Geological History and Origin of the Qinghai-Xizang Plateau*, p. 139–144.

112. Chen, W. (1981). In: *Proceedings of Symposium of Qinghai-Xizang (Tibet) Plateau, Vol. 1, Geology, Geological History and Origin of the Qinghai-Xizang Plateau*, p. 343.

113. Li, J., Li, B., Wnag, F., Zhang, Q., Wen. S., and Zhang, B., (1981). In: *Proceedings of Symposium of Qinghai-Xizang (Tibet) Plateau, Vol. 1, Geology, Geological History and Origin of the Qinghai-Xizang Plateau*, p. 111.

114. Guo, S.-X. (1981). In: *Proceedings of Symposium of Qinghai-Xizang (Tibet) Plateau, Vol. 1, Geology, Geological History and Origin of the Qinghai-Xizang Plateau*, p.201.

115. Xu, S. (1981). In: *Geological and Ecological Studies of the Qinghai-Xizang Plateau*, p. 247.
116. Molnar, P., and England, P. (1990). *Nature* **346**, p. 29.
117. Fort, M. (1996). *Palaeogeog. Palaeoclimatol. Palaeoecol.* **120**, p. 123.
118. Kroon, D., Steens, T., and Trolestra, S. R. (1991). In: *Proc. ODP Sci. Res. Vol. 116* (W. L. Prell and N. Niitsuma *et al.*, eds.), pp. 257–263. Ocean Drilling Program, College Station, Tex.
119. Weins, D. A., DeMets, C., Gordon, R. G., Stein, S., Argus, D., Engeln, J. F., Lundgren, P., Quible, D., Stein, C., Weinstien, S., and Woods, D. F. (1985). *Geophys. Res. Lett.*, **12**, p. 429.
120. Zuber, M. T. (1987). *J. Geophys. Res.* **92**, p. 4817.
121. Gordon, R. G., DeMets, C., and Argus, D. F.(1990). *Tectonics.* **9**, p. 409.
122. DeMets, C., Gordon, R. G., and Vogt, R. (1994). *Geophys. J. Int.* **119**, p. 893.
123. Sykes, L. R. (1970). *J. Geophys. Res.* **75**, p. 5041.
124. Royer, J.-Y., Gordon, R. G., and DeMets, C. (1994). *EOS*, 75, (supplement).
125. Cervany, P. F., Johnson, N. M., Tahirkeli, R. A. K., and Bonis, N. R. (1989). In: *Tectonics of the Western Himalayas* (L. L. Malinconico and R. J. Lillie, eds.) Geol. Soc. Am. Spec. Paper 232, Boulder, CO.
126. Copeland, P., Harrison, T. M., Hodges, K. V., Maréujol, P., Le Fort, P., and Pécher, A. (1991). *J. Geophys. Res.* **96**, p. 8475.
127. Zeitler, P. K. (1985). *Tectonics* **4**, p. 127.
128. Zeitler, P. K., Sutter, J. F., Williams, I. S., Zartman, R., and, Tahirkheli, R. A. K. (1989). In: *Tectonics of the Western Himalayas* (L. L. Malinconico and R.J. Lillie, eds.), pp. 1–22. Geol. Soc. Am. Spec. Paper 232, Boulder, CO.
129. Price, R. A. (1973). *Geol. Soc. Am. Abstr. Prog.* **5**, p. 772.
130. Jordan, T. E., Allemdinger, R. W., Damati, J. F., and Drake, R. E. (1993). *J. Geology* **101**, p. 135.
131. Burbank, D. W., and Reynolds, G. H. (1984). *Nature* **311**, p. 114.
132. Harrison, T. M., Yin, A., and Ryerson, F. J. (1996) in press (book being edited by Kevin Burke and Tom Crowley).
133. Windley, B. F. (1984). *The Evolving Continents*, John Wiley, New York, 399 pp.
134. Harrison, T. M., Ryerson, F. J., Le Fort, P., Yin, A., Lovera, O. M., and Catlos, E. J. (1997). *Earth Planet. Sci. Lett.*, **134**, p. E1.
135. Copeland P., Le Fort, P., Upreti, B. N., and Raï, S. (1997). Flexure of the main central thrust in the Kathmandu area due to ramping on the main boundary thrust. Proceedings of the Goldschmidt Conference, Tucson, Arizona, p. 42 (unpublished).

3

Variability in Age of Initial Shortening and Uplift in the Central Andes, 16–33°30'S

Teresa E. Jordan, James H. Reynolds III, and Johan P. Erikson

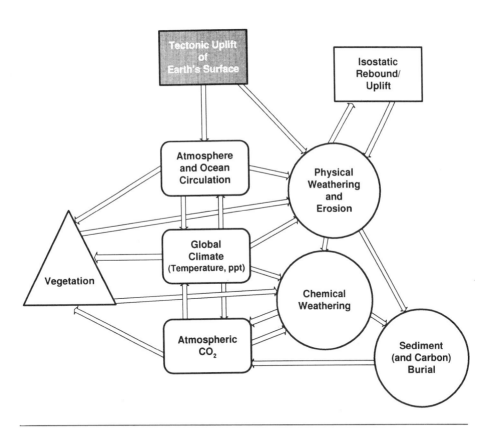

Teresa E. Jordan • Department of Geological Sciences, Cornell University, Ithaca, New York 14853–1504.   *James H. Reynolds III* • Department of Geosciences and Anthropology, Western Carolina University, Cullowhee, North Carolina 28723.   *Johan P. Erikson* • Department of Geology, University of California at Davis, Davis, California 95616.

Tectonic Uplift and Climate Change, edited by William F. Ruddiman. Plenum Press, New York, 1997.

1. INTRODUCTION

The north-trending Andean Cordillera is the highest landmass in the Southern Hemisphere. Lenters *et al.*[1] and Lenters and Cook[2] used general circulation model (GCM) experiments to show that the high elevation of the Andes combines with their narrow width to generate marked perturbations in South America's climate, even though the far-field effects on global climate are small. By extrapolation, topographic uplift of the Andean chain probably perturbed the paleoclimate of South America. This paper summarizes the history of that uplift between 16° and 33°30'S (Fig. 1), a zone that spans the tropics, a subtropical high-pressure belt, and the northern margin of the westerlies.[3]

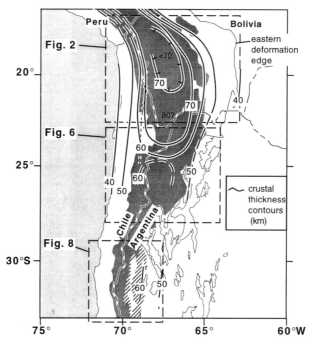

FIGURE 1. The Central Andean orogen above 3000 m elevation (dark shading) coincides with a region of thickened crust (expressed by contours of depth to Moho). Outlined regions east of the high-elevation ranges are also mountain ranges, uplifted during the Neogene. Cross-hatched area south of 29°S is the Pre-Cordillera thrust belt region and eastern flank of the Frontal Cordillera. Moho contours north of 28°S from compilation of Allmendinger *et al.*,[10] and rely primarily on refraction studies and broadband earthquake studies.[4,68-70] Gravity-based interpretations are indicated by the dashed 60-km contour near 25°S.[71] Incomplete contours south of 30°S are based on seismological studies of Regnier *et al.*[72] Thus the mapped contours mostly reflect the thickness of crust as defined by velocity properties, rather than crust as defined by density. Although elevation is directly related to the thickness of low-density (rather than low-velocity) crustal rocks, we consider crustal thicknesses defined by velocity to have less circularity in our analysis.

TABLE 1.   Principal Morphotectonic Units of the Central Andes

Region	Latitudes	Youngest magnetic arc	Foreland provinces	Initial uplift age (approximate)
1	16°–23°S (Fig. 3)	Western Cordillera — active volcanic chain	Altiplano — internally drained basin Eastern Cordillera — variably verging reverse and thrust faults Sub-Andean zone — thin-skinned thrust belt	25–26 million years
2	23°–28°S (Fig. 6)	Western Cordillera (Maricunga belt on western flank) — active volcanic chain	Puna — multiple internally drained basins; numerous stratovolcanos and calderas Eastern Cordillera — variably verging thrust faults Santa Bárbara System — variably verging reverse faults	16–17 million years
3	29°–31°30′S (Fig. 8)	Frontal Cordillera — inactive arc; reverse faults	Precordillera — thin-skinned thrust belt Sierras Pampeanas — reverse fault-bounded basement uplifts	20 million years
4	31°30′–33°30′S (Figs. 8 and 10)	Principal Cordillera — inactive arc; Aconcagua–La Ramada thrust belt	Precordillera — variably verging thin-skinned thrusts Sierras Pampeanas — reverse fault-bounded basement uplifts	19 million years

The high Andes mountains, in the 1700-km-long zone of interest, comprise a 350- to 400-km-wide Central Andean Plateau north of 27°S (see Fig. 1) and a 100-km-wide, rugged, ridge system to the south. Table 1 summarizes the locations and gross geologic aspects of the morphotectonic domains discussed here. In each of the four latitudinal belts discussed below, the basic elements are: a forearc zone, beginning at sea level (not discussed further), a magmatic arc (in some regions active, in others inactive for the last 6–10 million years) along the crest and western margin of the Cordillera, and east of the crest a broad foreland region dominated by structures indicative of crustal shortening.

High elevation in the Andes correlates closely with regions of unusually thick crust (Figs. 1 and 2a).[4,5] Between 17 and 24°S (Western Cordillera and Altiplano-Puna plateau) the crust is more than 70 km thick, whereas to the south, between 24° and 33°S (Puna Plateau, Frontal and Principal Cordilleras), it is at least 60 km thick (see Fig. 1).

We start with two premises concerning the topographic uplift history of the Central Andes:

1. Because elevation reflects isostatic adjustment to crustal thickness (Fig. 1) and lithosphere density structure (hence temperature), surficial uplift correlates in time with thickening crust or thinning mantle lithosphere. Isacks[6] calculated that the main contributor to high elevations in the Andes is thick crust (see e.g., Fig. 2a). Indeed, the predominant control by crustal thickness may be indicated by the fact that, from 14° to 28°S, the degree of mantle lithosphere thinning varies considerably along strike (based on Q in the lithosphere, other geophysical properties, and the chemistry of young volcanic rocks[7,8]), but the average elevation changes by less than 1 km (Fig. 1).

2. The main mode of crustal thickening in the Central Andes is shortening. Although Mesozoic and Cenozoic magmatic activity was an important ingredient in the evolution of the Andes, the volume of crust generated by magmatism and the rate of magmatic thickening of the crust are both grossly inadequate to explain the crustal volume and history.[6,9,10] In contrast, amounts of shortening, known best for the marginal thrust belts (Sub-Andean and Pre-Cordillera), but also estimated for the Eastern Cordillera, can account for 80–90% of the present crustal volume (especially apparent in Fig. 2a).[6,10–16] Therefore, we use history of crustal *shortening* as a proxy for history of crustal *thickening*, to determine the time(s) of principal elevation change.

One available data set that bears on past crustal thickness is the geochemistry of magmatic rocks, which evolve in response to the pressure conditions imposed by the thickness of overlying crust.[17–19] However, such geochemical data are limited to the ancient magmatic arc regions. In light of the above premises, our principal approach to determining the topographic uplift history of the Central Andes is to examine constraints on ages of crustal shortening from four study areas. One source of information about deformation ages is cross-cutting relations between structures and dated volcanic and sedimentary rocks. A

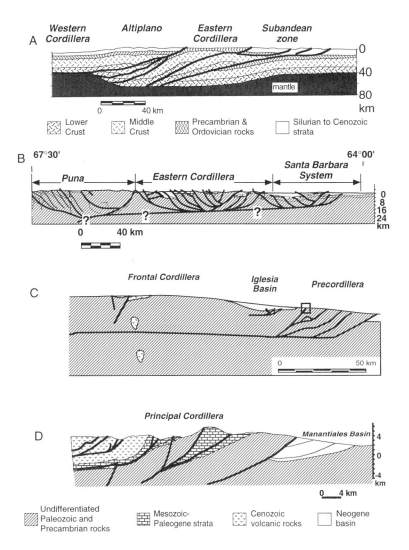

FIGURE 2. Cross sections from four regions discussed in text, at varying scales (all no vertical exaggeration). In a, b, and c, the faults shown are Cenozoic age contractional structures, some of which are reactivated older faults; in d, small offset Mesozoic normal faults with no Cenozoic slip are also visible [(a) from Schmitz[13]; (b) from Grier et al.[73] and Marrett et al.[43]; (c) from Allmendinger et al.[16]; (d) from Jordan et al.,[62] based partly on Ramos et al.[74]].

more regionally extensive data set comes from the ages of the strata found at the bases of foreland basin sequences. The results, detailed below, reveal that late Cenozoic Andean shortening began at least 6 million years earlier in the region between about 16° and 23°S than it did in areas farther south.

A vital step is to identify the initial surface datum from which uplift occurred. Unless the surface was at sea level immediately prior to major uplift, we need a description of the immediately preceding topography and/or crustal thickness, and their regional variations. As detailed below, the Neogene proves to be the time of principal crustal thickening. Thus to attempt to identify the initial datum, as part of the description of each region, we review the sparse data available on Oligocene crustal thickness. Two general observations are important in considering the Oligocene crustal thickness and topography. First, major segments of what is now the Central Andean Plateau were sufficiently low-lying in the late Cretaceous to be covered by shallow marine waters.[20-22] Second, Eocene deformational and magmatic activity, known as the Incaic event, produced some crustal thickening. South of 22°S, the Incaic effects visible at the surface were largely restricted to a zone west of the Western Cordillera,[23] and hence west of the zone with thick crust today.[4] However, north of 22°S, it contributed to thickening the crust of the Altiplano and Eastern Cordillera.[22]

2. OLIGOCENE CRUSTAL THICKNESS AND NEOGENE UPLIFT HISTORY

2.1. Region 1: Altiplano – Eastern Cordillera – Sub-Andean Zone (16 °–23 °S)

Little is known about the crustal thickness of the latest Oligocene and early Miocene based on available volcanic chemistry data, although Coira *et al.*[24] suggest that the crust underlying the Eastern Cordillera and eastern margin of Altiplano was more than 35–40 km thick. Indeed, Kennan *et al.*[25] and Sempere *et al.*[22] describe evidence of shortening in the Altiplano and Eastern Cordillera during the Eocene. Additional minor thickening of the Altiplano crust during the Paleogene occurred as a result of accumulation of at least 5 km of strata.[25]

In the west-verging Huarina thrust belt of the northern Altiplano (Fig. 3), Sempere *et al.*[26] interpret that strata at Salla, which bear Deseadan fauna and are unconformable over deformed Eocene strata, accumulated in a piggyback basin that was contemporaneous with the deformation. Tephrochronology and magnetostratigraphy can be interpreted to indicate that the Salla beds range in age from approximately 25 million years near the base of the section to about 20 million years near its top,[26,27] but MacFadden *et al.*[28] and Flynn and Swisher[29] use similar data to conclude that the age of the section ranges from about 28 to 25 million years.

Between 21° and 22°S latitude, preservation of late Oligocene and Miocene strata in the western part of the Eastern Cordillera (Tupiza area basins, Fig. 3) permits detailed reconstruction of the local deformation history. West-directed thrusting initiated in the latest Oligocene, and some combination of west- and east-vergent thrusting continued throughout the early and middle Miocene.[30-32]

In Region 1, the modern gravity field of the eastern flank of the Central Andean Plateau and the Chaco Plain is an excellent example of flexural subsidence in a foreland basin in response to crustal thickening by thrust faulting.[8,33,34]

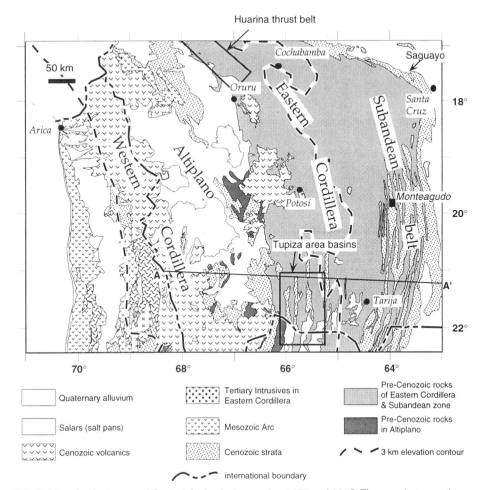

FIGURE 3. Geologic map of Central Andes between about 17° and 22°S. The area between the 3-km-contours constitutes the Central Andean Plateau. See Table 1 for general properties of the morphotectonic provinces that are labeled (modified from Pareja and Ballon[75] and Allmendinger et al.[10]). A-A' indicates line of cross section shown in Figure 2a.

Consequently, we hypothesize that accumulation of strata in the Sub-Andean foreland basin (subsequently deformed by thrust faulting) closely reflected the thickening and uplift of the contemporaneous mountain system. Strata began to accumulate in the foreland basin adjacent to the Altiplano in the Oligocene, consistent with the sparse data on initiation of thrusting. Near Santa Cruz (Quebrada Saguayo, Fig. 3), Deseadan fossils in strata 2 m above the base of the foreland basin section indicate that foreland basin accumulation began between 29 and 21 million years ago.[26] This age, dependent on the ages of the Deseadan land mammal stage, is completely consistent with Deseadan-bearing Salla strata, but very imprecise. Near Monteagudo (western part of Sub-Andean belt, near 20°S) (Fig. 4), an ash, fission-track dated from 24.4 ± 2.6 million years ago, occurs about 400 m above the base of the foreland basin section (Fig. 5).[35] One can only speculate about the age of the basal several hundred meters of section, but clearly thickening and uplift capable of generating a flexural basin began no later than the late Oligocene.

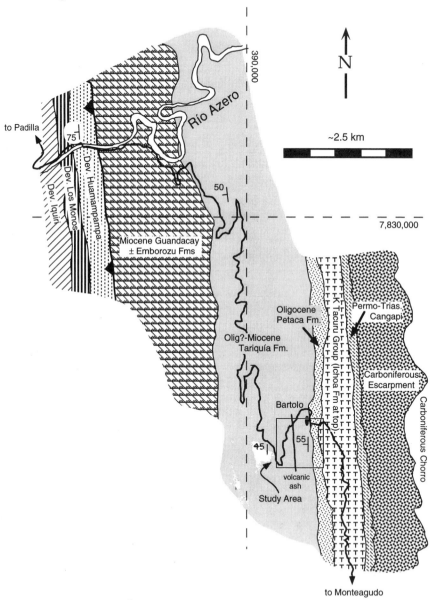

FIGURE 4.   Geologic map of region of study near Monteagudo, Bolivia, for which foreland basin strata reveal Miocene ages (see Fig. 5) (from Erikson and Kelley[35]). Abbreviations: Dev. = Devonian; Olig. = Oligocene; K = Cretaceous; Permo-Trias. = Permian and Triassic.

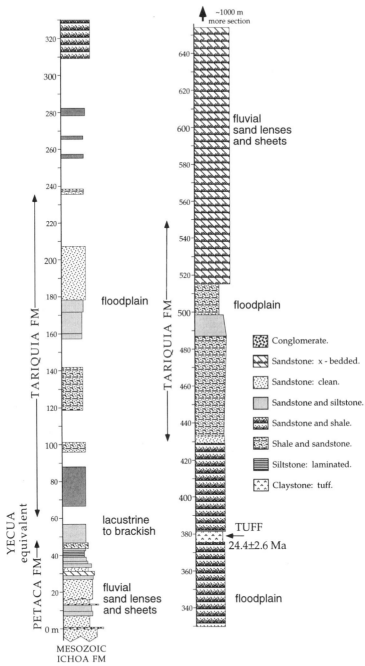

FIGURE 5. Stratigraphic terminology, lithologies, and depositional environments of strata in the lower part of the foreland basin section near Monteagudo, Bolivia (for locations, see Figs. 3 and 4). Lithostratigraphic terminology follows local usage of YPFB geologists, who consider the "Yecua equivalent" interval to be generally similar to, but distinguishable from, the Yecua Formation, which occurs widely to the east. The tuff horizon nearly 400 m above the base of the section was dated by fission-track analysis of zircons.[35] Abbreviations: FM = Formation; x-bedded = cross-bedded.

In summary, the high Andean Cordillera between about 16° and 23°S began a phase of crustal thickening and, by inference, uplift in the late Oligocene.[26] Prior to that late Oligocene uplift, what is now the Eastern Cordillera was probably already somewhat mountainous.[22,25] Crustal shortening and uplift continued during the early and middle Miocene in the Eastern Cordillera[30–32] and during the late Miocene to the present in the Sub-Andean belt.[14,36–38] Data are not yet adequate to conclude how much thickening and uplift occurred at progressive time steps in the Neogene.

2.2. Region 2: Maricunga Belt – Puna – Eastern Cordillera – Santa Bárbara System (23 °–28 °S)

Although the principal region of Eocene arc magmatism and Incaic deformation lies west of the Western Cordillera in this segment,[23] Andriessen and Reutter[39] attribute unroofing as far east as the Eastern Cordillera to this time interval (Fig. 6). Folds in the central Puna attributed to the Incaic event[40,41] are in fact very poorly dated and may well be of Neogene age. Chemical and isotopic

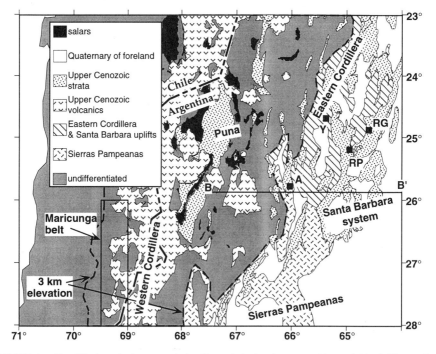

FIGURE 6.   Simplified geologic map of the Central Andes between 23° and 28°S. The region between the 3-km-contours is the Central Andean Plateau. Salars are internally drained salt pans. B-B′ indicates the position of the cross section shown in Fig. 2b. Foreland basin sections mentioned in the text are labeled A (Angastaco), Y (Yacones), RG (Río González), and RP (Río Piedras).

properties of late Oligocene and early Miocene volcanic rocks of the Maricunga belt (Fig. 6) suggest that the crust beneath what is now the westernmost plateau ranged in thickness from 35 km near 26°S to 50 km near 28°S.[19,42]

Comparison of the chemistry of 17 to 12-million-year-old volcanic rocks to that of earlier Miocene volcanic units in the Maricunga zone suggests that the crust thickened subtly in the interval between 20 and 17 million years.[19] Accumulation of the Atacama gravels on the western flank of the Central Andean Plateau indicates that there were high elevations to the east from about 16 to 12 million years ago.[42] However, Maricunga belt volcanic rocks dating from 11 to 7 million years ago are chemically more distinct from their predecessors than were those of the 17 to 12-million-year age range, suggesting significant thickening of the crust later than 12 million years ago. Volcanic chemistry suggests another thickening interval later than 7 million years ago.[19]

In the Puna region (Figs., 2b and 6), the oldest recognized Neogene shortening occurred between 17 and 13 million years ago, bracketed by cross-cutting relations of structures and volcanic units.[43] Even though such structural data are sparse, the sedimentary basins of the southern Puna provide a broad view of the regional distribution of crustal shortening. Late Oligocene redbeds seem to be regionally extensive, stretching across reverse-fault-bounded blocks and modern basins without interruption.[44] Overlying those redbeds, a regionally discontinuous evaporite–clastic sequence is strongly diachronous from one modern structural basin to the next. Ages of the units, thought to have accumulated in small but long-lived basins controlled only by local upper crustal structures, range from about 15 million years to the present.[44,45] The evaporitic nature of the strata indicates a climate and hydrology similar to today's, and implies that topographic uplift of the Puna prior to 15 million years ago had created internal drainage and a rainshadow like those that characterize the modern high plateau. In summary, structural shortening within the Puna began about 15 million years ago.

The foreland basin east of the Puna Plateau contains detritus shed from the Eastern Cordillera and from the ranges of the thick-skinned Santa Bárbara System (Figs. 2b and 6). Controls on subsidence in this part of the foreland basin have not been determined, but Whitman et al.[8] concluded that the eastern margin of the Andes in the belt is locally, rather than flexurally, compensated, implying that the subsidence history may not be simply linked to the shortening history of the Central Andean Plateau. Nevertheless, chronological data for this region are relatively complete. Reynolds et al.[46] reported on four sections with magnetic polarity stratigraphy in the foreland basin, exemplified by the Río González section (Fig. 7). Throughout the region the Anta Formation includes interbedded ashes (dates at multiple sites range between 14 and 15 million years ago), locally as much as 750 m above the base of the foreland basin section. Magnetic polarity zones below the dated horizons can be interpreted most simply to indicate an age of the base of the section near 16 million years.[46]

Sandstone provenance data from paleomagnetic sections at Río González[47] (J. R. Reynolds, R. Hernández, B. D. Idleman, J. M. Kotila, and R. V. Hilliard,

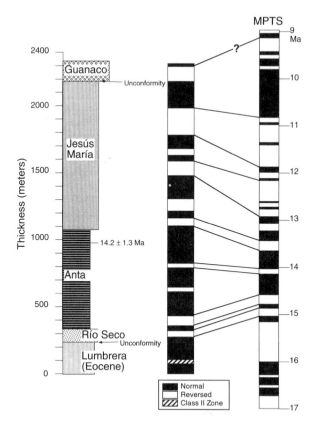

FIGURE 7. Formations (with resistance to erosion indicated by width of rectangle), magnetic polarity stratigraphy, and correlation to the magnetic polarity time scale of Cande and Kent[77] for strata in the foreland basin at Río González, Argentina (for location, see Fig. 6) (based on Reynolds *et al.*[46]). Independent age control is provided by a fission-track date on zircon of an intercalated volcanic ash.

unpublished), Río Piedras,[46,48] Río Yacones,[49] and Angastaco[50] (Fig. 6) provide insight to the sequence of foreland deformation.[49] Data from across the region indicate that the Puna was shedding sediment by 16–15 million years ago, western parts of the Eastern Cordillera were uplifted by 13–12 million years ago, and the eastern part of the Eastern Cordillera was a source area perhaps as early as 14(?)–13(?) million years ago. Thereafter, deformation in the predominantly west-verging Santa Bárbara System apparently migrated eastward, throughout late Neogene time. The Eastern Cordillera remained active, contemporaneous with thrusting in the Santa Bárbara System, until more recently than 6 million years ago.

In summary, the high Andean Cordillera between about 23° and 28°S exhibits no evidence of crustal thickening or uplift earlier than 20 million years ago. Beginning about 16–17 million years ago, multiple features (magmatic arc chemistry, the Atacama gravels, Puna basins and structures, and the foreland basin) independently reveal shortening. Thereafter, shortening within the Puna can be documented at many intervening stages until about 2–3 million years ago,[43] and shortening in the foreland continues to the present.[51]

2.3. Region 3: Flat-Slab Region: Frontal Cordillera – Pre-Cordillera – Sierras Pampeanas (29°–31°30′S)

Although the Sierras Pampeanas foreland basement uplifts (Fig. 8) are individually high ranges (typical peak elevations ranging from 3000 to 6000 m), it is the north-trending, continuous Andean Cordillera that is most likely to have

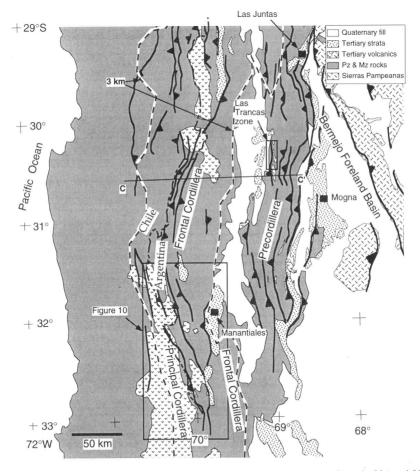

FIGURE 8. Simplified geologic map of the Central Andes between approximately 29° and 33°S. C-C′ indicates the position of the cross section shown in Fig. 2c. Foreland basin sections mentioned in the text are labeled. The Aconcagua–La Ramada thrust belt constitutes most of the region in the box corresponding to Fig. 10.

affected regional atmospheric circulation patterns. Thus we focus here entirely on the uplift history of the Frontal Cordillera and Pre-Cordillera.

The thickness of the crust beneath the Frontal Cordillera has not been measured directly, but seismological studies indicate a 60-km-thick crust beneath the Pre-Cordillera thrust belt (Fig. 1). Elastic modeling of gravity data suggests that the crust beneath the Frontal Cordillera is at least 55 km thick.[5] Late Oligocene to early Miocene volcanic rocks from 28° to 31°S (Fig. 8) are chemically similar to those in the modern southern volcanic zone, suggesting that the crust of what is now the High Andes (Frontal Cordillera) was 35–40 km thick.[5,17,18] No deformation or magmatism attributable to the "Incaic phase" of deformation is known at these latitudes.

Cross-cutting relations between volcanic rocks and high-angle reverse faults in the Frontal Cordillera (Fig. 2c) indicate deformation between 18 and 15 million years ago.[52] Arc chemistry implies thickening of the crust in about the same time interval. Kay *et al.*[18] showed that middle Miocene (16–11 million years ago) arc volcanic rocks are compositionally and chemically similar to modern magmatic rocks near 33°–34°S, where the crust is now 50–55 km thick, in contrast to their estimate of 35–40 km for the late Oligocene crust of the Frontal Cordillera.

Shortening of the westernmost Pre-Cordillera thrust belt (Las Trancas zone, Fig. 8) began somewhat earlier than in the adjacent Frontal Cordillera. Approximately 21 million years ago, redbeds accumulated across the region, with no apparent relationship to thrusting. However, 19.5 million years ago, conglomerates were derived from uplift of a thrust sheet and locally overlie that same thrust fault.[53]

In the Bermejo foreland basin (Fig. 8), many sections are well dated and there is extensive information concerning source histories of the clasts and subsidence histories. The base of the foreland basin is dated in most detail at the Mogna and Las Juntas (Figs. 8 and 9) sections, where magnetic polarity stratigraphy (calibrated with ash ages) indicates that the basin began to accumulate sediment around 18–19 million years ago[54,55] (J. Milana, F. Bercowksi, and T. Jordan, work in progress). At Las Juntas (Fig. 9), an undated eolian unit deposited in the foreland basin underlies 18-million-year-old strata, and probably correlates to eolian units known to overlie the 21-million-year-old redbeds.[56]

Thrusting in the Pre-Cordillera was interrupted in the interval from 19 to 16 million years ago. When it resumed, the locus of shortening migrated into the Central Pre-Cordillera. Most of the shortening occurred from about 15 million years ago to the present, with a peak in shortening rate around 10 million years ago.[53] In proximal parts of the foreland basin (Fig. 9), the accumulation of strata was slow until about 15 million years ago, when the rate increased markedly.

In summary, for the high Andean Cordillera between about 29° and 31°30'S, evidence of crustal thickening and related uplift is first seen about 20 million years ago. Thickening by faulting in the magmatic arc itself occurred between 18 and 15 million years ago, whereas the major interval of shortening in the foreland began about 15 million years ago.

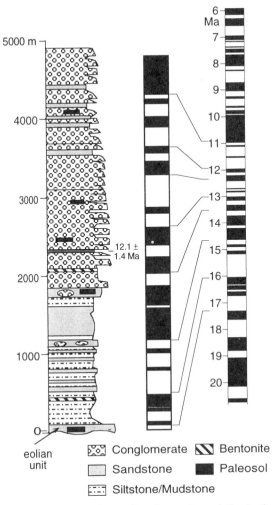

FIGURE 9. Formations, magnetic polarity stratigraphy, and correlation to the magnetic polarity time scale of Cande and Kent[76] for strata in the Bermejo foreland basin at Las Juntas, Argentina (for location, see Fig. 8) (modified from Reynolds *et al.*[54]). Independent age control is provided by a fission-track date of an intercalated volcanic ash.

2.4. Region 4: Principal Cordillera – Frontal Cordillera – Pre-Cordillera (31°30′–33°30′S)

Oligocene and early Miocene magmatic rocks of the western flank of the Andean crestline are isotopically primitive, with the most MORB-like properties of any units erupted from the Cretaceous to Recent[57,58] (A. Kurtz and S. M. Kay, unpublished data), suggesting crustal thinning and mantle upwelling at that time.[57] These rocks occupy a belt whose present crustal thickness exceeds 55 km (Fig. 1).[5]

Unlike the area immediately to the north, the style of deformation in the main Andean Cordillera (Principal Cordillera) (Figs. 8 and 10) is at least partly thin-skinned, including the Aconcagua thrust belt in the south and the La Ramada thrust belt in the north (see Fig. 2d). Consequently, there is likely to have been more local shortening than in Region 3.[59,60] Shortening farther east, in the thick-skinned Frontal Cordillera and the complex Pre-Cordillera thrust belt (Fig. 8), also must have caused uplift.

Although Cerro Aconcagua, the highest peak in the Western Hemisphere (Fig. 10), began as a stratovolcano, it owes much of its elevation to uplift in the Aconcagua thrust belt.[61] Ramos *et al.*[61] suggest that K/Ar ages of a stock and its wall rocks in the western part of the thrust belt are due to uplift and denudation caused by initial thrusting around 21 million years ago. Kurtz[58] also suggested that a pluton on the western flank of the Andes near 33°30′S was slowly denuded between 22 and 16 million years ago. More direct evidence, cross-cutting relations of structures and volcanic rocks, suggests that thrusting in the central and eastern parts of the thrust belt was underway by 15 million years ago and continued until about 9 million years ago.[61] The chemistry of volcanic rocks at Aconcagua that are from 15 to 8 million years old suggests that the crust was like today's crust near 34°S in the southern volcanic zone (estimated to be approximately 50 km thick)[5,18] Rare-earth element compositions of late Miocene and Pliocene plutonic rocks near 34°S indicate that the magmas must have equilibrated in crust that was thicker than 60 km.[58]

The Manantiales foreland basin is associated with the Principal Cordillera thrust system (Figs. 2d, 10). Magnetic stratigraphy, calibrated by ages of interbedded ashes, shows that accumulation in the basin began around 19 million years ago (Fig. 11). The history of the basin suggests that thrusting was active as late as 9 million years ago.[62,63] Irigoyen *et al.*[64] described preliminary data suggesting that the Frontal Cordillera at 33°S began uplift around 9 million years ago. Bercowski *et al.*[65] showed that accumulation in the foreland basin east of the Pre-Cordillera began no later than 15 million years ago.

In summary, for the high Andean Cordillera between about 31°30′ and 33°30′S, the first evidence of crustal shortening and related uplift is seen about 19 million years ago. Shortening by the Aconcagua–La Ramada thrust belt was responsible for only part of the total thickening and, by inference, uplift of the zone. Crustal thickening expressed in the Frontal Cordillera and Pre-Cordillera also contributed, but the timing is not well known.

3. CONCLUSIONS

Uplift of the Central Andes between 16° and 28°S occurred in two stages, widely separated in time: (1) Parts of the Central Andean Plateau area were near sea level in the Cretaceous; (2) Eocene (Incaic) deformation accounted for perhaps one-quarter to one-half of the Cenozoic thickening and uplifting in limited regions, largely in the Western Cordillera but at least locally in the plateau region

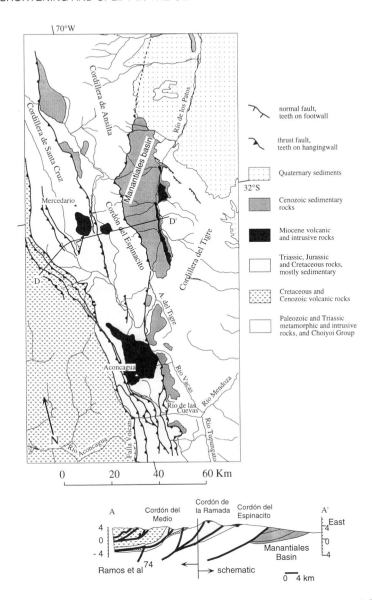

FIGURE 10.   Geologic map of the Aconcagua–La Ramada thrust belt in the Principal Cordillera, and neighboring parts of the frontal Cordillera (see Fig. 8 for location) (from Jordan et al.[62]). The Manantiales basin is the foreland basin to the La Ramada thrust belt. D-D' indicates position of cross section shown in Fig. 2d.

(Altiplano and part of the Eastern Cordillera); and (3) a phase of late Oligocene and younger thickening–uplift accounted for most of the Central Andean thickening and elevation, including all that east of the plateau.

North of 23°S there is firm evidence that the second and more important stage of crustal shortening and thickening began in the late Oligocene (approxi-

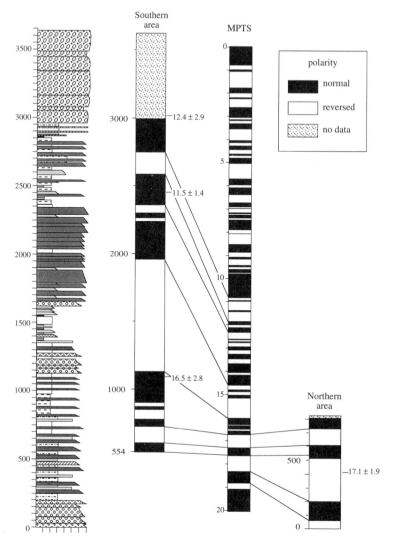

FIGURE 11.   Lithologies, magnetic polarity stratigraphy, and correlation to the magnetic polarity time scale of Cande and Kent[77] for strata of the Chinches Formation in the Manantiales foreland basin, Argentina (for location, see Fig. 10) (simplified from Jordan et al.[62]). Independent age control is provided by four fission-track dates of intercalated volcanic ashes.

mately 26 million years ago). South of 23°S, there is firm evidence that shortening began no earlier than 20 million years ago. Indeed, several lines of evidence suggest that important shortening between 23° and 28°S began 16–17 million years ago, whereas from 29° to 33°30′S shortening began 20 million years ago. Thus topographic uplift of the Andean crestline was diachronous in the Central Andes, and 6 to 10 million years earlier from 17° to 23°S (in the tropics) than from 23° to 33°30′S (in the subtropical high zone and westerlies). The exact latitude across which the time of initial uplift changed from the late Oligocene to the early or middle Miocene is uncertain from the available data.

Although many authors have concluded that the reorganization of plates and acceleration of convergence 26 million years ago caused the initiation of shortening along the entire Central Andean margin,[66,67] response to that acceleration is dissimilar along strike. Shortening and uplift began at least 6 million years earlier in the north than in the south, and may be youngest in a middle zone (23°–28°S), compared to neighboring zones to its south and north.

Acknowledgments

We thank Richard W. Allmendinger, Suzanne Mahlburg Kay, and Bryan Isacks (Cornell University) for sharing their knowledge of the Central Andes and data bases. We thank Constantino Mpodozis and Andres Tassara (SERNAGEO-MIN, Chile), T. Charrier (Universidad de Santiago, Chile), E. Scheuber (Freie Universität Berlin, Germany), and T. Sempere (ORSTOM) for tremendous assistance finding not-yet-published constraints on pre-Neogene crustal thicknesses, and T. Sempere, S. Tawackoli (Freie Universität Berlin, Germany) and B. Horton (University of Arizona) for preprints of their studies in Bolivia. Rick Allmendinger generously assisted in preparation of the manuscript and figures. We gratefully acknowledge the many contributions of YPF S.A. (Argentina) and YPFB (Bolivia) to our studies of the foreland basins. Supported in part by NSF award GER 9022811 to Jordan, and by Petroleum Research Fund awards PRF #24348-B2 and PRF #28363-B8 to Reynolds.

REFERENCES

1. Lenters, J. D., Cook, K. H., and Ringler, T. D. (1995). *J. Clim.* **8**, p. 2113.
2. Lenters, J. D., and Cook, K. H. (1995). *J. Clim.* **8**, p. 2988.
3. Schwerdtfeger, W. (1976). In: *Climates of Central and South America, Vol. 12* (W. Schwerdtfeger, ed.), pp. 1–12. Elsevier, Amsterdam.
4. James, D. E. (1971). *J. Geophys. Res.* **76**, p. 3246.
5. Tassara, A., and Yañez, G. (1996). In: *Troisieme Symposium International sur la Geodynamique andine*, ORSTOM, Saint-Malo, France, p. 115.
6. Isacks, B. L. (1988). *J. Geophys. Res.* **93**, p. 3211.
7. Kay, S. M., Coira, B., and Viramonte, J. (1994). *J. Geophys. Res.* **99**, p. 24323.
8. Whitman, D., Isacks, B. L., and Kay, S. M. (1996). *Tectonophysics* **259**, p. 29.
9. Francis, P. W., and Hawkesworth, C. J. (1994). *J. Geol. Soc. Lond.* **151**, p. 845.
10. Allmendinger, R. W., Isacks, B. L., Jordan, T. E., and Kay, S. M. (1997). *Ann. Rev. Earth Planet. Sci.* **25**, p. 139.
11. Roeder, D. (1988). *Tectonics* **7**, p. 23.
12. Sheffels, B. M. (1990). *Geology* **18**, p. 812.
13. Schmitz, M. (1994). *Tectonics* **13**, p. 484.
14. Baby, P., Herail, G., Salinas, R., and Sempere, T. (1992). *Tectonics* **11**, p. 523.
15. Baby, P., Rochat, P., Hérail, G., Mascle, G., and Paul, A. (1996). In: *Troisieme Symposium International sur la Geodynamique andine*, ORSTOM, St. Malo, France, p. 281.
16. Allmendinger, R. W., Figueroa, D., Snyder, D., Beer, J., Mpodozis, C., and Isacks, B. L. (1990). *Tectonics* **9**, p. 789.

17. Kay, S. M., Maksaev, V., Moscoso, R., Mpodozis, C., and Nasi, C. (1987). *J. Geophys. Res.* **92**, p. 6173.
18. Kay, S. M., Mpodozis, C., Ramos, V. A., and Munizago, F. (1991). In: *Andean Magmatism and Its Tectonic Setting, Vol. 265* (R. S. Harmon and C. W. Rapela, eds.), pp. 113–137. Geological Society of America, Boulder, CO.
19. Kay, S. M., Mpodozis, C., Tittler, A., and Cornejo, P. (1994). *Int. Geol. Rev.* **36**, p. 1079.
20. Salfity, J., Marquillas, R., Gardeweg, M., Ramirez, C., and Davidson, J. (1985). *IV Congreso Geológico Chileno* **4**, p. 654.
21. Sempere, T. (1995). In: *Petroleum Basins of South America, Vol. 62* (A. J. Tankard, S. R., Suárez, and H. J. Welsink, eds.), pp. 207–230. American Association of Petroleum Geologists, Tulsa, Okla.
22. Sempere, T., Butler, R. F., Richards, D. R., Marshall, L. G., Sharp, W., and Swisher, C. C., III, (1997). *Geol. Soc. Am. Bull.* **109**, p. 709.
23. Scheuber, E., Bogdanic, T., Jensen, A., and Reutter, K.-J. (1994). In: *Tectonics of the Southern Central Andes: Structure and Evolution of an Active Continental Margin* (K.-J. Reutter, E. Scheuber, and P. J. Wigger, eds.), pp. 121–140. Springer-Verlag, New York.
24. Coira, B., Kay, S. M., and Viramonte, J. (1993). *Int. Geol. Rev.* **35**, p. 677.
25. Kennan, L., Lamb, S., and Rundle, C. (1995). *J. S. Am. Earth Sci.* **8**, p. 163.
26. Sempere, T., Hérail, G., Oller, J., and Bonhomme, M. G. (1990). *Geology* **18**, p. 946.
27. McRae, L. E. (1990). *J Geol.* **98**, p. 479.
28. MacFadden, B. J., Campbell, K. E., Jr., Cifelli, R. L., Siles, O., Johnson, N. M., Naeser, C. W., and Zeitler, P. K. (1985). *J. Geol.* **93**, p. 223.
29. Flynn, J. J., and Swisher, C. C., III (1995). In: *Time Scales and Global Stratigraphic Correlation*, Special Publication No. 54, pp. 317–333. SEPM.
30. Hérail, G., Oller, J., Baby, P., Blanco, J., Bonhomme, M. G., and Soler, P. (1996). *Tectonophysics* **259**, p. 201.
31. Tawackoli, S., Jacobshagen, V., Wemmer, K., and Andriessen, P. A. M. (1996). In: *Troisieme Symposium International sur la Geodynamique andine*, ORSTOM, Saint-Malo, France, p. 505.
32. Horton, B. (1996). In: *Troisieme Symposium International sur la Geodynamique andine*, ORSTOM, Saint-Malo, France, p. 383.
33. Watts, A. B., Lamb, S. H., Fairhead, J. D., and Dewey, J. F. (1995). *Earth Planet. Sci. Lett.* **134**, p. 9.
34. Lyon-Caen, H., Molnar, P., and Suárez, G. (1985). *Earth Planet. Sci. Lett.* **75**, p. 81.
35. Erikson, J. P., and Kelley, S. A. (1995). Late Oligocene initiation of foreland-basin formation in Bolivia based on newly dated volcanic ash. In: *IX Congreso Latinoamericano de Geología*. Ministerio de Energia y Minas, Caracas, Venezuela.
36. Reynolds, J. H., Hernández, R. M., Idelman, B. D., Naeser, C. W., and Guerstein, P. G. (1996). *Geol. Soc. Am. Abstr. with Prog.* **28**, p. A-59.
37. Reynolds, J. H., Idelman, B. D., Hernández, R. M., and Naeser, C. W. (1993). *Geol. Soc. Am. Abstr. with Prog.* **25**, p. 473.
38. Hernández, R. M., Reynolds, J., and Disalvo, A. (1996). *Boletín de Informaciones Petroleras* (*Tercera Epoca*) **12**, pp. 80–93. YPF, Buenos Aires, Argentina.
39. Andriessen, P. A. M., and Reutter, K.-J. (1994). In: *Tectonics of the Southern Central Andes: Structure and Evolution of an Active Continental Margin* (K.-J. Reutter, E. Scheuber, and P. J. Wigger, eds.), pp. 141–153. Springer-Verlag, New York.
40. Schwab, K. (1985). *IV Congreso Geológico Chileno* **1**, pp. 2-138–2-158.
41. Reutter, K.-J., Geise, P., Götze, H.-J., Scheuber, E., Schwab, K., Schwarz, G., and Wigger, P. (1988). In: *The Southern Central Andes — Lecture Notes in Earth Sciences* (H. Bahlburg, ed.), pp. 231–161. Springer-Verlag, New York.
42. Mpodozis, C., Cornejo, P., Kay, S. M., and Tittler, A. (1995). *Rev. Geol. Chile* **22**, p. 273.
43. Marrett, R. A., Allmendinger, R. W., Alonso, R. N., and Drake, R. E. (1994). *J. S. Am. Earth Sci.* **7**, p. 179.
44. Vandervoort, D. S., Jordan, T. E., Zeitler, P. K., and Alonso, R. N. (1995). *Geology* **23**, p. 145.
45. Vandervoort, D. S. (1993). Non-marine evaporite basin studies, southern Puna plateau, Central Andes: (Ph.D. Thesis), Cornell University, Ithaca, NY.
46. Reynolds, J. H., Idelman, B. D., Hernández, R. M., and Naeser, C. W. (1994). *Geol. Soc. Am. Abstr. with Prog.* **26**, p. 503.
47. Hilliard, R., Kotila, J. M., and Reynolds, J. H. (1996). *Geol. Soc. Am. Abstr. with Prog.* **28**, p. 442.

48. Galli, C. I., Hernández, R. M., and Reynolds, J. H. *Rev. Assoc. Geol. Arg.* (unpublished).
49. Kotila, J. M., Hilliard, R., Foldesi, C. P., Butz, D. J., and Reynolds, J. H. (1996). *Geol. Soc. Am. Abstr. with Prog.* **28**, p. 442.
50. Butz, D. J., Foldesi, C. P., and Reynolds, J. H. (1995). *Geol. Soc. Am. Abstr. with Prog.* **27**, p. 39.
51. Jordan, T. E., Isacks, B. L., Allmendinger, R. W., Brewer, J. A., Ramos, V. A., and Ando, C. J. (1983). *Geol. Soc. Am. Bull.* **94**, p. 341.
52. Maksaev, V., Moscoso, R., Mpodozis, C., and Nasi, C. (1984). *Rev. Geol. Chile* **21**, p. 12.
53. Jordan, T. E., Allmendinger, R. W., Damanti, J. F., and Drake, R. (1993). *J. Geol.* **101**, p. 135.
54. Reynolds, J. H., Jordan, T. E., Johnson, N. M., Damanti, J. F., and Tabbutt, K. T. (1990). *Geol. Soc. Am. Bull.* **102**, p. 1607.
55. Milana, J. P. (1991). Sedimentología y magnetoestratigrafía de formaciones cenozoicas en el area de Mogna y su inserción en el marco tectosedimentario de la Precordillera Oriental: (Ph.D. Thesis), Universidad Nacional de San Juan.
56. Jordan, T. E., Drake, R. E., and Naeser, C. (1993). *XII Congreso Geológico Argentino y II Congreso de Exploración de Hidrocarburos* **2**, p. 132.
57. Nyström, J. O., Parada, M. A., and Vergara, M. (1993). Sr–Nd isotope compositions of Cretaceous to Miocene volcanic rocks in Central Chile: a trend towards a MORB signature and a reversal with time. In: *Second International Symposium on Andean Geodynamics*, pp. 411–414, Oxford, UK.
58. Kurtz, A. C. (1996). Geochemical evolution and uplift history of Miocene plutons near the El Teniente copper mine, Chile (34°S-35°S): (Masters Thesis), Cornell University, Ithaca, NY.
59. Ramos, V. A. (1985). *IV Congreso Geológico Chileno* **1**, pp 2-104–2-108.
60. Cristallini, E. O., Mosquera, A., and Ramos, V. A. (1994). *Associatión Geológica Argentina, Revista* **49**, p. 165.
61. Ramos, V. A., Perez, D., and Aguirre-Urreta, M. B. (1990). *XI° Congreso Geologico Argentino* **2**, p. 361.
62. Jordan, T. E., Tamm, V., Figueroa, G., Flemings, P. B., Richards, D., Tabbutt, K., and Cheatham, T. (1996). *Rev. Geol. Chile* **23**, p. 43.
63. Perez, D. J. (1995). Evolución geológica de la región del Cordón del Espicacito, Provincia de San Juan, Argentina: (Ph.D. Thesis), Universidad de Buenos Aires, Buenos Aires.
64. Irigoyen, M. V., Brown, R. L., and Ramos, V. (1995). Magnetic polarity stratigraphy and sequence of thrusting: 33°S latitude, Mendoza province, Central Andes of Argentina (abst.). In: *Andean Thrust Tectonics Symposium*, pp. 16–17. International Union of Geological Sciences, International Program of the Lithosphere, and Asociación Geológica Argentina, San Juan, Argentina.
65. Bercowski, F., Ruzycki, L., Jordan, T., Zeitler, P., Caballero, M. M., and Perez, I. (1993). *XII Congreso Geológico Argentino y II Congreso de Exploración de Hidrocarburos* **1**, p. 212.
66. Pilger, R. J., Jr. (1984). *J. Geol. Soc. Lond.* **141**, p. 793.
67. Jordan, T. E., and Gardeweg, P. M. (1989). In: *The Evolution of the Pacific Ocean Margins* (2. Ben-Avraham, ed.), pp. 193–207. Oxford University Press, New York.
68. Wigger, P., Schmitz, M., Araneda, M., Asch, G., Baldzuhn, S., Giese, P., Heinsohn, W.-D. Martínez, E., Ricaldi, E., Röwer, P., and Viramonte, J. (1994). In: *Tectonics of the Southern Central Andes: Structure and Evolution of an Active Continental Margin* (K.-J. Reutter, E. Scheuber, and P. J. Wigger, eds.), pp. 23–48. Springer-Verlag, Berlin.
69. Beck, S. L., Zandt, G., Myers, S. C., Wallace, T. C., Silver, P. G., and Drake, L. (1996). *Geology* **24**, p. 407.
70. Zandt, G., Velasco, A. A., and Beck, S. L. (1994). *Geology* **22**, p. 1003.
71. Götze, H.-J., Lahmeyer, B., Schmidt, S., and Strunk, S. (1994). In: *Tectonics of the Southern Central Andes: Structure and Evolution of an Active Continental Margin* (K.-J. Reutter, E. Scheuber, and P. J. Wigger, eds.), pp. 7–23. Springer-Verlag, Berlin.
72. Regnier, M., Chiu, J.-M., Smalley, R., Jr., Isacks, B. L., and Araujo, M. (1994). *Bull. Seismol. Soc. Am.* **84**, p. 1097.
73. Grier, M. E., Salfity, J. A., and Allmendinger, R. W. (1991). *J. S. Am. Earth Sci.* **4**, p. 351.
74. Ramos, V. A., Oegarra, M., and Cristallini, E. (1996). *Tectonophysics* **259**, p. 185.
75. Pareja, L., and Ballon, A. R. (1978). Mapa Geológica de Bolivia (escala 1:1,000,000). YPFB and GEOBOL, Bolivia.
76. Cande, S. C., and Kent, D. V. (1995). *J. Geophys. Res.* **100**, p. 6093.
77. Cande, S. C., and Kent, D. V. (1992). *J. Geophys. Res.* **97**, p. 13917.

4

Late Neogene Uplift in Eastern and Southern Africa and Its Paleoclimatic Implications

Timothy C. Partridge

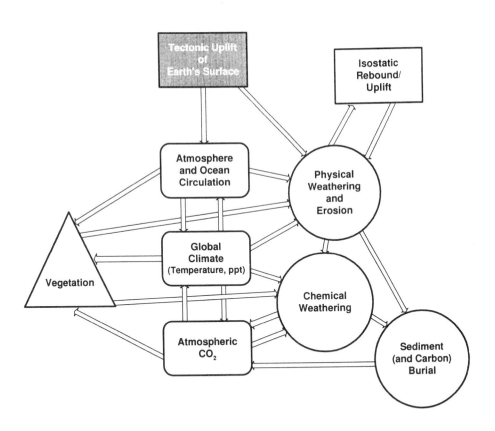

Timothy C. Partridge • Climatology Research Group, University of the Witwatersrand, WITS 2050, Johannesburg, South Africa.

Tectonic Uplift and Climate Change, edited by William F. Ruddiman. Plenum Press, New York, 1997.

1. INTRODUCTION: PLATE TECTONIC SETTING

With the exception of the Mediterranean and Red Sea littorals, Africa is one of the few continents whose underlying plate is entirely bounded by passive margins. The absolute motion of the African plate has, indeed, been relatively small since the end of the Cretaceous, amounting to no more than 14° of northward movement.[1] However, the rate of drift toward the Eurasian and Indian–Arabian boundaries has not been constant through this period: its progress was significantly influenced by global patterns of plate movement, especially during the Neogene.[2]

Pollitz[3] links a late to end Miocene (~6 million years ago) change in the absolute motion of the African plate to: (1) the opening of the modern Gulf of Aden and the Red Sea (5–4 million years ago); (2) a change in the orientation of principal stresses on the European platform (8–4 million years ago); (3) a slowdown in spreading rate along the southern Mid-Atlantic ridge (8–4 million years ago); and (4) the initiation of the extensional phase of the East African Rift (7–4 million years ago). The first three correlations associate the change in the absolute motion of the plate with relative-motion changes along its boundaries with the Indian, Eurasian, and (North-South?) American plates, respectively. The last correlation, based primarily on the chronology of the Kenya Rift, relates it to the ongoing breakup of the African plate into separate Nubian and Somalian plates, which the current model of global plate motions[4] fails to resolve.

In a recent reevaluation of motions within the African plate Jestin *et al.*[5] have emphasized the division of the African plate into western (Nubian) and eastern (Somalian) sectors separated by the tectonically active East African Rift System. They have proposed a model for these subplates in which the Euler pole of rotation for the Somalia plate is located about 56°S and 20°E. Around this pole the velocity of the southern African portion of the Somalia plate is predicted as some 3–4 mm/year in an eastward direction. Ben Avraham *et al.*[6] have noted however, that early results of space-geodetic very long baseline interferometry suggest that absolute motion in the vicinity of Hartebeesthoek Radio-Astronomical observatory in South Africa may be as high as 8 mm/year in a southeasterly direction, implying that this site may be located on an extension of the Nubian plate, rather than of the Somalian. Until further studies resolve this apparent anomaly, it is probably as well to bear in mind the conclusion of Gordon and Stein[7] that the southern African area represents a "wide plate boundary zone."

In connection with changes in stresses on the European platform and the possible relaying of Pacific–North American kinematic changes to the African plate via coupling along the North American–African plate boundary, Sloan and Patriat[8] have documented a change in axial orientation, spreading direction, and spreading rate that occurred along the Mid-Atlantic ridge at approximately anomaly 4 time (7.01 million years ago on a polarity timescale calibrated by Patriat). The African plate is, of course, directly coupled to the Eurasian plate in the Alpine–Mediterranean belt. If, as speculated above, the changes in Pacific and North American plate motion were induced by forces acting around the Eurasian–Pacific–North American triple junction, then alterations in the balance of forces along the Eurasian–African plate boundary may have been more directly responsible for the change 6 million years ago in African absolute motion. These events along the Eurasian–African boundary include the transformation of the larger Indo–Australian plate to the Indian plate and the latter's interaction with the Arabian plate.

During the terminal Miocene (6.0–5.0 million years ago) the link between the deep Mediterranean Basin and the Atlantic Ocean was repeatedly closed by a combination of glacioeustatic lowering of sea level and tectonic movements at the western end of the Mediterranean connected to its gradual narrowing during the late Cenozoic, as Africa drifted northward.[9,10]

The above evidence is interpreted as documenting a worldwide sequence of episodic tectonic changes during late Miocene to earliest Pliocene times, between roughly 10 and 5 million years ago. These changes appear mostly to follow temporally upon, and may therefore be a geodynamic consequence of, an important tectonic episode involving the culmination of major uplift along the Indian–Eurasian plate boundary around 8 million years ago.[11] Perhaps the analogy of a tectonic "domino effect" can be drawn, but in the absence of precise age constraints, it may not be appropriate to speculate on which of these tectonic episodes has causal priority. Because the system of lithospheric plates is a rigidly interconnected global entity on a closed surface, any local tectonic mechanism that allows all related episodic changes to proceed worldwide is, in a sense, a cause of them.

The evidence reviewed here, particularly that for incipient extension of the East African Rift System into southern Africa, goes some way toward explaining the recent stress regime within sub-Saharan Africa and the differing tectonic responses in the two regions. Zoback's[12] global synthesis has shown that large regions in the interiors of plates are characterized by uniform compressive stress orientations, which are produced by forces acting on the plate boundaries (e.g., drag resistance and ridge-push). In these interior areas maximum principal stresses are horizontal. In the southwestern part of southern Africa an east–west horizontal stress regime prevails, which has been dominated by compressional forces originating in the Mid-Atlantic Ridge.[12] This regime is manifested in strike-slip faulting along major east–west trending fracture systems. That the crust is widely at the point of failure is indicated by continuing seismic activity along these dislocations, most of which originated through mechanisms of inversion tectonics during Gondwana rifting, when faults generated in compressional regimes were reexploited in reverse mode during continental separation.[13]

In contrast, in the vicinity of the East African Rift, the stress field is dominated by buoyancy forces (swell–push), which have been generated by the upwelling of asthenospheric material and thinning of the lithosphere.[12] This has resulted in the rotation of the axis of maximum stress into the vertical plane, with S_2 orientated north-northwest–south-southeast, giving rise to a predominance of normal faulting. In the area between these two fundamentally differing regions, S_1 remains vertical, but S_2 assumes a north-west–southeast orientation.

2. REGIONAL TECTONIC RESPONSES

Tectonic responses in Africa must be gauged on a variety of evidence. Because of an absence of Cretaceous and younger marine sediments over most of the plateau areas, and also because very little crustal shortening has occurred during the Cenozoic, conventional criteria for inferring uplift can seldom be invoked. Constructional volcanic episodes (as along parts of the East African Rift) provide unambiguous evidence. On its own, rift faulting does not necessarily imply uplift of adjoining areas; ancillary evidence for upwarping of the rift shoulders (deformed geological contacts and planation surfaces) and local altitudinal influences on vegetation belts must be provided. The significance of the distribution of planation surfaces has long been a matter of debate, but, in the African context, is now less controversial because of the availability of better identification criteria and chronological control. Offshore sedimentary responses are also more precisely constrained.

The plate tectonic history and regional patterns of stress outlined in Section 1 have given rise to widely differing responses in eastern and southern Africa. The late Neogene component of drift of the African plate was too small to have caused any significant changes in climate. Slow, progressive crustal separation along the Rift Valley of East Africa has been linked to major uplift and volcanism during the Neogene. Evolution of the rift system was manifested in the development of a large number of discrete fault basins along its eastern and western branches, the more or less concurrent rise of two major domal structures (the Afar Plateau and the East African Plateau), and the superimposition of additional relief upon these elevated areas through the rise of the rift shoulders and the development of volcanic massifs, chiefly along the eastern branch of the rift system. The overall effect of these events was the augmentation of relief adjoining the rifts by more than a kilometer in many areas, while the topography within the confines of the rifts themselves became fragmented. Although significant locally, lava and ash outpourings and the generation of volcanic aerosols (gases and dust) were small by global standards and could hardly have given rise to important climatic effects.

Erosional processes operating on the newly uplifted rift margins would, given sufficient time, have reduced their environmental significance, at least locally; landscape degradation is, however, a slow and progressive process and is thus unlikely to have been associated with rapid changes in local environments in comparison with those wrought by periodic faulting movements. The generation of new relief through tectonic and volcanic activity seems, on the whole, to have counterbalanced erosion. Progressive sedimentation within the rift basins would, however, have tended to reduce the topographic fragmentation generated during recurrent faulting.

In southern Africa, Neogene tectonism followed a different style. There, large-scale epeirogenic uplift and flexuring with the eastern coastal hinterland were associated with a lesser degree of topographic fragmentation than in East Africa, except in some local areas. Late Neogene uplifts were somewhat smaller than their East African counterparts, with correspondingly fewer impacts on regional climate.

A link between the two regions is provided by the extensive areas of anomalously elevated topography with which they are associated. Named the African Superswell by Nyblade and Robinson,[14] this high-lying ground is present not only within the limits of the continent, where it carries across it a series of widely preserved and readily distinguishable planation surfaces, but extends into the surrounding oceans to the edge of the African plate (Fig. 1). As an entity it is therefore almost an order of magnitude larger than the Tibetan Plateau. Over most of the Superswell, the positive anomalies, defined in relation to global-mean continental and ocean-floor elevations, exceed 500 m; in fact, within large parts of East Africa, the anomaly is greater than 1000 m, and this is also the case for elevated areas of the eastern hinterland of southern Africa.

The broad, low-amplitude nature of the Superswell suggests that its origin should be sought in a buoyancy residing deep within the earth's mantle, a supposition confirmed by the recent studies on global tomography of Su *et al.*[15] and the seismic studies of Tanaka and Hamaguchi.[16] The smaller, high-amplitude anomalies superimposed upon the Superswell, such as the raised flanks of the East African Rift System, are linked to more localized mantle and lithospheric sources. The deep-seated buoyancy reflected by the Superswell is thought by Nyblade and Robinson[14] to be attributable to the movement of the African plate over numerous hotspots during the last 200 million years. The mushrooming of mantle-plume heads beneath the lithosphere can produce uplifts of about 1 km over diameters of 2000 km, indicating that large regional effects can be generated by movement over a relatively small number of hotspots. The presence of multiple hotspot tracks across southern Africa, however, suggests to Nyblade and Robinson[14] that the observed uplift may have been caused by thermal alteration of the lithosphere by plume tails rather than by plume heads. The presence of plume-derived material with temperatures elevated by no more than 10°–20°C could conceivably have provided the requisite density contrast. However, evidence for such pervasive thermal alteration of the lithosphere beneath both eastern and southern Africa is not unequivocally forthcoming from seismic studies, although measurements do suggest that, within the Kalahari Craton and the surrounding mobile belts, heat flow is raised sufficiently to account for regional uplifts of up to 500 m.[14]

The extensive negative Bouguer gravity anomalies within much of eastern and southern Africa (Fig. 2)[17] indicate the probable presence of widespread areas of low-density mantle lithosphere (with density deficiencies of 50–100 kg/m³), but these low-density areas are too large to be attributed to lithospheric heating alone; rather, they imply extensive thinning of the mantle lithosphere and its replacement by hot asthenospheric material, a circumstance that, again, is not firmly substantiated by the seismic evidence, although Bloch *et al.*[18] and Clouser and Langston[19] see evidence for a reduction in lithospheric thickness by some 50 km beneath parts of southern Africa, which may be a result of basal alteration or replacement. If sufficiently recent, these changes need not necessarily be associated with surface thermal anomalies. What is now beyond doubt is that southern Africa is roughly centered over the largest negative anomaly in *S*-wave

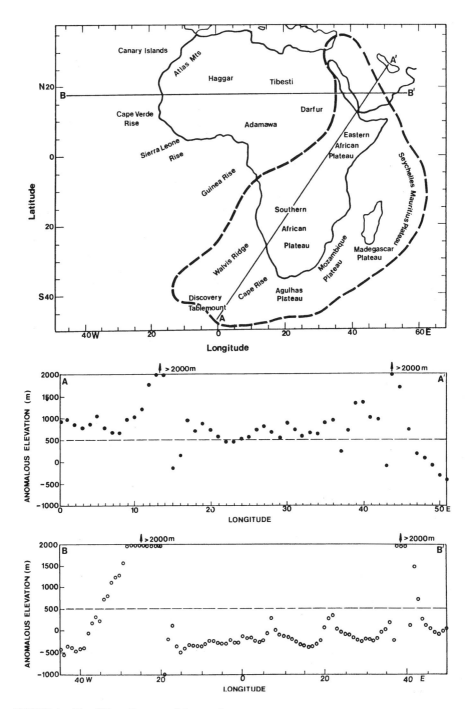

FIGURE 1.   The African Superswell (upper diagram, outlined by bold dashed line). A-A' and B-B' (lower two diagrams) are cross sections along the two transects (after Nyblade and Robinson[14]).

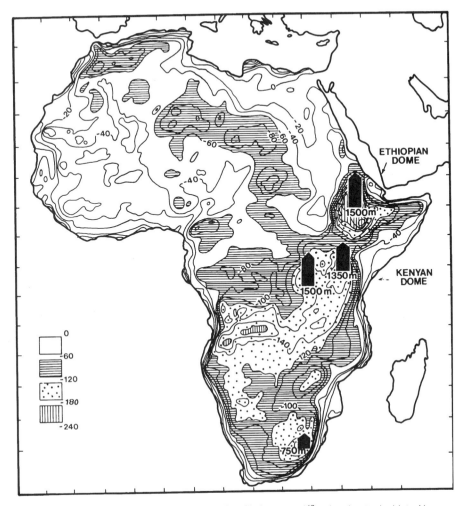

FIGURE 2. Bouguer gravity map of Africa (after Sletlene *et al.*[17]), showing typical late Neogene uplifts associated with the principal negative Bouguer gravity anomalies that characterize the eastern hinterland of the continent.

velocity within the lower mantle.[15] Moreover, seismic studies by Vinnik *et al.*[20] have detected the top of a "highly unusual" slow velocity layer below part of the Kaapvaal Craton of southern Africa at a depth of about 380 km, close to the mantle transition zone. It is entirely possible that this slow anomaly represents a mantle-plume head, similar to the East African plume, rooted in the larger deep mantle anomaly revealed in the tomographic studies.[21] Support is given to this hypothesis by the occurrence, in 1983, of a small volcanic eruption within the mountains in Lesotho. Recent unpublished studies have shown that the lava erupted has a deep-seated source and is probably the first manifestation of renewed volcanism in the area since the Cretaceous.

The timing of uplift within different regions of the African Superswell has been a matter of considerable debate and is considered further below.

3. EAST AFRICA

3.1. Uplift and Volcanism

The East African Rift System covers more than 3200 km from the Afar triple junction, formed by the Red Sea and the Gulf of Aden, to the lower Zambezi River valley in southern Africa. South of the Turkana depression in Kenya, the rift bifurcates into eastern (Kenya) and western (Gregory) branches around the Nyanza Craton, which coincides, in part, with the uplifted East African Plateau (Fig. 3).[22] Along both the Gregory and the Kenya rifts, significant additional relief

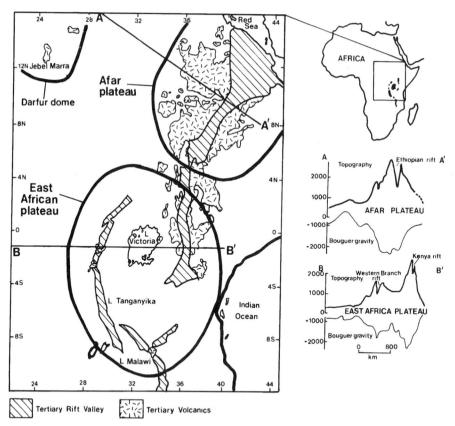

FIGURE 3. East Africa, showing the location of uplifted domes or plateaus and profiles of gravity and topography. The heavy solid lines enclose elevations greater than 800 m (after Ebinger *et al.*[22]).

has been superimposed upon the broad, domal topography of the East African Plateau. Much of this local relief is due to rift-shoulder uplift, but a significant part of that forming the Kenya Dome around the eastern rift is constructional and was produced by the eruption of some 140,000 km^3 of volcanics since the Miocene.[23]

Recent models of rift evolution can satisfactorily explain the complex suite of ecological and topographic features that make up the East African Rift System and have been reviewed by Partridge et al.[24] The pattern of broad negative Bouguer gravity anomalies that underlie the Afar and East African plateaus (Figs. 2 and 3) suggests that the uplift of the former is isostatically compensated while that of the latter may be overcompensated as a result of ongoing convection within the asthenosphere.[22] This ascent of asthenopheric material has given rise to a mushroom-shaped mantle plume beneath the apex of the Kenya Dome.[25] In both the Afar and Kenya rifts considerable lithospheric thinning is evident, but crustal extension probably amounts to no more than 20–30 km.

The pattern and timing of vertical tectonic events associated with East African Rift evolution are far from simple. In the 1000-km-wide Afar Plateau in the north, the eruption of flood basalts occurred in the Paleogene between 49 and 33 million years ago[26–28]; faults bounding the Ethiopian Rift began to develop in late Oligocene–early Miocene times,[26] and by the mid-Miocene the rift was well established.[29] Some uplift preceded this early faulting, and an episode of alkali basalt and trachyte volcanism occurred between 18 and 11 million years ago.[28] However, pollen recovered from lignites sandwiched between late Miocene (8 million years ago) basalt flows near Gondar display spectra similar to those of contemporary lowland rain forest,[30] indicating that elevation of at least the northwestern part of the Afar Plateau to its present altitude of ca. 2000 m has occurred since that time (Fig. 4). Baker et al.[31] referred the major component of this movement (1000–1500 m) to the late Pliocene; Adamson and Williams[29] cite more specific evidence, which links a change from lacustrine to fluviatile sedimentation in the Middle Awash Valley of the Ethiopian Rift, dated to between 4.0 and 3.8 million years ago, to tectonic disruption of early Pliocene lake systems following continued movements along the western fault scarp of the Ethiopian Rift. This faulting was associated with both the lowering of the Awash Graben and the continued rise of the Afar Plateau and served to accentuate climatic contrasts between lowland and upland through the medium of both altitudinal and rainshadow effects. Denys et al.[32] bracket a major phase of tectonic movement along the western scarp of the Ethiopian Rift between 2.9 and 2.4 million years ago, over which interval there was a massive influx of detrital sediments into the Hadar Basin.

The geological events of this period were certainly of a magnitude sufficient to induce major environmental changes: Bonnefille et al.[33] have noted that pollen spectra from the Hadar hominid site within the Ethiopian Rift indicate that the vegetation present between 3.3 and 2.9 million years ago has no analogues in the subdesertic steppe flora that characterize the area today and is most similar to modern vegetation communities occurring between elevations of 1600 and 2200 m

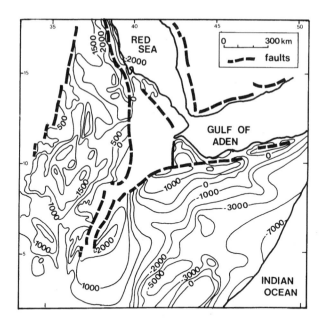

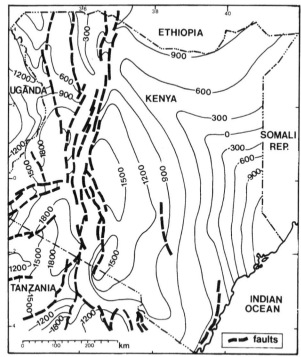

FIGURE 4.   (Top) Isohypsals of the Precambrian basement in Ethiopia. (Bottom) Isobases of the sub-Miocene erosion surface in Kenya (equivalent to the Post-African 1 surface in southern Africa). Elevations in meters (after Baker *et al.*[31]).

and receiving two to three times the present rainfall. They favor downfaulting within the graben of ca. 1000 m as the most likely cause of these changes. WoldeGabriel et al.[27] believe that graben subsidence may, in fact, have been as great as 2 km on the evidence of a marker tuff (the Munesa Crystal Tuff dated at 3.5 million years ago), which is exposed on both rift margins and is present also in a geothermal well beneath the rift floor. However, graben lowering on this scale could not have occurred without major concomitant uplift of the rift flanks as envisaged by Adamson and Williams.[29]

The available evidence is thus strongly suggestive of at least 1000 m of uplift within the Afar Plateau, with concurrent graben subsidence within the Ethiopian Rift, beginning later than 3.5 million years ago (probably around 2.9 million years ago). The greater part of these tectonic adjustments can, on the available evidence, probably be referred to the period prior to 2.4 million years ago, when the large influx of sediments into the Hadar Basin ceased. The arguments advanced by Molnar and England[34] against the use of sedimentological and paleobotanical data to infer uplift do not apply in this case in that the evidence points to *local* changes, which are most compatible with graben development, rather than to a regional response.

Averaging about 1200 m in elevation, the East African Plateau to the south supports narrower belts of more elevated topography associated with the Kenya and Western rift systems. The rift-flank areas are some 100–200 km wide and have been uplifted above the surrounding plateau by ca. 1000 m. Along the Kenya Rift, volcanism and regional uplift began in the early Miocene.[23,35,36] The doming brought about by these events probably dates to around 20 million years ago and amounted to no more than about 300–500 m on the basis of deformations of planation surfaces associated with deposits of well-established age.[31,35] Rapid sedimentation within the Tana Basin resulted. Unlike the Western Rift, where the elevated rift flanks are almost exclusively the result of uplift, a considerable proportion of the topography was associated with the growth of major volcanic edifices, particularly toward the northern end of the rift. In total, some 140,000 km^3 of volcanic material have erupted in the vicinity of the Kenya Rift since the early Miocene[23]; and the tempo of volcanic activity increased with time: flood phonolites were erupted from about 15 million years ago and were followed by trachytes and basalts from about 10 million years ago[36] (Fig. 5).[38–40] Important new volcanic centers were added around 5 million years ago. Major uplift, concentrated in the vicinity of the Kenya Dome, occurred above a deep anomalous mantle zone identified in teleseismic studies.[25] This phase of movement, documented by the deformation of well-dated planation surfaces, by the faulting and deformation of deposits of known age, and by renewed basin sedimentation, occurred later than about 3 million years ago and reached a maximum during the Plio–Pleistocene interval[31,35–37,41–43] (see Fig. 5). In the course of these movements the rift shoulders were raised by 1200–1500 m in places. The rift itself was characterized by the formation of a number of discrete fault basins, some containing relief of as much as 1800 m between rift floor and rift shoulder.

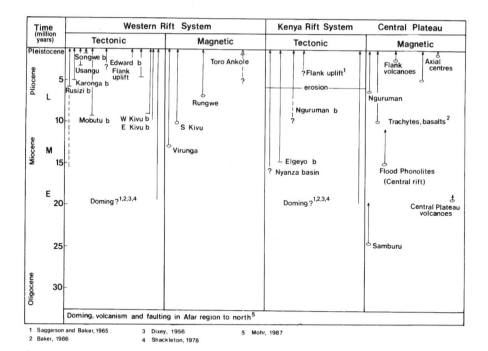

1 Saggerson and Baker, 1965   3 Dixey, 1956   5 Mohr, 1987
2 Baker, 1986   4 Shackleton, 1978

FIGURE 5. Chronological constraints on vertical movements, volcanic activity, and basinal subsidence within the East African Plateau region. Lines indicate approximate time span of activity; dashes and question marks are used where dating is uncertain (after Ebinger[37]). Sources: (1) Saggerson and Baker[35]; (2) Baker[36]; (3) Dixey[38]; (4) Shackleton[39]; (5) Mohr.[40]

The evolution of the Western Rift differed from that of the Kenya Rift in that early Miocene doming around the incipient rift structure was of smaller amplitude and volcanic activity began later (12–10 million years ago) and was restricted mainly to fault-bounded basins.[37,42,43] Uplift of the rift shoulders averaged about 1500 m but reached 4300 m in the Ruwenzori Mountains; on the evidence of sedimentary and volcanic sequences, the major part of these movements occurred in the period from 3 to 2 million years ago. Figure 6 represents a reconstruction of the three major stages of topographic and rift evolution in East Africa.[43] By the time of maximum uplift, the rift had become segmented into numerous separate basins, some of which had floors that extended below sea level. In large parts of both the Kenya and Western rifts this fragmentation persisted until about 2 million years ago, after which progressive sedimentation largely obliterated the effects of local rift structures. Uplift, particularly in the Ruwenzori area, continued well into the Pleistocene, and the ponding of Lake Victoria can, in fact, be linked to upwarping at the northern end of the Albertine Rift in the terminal Pleistocene. The typical amplitude of late Neogene uplifts in the eastern hinterland of Africa was indicated earlier in Fig. 2.

      In summary, an early phase of faulting and uplift occurred in the Afar Plateau area in Oligocene to early Miocene times and in the vicinity of the East

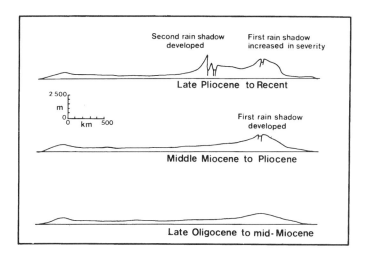

FIGURE 6. Schematic cross sections across the East African rifts at the equator, showing three stages in the evolution of the associated topography (after Pickford *et al.*[43]).

African Plateau during the early Miocene. Vertical movements totaling 300–500 m were involved in both areas. Along the Kenya Rift the subsequent growth of large volcanoes, accompanied by further tectonic movements, created sufficient relief by 7 million years ago to produce important local climatic effects. In both the Afar and East African plateau areas the major period of subsidence of the rift floors, with concomitant uplift of the rift shoulders by some 1000–1500 m, took place later than about 3 million years ago and reached a maximum in the Plio–Pleistocene interval.

3.2. Climatic Effects of Uplift

Over much of eastern and southern Africa, late Cenozoic tectonics were an important modifier of more pervasive climatic trends. Not only did increasing relief significantly influence local patterns of precipitation (through orographic and rainshadow effects), but the rising interior plateaus became cooler and exerted an important influence on atmospheric dynamics. This is particularly evident in East Africa, where the local effects of topographic fragmentation along the rift shoulders must be separated from the regional impact of plateau uplift and from changes of global influence.

Although records from Miocene sites in East Africa are few and discontinuous, some useful conclusions can be drawn for the latter part of this period. Pickford *et al.*[43] have drawn attention to the fact that, by about 7 million years ago, the flanks of the Kenya Rift had been elevated to a sufficient altitude to produce rainshadow effects within the rift itself and between it and the Indian Ocean. Changes associated with the passing of this threshold may well have given rise to the faunal changes documented by Hill[44] in sedimentary sequences of the

Tugen Hills, to the west of Lake Baringo. A substantial positive shift in the $\delta^{18}O$ of paleosol carbonates from East African Rift sites between 8.5 and 6.5 million years ago has been recorded by Cerling[45] and is interpreted as reflecting an increase in the abundance of C_4 (grassland) biomass over that interval. This picture would certainly accord with the growing influence of the rift shoulders during the latter part of the Miocene, as recognized by Pickford and his co-workers, rather than with a response to climatic signals of a global nature alone. In a recent study Cerling et al.[46] argue that $\delta^{18}O$ decreased on a global basis between 7 and 5 million years ago, probably as the result of a decrease in atmospheric CO_2. Molnar et al.[11] have presented persuasive evidence for major rapid uplift of the Tibetan Plateau around 8 million years ago, which they consider to be the cause of the intensification of the Asian monsoon, as revealed in a number of oceanic sequences. They believe that the rapid exhumation of silicate rocks, and the increased chemical weathering precipitated by these events, may have played a major role in this change in atmospheric chemistry.

Evidence of important climatic changes in East Africa coinciding with the Messinian salinity crisis is not forthcoming, at least at present, from the patchy terrestrial record, but data from cores off the west coast of North Africa provide important information on the onset of hyperarid conditions within that part of the continent. Sediment studies by Tiedemann et al.[47] indicate an increase in siliciclastic dust within these cores beginning at about 6 million years ago, but increasing significantly between 4.6 and 3.7 million years ago, reflecting substantial enlargement of the summer dust plume. The authors interpret these changes as resulting from major intensification of the African Easterly Jet-Saharan Air Layer circulation system over North Africa, the desiccating effects of which were further augmented by an increase in the northeast trade winds from about 3.2 million years ago onward. The effects of the gradual tectonic narrowing of the Mediterranean that culminated in the Messinian events between 5.6 and 4.8 million years ago are not manifested in these or other deep-sea records.[48]

A number of influences of global significance may have contributed to North African aridification in the late Miocene and early Pliocene: uplift of the Tibetan Plateau, which apparently included a significant late Miocene component,[11] would have created a flow of dry, northeasterly winds over the central and eastern parts of North Africa and accentuated atmospheric subsidence over its western regions.[49] At the same time global CO_2 values appear to have decreased significantly between the late Miocene and the early Pliocene[46,50]; this decrease, too, may have had as its primary forcing mechanism the increased weathering of silicate rocks on a worldwide basis as a result of Tibetan Plateau uplift.[51] These global influences may have been accentuated by a reduction in moisture, brought from the Indian Ocean by the southeast trade winds, as a result of the rise of the East African Plateau. The aeolian sands, which blanket areas currently occupied by equatorial rain forest (e.g., the Series des Sables Ocres of Zaire), were probably first distributed over a mid-Tertiary land surface during the Mio–Pliocene and may conceivably reflect a central African response to these influences. A return to more humid conditions has clearly occurred during the Pleistocene.

A substantial body of evidence summarized in the previous section now points to the period between about 3 and 2 million years ago as one in which some of the most dramatic tectonic events of the Neogene occurred in East Africa. Because this interval spans a period during which major global climatic changes occurred in response to the growth of ice sheets in the Northern Hemisphere, it is not entirely surprising that some of the East African signals of major environmental change around this time are stronger than would be expected in the tropics, where the response to changes in the high latitudes is frequently ambiguous. As Pickford[52] has pointed out, the most important factor in the evolution of East African climates and faunas was the uplift of the mountain ranges flanking the rifts during the Pliocene and Pleistocene, but these changes were, in turn, themselves affected by global-scale changes related to Neogene cooling, which culminated in the glaciations of the last 3 million years.

Some of the best evidence for the advent of a cooler and drier climate in East Africa around 2.5 million years ago comes from the work of Gasse[53] and Bonnefille[54] on the sediments of Lake Gadeb on the Afar Plateau. Here diatomaceous sediments between tuffs dated at 2.51 and 2.35 million years ago have yielded pollens that reflect the cool, montane conditions now found over 1200 m above Lake Gadeb's 2300-m elevation. The pollen spectra indicate a temperature decline of between 4° and 6°C around 2.5 million years ago.[54] At the hominid site of Hadar, within the Ethiopian Rift, pollen from deposits radio-metrically bracketed between 3.3 and 2.9 million years ago is interpreted as representative of an "Afromontane" flora characteristic of present highland areas and differing markedly from the subdesertic steppe vegetation that characterizes this part of the rift today.[33] Possible altitudinal changes as a result of rift tectonics have been discussed previously. These effects apart, the evidence suggests the existence of a climate significantly moister than today's or that which followed the 2.5-million-year cooling.

Palynological data from the delta of the Omo River in Ethiopia indicate an expansion of grasslands at this time,[55] and macrobotanical studies by Bonnefille and Letouzey[56] show that fossilized wood and fruits disappeared more or less simultaneously from the lower Omo area. The work of Wesselman[57] and Vrba[58,59] on the Omo mammalian faunas also indicates a shift to greater aridity about 2.5 million years ago: Vrba's work records a pronounced increase in open grassland species of bovids, whereas Wesselman's analyses of a number of micromammalian assemblages spanning this interval show the increasing domi-nance of species adapted to grassland environments by around 2.4 million years ago (recent recalibration of the Plio–Pleistocene paleomagnetic timescale by Hilgen[60,61] and Shackleton et al.[62,63] indicates that these dates for the Omo area should now be increased by about 5%). Further east, in the Turkana Basin, the replacement of *Hipparion* by *Equus* and other species, and notable turnovers in bovid taxa, together constitute the widespread pulse around 2.7–2.5 million years ago associated with an increase in areas of open, dry grassland.[64] In gastropod assemblages and pedogenic carbonate horizons from deposits of Lake Turkana, changes in stable isotope ratios documented by Abell[65] and Cerling et al.[66] show a trend toward reduced precipitation between tuffs dated radiometrically at 3.2

and 1.9 million years ago. The most dramatic increase in both $\delta^{18}O$ and $\delta^{13}C$, indicating a major expansion of grasslands, occurred, however, in the Turkana Basin around 1.8 million years ago[45]; this event suggests that the largest spread of C_4 grasslands in Africa occurred later than suggested by some other lines of evidence.

4. SOUTHERN AFRICA

4.1. The Interplay between Tectonism and Climate during the Cenozoic

The view has long been held that, except in the Rift Valley region of East Africa and in the Atlas belt, vertical uplift within Africa has been negligible since the Paleogene.[67] This view has been challenged on the basis of recent evidence that large-scale uplift of the eastern hinterland of southern Africa occurred during the late Neogene.[24,68] Much of the evidence for the timing and extent of these movements comes from the legacy of planation surfaces that have been formed since Africa separated from the other continents of the Gondwanaland mosaic. At the time of rifting, southern Africa stood relatively high (in the vicinity of 2000–2500 m AMSL).[24,68] This resulted in rapid denudation of much of the subcontinent to the base level of the newly created marginal oceans; erosion was aided by the humid, tropical climates of the Cretaceous, which favored deep weathering and the accelerated removal of the regolith. By the end of the Cretaceous, the marginal escarpment created by rifting had receded some 120 km inland from the south-eastern coast[69] and approximately 50 km from the west coast. Major rivers such as the Orange and Limpopo developed conspicuous nickpoints inland of the receding escarpment (e.g., the Aughrabies Falls); above these, denudation of the continental interior proceeded in parallel with that inland of the coastal margins. In the process, up to about 3000 m of material was eroded from the coastal hinterland, with lesser thicknesses being removed from inland areas.[70] The end result of this postrifting cycle of denudation was the formation, by the end of the Cretaceous, of coeval surfaces of low relief separated by the step of the Great Escarpment.

A well-integrated network of large rivers, whose legacy is now preserved in the high-level Upper Cretaceous gravels of the western interior, developed on this vast pediplain under the humid Cretaceous climate. While there is evidence to suggest that limited tectonic movements occurred during the long duration of this cycle, notably during the Valanginian–Hauterivian and Coniacian–Santonian intervals, the local effects of these were largely elided, through time, into a single, continuous surface of low relief, known as the African surface. This surface preserves the remains of the crater-facies of diatremes of end-Cretaceous age,[71] and its surviving remnants are extensively armored by duricrusts, beneath which deep kaolinization is ubiquitous. There is no convincing evidence, contra Gilchrist and Summerfield,[72] for the development of an inward-migrating flexural bulge up to 600 m in amplitude, which they argue, largely on theoretical grounds, would have formed in response to Cretaceous denudation of the coastal hinterland and concomitant sedimentary loading of the continental shelf.[24]

The extensive plains, punctuated by a few mountain massifs (e.g., the Namaqua and Lesotho Highlands and the Cape Fold Mountains), which characterized the late Cretaceous topography of southern Africa, bear the imprint of the major climatic changes that ushered in the Cenozoic. Particularly important evidence for these changes are the extensive silcrete duricrusts that cap remnants of the African surface west of 29°E longitude and extend into southeastern Zimbabwe. Both palynological evidence from diatremes in the west of southern Africa[73] and the morphology of the silcrete itself, including relict columnar or prismatic structure indicative of a saline soil environment, suggest that significant desiccation occurred at the end of the Cretaceous.[74] Indeed, a substantial body of evidence indicates a major change in atmospheric chemistry, associated with desiccation and cooling on a global scale, at that time; there are good grounds for inferring that this temperature decline was followed by a prolonged interval of desiccation in the earliest Cenozoic. The presence of some 50–60 m of cross-bedded aeolian sands in southern Namibia indicates that these changes were associated with hyperarid conditions, which may have persisted until about 50 million years ago.[75,76] In contrast, mesic conditions appear to have characterized the southeastern hinterland during this period,[77] suggesting that the current east–west climatic gradient across southern Africa was established (at least temporarily) as far back as the early Cenozoic.

Little is known about local paleoclimates during the latter part of the Eocene and the Oligocene, owing to an absence of continental sediments that can be assigned with certainty to this period (sedimentation had almost certainly been occurring within the interior Kalahari Basin since the Campanian, when evidence from the southern rim of the basin indicates that a major southward-flowing drainage was disrupted by the first basining movements, but a detailed chronostratigraphy has yet to be worked out for this important terrestrial sequence, which in places comprises more than 400 m of sediment). Although evidence exists in support of localized uplift in the Eocene–Oligocene interval,[78] no regional-scale movements were apparent in southern Africa until the Neogene.

Data on the extent of Neogene movements come from several sources: deformation of remnants of the African surface, and the younger, less perfectly planed Post-African I surface (Fig. 7) provides one line of evidence. Undisturbed gradients across such areas of advanced planation are usually less than 1 m/km and seldom rise to more than twice this value. In the coastal hinterland of Natal, however, gradients across accordant remnants of the African surface can be reconstructed from remnants of well-developed laterite duricrusts and deeply kaolinized weathering profiles; these commonly attain gradients of 30 m/km, and are only marginally less steep in profiles across surviving areas of the Post-African I surface. Divergence between the two surfaces across the crest of the uplift, which is located some 80 km from the coast, indicates that maximum vertical movements after the end of the African cycle were of the order of 250 m; subsequent deformations of the Post-African I surface ranged from 600 to 900 m (Fig. 7).

Supporting evidence for these deformations is provided by the long profiles of major rivers of the east coast (Fig. 8),[79] which are convex upward, clearly reflecting the influence of recent uplift and warping. Near Port Elizabeth on the

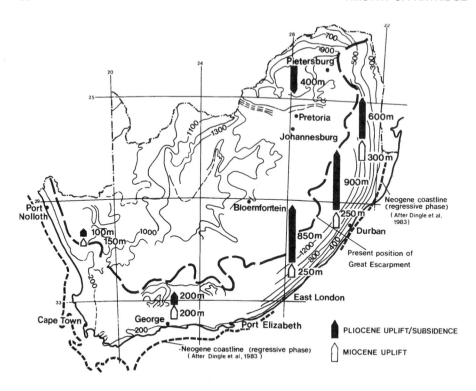

FIGURE 7.   Generalized contours in meters on the Post-African I (early Miocene) erosion surface in South Africa. Open arrows indicate amplitude of early Miocene uplift; solid arrows show Pliocene uplift or subsidence. The position of the Great Escarpment is shown by a long broken line (after Partridge and Maud[68]).

southeast coast, late Miocene and early Pliocene marine deposits have been raised to elevations of up to 330 m along the southeastern flank of the axis of uplift, about halfway between the coast and the zone of maximum uplift.[80,81] Allowing a concomitant sea-level stand of 80–90 m,[82] this evidence points to a total uplift of some 600 m within the last 5 million years. The offshore response was an increase in the sedimentation rate, above a coeval acoustic reflector, by about 50% over the average since continental rifting, 145 million years ago.[83]

Apatite fission-track analyses undertaken at localities both landward and seaward of the Great Escarpment of southern Africa[70] have yielded results that in no way contradict the erosional and tectonic history outlined above, despite claims to the contrary. R. W. Brown's comment (Brown,[70] p. 77) that: "these data indicate that some areas mapped by Partridge and Maud (1987) as remnants of an assumed Jurassic surface were at burial depths equivalent to > 120°C until the early Cretaceous" reflects a serious misrepresentation of our 1987 review. In this we stress that *no* areas referable to the prerifting ("Gondwana") land surface are currently preserved in the southern African landscape and that erosion, *well*

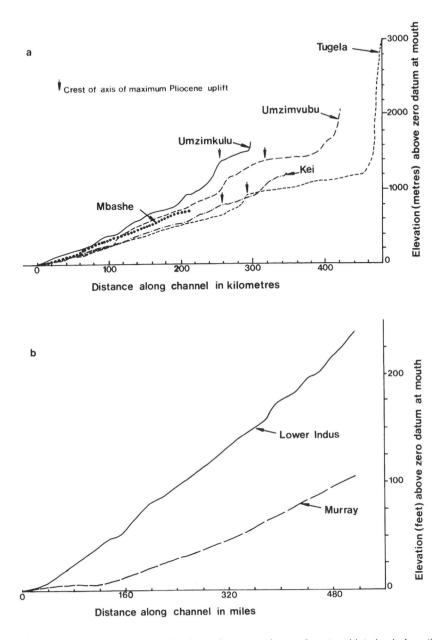

FIGURE 8.   (a) Long profiles of major rivers that cross the southeastern hinterland of southern Africa; (b) typical river long profiles in areas unaffected by recent uplift (after Leopold et al.[79]).

in excess of a kilometer in some areas, occurred both seaward and inland of the Great Escarpment, mostly during the Cretaceous; this led ultimately to the formation of the duricrust-armored African surface by the early Tertiary. The reestablishment of high elevations within the continental interior was largely the result of Neogene uplifts, which accentuated the relief along the Great Escarpment and initiated a new phase of erosion in the coastal hinterland.

4.2. Late Neogene Climatic Responses

The climatic effects of the Neogene uplifts documented above were undoubtedly considerable, and tended to accentuate the patterns induced by contrasting current regimes along the eastern and western seaboards of southern Africa. Temporary aridity within the western hinterland, which ushered in the Cenozoic, was replaced by more mesic conditions and a woodland vegetation in the early Miocene[84]; a return to aridity in this area occurred somewhat later than 13 million years ago with the permanent establishment of the East Antarctic ice sheet and the consequent initiation of cold upwelling in the Benguela Current System of the Atlantic seaboard. With the rise of the southeastern hinterland, the meridional precipitation gradient reestablished during the mid-Miocene was strengthened through the orographic capture of moisture advected from the western Indian Ocean, especially in those areas proximal to the warm Mozambique and Agulhas currents. Precipitation was concentrated on the windward side of the uplifted areas, with concomitant rainshadow effects in areas to the west. Thus, despite widespread evidence of somewhat warmer and more mesic conditions during the Pliocene,[85] the rejuvenation of local drainage networks in the semiarid western interior of the subcontinent (e.g., the Carnarvonleegte System near Kenhardt in the Northern Cape)[84] was limited both in time and space. The apparently minor influence of Pliocene warming on southern African climates may also be attributable to markedly reduced surface temperatures, consequent upon uplift.

Although the timing of various pulses of Neogene uplift is not yet well bracketed, the evidence now available suggests that the later phases of movement may have occurred close to the onset of a further important interval of global cooling, which began about 2.8 million years ago. This event is prominent in the global oceanic record[86,87] and represents a significant period of cooling, which culminated in the first major extension of the Northern Hemisphere ice sheets; this led to the rafting of debris into the temperate latitudes of the North Atlantic during Isotope Stage 104.[88] At the same time cycles dominated by orbital precession at 19- to 23-thousand-year frequency were replaced by 41-thousand-year glacial–interglacial couplets controlled by orbital obliquity. Fluxes of aeolian dust in cores off the east and west coasts of North Africa reflect concurrent changes in the intensity of seasonal winds carrying dust from sources in Arabia and East Africa.[87] Terrestrial responses included the aridification of large tracts of sub-Saharan Africa, including the development of extensive dune systems on the Miocene (Post-African I) land surface. In southern Africa discontinuous dunes penetrated from the southern limits of the present Kalahari Basin as far south as the Cape Fold Mountains.[68,89] This so-called mega-Kalahari is, in fact, the largest single body of windblown sand in the world.

Similarly profound changes, probably brought about by the combined effects of uplift and global cooling, can be discerned in the hominid-bearing cave deposits of the Transvaal, which fall within the area affected directly by Pliocene uplift of the southeastern hinterland. Several data sets provide independent lines of evidence pointing to increased aridification at this time. At the Makapansgat Limeworks and at Sterkfontein, the advent of cyclic sedimentation with a significant increase in the proportion of coarse debris occurs within the upper *Australopithecus africanus* levels — Member 4 in each case.[90,91] Aridification, with a more seasonal distribution of rainfall, is indicated. Vrba's[92,93] analyses of the bovid faunas from Sterkfontein, Swartkrans, and Kromdraai show that this transition was also associated with a change from species indicative of a relatively greater bush cover to others reflecting a significantly higher proportion of grass. On the basis of paleontological comparisons she places this change, which corresponds with a major shift in the composition of the antelope fauna, some time between 2.6 and 2.0 million years ago.[59] Simultaneously, changes in local hominid populations from those present in Member 4 at Sterkfontein to the admixture of robust forms and early representatives of genus *Homo* accompanied this faunal turnover.

5. CONCLUSIONS

Evidence from both eastern and southern Africa indicates that, contrary to earlier views, the anomalous hypsometry of these areas was not inherited from the time of continental rifting, during the late Jurassic and early Cretaceous, but was caused by uplift movements, which occurred chiefly during the Neogene. As Partridge et al.[2] have noted, the present altitudinal distribution of late Cretaceous to Miocene sediments on the African continent argues for major uplift during the Cenozoic, which gives the eastern and southern parts of the continent the character of a relatively high plateau. Vertical movements as large as 1 km were common, and were considerably exceeded in areas of East Africa affected by the mantle-plume activity that gave rise to the Rift Valley System and its associated volcanoes. Age control is not always accurate, but there is an overwhelming weight of evidence that, in both parts of the continent, most of the movements were concentrated in the late Neogene.

A complex legacy of climatic changes has resulted in which the influence of local topographic effects is frequently difficult to separate from regional or even global influences. Model studies are needed to assess the influence of the uplift of an area far larger (but lower) than the Tibetan Plateau on global patterns of atmospheric circulation and climate; there can be little doubt that these were considerable. Perhaps the most noteworthy results of the interplay between these tectonic events and global climate change were the imposition of regional climatic stimuli and the creation of local ecological mosaics favorable to speciation and rapid faunal turnover. The emergence of our earliest ancestors on the African continent may have owed more to these influences than has hitherto been acknowledged.

REFERENCES

1. Smith, A. G. and Briden, J. C. (1982). *Mesozoic and Cenozoic Palaeocontinental Maps.* Cambridge University Press, Cambridge.
2. Partridge, T. C., Bond, G. C., Hartnady, C. J. H., deMenocal, P. B., and Ruddiman, W. F. (1995). In: *Paleoclimate and Evolution, with Emphasis on Human Origins* (E. S. Vrba, L. H. Burckle, G. H. Denton, and T. C. Partridge, eds.), pp. 8–23. Yale University Press, New Haven.
3. Pollitz, F. F. (1991). *Tectonophysics* **194**, p. 91.
4. DeMets, C., Gordon, R. G., Argus, D. F. and Stein, S. (1990). *Geophys. J. Intern.* **101**, p. 425.
5. Jestin, F., Huchon, P., and Gaulier, J. M. (1994). *Geophys. J. Intern.* **116**, p. 637.
6. Ben-Avraham, Z., Hartnady, C. J. H., and le Roex, A. P. (1995). *J. Geophys. Res.* **100**, p. 6199.
7. Gordon, R. G., and Stein, S. (1992). *Science* **256**, p. 333.
8. Sloan, H., and Patriat, P. (1992). *Earth Planet. Sci. Lett.* **113**, p. 323.
9. Stanley, D. J., and Wezel, R.-C. (eds.) (1985). *Geological Evolution of the Mediterranean Basin.* Springer-Verlag, Berlin.
10. Hodell, D. A., Elmstrom, K. M., and Kennett, J. P. (1986). *Nature* **320**, p. 411.
11. Molnar, P., England, P., and Martinod, J. (1993). *Rev. Geophys.* **31**, p. 357.
12. Zoback, M. L. (1992). *J. Geophys. Res.* **97**, p. 11703.
13. de Wit, M. J., and Ransome, I. G. D. (1992). *Inversion Tectonics of the Cape Fold Belt, Karoo and Cretaceous Basins of Southern Africa.* Balkema, Rotterdam.
14. Nyblade, A. A., and Robinson, S. W. (1994). *Geophys. Res. Lett* **21**, p. 765.
15. Su, W. J., Woodward, R. L., and Dziewonski, A. M. (1994). *J. Geophys. Res.* **99**, p. 6945.
16. Tanaka, S., and Hamaguchi, H. (1992). *Tectonophysics* **209**, p. 213.
17. Sletlene, L., Wilcox, L. E., Blouse, R. S., and Sanders, J. R. (1973). A Bouger Gravity Map of Africa. DMAAC Technical Paper 73-003, Defense Mapping Agency, St. Louis.
18. Bloch, S., Hales, A. L., and Landisman, M. (1969). *Bull. Seismol. Soc. Amer.* **59** p. 1599.
19. Clouser, R. H., and Langston, C. A. (1990). *J. Geophys. Res.* **95**, p. 17403.
20. Vinnik, L. P., Green, R. W. E., and Nicolaysen, L. O. (1994). *EOS* **75**, p. 230.
21. Hartnady, C. J. H., and Partridge, T. C. (1995). *Proceedings of the Centennial Geocongress of the Geological Society of South Africa,* Vol. 1, p. 456.
22. Ebinger, C. J., Bechtel, T. D., Forsyth, D. W., and Bowin, C. O. (1989). *J. Geophys. Res.* **94**, p. 2883.
23. Williams, L. A. J. (1978). In: *Petrology and Geochemistry of Continental Rifts,* Vol. 1, *Proceedings of NATO Advanced Study Institute on Paleorift Systems,* p. 36 (H. J. Neumann and I. B. Ramberg, eds.). NATO Advanced Study Institute, Reidel, Dordrecht.
24. Partridge, T. C., Wood, B. A., and de Menocal, P. B. (1995). In: *Paleoclimate and Evolution with Emphasis on Human Origins.* (E. S. Vrba, G. H. Denton, T. C. Partridge, and L. H. Burckle, eds.), pp. 331–335. Yale University Press, New Haven.
25. Achauer, U., Maguire, P. K. H., Mechie, J., Green, W. V., and the KRISP Working Group. (1992). *Tectonophysics* **213**, p. 257.
26. Davidson, A., and Rex, D. C. (1980). *Nature* **283**, p. 657.
27. WoldeGabriel, G., Aronson, J. L., and Walter, R. C. (1990). *Geol. Soc. Am. Bull.* **102**, p. 439.
28. Ebinger, C. J., Yemano, T., WoldeGabriel, G., Aronson, J. L., and Walter, R. C. (1993). *J. Geol. Soc. London* **150**, p. 99.
29. Adamson, D. A., and Williams, M. A. J. (1987). *J. Human Evol.* **16**, p. 597.

30. Yemane, K., Bonnefille, R., and Faure, H. (1985). *Nature* **318**, p. 653.
31. Baker, B. H., Mohr, P. A., and Williams, L. A. J. (1972). The geology of the Eastern Rift System of Africa. Geological Society of America, Special Paper, No. 136.
32. Denys, C., Chorowicz, J., and Tiercelin, J. J. (1986). In: *Sedimentation in the African Rifts* (L. E. Frostick *et al.*, eds.), pp. 363–372. Special Publication of the Geological Society of London, No. 25.
33. Bonnefille, R., Vincens, A., and Buchet, G. (1987). *Palaeogeog., Palaeoclim. Palaeoec.* **60**, p. 249.
34. Molnar, P., and England, P. (1990). *Nature* **346**, p. 29.
35. Saggerson, E. P., and Baker, B. H. (1965). *Quar. J. Geol. Soc. London* **121**, p. 51.
36. Baker, B. H. (1986). In: *Sedimentation in the African Rifts* (L. E. Frostick *et al.*, eds.), pp. 45–57. Special Publication of the Geological Society of London, No. 25.
37. Ebinger, C. J. (1989). *Geol. Soc. Am. Bull.* **101**, p. 885.
38. Dixey, F. (1956). *The East African Rift System*. Supplementary Bulletin, Overseas Geological Mineral Resources, No. 1, H.M. Stationery Office, London.
39. Shackleton, R. M. (1978). In: *Geological Background to Fossil Man* (W. W. Bishop, ed.), pp. 20–28. Scottish Academic Press, Edinburgh.
40. Mohr, P. A. (1987). *EOS* **68**, p. 721.
41. Fairhead, J. D. (1986). In: *Sedimentation in the African Rifts* (L. E. Frostick *et al.*, eds.), pp. 19–27. Special Publication of the Geological Society of London, No. 25.
42. Ebinger, C. J., Deino, A. L., Drake, R. E., and Tesha, A. L. (1989b). *J. Geophys. Res.* **94**, p. 15, 785.
43. Pickford, M., Senut, B., and Hadoto, D. (1993). *Geology and Palaeobiology of the Albertine Rift Valley, Uganda-Zaire, Vol. 1 Geology*. International Centre for Training and Exchanges in the Geosciences (CIFEG), Occasional Paper 1993/24, Orleans, France.
44. Hill, A. (1987). *J. Human Evol.* **16**, p. 583.
45. Cerling, T. E. (1992). *Palaeogeog., Palaeoclima., Palaeoec.* **97**, p. 241.
46. Cerling, T. E., Wang, Y., and Quade, J. (1993). *Nature* **361**, p. 344.
47. Tiedemann, R., Sarnthein, M., and Stein, R. (1989). *Ocean Drilling Program: Scientific Results* **108**, p. 241.
48. Ruddiman, W. F., Sarnthein, M., Backman, J., Baldauf, J. G., Curry, W., Dupont, L. M., Janecek, T., Pokras, E. M., Raymo, M. E., Stabell, B., Stein, R., and Tiedemann, R. (1989). *Ocean Drilling Program: Scientific Results* **108**, p. 463.
49. Ruddiman, W. F., and Kutzbach, J. E. (1989). *J. Geophys. Res.* **94**, p. 18409.
50. Cerling, T. E. (1991). *Am. J. Sci.* **291**, p. 337.
51. Raymo, M. E., Ruddiman, W. F., and Froelich, P. N. (1988). *Geology* **16**, p. 649.
52. Pickford, M. (1990). *Human Evol.* **5**, p. 1.
53. Gasse, F. (1980). *Rev. Algologique* **3**, p. 1.
54. Bonnefille, R. (1983). *Nature* **303**, p. 487.
55. Bonnefille, R. (1976). In: *Earliest Man and Environments in the Lake Rudolf Basin* (Y. Coppens *et al.*, eds.), pp. 421–431. University of Chicago Press, Chicago.
56. Bonnefille, R., and Letouzey, R. (1976). *Adansonia* **16**, p. 65.
57. Wesselman, H. B. (1985). *S. Afr. J. Sci.* **81**, p. 260.
58. Vrba, E. S. (1985). *S. Afr. J. Sci.* **81**, p. 263.
59. Vrba, E. S. (1995). In: *Paleoclimate and Evolution with Emphasis on Human Origins* (E. S. Vrba *et al.*, eds.), pp. 385–424. Yale University Press, New Haven.
60. Hilgen, F. J. (1991). *Earth Planet. Sci. Lett.* **104**, p. 226.
61. Hilgen, F. J. (1991). *Earth Planet. Sci. Lett.* **107**, p. 349.
62. Shackleton, N. J., Berger, A., and Peltier, W. R. (1990). *Trans. Roy. Soc. Edinburgh: Earth Sci.* **81**, p. 251.
63. Shackleton, N. J., Crowhurst, S., Hagelberg, T., Pisias, N. G., and Schneider, D. A. (1995). *Ocean Drilling Program: Scientific Results*, Vol. 138.
64. Feibel, C. S., Harris, J. M., and Brown, F. H. (1991). In: *Koobi Fora Research Project*, Vol. 3 (J. M. Harris, ed.), pp. 321–346. Clarendon Press, Oxford.
65. Abell, P. I. (1982). *Nature* **297**, p. 231.
66. Cerling, T. E., Hay, R. L., and O'Neil, J. R. (1977). *Nature* **267**, p. 137.
67. Gilchrist, A. R., Kooi, H., and Beaumont, C. (1994). *J. Geophys. Res.* **99**, p. 12,211.

68. Partridge, T. C., and Maud, R. R. (1987). *S. Afr. J. Geol.* **90**, p. 179.
69. Matthews, P. E. (1978). *Petros* **8**, p. 26.
70. Brown, R. W. (1991). *Geology* **19**, p. 74.
71. Moore, A. E. (1979). The geochemistry of the olivine melilitites and related rocks of Namaqualand/Bushmanland, South Africa. Ph.D. Thesis, Department of Geology, University of Cape Town.
72. Gilchrist, A. R., and Summerfield, M. A. (1990). *Nature* **346**, p. 739.
73. Scholtz, A. (1985). *Ann. S. Afr. Museum* **95**, p. 1.
74. Partridge, T. C., and Maud, R. R. (1989). *S. Afr. J. Sci.* **85**, p. 428.
75. Ward, J. D., Seely, M. K., and Lancaster, I. N. (1983). *S. Afr. J. Sci.* **79**, p. 175.
76. Ward, J. D., and Corbett, I. (1990). In: *Namib Ecology: 25 Years of Namib Research* (M. K. Seely, ed.), pp. 17–25. Pretoria: Transvaal Museum.
77. Dingle, R. V., Siesser, W. G., and Newton, A. R. (1983). *Mesozoic and Tertiary Geology of Southern Africa.* Balkema, Rotterdam.
78. McCarthy, T. S., Moon, B. P., and Levin, M. (1985). *S. Afr. Geograph. J.* **67**, p. 160.
79. Leopold, L. B., Wolman, M. G., and Miller, J. P. (1964). *Fluvial Processes in Geomorphology.* Freeman, San Francisco.
80. King, L. C. (1972). *Trans. Geol. Soc. S. Afr.* **75**, p.159.
81. le Roux, F. G. (1989). The lithostratigraphy of Cenozoic deposits along the southeast Cape coast as related to sea-level changes. M.Sc. Thesis, Department of Geology, University of Stellenbosch.
82. Haq, B. U., Hardenbol, J., and Vail, P. R. (1987). *Science* **235**, p. 1156.
83. Martin, A. K. (1987). *S. Afr. J. Sci.* **83**, p. 716.
84. de Wit, M. C. J. (1993). Cenozoic evolution of drainage systems in the north-western Cape. Ph.D. Thesis, Department of Geology, University of Cape Town.
85. Dowsett, H. J., and PRISM Project Members. (1994). *EOS* **75**, p. 206.
86. Shackleton, N. J. (1995). In: *Paleoclimate and Evolution with Emphasis on Human Origins* (E. S. Vrba *et al.*, eds.), pp. 242–248. New Haven, Yale University Press.
87. deMenocal, P. B., and Bloemendal, J. (1995). In: *Paleoclimate and Evolution with Emphasis on Human Origins* (E. S. Vrba *et al.*, eds.), pp. 262–288. Yale University Press, New Haven.
88. Shackleton, N. J., Backman, J., Zimmerman, H., Kent, D. V., Hall, M. A., Roberts, D. G., Schnitker, D., Baldauf, J. G., Desprairies, A., Homrighausen, R., Huddlestun, P., Keene, J. B., Kaltenback, A. J., Krumsiek, K. A. O., Morton, A. C., Murray, J. W., and Westberg-Smith, J. (1984). *Nature* **307**, p. 620.
89. Partridge, T. C. (1992). *Quatern. Intern.* **17**, p. 105.
90. Partridge, T. C. (1985). *S. Afr. J. Sci.* **81**, p. 245.
91. Partridge, T. C. (1986). *S. Afr. J. Sci.* **82**, p. 80.
92. Vrba, E. S. (1974). *Nature* **250**, p. 19.
93. Vrba, E. S. (1975). *Nature* **254**, p. 301.

III

General Circulation Model Studies of Uplift Effects on Climate

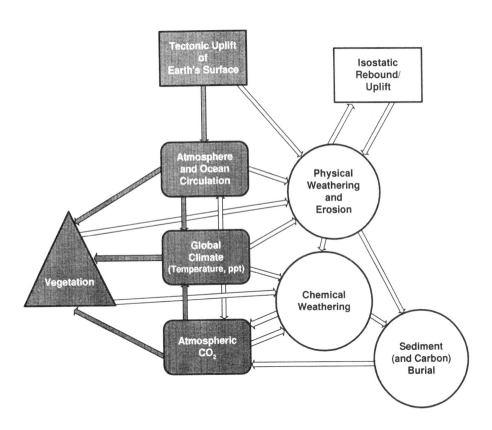

5

Mountains and Midlatitude Aridity

Anthony J. Broccoli and Syukuro Manabe

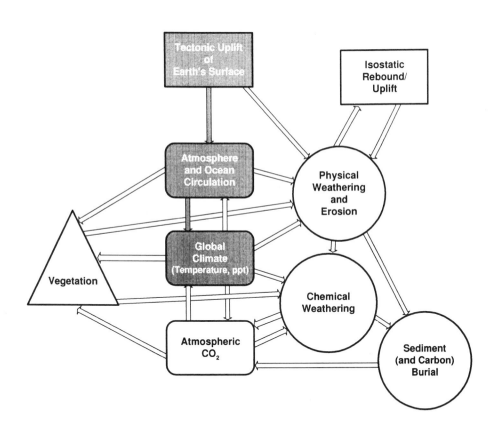

Anthony J. Broccoli and Syukuro Manabe • Geophysical Fluid Dynamics Laboratory/NOAA, Princeton University, Princeton, New Jersey 08542.

Tectonic Uplift and Climate Change, edited by William F. Ruddiman. Plenum Press, New York, 1997.

1. INTRODUCTION

The geological record provides evidence that suggests a link between mountain uplift and changes in climate over substantial regions of the world.[1] One approach to understanding the effect of mountain uplift on the evolution of climate is to investigate the role of present-day orography in determining the modern distribution of climates. A better understanding of the mechanisms by which the global distribution of mountains affects today's climate may shed light on the changes in climate that would occur in response to changes in orography. Unfortunately, even to understand the role of modern orography presents a difficult task that is not likely to be accomplished using observations alone, because the interaction among the many components of the climate system limits our ability to isolate mountain effects from other influences such as the location, size, and shape of the continents or the spatial distribution of incoming solar radiation.

To overcome this difficulty, numerical models of the Earth's climate can be employed to provide a framework in which it is possible to isolate the effects of mountains on the distribution of climate. They can be used to perform a controlled experiment in which climate is simulated with and without orography, just as scientists in other disciplines would conduct controlled experiments in the laboratory. Provided that the climate model is realistic enough to represent the relevant processes, results from such experiments may be expected to elucidate the mechanisms that operate in the real climate system. Thus climate modeling is a powerful tool for exploring the effects of orography on climate.

In this chapter we will present some recent results from a series of experiments using a climate model developed at the Geophysical Fluid Dynamics Laboratory (GFDL) of the National Oceanic and Atmospheric Administration.[2,3] These results indicate that the existence of midlatitude arid climates may be linked to the modern distribution of orography, making it possible that the geologically recent expansion of such dry regions is associated with mountain uplift.

2. DISTRIBUTION OF ARID CLIMATES

Most of the Earth's arid climates (i.e., those in which potential evapotranspiration exceeds precipitation on an annual basis) are found in the lower and middle latitudes. Widespread subtropical dry regions exist in northern Africa, the Middle East, Pakistan and northwestern India, the southwestern United States and Mexico, coastal Peru and Chile, southern Africa, and Australia. All of these regions lie in belts between 15° and 35° latitude on either side of the equator, and their existence is explained by their location beneath the subsiding branch of the Hadley circulation. But other sizable dry regions exist across the interior of Asia from Turkestan east to the Gobi Desert, in the western interior of North America from the Great Basin to the prairies of Canada and the Great Plains, and east of the Andes in south central South America. These dry regions are not as readily explained by a simple schematic model of global circulation, since they are located beneath the traveling disturbances of the midlatitude westerlies.

Midlatitude aridity in the Northern Hemisphere is often explained by the large distance between oceanic moisture sources and the interiors of North American and Asia, with the dryness accentuated in some locations by the presence of mountain barriers upwind.[4-6] If distance from oceanic moisture sources is its primary cause, then midlatitude aridity (although perhaps less intense) would be expected even in the absence of orography. This hypothesis can be tested with atmospheric general circulation models (GCMs) using the methodology described in the introduction to this chapter.

3. REVIEW OF PREVIOUS CLIMATE MODELING WORK

In the early days of climate modeling, a number of studies were performed in which GCMs were used to simulate the Earth's climate with and without mountains.[7-11] Some information resulting from these and other studies hints at the possibility that midlatitude aridity may be related to the modern distribution of orography. In a paper that primarily concerned changes in the boreal wintertime atmospheric circulation, Manabe and Terpstra[10] briefly discussed the differences in precipitation distribution based on simulations of the January climate with and without mountains. They found a zonal belt of moderate precipitation stretching across the midlatitudes of the Northern Hemisphere in the case without mountains, and interruptions in this zonal belt over the interiors of Eurasia and North America in their experiment with mountains. In their investigation of the role of mountains in the south Asian summer monsoon circulation, Hahn and Manabe[11] found that the monsoon circulation was greatly altered in their experiment without mountains, and that the presence of the Tibetan Plateau was required for the northward expansion of the monsoon over India. Using realistic orography and incorporating seasonal variation, Manabe and Holloway[12] found that their model simulated the midlatitude dryness of the Eurasian interior. Based on the results of the winter simulations of Manabe and Terpstra[10] and the summer simulations of Hahn and Manabe,[11] they suggested that the Tibetan Plateau plays a major role in maintaining this dryness.

Interest in mountain effects on global circulation and climate resurfaced in the latter half of the 1980s.[13-15] Figuring prominently in this renewed interest was the suggestion that changes in climate during the past 30 to 40 million years may be associated with large-scale uplift of the Tibetan Plateau and the western United States.[16] Kutzbach et al.[15] used the Community Climate Model of the National Center for Atmospheric Research (NCAR CCM) to run perpetual January and July integrations with and without mountains in order to study changes in climate in both the summer and winter seasons. Substantial similarity

exists between their January results and those of Manabe and Terpstra,[10] as well
as between the response of the model to changes in orography and climatic
changes in the geological record.[16]

The remainder of this chapter will focus on the results obtained by Manabe
and Broccoli[2] and Broccoli and Manabe[3] using the GFDL climate model. We
will examine specifically the ability of the model to simulate midlatitude arid
climates. We will also investigate the mechanisms, both atmospheric and hydrologic, by which those climates are maintained.

4. MODEL DESCRIPTION

The version of the GFDL climate model used in this study consists of two
basic units: (1) a general circulation model of the atmosphere, and (2) a heat and
water balance model over the continents. These components are fully interactive.
To reduce the computational burden, these components are not coupled to an
ocean model; instead, the geographical distribution of sea surface temperature and
sea ice is prescribed, varying seasonally, in a manner consistent with observations.

The atmospheric model employs the spectral transform method, in which the
horizontal distributions of the primary variables are represented by spherical
harmonics. The present model retains 30 zonal waves, adopting the so-called
rhomboidal truncation. The spacing of the transform grid is 2.25° latitude by 3.75°
longitude. Normalized pressure is used as the model's vertical coordinate, with
nine unevenly spaced levels used for finite differencing. Gordon and Stern[17]
provide further discussion of the dynamical component of this model.

Solar radiation at the top of the atmosphere is prescribed, varying seasonally
but not diurnally. Computation of the flux of solar radiation is performed using
a method similar to that of Lacis and Hansen,[18] except that the bulk optical
properties of various cloud types are specified. Terrestrial radiation is computed
as described by Stone and Manabe.[19] For the computation of radiative transfer,
clouds are prescribed, varying only with height and latitude. The mixing ratio of
carbon dioxide is assumed constant everywhere, and that of ozone is specified as
a function of height, latitude, and season.

Over the continents, surface temperatures are computed from a heat balance
with the requirement that no heat is stored in the soil. Both snow cover and soil
moisture are predicted. A change in snow depth is predicted as the net contribution
from snowfall, sublimation, and snowmelt, with the latter two determined from the
surface heat budget. Soil moisture is computed by the "bucket method." The soil is
assumed to have a water-holding capacity of 15 cm. If the computed soil moisture
exceeds this amount, the excess is assumed to be runoff. Changes in soil moisture are
computed from the rates of rainfall, evaporation, snowmelt, and runoff. Evaporation from the soil is determined as a function of soil moisture and the potential
evaporation rate (i.e., the hypothetical evaporation rate from a completely wet soil).
Further details of the hydrologic computations can be found in Manabe.[20]

An additional characteristic of the model is a parameterization of the drag that results from the breaking of orographically-induced gravity waves. Parameterizations of this kind have been found to improve the performance of atmospheric GCMs both for weather forecasting and climate simulation.[21,22] In the current experiments, gravity wave drag is used in all integrations where orography was present, and it substantially improves the simulation of midlatitude aridity. A description of this parameterization and a discussion of its impact on the simulation of climate are contained in Appendix A of Broccoli and Manabe.[3]

5. EXPERIMENTAL DESIGN

Three numerical integrations are run with the model as described in the previous section. In the first of these, the mountain (M) integration, realistic geography and topography are used as depicted in Fig. 1. Because topography in a spectral model must be represented mathematically by a finite series of spherical harmonics, it is relatively smooth. Despite this smoothness, most of the large-scale features of the global topography are represented, such as the Tibetan Plateau, the Rocky and Andes cordilleras, the east African highlands, and the Greenland and Antarctic ice sheets. A second integration, the no-mountain (NM) integration, uses the same geographical distribution of land and sea as the M integration, but with flat continents. Comparing the results from these two integrations provides the response of the climate system (as represented by the model) to the presence of orography. A wide variety of interactions or feedbacks is included in the model, and thus the response takes into account the effects of these interactions. A third integration, known as the fixed soil moisture (FSM) integration, is performed to evaluate one of these feedbacks, specifically that between the land surface and the atmosphere. The FSM integration is identical to the M integration except that the soil moisture value at each gridpoint is prescribed, varying seasonally, rather than predicted from a water balance. (The soil moisture values are based on the climatological values from the NM integration using the procedure employed by Delworth and Manabe[23] in their prescribed soil moisture experiment.) By utilizing the results from the FSM integration, it is possible to determine to what extent the interaction between the land surface and the atmosphere enhances the mountain-induced changes in climate.

All integrations were initiated from an isothermal, resting atmosphere and integrated until the simulated seasonal cycle of climate reached a quasi-equilibrium. Even though this initial state is very unrealistic, the atmospheric component of the model quickly adjusts and loses its memory of the initial conditions. Since sea surface temperature and sea ice were prescribed, no thermal adjustment of the ocean temperature was required, so only a relatively short spin-up period of several years was required for quasi-equilibrium to occur. A period of 3 years subsequent to the achievement of quasi-equilibrium was retained for analysis, and

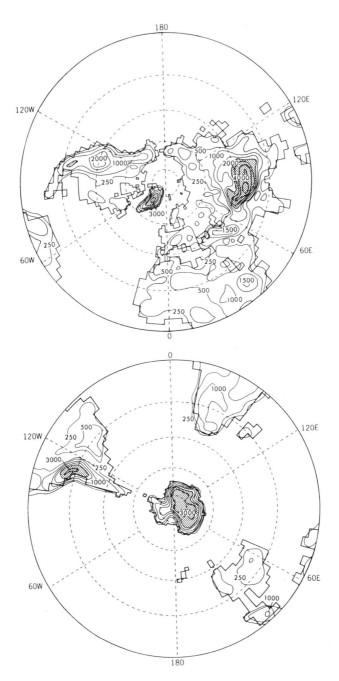

FIGURE 1. Surface elevation (m) of topography used for the M integration: (Top) Northern Hemisphere. (Bottom) Southern Hemisphere. Stippled areas indicate elevations >2000 m. Contours at 250, 500, 1000, 1500, 2000, 3000, 4000, and 5000 m. Owing to spectral truncation, contours extend over oceans as model elevation is not constrained to be zero at ocean gridpoints.

the results presented herein utilize data from those 3-year periods unless otherwise noted.

6. SIMULATED DISTRIBUTION OF DRY CLIMATES

The climate model output is used to determine the Köppen climate classification at each gridpoint for the M and NM integrations. Climate classification systems such as this are useful, albeit arbitrary, in providing an indication of the overall character of the simulated climates. In the Köppen system, monthly mean temperature and precipitation are used to classify climate in a manner intended to correspond to the prevailing natural vegetation. The Köppen scheme is a commonly used climate classification that has been used previously to examine climates simulated by GCMs.[12,24] A detailed description of the Köppen scheme can be found, for example, in Lamb.[25]

Köppen category B denotes desert and steppe climates, where precipitation is insufficient to meet the demands of forest vegetation at the prevailing temperatures. The spatial distributions of the B climates simulated by the M and NM integrations are compared to the observed distribution in Fig. 2. In the NM integration, dry climates are confined to zonal (or east–west) belts between 15° and 35° latitude in each hemisphere. These subtropical belts of aridity are nearly continuous, with only the southeastern parts of Asia and North America experiencing more moist climates. Midlatitude dryness is almost completely absent in the NM integration. This contrasts sharply with the results from the M integration, in which the subtropical dry regions are not as extensive and Köppen's category B climates occur across the midlatitude interior of Eurasia and, to a lesser extent, North America. Comparison with the observed distribution indicates that the M integration is in good agreement over the midlatitude Northern Hemisphere, implying that the model's simulation of temperature and precipitation in these regions is realistic.

The geographical distribution of annual mean soil moisture (Fig. 3) provides another measure of model aridity. In the NM integration, little east–west variation of soil moisture occurs, although the east coasts of North America and Asia are somewhat wetter than the remainder of those continents (as noted in the distribution of B climates). The pattern of soil moisture contours in the NM integration is largely zonal, with values in excess of 80% of saturation along the northern coasts of the continents and below 20% in the subtropics. This zonal symmetry is not present in the M integration, in which substantial east–west variation exists across both Eurasia and North America. Of particular interest to this study are two midlatitude regions of soil dryness. One extends eastward from the Caspian Sea across central Asia into northwest China, while another lies just east of the Rocky Mountains in Canada and the northern United States.

There is a great deal of similarity between the areas of low soil moisture and the regions where Köppen's category B climates are simulated. The reason for this

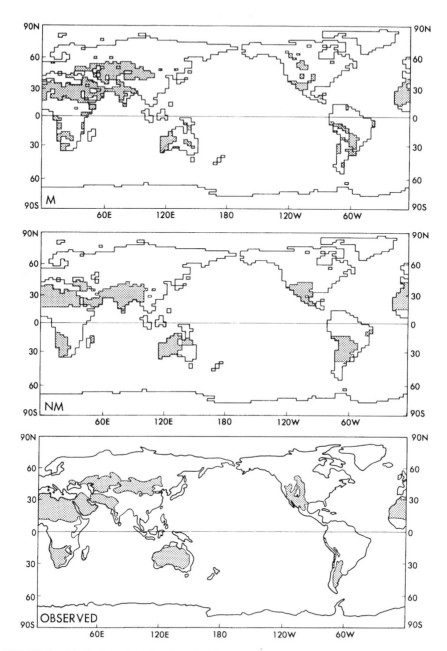

FIGURE 2.   Distribution of arid and semiarid climates (indicated by stippling) according to the
Köppen climate classification: (Top) M integration. (Center) NM integration. (Bottom) Observed.

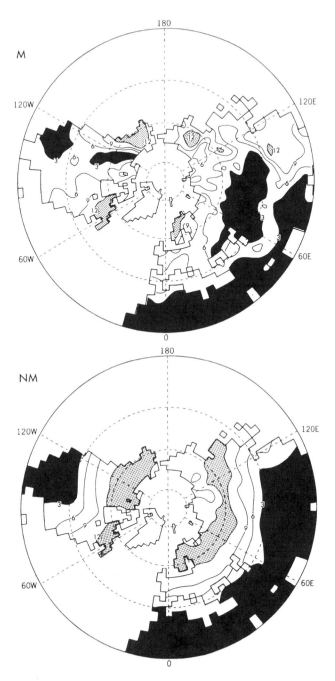

FIGURE 3. Annual mean soil moisture (cm): (Top) M integration. (Bottom) NM integration. Stippling indicates soil moisture > 12 cm; solid black indicates soil moisture < 3 cm. The field capacity of soil moisture in both integrations is 15 cm everywhere. A 1–2–1 smoothing has been applied in both directions to reduce grid scale variability.

similarity is evident when the method of computation of the Köppen classification is compared with the hydrologic component of the climate model. The Köppen scheme classifies arid and semiarid climates on the basis of an empirical relationship involving temperature and precipitation, using temperature as a surrogate for the potential evaporation from the land surface. The B climates are those in which potential evaporation substantially exceeds precipitation. In the climate model, a balance must exist (on an annual mean basis) between the inflow to the soil moisture "bucket" (precipitation) and the outflow (evaporation and runoff). The evaporation E is related to the potential evaporation E_p by $E = \beta E_p$, where $\beta = \min(w/w_k, 1.0)$, with w the soil moisture and w_k an empirical value (chosen to be 11.75 cm, or 75% of field capacity) below which evaporation is limited by soil moisture. The potential evaporation is determined from a surface heat budget, and is thus positively correlated with temperature. Thus for $w < w_k$, the actual evaporation is smaller than the potential evaporation by a factor of w/w_k. Given the balance requirement between precipitation and evaporation (if runoff is assumed to be small), this implies that potential evaporation exceeds precipitation wherever $w < w_k$. Thus the regions where soil moisture is well below the critical value (w_k) of 75% of field capacity experience potential evaporation that greatly exceeds precipitation.

The contrast between the largely zonal pattern of the NM integration and the more complex pattern of the M integration that appears in the Köppen classification and soil moisture distributions is also present for annual mean precipitation (Fig. 4). Without orography, precipitation is lightest over the Arctic Ocean and in two subtropical regions: southwestern North America and the adjacent Pacific, and northern Africa and the nearby Atlantic. Precipitation is relatively heavy in a band that circles the hemisphere between 45° and 65°N, and also at lower latitudes over the western North Atlantic and western North Pacific. In the M integration, more east–west variability occurs at midlatitudes than in the NM integration, and precipitation is light over the same continental interior regions where soil moisture is low. Annual precipitation rates of less than 1 mm/day extend across central Asia and in a band just east of the Rocky Mountains in North America.

Three regions (outlined in Fig. 5) were selected for more detailed analysis: west central Asia, east central Asia, and the Canadian prairie. To explore the seasonal variations of the difference in precipitation between the M and NM integrations, the annual cycle of precipitation is computed for each of these regions (Fig. 6). In all three areas, precipitation in the M integration is substantially lower than in the NM integration throughout the entire year. The seasonal cycles of precipitation for the M integration are somewhat different in each region, with a late winter/early spring maximum in west central Asia, a summer maximum in east central Asia, and a spring maximum in the Canadian prairie. In the NM integration, the heaviest precipitation in all three regions occurs during the period March through June, with a secondary autumn maximum in west central Asia and the Canadian prairie. Although there is some disagreement between the annual march of precipitation from the M integration and the

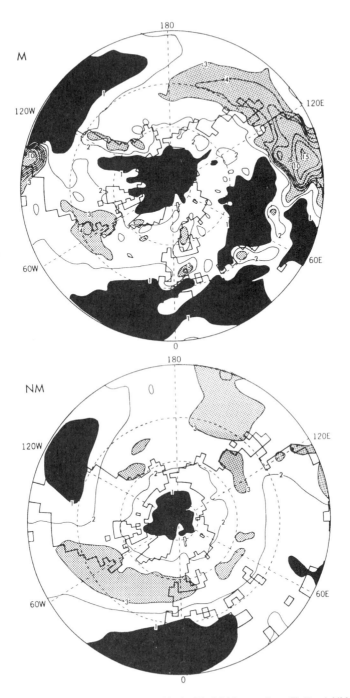

FIGURE 4.   Annual mean precipitation (mm/day): (Top) M integration. (Bottom) NM integration. Contours at 1, 2, 3, 4, 5, 6, 8, 10, 15, 20, 30 mm/day. Smoothing as in Fig. 3.

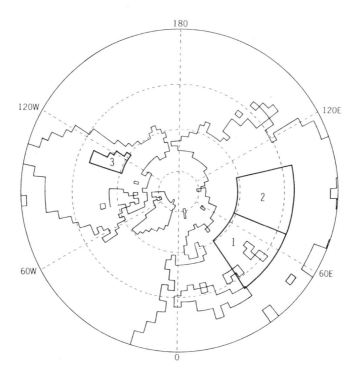

FIGURE 5.   Areas used in computation of regional precipitation: (1) west central Asia; (2) east central Asia; (3) Canadian prairie. Only land points within these regions were used in the computations.

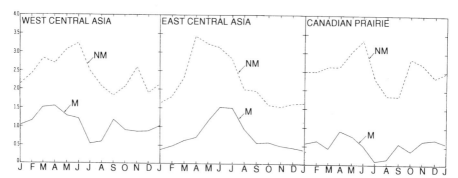

FIGURE 6.   Seasonal variation of monthly precipitation (mm/day) from the M and NM integrations for three regions: (Left) west central Asia. (Center) east central Asia. (Right) Canadian prairie.

observed precipitation climatology for west central Asia and the Canadian prairie, these differences are generally small when compared to the differences between the M and NM integrations.

7. LARGE-SCALE ATMOSPHERIC CIRCULATION

A substantial decrease in both soil moisture and precipitation is simulated over portions of the interiors of Eurasia and North America in response to the modern distribution of orography. How do mountains alter the atmospheric circulation to produce this drying? A relatively simple and well-understood mechanism probably contributes to the dryness of the western interior of North America. As the prevailing westerly winds in the lower troposphere encounter the Rocky Mountains, air is forced upward. The reduction of pressure as the air ascends results in a cooling, so that much of the moisture contained in the air condenses to form clouds and precipitation. Once the air crosses the highest elevations and starts downward, the increasing pressure warms the moisture-depleted air. (The well-known chinook winds of the Rocky Mountain foothills result from a similar mechanism on a smaller scale.) The ascending moist air on the upwind side results in orographic precipitation, while the subsiding dry air on the downwind side inhibits precipitation. Although this so-called rainshadow effect plays an important role in the dryness of interior North America, it cannot satisfactorily explain the extensive dryness of central Asia, as that continent lacks an extensive meridional barrier such as the Rocky Mountains. Other mechanisms must be involved in maintaining the arid climate of the Eurasian interior, and these mechanisms contribute to the dryness of western interior North America as well.

To identify one such mechanism, consider the processes responsible for the precipitation that falls in midlatitudes. Much of this precipitation is associated with the passage of transient disturbances (i.e., extratropical cyclones) with timescales of about 3–5 days. As discussed extensively by James,[26] such disturbances are not uniformly distributed throughout the midlatitudes of the Northern Hemisphere. Instead, they tend to be organized along "storm zones" or "storm tracks" that extend across the North Pacific and North Atlantic oceans. These locations have a specific relationship to the preferred positions for waves in the upper-tropospheric westerlies that become evident when the circulation is averaged over time periods of many seasons. The major storm tracks parallel the upper-tropospheric jet streams that occur between the prominent cold season stationary trough axes over eastern Asia and eastern North America and their corresponding downstream ridges.[27–29]

Theoretical studies[30,31] have linked the positions of these stationary waves to the spatial distribution of the underlying topography, with the ridge position (anticyclonic flow) in the vicinity of the topographic ridge and the trough position (cyclonic flow) downstream. Using a steady linear primitive equation model, Nigam et al.[32] found that orographic forcing was responsible for about two-thirds of the overall amplitude of winter stationary waves in the upper troposphere in the Northern Hemisphere. Thus it can be argued that mountains may contribute to the existence of midlatitude dry regions by inducing stationary waves that largely determine the favored locations for extratropical disturbance activity. If this argument is correct, there should be a clear relationship between the distribution of midlatitude precipitation and stationary wave position. This hypothesis can be evaluated by examining the simulated atmospheric circulation from the M integration and comparing it to the corresponding circulation features in the NM integration.

The large seasonal changes in climate that occur in the midlatitude Northern Hemisphere require an examination of the entire seasonal cycle of precipitation and circulation. To facilitate this approach, longitude–time sections have been constructed by computing meridional averages of precipitation and 500-mb geopotential height for the belt from 30°–50°N. The 500-mb geopotential height is used to depict the atmospheric circulation, as winds at this level (to a close approximation) blow parallel to the geopotential height contours, with lower heights to the left. The geopotential height is expressed as a departure from the zonal mean to emphasize the seasonal variations in the position of stationary waves and eliminate the large thermally induced seasonal cycle in zonal mean geopotential height. Troughs and ridges appear as relative minima and maxima, respectively.

The longitude–time section for the M integration (Fig. 7, left) indicates the existence of waves that maintain an almost constant longitude for substantial portions of the seasonal cycle. Of most importance to this study are the ridge–trough couplets that occur over North America through most of the year and the eastern half of Eurasia during the cold season, with the trough axes along or just east of the east coasts of these continents. (Note that the Tibetan Plateau and Rocky Mountains are at longitudes of 75–100°E and 240–255°E, respectively.) For each of these features, the heaviest precipitation tends to occur along and just east of the trough axis, with much lighter precipitation occurring upstream in the vicinity of and just downstream of the ridge. This is consistent with the hypothesis, based in part on theoretical studies of the wintertime Northern Hemisphere circulation, that orographically induced stationary waves organize the distribution of midlatitude precipitation and contribute to the existence of midlatitude dry regions.

In the warm season there is a substantial weakening and westward shift of the east Asian trough as the Asian monsoon circulation becomes a more dominant influence. This is accompanied by a corresponding westward shift in both the wet and dry regions. In contrast, there is very little evidence of a similar circulation change over North America, as expected given the absence of a strong monsoon circulation there.

The stationary waves from the NM integration (Fig. 7, right) have much smaller amplitude and less seasonal persistence than those of the M integration. Although the precipitation distribution in the NM integration is not perfectly uniform, it is much more amorphous and lacks the relationship to the stationary

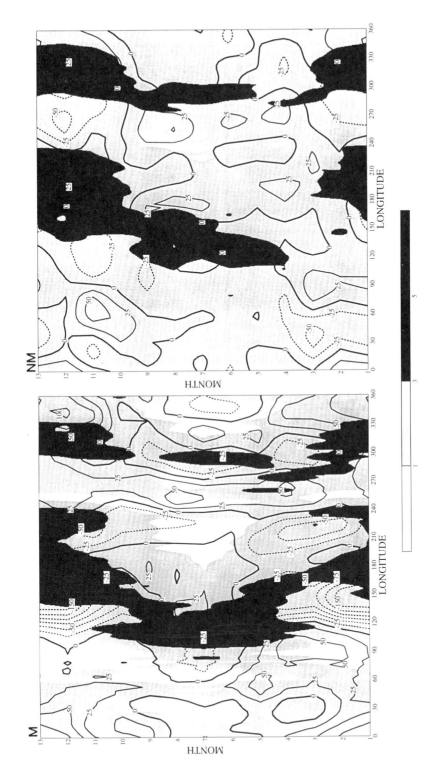

FIGURE 7. Longitude–time section of 500-mb geopotential height departure from zonal mean (m; indicated by contours with negative contours dashed) and precipitation (mm/day; indicated by shading) averaged from 30° to 60°N. For reference, the Tibetan Plateau and Rocky Mountains are at longitudes of 75–100°E and 240–255°E, respectively.

wave pattern that is so prominent in the M integration. The absence of midlatitude dryness over the continents is evident from the reduction in the prominence of the unshaded regions, indicative of precipitation rates less than 1 mm/day. There remains some tendency for the oceans to be wetter than the continents in winter, and for the east coasts of the continents to be wetter than the west coasts in summer.

To further examine the relationships between circulation and precipitation in the M and NM integrations, a more detailed examination is warranted. Because the longitude–time section from the M integration provides evidence of important seasonal changes, particularly over Eurasia, the examination will focus on the boreal winter and summer seasons.

7.1. Winter

Various diagnostics of the atmospheric circulation at the 500-mb level (i.e., 5–6 km above sea level) are used to examine the circulation in boreal winter (December–January–February). The stationary waves can be seen as the meandering of the 500-mb geopotential height contours. Isotachs, or contours of wind speed, enable the identification of jet streams, and the root-mean-square (rms) of the daily geopotential height (band-pass-filtered to retain fluctuations with timescales between 2.5 and 6 days) provides a measure of storminess. The 500-mb level is chosen as Blackmon et al.[29] and Blackmon and Lau[33] have found that the rms of band-pass-filtered heights at that level corresponds well with synoptic-scale disturbances. While higher tropospheric levels may be equally or more suitable for identifying the stationary wave pattern and jet stream locations, comparison of upper tropospheric maps with those from the 500-mb level suggests that the features are similar.

When compared to the NM integration, the amplitude of the stationary waves in the M integration is substantially larger, as noted earlier from the longitude–time sections. The geopotential heights and isotachs (Fig. 8, top) from the NM integration follow a nearly circular pattern, with only low-amplitude waves and a circumpolar jet axis centered between 45° and 50°N. In contrast to this zonally symmetric picture, stationary waves are present in the M integration with troughs situated downstream of the Tibetan Plateau and Rocky Mountains. Each of these troughs is associated with a wind maximum located downstream, a feature simulated by Bolin[30] and Cook and Held[34] in response to idealized topography. The increase in the amplitude of stationary waves in the presence of orography is similar to that found by Manabe and Terpstra[10] and Kutzbach et al.[15]

Contrasting patterns between the M and NM integrations also appear in the rms of band-pass-filtered 500-mb heights (Fig. 8, center). In the NM integration, storminess is high in a circumpolar band between 40° and 60°N, nearly coincident with the maximum 500-mb winds. In the M integration, the bands of high storminess are more narrowly confined to maxima stretching across the North Pacific and North Atlantic. These storm tracks closely parallel the 500-mb wind

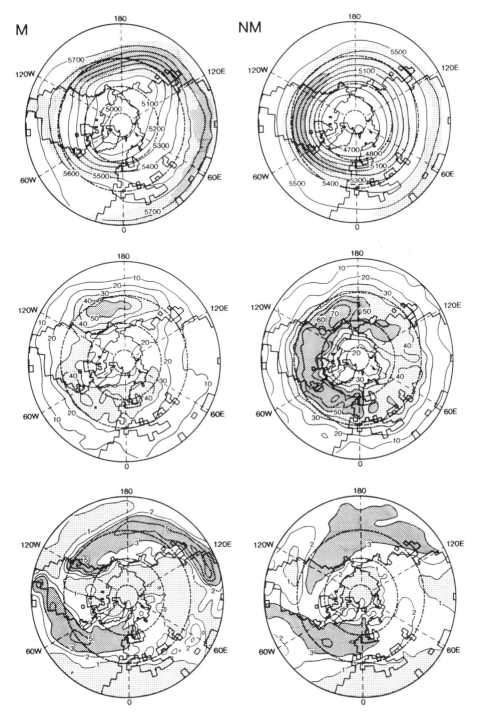

FIGURE 8. December–January–February circulation and precipitation from the (left) M integration and (right) NM integration: (Top) 500-mb geopotential height (dm). Light stippling indicates 500-mb winds > 15 m/s; dense stippling > 25 m/s. (Center) Root-mean-square (rms) of band-pass-filtered 500-mb geopotential height (m). The bandpass filter selects disturbances between 2.5 and 6 days. Light stippling indicates values > 30 m; dense stippling > 50 m. (Bottom) Precipitation (mm/day). Light stippling indicates precipitation < 1 mm/day; dense stippling > 3 mm/day. Contour interval as in Fig. 4, and smoothing as in Fig. 3.

speed maxima that occur downstream of the trough axes, in a relationship very similar to observations.[28] An association between storm tracks and jet axes is expected owing to the relationship between vertical wind shear and storm development, also noted by Manabe and Terpstra[10] in their GCM simulation. Substantially lower rms values occur over the regions of midlatitude aridity in the M integration than in the same locations in the NM integration, indicating a reduction in synoptic disturbance activity.

To see the relationship between precipitation and these circulation features, the bottom of Fig. 8 contains maps of winter precipitation from the M and NM integrations. There is some similarity between these distributions and their annual mean counterparts (Fig. 4). A zonal band of moderate precipitation (~ 2–3 mm/day) is evident over the continents between 40° and 60°N in the NM integration, with the heaviest amounts (>3 mm/day) over the western continental midlatitudes. This rain belt nearly coincides with the axis of maximum 500-mb winds and storminess. Elsewhere, the tendency for winter precipitation to be heavier over the oceans is also apparent, as discussed earlier in this section. In the M integration, much larger spatial variations occur in middle latitudes, with precipitation rates less than 1 mm/day east of the Canadian Rockies and across a vast region in eastern and central Asia. Precipitation over these regions is as little as one-third of that in the NM integration. Much heavier amounts of more than 4 mm/day occur over southeastern portions of Asia and North America near the western ends of the oceanic storm tracks.

The circulation maps further support the relationship between midlatitude precipitation and stationary waves during boreal winter that was evident in the longitude–time section from the M integration. Precipitation tends to be heaviest in the areas downstream of the stationary wave troughs, such as from the east coast of North America east–northeastward across the North Atlantic, and from China across Japan into the North Pacific. These regions of heavy precipitation are associated with high synoptic disturbance activity and relatively strong midtropospheric winds. In areas upstream of the troughs, synoptic disturbance activity is relatively weak and precipitation is light. In the more zonally symmetric NM integration, stationary wave amplitudes are small, so that precipitation and disturbance activity are associated with a circumpolar jet axis that is continuous around the hemisphere.

7.2. Summer

The 500-mb circulation in the NM integration maintains its generally zonal character during boreal summer (Fig. 9, right). Although the associated jet stream is not as prominent, synoptic disturbance activity continues at a moderate level in a circumpolar belt between 40° and 60°N. The midlatitude rainbelt associated with these features is quite prominent over the continents, and is located toward the poleward side of this belt. Summer also brings relatively heavy rainfall at lower latitudes over the east coasts of North America and Asia. These are regions that lie under southerly surface flow on the west side of the oceanic

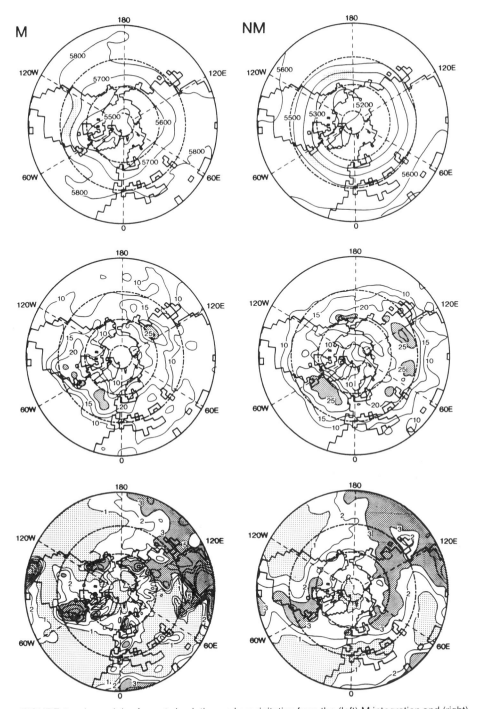

FIGURE 9. June–July–August circulation and precipitation from the (left) M integration and (right) NM integration: (Top) 500-mb geopotential height (dm). Light stippling indicates 500-mb winds >15 m/sec; dense stippling >25 m/sec. (Center) Root-mean-square (rms) of band-pass-filtered 500-mb geopotential height (m). The band-pass filter selects disturbances between 2.5 and 6 days. Stippling indicates values >25 m. (Bottom) Precipitation (mm/day). Light stippling indicates precipitation <1 mm/day; dense stippling >3 mm/day. Contour interval as in Fig. 4, and smoothing as in Fig. 3.

subtropical anticyclones, which reach their northernmost latitudes during the season.

In contrast to the NM integration, summer brings substantial changes to the stationary wave pattern in the M integration (Fig. 9, left), in terms of both wave amplitudes and positions, as discussed earlier in conjunction with the longitude–time sections. While a trough remains over eastern North America, the east Asian trough is not present. This is probably a consequence of the seasonal retreat of the westerlies, which are now far enough north that they no longer encounter the Tibetan Plateau. Despite this change, extensive dryness still prevails over central Asia, particularly west of 100°E, where the axis of a weak trough is located. The absence of this trough in the NM experiment suggests that it is the result of orographic forcing, either direct or indirect. The summer precipitation difference between the M and NM integrations in this region is substantial, with much drier conditions prevailing in the M experiment.

Previous work points to changes in the south Asian monsoon, the dominant circulation over much of the Asian continent in boreal summer, as a possible source of the reduction in summer precipitation over central Asia. In their GCM simulations with and without orography, Hahn and Manabe[11] found that mountains played a large role in the south Asian monsoon. They found that a center of rising motion was present above the Tibetan Plateau and its southern slope in the integration with mountains, but absent from the same location in the integration without mountains. Their result suggests that some of the differences in central Asian summer precipitation between the NM and M integrations could be due to the changes in the monsoon circulation.

Rising motion in the atmosphere is generally associated with the convergence of air in the lower troposphere and a corresponding divergence at upper levels. The irregularity of the surface often creates small-scale circulations that make it difficult to diagnose low-level convergence, so upper-level divergence is frequently used as an indication of the large-scale circulation. One way of diagnosing the upper-level divergence is to compute the velocity potential, as the divergent component of the flow is along the gradient of velocity potential. Thus a minimum (maximum) in the velocity potential is associated with large-scale divergence (convergence) at that level. The velocity potential at the 205-mb level (an altitude of ~ 12 km) from the NM integration (Fig. 10, center) indicates a very large-scale dipole pattern, with an elongated minimum over the western tropical Pacific and southeast Asia and a maximum over the tropical Atlantic. The heavy precipitation falling from southeast Asia eastward along the equatorial region suggests the importance of the latent heat of condensation in maintaining the upward vertical motion associated with this circulation. A similar dipole pattern is also present in the M integration (Fig. 10, top), but the primary center of the elongated minimum is shifted west-northwestward to southeast Asia, where precipitation is very heavy. In the difference map (Fig. 10, bottom) a strong center of negative values is located above the southeast portion of the Tibetan Plateau, indicative of increased upper divergence at that location. An area of enhanced precipitation coincides with this center, consistent with the orographically induced enhancement of the south Asian monsoon first noted by Hahn and Manabe.[11]

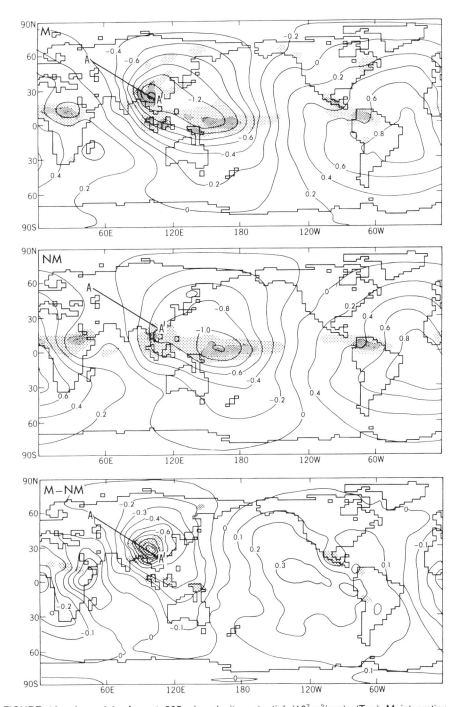

FIGURE 10. June–July–August 205-mb velocity potential ($10^7\,\mathrm{m^2/sec}$): (Top) M integration. (Center) NM integration. (Bottom) M – NM. Light stippling indicates precipitation between 5–10 mm/day; dense stippling precipitation >10 mm/day. In bottom panel, stippling indicates precipitation increase >4 mm/day. The lines AA′ in the top and center panels depict the orientation of the cross section shown in Fig. 12.

The intense latent heating above the southeast portion of the Tibetan Plateau produces a distinctive warm core structure, as is often observed in low-latitude circulations driven by latent heating. In such circulations, a low-level cyclonic circulation (the "monsoon low" or "south Asian low" in this case) is located beneath a divergent anticyclonic circulation in the upper troposphere. The lower tropospheric cyclonic circulation is evident in a map of wind vectors at the 830-mb level from the M integration (Fig. 11, top), although proximity to the elevated surface produces irregularities in the low-level flow. Further evidence of the vertical circulation associated with the summer monsoon appears in a cross section of vertical pressure velocity (Fig. 12, top) extending from northwest to southeast across the Tibetan Plateau. Intense upward motion (indicated by negative values) occurs above the southeast portion of the plateau and its vicinity as southwesterly winds in the lower troposphere (Fig. 11, top) encounter its southern slope. Meanwhile, subsidence occurs to the northwest of the Tibetan Plateau.

The relatively dry air that subsides northwest of the Tibetan Plateau in the M integration is drawn south and east across much of central Asia by the low-level cyclonic circulation. To the south, the northward flow of air from the Indian Ocean and subcontinent does not penetrate beyond the Tibetan Plateau. This contrasts with the integration without mountains, where strong lower tropospheric westerlies occur over middle latitudes, extending equatorward to the latitude of the Tibetan Plateau (Fig. 11, bottom). The south Asian low is much weaker in the NM case, and strong centers of time-averaged vertical velocity are absent from south Asia (Fig. 12, bottom).

An active storm track, indicated by the rms of band-pass-filtered 500-mb height (Fig. 9, center), is associated with the midlatitude westerlies of the NM integration. The disturbances propagating along this track bring substantial precipitation to central Asia. By contrast, the low-level cyclonic circulation surrounding the Tibetan Plateau in the M integration weakens the westerlies in the region north of the plateau, so that the strongest winds in the lower troposphere are weaker and shifted poleward. A corresponding change in the storm track occurs, with the band of maximum synoptic disturbance activity weakened and shifted northward. The combination of subsidence, the low-level flow of dry air, and this reduction of storminess is unfavorable for summer precipitation over central Asia.

8. WATER VAPOR TRANSPORT

The impact of orography on midlatitude aridity can be viewed from another perspective by examining the annual mean vertically integrated water vapor transport from the M and NM integrations. Assuming that no net change in moisture storage in the atmosphere or soil occurs, a condition generally met when averaging over an annual cycle, the divergence of this transport is equal to the

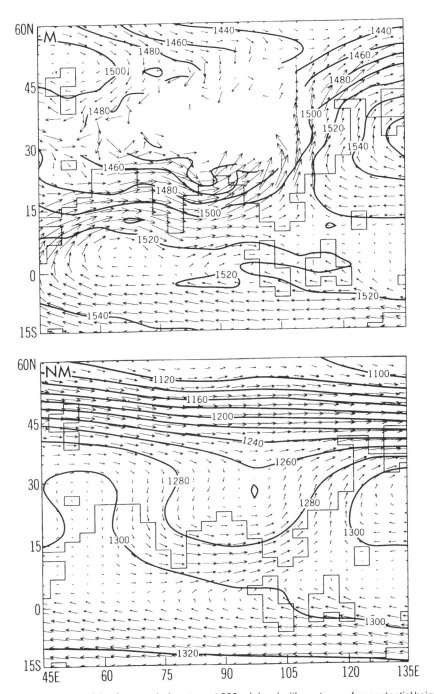

FIGURE 11. June–July–August wind vectors at 830-mb level with contours of geopotential height at 850 mb superimposed. No vectors or contours are plotted if the appropriate level is below the surface: (Top) M integration. (Bottom) NM integration.

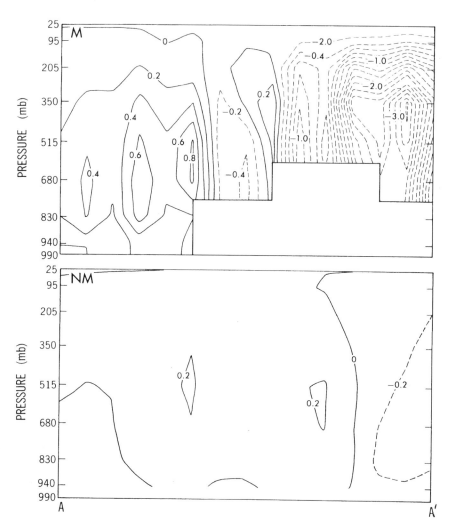

FIGURE 12.   Cross section of vertical pressure velocity (dyne/cm² sec¹) along the line depicted in Fig. 10. A 1–2–1 smoothing has been applied to the geographical distribution of vertical pressure velocity at each level before forming the cross section. Dashed contours indicate negative values (i.e., upward motion): (Top) M integration. (Bottom) NM integration.

difference between evaporation and precipitation. Owing to data-processing limitations, the moisture transport was computed using only one year of data from each integration, but given the large changes in boundary conditions and the model's substantial response to those changes, it is likely that the differences are representative of what would have been obtained from a sample of longer duration.

In the NM integration (Fig. 13, bottom), a strong annually averaged flow of moisture prevails throughout the belt from 40° through 60°N, carrying an ample

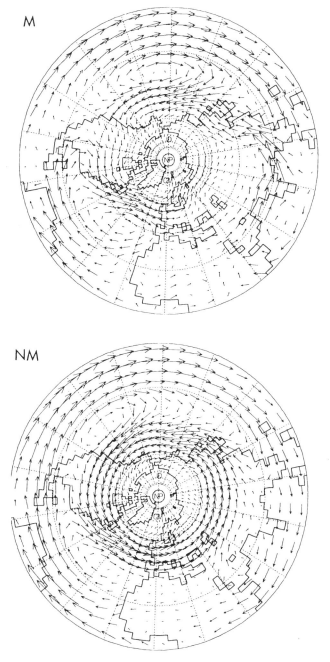

FIGURE 13. Annual average vertically integrated water vapor transport vectors: (Top) M integration. (Bottom) NM integration.

supply of moisture from oceanic sources into the continents. Anticyclonic circulations centered over the eastern North Pacific, central North Atlantic, and northern Arabian Sea carry moisture from the tropics into this westerly transport stream. In addition, moisture from the west penetrates deep into the interiors of North America and Eurasia. The high soil moisture values of the NM integration may promote this penetration by allowing moisture to be recycled through the land surface, since evaporation from the relatively wet soil takes place at a substantial fraction of the potential rate. Strong westerly transport also occurs across the North Atlantic and North Pacific in the M integration (Fig. 13, top) in a manner similar to the NM integration, but the moisture flux weakens as it enters North America and Europe and bypasses the continental interior. Thus the effect of orography is to substantially reduce the water vapor transport into the interior regions of both North America and Eurasia.

9. ROLE OF SOIL MOISTURE FEEDBACK

The effect of the interaction between soil moisture and the atmosphere contributes to the response of the model to orography in the results presented thus far. In the context of the present experiment, this interaction, or feedback, might operate in the following manner. The results presented in the previous sections have shown that the presence of orography induces a decrease in precipitation for the interiors of Eurasia and North America. By altering the water budget, this reduction in precipitation induces a decrease in soil moisture. If there is sufficient radiative energy available at the surface, the reduction in soil moisture would lead to a decrease in evaporation from the surface. This lowers the atmospheric humidity, leading to an additional decrease in precipitation. As described in this example, the feedback would be positive, meaning that the initial response (the decrease in precipitation) is enhanced by the feedback process.

Among the virtues of climate models is that they allow the importance of such processes to be evaluated quantitatively. Previous climate model studies[35,36,23] have demonstrated that the interaction between soil moisture and the atmosphere can be quite important. To explore the role of soil moisture feedback in the response of the present model to orography, the FSM integration is included in this study. As described in Section 5, this integration is identical to the M integration except that it uses prescribed soil moisture values (varying seasonally and spatially) taken from the NM integration. Differences between the FSM and NM integrations represent the effects of orography *without* soil moisture feedback, while differences between the M and NM integrations include soil moisture feedback.

The difference in annual mean precipitation between the M and NM integrations (Fig. 14, top) illustrates the large reduction of precipitation over the midlatitude continental interiors in a belt between approximately 40° and 60°N. A vast area experiences a decrease in precipitation rate larger than 1 mm/day,

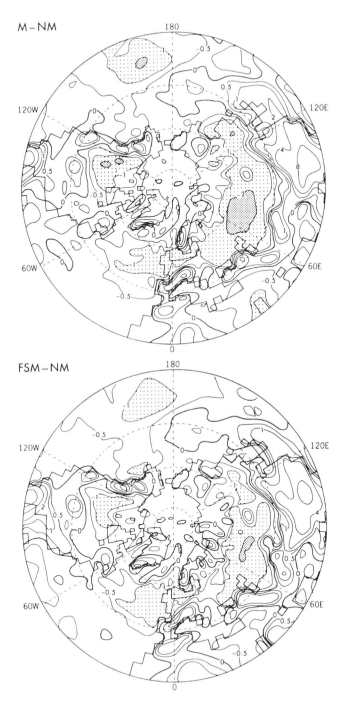

FIGURE 14. Annual mean precipitation difference (mm/day): (Top) M − NM. (Bottom) FSM − NM. Contours at −2, −1, −0.5, 0, 0.5, 1, 2, 4, 8 mm/day. Smoothing as in Fig. 3.

with decreases of more than 2 mm/day in central Asia and in small areas of western Canada. This pattern is largely reproduced in the annual mean precipitation differences between the FSM and NM integrations (Fig. 14, bottom), although the magnitudes of the precipitation reduction are somewhat smaller. This indicates that the effect of soil moisture feedback makes a modest contribution to the annual mean response.

This conclusion is confirmed by once again focusing on the three regions depicted earlier in Fig. 6 and computing differences in annual precipitation for each of them. These statistics are presented in Table 1. The M − NM differences represent the total effect of orography on precipitation, and the FSM − NM differences represent the effect of orography without soil moisture feedback. For the three regions, the contribution of soil moisture feedback accounts for 26–34% of the overall reduction in precipitation.

On a seasonal basis, however, this feedback mechanism is more important. Examination of the seasonal cycle of precipitation differences (Fig. 15) indicates a distinct seasonal cycle in the importance of soil moisture feedback. In all three regions, soil moisture feedback plays an insignificant role during the winter months, as the reduction in precipitation owing to orography is almost as large without soil moisture feedback as it is with that feedback included. As surface temperatures are quite low and little radiative energy is available during the winter, potential evaporation is small and the effect of soil moisture on the atmosphere is limited. Soil moisture feedback has little or no effect on the stationary wave pattern, as evidenced by the similarity of the 500-mb-height maps for the FSM integration (not shown) and the M integration.

The importance of soil moisture feedback increases during the spring, becoming most significant during the summer months, when surface temperatures and potential evaporation are relatively high. Soil moisture feedback accounts for a majority of the orographically induced summer reduction in precipitation, with contributions of 46, 55, and 87% in the west central Asia, Canadian prairie, and east central Asia regions, respectively. During autumn the contribution of soil moisture feedback to the change in precipitation decreases rapidly in association with the decrease in incoming solar radiation at the surface.

TABLE 1. Differences in Annual Mean Precipitation
Rate (cm/day) for the regions depicted in Fig. 6 [a]

Region	M − NM	FSM − NM
West central Asia	−0.140	−0.093
East central Asia	−0.148	−0.097
Canadian prairie	−0.201	−0.149

[a] The M − NM difference represents the total effect of orography on precipitation, while the FSM − NM difference represents the effect of orography without soil moisture feedback.

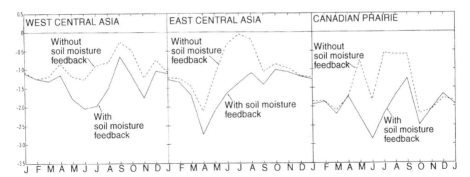

FIGURE 15. Seasonal variation of the difference in monthly precipitation (mm/day) between the M and NM integrations (solid lines), representing the effect of orography with soil moisture feedback, and between the FSM and NM integrations (dashed lines), representing the effect of orography without soil moisture feedback: (Left) west central Asia. (Center) east central Asia. (Right) Canadian prairie.

10. PALEOCLIMATIC IMPLICATIONS

The current set of integrations was not designed to correspond to any period in Earth history, as there were no periods in the Earth's past when the current distribution of continents occurred simultaneously with uniform low relief. Nonetheless, it may be possible to use these integrations to infer the qualitative changes in climate that might be associated with relatively recent orographic uplift. For example, since the continental interiors of Eurasia and North America are drier in the M integration than in the NM integration, one could infer that orographic uplift near these regions would also lead to drying. Recent simulations using more realistic scenarios of continental drift and tectonic evolution confirm the importance of Tibetan Plateau uplift to the drying of central Asia (A. Bush, personal communication, 1996).

Paleoclimatic data assembled by Ruddiman et al.[1] for comparison with their climate modeling experiments provides a source of evidence for changes in climate over the midlatitude continental interiors. Drier winters in the northern Great Plains are indicated by the gradual transition from forest to grassland during the late Miocene to Pliocene. A transition from forest to steppe in northwest China, an expansion of deserts in the Asian interior, and increasing fluxes of windborne dust into the North Pacific Ocean suggest a trend toward drier climates in the Eurasian interior. These changes are qualitatively consistent with those occurring in our GCM experiments in response to the presence of orography.

There are some unresolved issues in this interpretation. The gradual decrease of atmospheric CO_2 over the last 10 million years may have favored plants using the C_4 metabolic pathway (mainly grasses) at the expense of those using the C_3 pathway (mainly trees), as discussed in Chapter 13. Thus some of the expansion of grassland in interior North America and Eurasia over this time period could be due to the direct effect of CO_2 rather than to climate. Also, the timing and extent of recent uplift is crucial to the argument that changes in the paleoclimate record support the results of our climate modeling experiments, and there is some controversy about the evidence for recent uplift in certain regions (Chapter 1).

The resolution of these issues is beyond the scope of this chapter and is addressed elsewhere in this volume. However, the consistency between the changes in climate simulated by our model and those inferred from the geological record supports the connection between mountains and midlatitude aridity, provided that the uplift of western North America and the Tibetan Plateau region is relatively recent and the changes in prevailing vegetation are due to climate.

11. SUMMARY AND CONCLUDING REMARKS

Motivated by previous GCM experiments, an atmospheric GCM with interactive hydrology was used to perform integrations with and without mountains. Substantial differences in precipitation and soil moisture were simulated over the interiors of North America and Eurasia in response to orography. Several mechanisms have been identified that produce this response.

For most of the year, large-amplitude stationary waves in the midtropospheric circulation occur in response to the Tibetan Plateau and Rocky Mountains, with anticyclonic flow near the longitude of the orographic ridge and cyclonic flow downstream. The midlatitude dry regions are located in the zone of large-scale subsidence and infrequent storm development that is found upstream of the troughs of these waves, while ascent and storminess occur further downstream. In the absence of orography, the atmospheric circulation is more zonally symmetric and a belt of storminess circles the Northern Hemisphere between 40° and 60°N, producing heavier precipitation over the midlatitude continental interiors. In North America, the rainshadow effect associated with the forced ascent of air by the meridional barrier of the Rocky Mountains and its subsequent descent also contributes to the aridity of the continental interior.

The North American interior remains downstream of a stationary wave during the summer, but a seasonal change in the midlatitude westerlies allows the South Asian monsoon to be the dominant circulation influencing the climate of interior Eurasia. The Tibetan Plateau exerts an important influence on the monsoon circulation, strongly influencing the dryness of interior Asia. With orography present, the rising branch of the Asian monsoon circulation is centered over the southeast portion of the plateau, and the associated low-level cyclonic circulation brings a flow of relatively dry, subsiding air southwestward across central Asia. In addition, a trough in the westerlies is situated to the north of the Tibetan Plateau. This trough, which is not present in the integration without mountains, organizes precipitation to its east and relatively dry conditions to its west across central Asia.

A substantial change in the transport of water vapor into the continental interiors occurs in response to orography. Without mountains, a strong westerly flow of moisture exists throughout the Northern Hemisphere midlatitudes, with tropical moisture drawn into this flow by anticyclonic circulations over the subtropical North Pacific, North Atlantic, and Arabian Sea. In contrast, the westerly transport of moisture in the M integration weakens as it reaches the west coasts of Europe and North America. Thus although the sizes and locations of the continents are identical in the two integrations, substantial differences exist in the transport of water vapor across the continental interiors. The results from the NM integration suggest that the recycling of atmospheric water vapor through precipitation and subsequent evaporation from the land surface can promote the penetration of moisture into these interior regions.

Soil moisture feedback is responsible for between one-fourth and one-third of the orographically induced reduction in annually averaged precipitation over the midlatitude arid regions. The contribution of this feedback undergoes a large seasonal variation. In winter, when little radiative energy is available at the surface, the soil moisture feedback effect is minimal, whereas it contributes more than half of the total response in summer. A portion of this feedback arises because the land surface provides a memory that allows the decrease in cold season precipitation owing to orographic stationary wave forcing to contribute to a reduction in precipitation during subsequent months.

Our analysis suggests an alternative to the traditional explanation that physical distance from oceanic moisture sources is a major cause of the dry climates of the midlatitude continental interiors. The relative wetness of the continents in the NM experiment indicates that large continents alone may not ensure the existence of midlatitude arid regions in their interiors. Therefore, we believe that orographically induced changes in atmospheric circulation and feedbacks involving the land surface may be quite important in determining the flow of moisture into continental interiors. Although rainshadow effects undoubtedly contribute to the dryness of western interior North America and, perhaps to a lesser extent, the Gobi Desert, no sizable mountain barriers lie upstream of the vast area of dryness in central Eurasia extending from the Black and Caspian seas eastward to the Tien Shan mountains. Thus we suggest that mountains are responsible for the existence of these dry regions, acting through large-scale effects on the atmospheric circulation.

Paleoclimatic evidence (i.e., vegetation changes, windborne dust) of less aridity in the northern Great Plains and central Asia during the late Tertiary supports this possibility. Provided that the uplift of the Tibetan Plateau and western North America is geologically recent, the consistency of our results with the paleoclimatic evidence lends credence to our hypothesis for the maintenance of midlatitude aridity.

There are reasons for caution regarding our results. Our experiments have limited the interaction between the atmosphere and ocean by prescribing the modern distribution of sea surface temperature. It is quite likely that the dramatic changes in atmospheric circulation we have simulated in the NM integration would alter the exchanges of heat, fresh water, and momentum between the atmosphere and ocean, thus potentially altering the sea surface temperature distribution and, perhaps, the ocean circulation. To fully explore this possibility will require the use of a coupled atmosphere–ocean model, a worthwhile topic of future investigation (see Chapter 6). Meanwhile, the consistency of our results with those from other studies of orographic effects using different GCMs and simple linear models suggests that they are relatively robust. This gives us some confidence that the mechanisms we have identified are largely responsible for midlatitude aridity.

REFERENCES

1. Ruddiman, W. F., Prell, W. L., and Raymo, M. E. (1989). *J. Geophys. Res.* **94**, p. 18379.
2. Manabe, S., and Broccoli, A. J. (1990). *Science* **247**, p. 192.
3. Broccoli, A. J., and Manabe, S. (1992). *J. Clim.* **5**, p. 1181.
4. Haurwitz, B., and Austin, J. M. (1944). *Climatology*, McGraw-Hill, New York.
5. Crutchfield, H. J. (1974). *General Climatology*, Prentice-Hall, Englewood Cliffs, NJ.
6. Trewartha, G. T., and Horn, L. H. (1980). *An Introduction to Climate*, McGraw-Hill, New York.
7. Mintz, Y. (1965). Very long-term global integration of the primitive equations of atmospheric motion, *WMO Tech. Note No. 66*, p. 141.
8. Kasahara, A., and Washington, W. M. (1971). *J. Atmos. Sci.* **28**, p. 657.
9. Kasahara, A., Sasamori, T., and Washington, W. M. (1973). *J. Atmos. Sci.* **30**, p. 1229.
10. Manabe, S., and Terpstra, T. B. (1974). *J. Atmos. Sci.* **31**, p. 3.
11. Hahn, D. G., and Manabe, S. (1975). *J. Atmos. Sci.* **32**, p. 1515.
12. Manabe, S., and Holloway, J. L., Jr. (1975). *J. Geophys. Res.* **80**, p. 1617.
13. Lau, N.-C. (1986). *Proceedings of International Symposium on the Qinghai-Xizang Plateau and Mountain Meteorology*, American Meteorological Society, p. 241.
14. Tokioka, T., and Noda, A. (1986). *J. Meteor. Soc. Japan* **64**, p. 819.
15. Kutzbach, J. E., Guetter, P. J., Ruddiman, W. F., and Prell, W. L. (1989). *J. Geophys. Res.* **94**, p. 18393.
16. Ruddiman, W. F., and Kutzbach, J. E. (1989). *J. Geophys. Res.* **94**, p. 18409.
17. Gordon, C. T., and Stern, W. F. (1982). *Mon. Wea. Rev.* **110**, p. 625.
18. Lacis, A. A., and Hansen, J. E. (1974). *J. Atmos. Sci.* **31**, p. 118.
19. Stone, H. M., and Manabe, S. (1968). *Mon. Wea. Rev.* **96**, p. 735.
20. Manabe, S. (1969). *Mon. Wea. Rev.* **97**, p. 739.
21. McFarlane, N. A. (1987). *J. Atmos. Sci.* **44**, p. 1775.
22. Palmer, T. N., Shutts, G. J., and Swinbank, R. (1986). *Quart. J. Royal. Meteor. Soc.* **112**, p. 1001.
23. Delworth, T., and Manabe, S. (1989). *J. Clim.* **2**, p. 1447.
24. Guetter, P. J., and Kutzbach, J. E. (1990). *Climatic Change* **16**, p. 193.
25. Lamb, H. H. (1972). *Climate: Present, Past and Future*, Methuen, London.
26. James, I. N. (1994). *Introduction to Circulating Atmospheres*, Cambridge University Press, Cambridge.
27. Palmen, E., and Newton, C. W. (1969). *Atmospheric Circulation Systems*, Academic Press, New York.
28. Lau, N.-C. (1988). *J. Atmos. Sci.* **45**, p. 2718.
29. Blackmon, M. L., Wallace, J. M., Lau, N.-C., and Mullen, S. L. (1977). *J. Atmos. Sci.* **34**, p. 1040.
30. Bolin, B. (1950). *Tellus* **2**, p. 184.

31. Held, I. M. (1983). In: *Large Scale Dynamical Processes in the Atmosphere* (B. J. Hoskins and R. P. Pearce, eds.), pp. 127–168. Academic Press, London.
32. Nigam, S., Held, I. M., and Lyons, S. W. (1988). *J. Atmos. Sci.* **45**, p. 1433.
33. Blackmon, M. L., and Lau, N.-C. (1980). *J. Atmos. Sci.* **37**, p. 497.
34. Cook, K. H., and Held, I. M. (1992). *J. Atmos. Sci.* **49**, p. 525.
35. Rind, D. (1982). *Mon. Wea. Rev.* **110**, p. 1487.
36. Yeh, T.-C., Wetherald, R. T., and Manabe, S. (1983). *Mon. Wea. Rev.* **111**, p. 1013.

6

The Effects of Uplift on Ocean–Atmosphere Circulation

David Rind, Gary Russell, and William F. Ruddiman

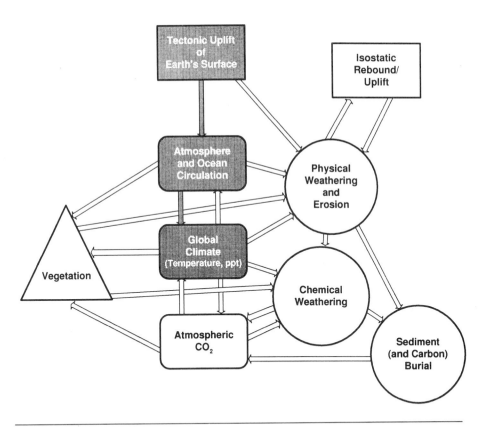

David Rind and Gary Russell • Goddard Space Flight Center, Institute for Space Studies, New York, New York 10025   *William F. Ruddiman* • Department of Environmental Sciences, University of Virginia, Charlottesville, Virginia 22903.

Tectonic Uplift and Climate Change, edited by William F. Ruddiman. Plenum Press, New York, 1997.

1. INTRODUCTION

The potential effect of uplift on climate and circulation during the Cenozoic has several different interacting components. First, changing topography alters the standing wave structure in the atmosphere, and hence the monthly mean longitudinal distribution of winds and temperatures. The current position of the standing long-wave troughs in the atmosphere, which in winter are located along the east coasts of North America and Asia, with a smaller trough in central Europe, depend upon the current distribution of topography (including the land–ocean distribution). These trough locations guide the position of storms and storm tracks, and the resulting wind circulation; for example, were a trough to be located over western Europe today, the airflow into the region would come down from Greenland and conditions would be much colder. (That such a circulation is hypothesized for the Little Ice Age climate extreme—a time when the topographic distribution was similar to today's—indicates that the mean trough positions are affected by other factors as well, most likely including the latitudinal temperature gradient).

The effects of altered topography on the atmosphere have potential consequences for the ocean as well. The surface wind field drives surface ocean currents; today, for example, the predominant surface winds in the North Atlantic from $30°-60°N$ are from southwest to northeast, driving the North Atlantic Drift to provide the warm water that influences the climate of western Europe. This circulation results from the mean position of the standing wave trough over eastern North America. A similar circulation pattern exists in the North Pacific, affecting the Kuroshio Current and its eastward extension, owing to the standing wave trough over eastern Asia. Storms following these circulation features in both oceans produce precipitation, which influences the salinity at high latitudes, and hence production of deep water, which drives the large-scale ocean circulation. A change in topography can thus affect both the surface and thermohaline circulations of the oceans, altering ocean temperatures and climate downstream.

Finally, changes in both the atmosphere and ocean circulation can potentially affect the climate of the globe as a whole. Changes in circulation imply possible changes in poleward heat and moisture transports, affecting the latitudinal distribution of water vapor, sea ice, and cloud cover, all of which help determine the net radiative balance of the planet. Hartmann[1] noted that the quasi-stationary planetary scale waves affect the distribution of precipitation between land and ocean, giving rise to planetary albedo changes via vegetation–albedo feedback and affecting cloud cover through the contrast in clear-sky albedo between land and ocean. In this way, more than regional climates may be affected by topography; the global mean state may be altered.

All of these possibilities are especially relevant to the climate of the late Cretaceous and early Cenozoic, which was much warmer than today's, especially at high latitudes. It is intriguing to speculate that at least some of this difference was associated with altered topography, in addition to possible changes in atmospheric CO_2. The high topography of both the American west and the Himalayas was established during the last 70 million years. To relate these changes to climate, it would be useful to understand the chronology of these events in relationship to Cenozoic climatic decline, in general, and intra-Cenozoic climatic fluctuations.

Unfortunately, the chronology of topographic changes during the Cenozoic is not firmly established. One theory holds that the high terrain of the American West (the Sierra Mountains, the Basin and Range Province, the Colorado Plateau, and the Rocky Mountains and the High Plains) was created during the Laramide orogeny from 75 to 45 million years ago. Compressive deformation is usually required to generate the crustal shortening that creates high mountains and plateaus, and the Laramide orogeny is the last such compressive event to affect western North America on large spatial scales.[2,3] Recent paleobotanical work on leaf margin types supports this view.[4]

The other view is that much of the uplift of the American West dates from the last 20 million years or so of the Cenozoic. This is largely based on evidence of strong erosional dissection of the highest terrain during the late Cenozoic, and in part on paleobotanical assemblage data. This view emerges prominently in the introductory sections of several chapters contributed by geologists to the DNAG volume on the American Cordillera.[5] Late Cenozoic uplift of the American West is attributed to several possible causes, including deep-seated heating processes involving the upper asthenosphere and lower lithosphere.

The situation for the Tibetan–Himalayan plateau–mountain complex in southern Asia is better constrained. Uplift postdates the initial collision of India and Asia 55–40 million years ago. The area experienced major uplift during the middle and late Cenozoic, beginning near 25–20 million years ago (Chapter 2). This uplift probably occurred later than much of that in the American West.

Given the state of this ongoing debate, we decided to run a sensitivity study of the effects of uplift on climate with a general circulation model (GCM) considering only changes in southern Asia. Additional studies altering the topography of the American West should be done as well, and will be the subject of further experiments.

2. PREVIOUS MODELING STUDIES

Several previous GCM studies have compared "no mountain" versus "mountain" cases, which contrast full orography versus no orography. With such a simulation, Manabe and Terpstra[6] found that without mountains, the atmospheric circulation is almost devoid of standing wave energy, with transient wave energy taking its place. This resulted in a longitudinally less diverse climate, less focused storm track paths, and more zonal flow over the oceans. Similar results were obtained by Hay et al.[7] and by Rind and Chandler,[8] who reduced the topography in the American West and Tibetan region to 900 m. In terms of broad plateau-scale orography, these experiments are geologically most comparable to a comparison of the mid-Cretaceous (low orography 100 million years ago) versus the full orography of today.[9] Ruddiman and Kutzbach[10] and Kutzbach *et al.*[11] ran, in addition, a half-mountain case, with similar, although less extreme, changes.

Kutzbach *et al.*[12] also explored the possibility of separate uplift by running tests with full orography in western North America and with no orography in Tibet, as well as the converse. They found that Tibetan uplift controlled monsoonal and jet stream dynamics in southern Asia and also affected jet stream meander amplitudes over North America. Uplift in western North America primarily affected climate in North America and immediately adjacent sectors, with some downstream effects as far east as Europe (but not over Asia). Uplift in both regions led to the same reinforcing pattern of westerly waves over North America.

Two papers in this volume also implicitly acknowledge the possibility of uplift in North America being earlier than that in Tibet. Ruddiman *et al.* (Chapter 9) and Kutzbach *et al.* (Chapter 7) utilize a sensitivity test in which Tibet is placed at one-quarter of its modern elevation and North America at three quarters of the present, as a plausible simulation for conditions 20 million years ago.

In terms of the possible effects of uplift, these atmospheric GCM experiments were only able to investigate the first of the three consequences discussed in the introduction: the impact on the atmospheric circulation itself. In order to begin to address the question of potential effects on the other two processes, ocean circulation and climate, the ocean must be allowed to respond to atmospheric changes. One such experiment is discussed below.

3. MODEL AND EXPERIMENTAL PROCEDURE

To investigate the coupled atmosphere–ocean response to uplift, we use the GISS coupled atmosphere–ocean GCM.[13] The model consists of a fully dynamic ocean and atmosphere, run at $4° \times 5°$ horizontal resolution, with 9 vertical levels in the atmosphere and 13 in the ocean. It does not use flux corrections (salt and heat added to the ocean), and its deep-water production is somewhat weaker than is currently observed. Overall, the primary deficiencies are warmer-than-observed sea surface temperatures along the west coasts of continents in the tropics and somewhat colder temperatures at high latitudes, associated with the reduced ocean heat transports. A full description of the coupled model results, including a comparison with observations, is presented in Russell *et al.*[13]

To explore the effects of uplift on the ocean circulation, it is necessary to run the model in a synchronous, fully coupled mode. Since uplift will change the proportion of standing to transient eddy energy, using time-averaged forcing from

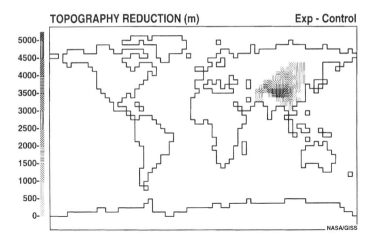

FIGURE 1. Topography reduction employed in the experiment.

the atmospheric model to spin-up the ocean and achieve equilibrium (a standard procedure) is not a practical option because time-averaging removes transient eddy effects. Since the coupled model time-step is determined by the atmospheric requirements (7.5-min time-step), equilibrium simulations are not currently possible. Therefore in the following experiment, we simply investigate "tendencies" induced by the changed topography.

The coupled model starts from observed initial conditions and is spun-up for 16 model years. Experiments run for 100+ years show that most of the deviation from the initial conditions occurs during this time period, with only a slow drift apparent thereafter. Starting from year 17, two simulations are run: one with full topography (the control run), and the other with the topography in southern Asia reduced to at most 300 m, a maximum change of almost 5 km (see Figure 1). Both simulations are continued through year 40, and comparisons are made for each of the same 8-year intervals for experiment and control. We are particularly interested in trends that developed during the first 8 years, and continued/intensified for the remainder of the simulations; unless otherwise indicated, this was the case for the results discussed below.

Note that while these experiments are perforce to investigate uplift effects during the Cenozoic, no attempt is made to include other boundary conditions relevant to that time, i.e., current values are used for land ice, land–ocean configuration (including the Panamanian Isthmus), ocean bottom topography, and CO_2 levels. It is conceivable that results might be altered by changing some of these parameters (many of which are not well known, independent of chronology uncertainties). However, as shown below, the impacts of the experiment as run are direct, and it is likely that some portion of the resultant change would exist regardless of the background state.

4. RESULTS

In this section, we show the changes induced by the altered topography for years 33–40. In all cases, we concentrate on those differences that arose during the first 8 years of the experiment after initial spin-up (years 17–24) and were maintained or intensified for the rest of the simulation. The changes explicitly discussed are all highly significant relative to the model's unforced interannual variation.

4.1. Atmospheric Changes

Several factors influence the following results, as discussed, for example, by Ruddiman and Kutzbach[14]: topography alters the flow pattern, inducing meridional components to the zonal flow, directly affecting the standing wave distribution; the high topography is associated with smaller air densities, allowing for greater seasonal temperature responses, with subsequent sinking air in winter and rising air in summer; and the topography itself induces vertical motions as air ascends or descends the plateaus. In winter the Asian topography is associated with sinking air and higher pressure, while in summer there is rising air and lower pressure. Consistent with the previous modeling experiments, therefore, we should expect increased topography to produce more local anticyclonic (clockwise) flow in winter and cyclonic (counterclockwise) flow in summer.

The differences between the control and the experiment for the 500-mb geopotential height field are shown in Fig. 2. (Note that since the experiment has reduced topography, subtracting the experiment from the control is equivalent to uplift). From December through February (Fig. 2a) (henceforth called winter) the increased topography results in a broad area of higher geopotential heights across most of Eurasia and the North Atlantic, extending to North America. Accompanying this change is a decrease in geopotential height over the North Pacific.

The effects from June through August (Fig. 2b) (henceforth called summer) are somewhat different, since the mean wavelength of the planetary long waves is shorter in that season, associated with the reduced zonal wind velocities, and there is a cyclonic response in the vicinity of the region where the plateau was removed. Nevertheless, increased heights prevail across Europe and the North Atlantic.

Shown in Fig. 3a, b are the corresponding changes in sea-level pressure for the two seasons. Associated with the relative ridging at 500 mb over the North Atlantic and Eurasia in winter, sea-level pressure increases over the same wide area. In the eastern North Atlantic, this represents an intensification and northward extension of the subtropical high. Over the North Pacific, the Aleutian low is intensified, and the subtropical high is also slightly stronger. In summer, lower sea-level pressure is established over southern Asia and northeastern Africa. These effects are consistent with the expectations noted above.

With the change in sea-level pressure distribution, and its gradients, surface wind velocities also are affected (Fig. 4a, b). The increased sea-level pressure over the North Atlantic and most of Eurasia results in a large-scale increase in the anticyclonic turning of the wind, with southwest flow at higher latitudes, and

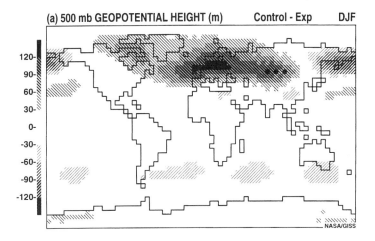

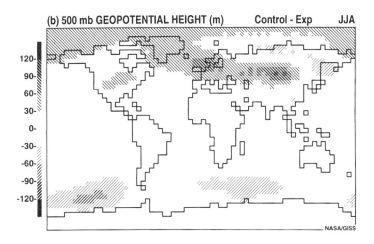

FIGURE 2. Five hundred mb-geopotential height changes: (a) December through February; (b) June through August. Results in this and all other figures are averages for years 33–40.

northeast flow further south. There is also some small intensification of the easternmost portion of the Icelandic low, visible through the increased cyclonic tendency east of Greenland (less than 2 mb difference, and therefore not evident in the sea-level pressure change presentation). Over the northwestern Pacific, the situation is reversed: there is more cyclonic flow owing to lower sea-level pressure (in winter). Over the eastern extratropical North Pacific, increased sea-level pressure produces a more anticyclonic circulation.

Changes in the surface wind alter the local heat advection and are responsible for perturbing the atmospheric temperatures. The surface air temperature differences are given in Fig. 5a, b. Where the wind is now more from the south (north) temperatures in general have warmed (cooled). Note in particular the

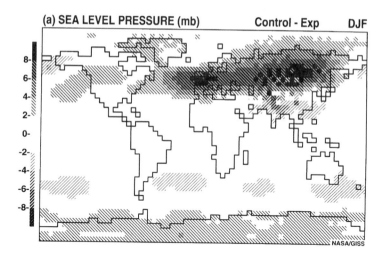

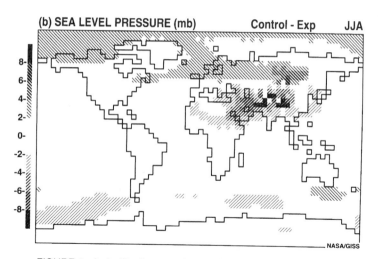

FIGURE 3. As in Fig. 2 except for sea-level pressure changes.

extreme warming over the northeast Atlantic during winter, with cooling over the Northern Pacific.

In regions of higher (lower) pressure, precipitation tends to decrease (increase) (Fig. 6a, b), with an extensive region of reduced precipitation across Eurasia in both seasons.

4.2. Oceanic Changes

The sea-level pressure and surface wind forcing changes described above are somewhat different for the two major Northern Hemisphere ocean basins, so the

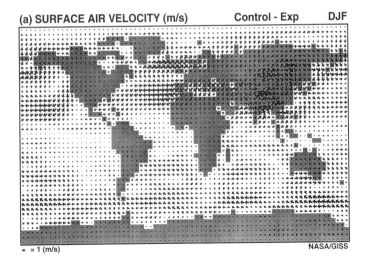

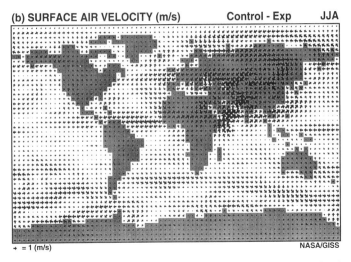

FIGURE 4. Surface wind velocity changes: (a) December through February; (b) June through August.

results for the North Atlantic and North Pacific oceans are discussed separately. There are also changes in the Southern Hemisphere, which are discussed at the end of this section.

4.2.1. Atlantic Ocean

The effect on the ocean surface temperature is given in Fig. 7a, b. Again warming prevails throughout much of the North Atlantic, although given the greater heat capacity of the ocean, the temperature changes are smaller than in the atmosphere. The ocean temperature changes are potentially associated with

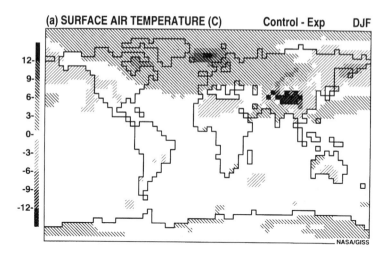

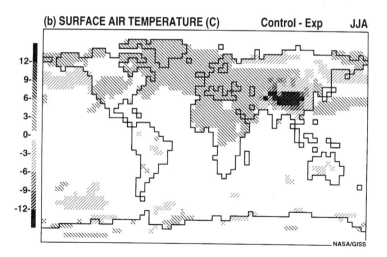

FIGURE 5. As in Fig. 2 except for surface air temperature.

two factors: (1) changes in heat exchange with the atmosphere, owing to altered latent or sensible heat fluxes, or altered solar radiation absorption; or (2) changes in the ocean currents driven by the surface wind field. With warmer North Atlantic sea surface temperatures and an increased anticyclonic wind flow, latent and sensible heat loss to the atmosphere decrease in some regions and increase in others, with magnitudes on the order of $5-20\,\mathrm{Wm}^{-2}$ (Fig. 8a). Absorbed solar radiation in this region generally increases by $5-10\,\mathrm{Wm}^{-2}$ (Fig. 8b). There is no consistent forcing over the basin to provide for the large-scale North Atlantic temperature changes shown in Figure 7, although in some areas reduced surface

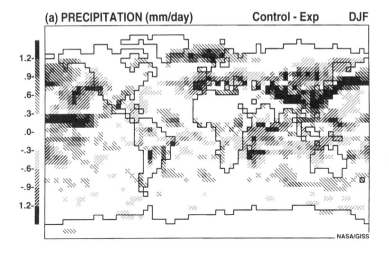

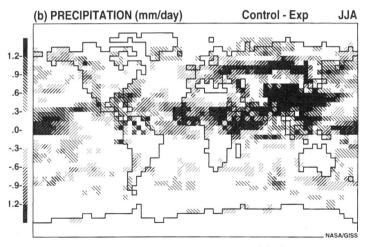

FIGURE 6. As in Fig. 2 except for precipitation.

fluxes and increased solar absorption are both contributing to the warming (e.g., 30°–50°N, 70°–80°N). Note that since the prevailing pressure pattern for this region is one of higher pressure, an increase in anticyclonic flow represents a reduction of the wind forcing, which can lead to reduced fluxes from the ocean surface to the atmosphere despite the warmer temperatures. Since the atmosphere is generally warming more than the ocean, a decrease in sensible heat flux is to be expected. The increased sea-level pressure also leads to slightly less cloud cover, increasing solar radiational heating.

The ocean currents above the thermocline are partially driven by the surface wind field and partially by the thermohaline circulation; hence their change (Fig.

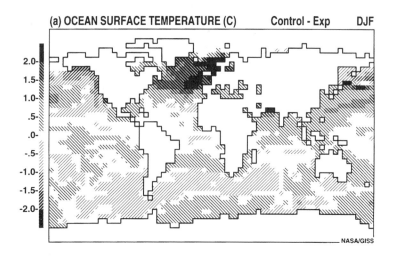

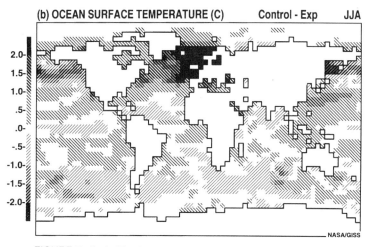

FIGURE 7. As in Fig. 2 except for ocean surface temperature.

9a) is at least partly associated with altered wind forcing. Comparisons of Figs. 4 and 9 indicate that the increased anticyclonic wind flow in the high-latitude North Atlantic results in an anticyclonic tendency for the underlying ocean currents as well. In particular, there is greater eastward flow south of Greenland and intensification of the northeastern portion of the North Atlantic Drift. This represents an amplification of the normal surface currents in the model of about 10%. The change in ocean heat transports associated with this altered circulation is shown in Fig. 9b. In the northeastern North Atlantic, where the change in ocean surface temperature is most extreme, there is an increase in poleward heat flux.

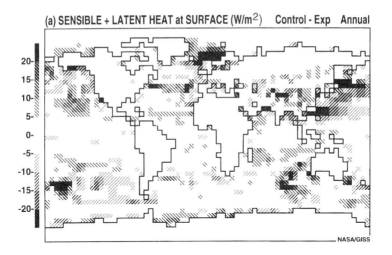

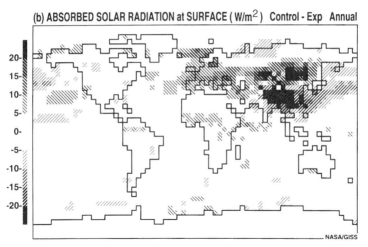

FIGURE 8. Annual average change in: (a) sensible plus latent heating fluxes; (b) absorbed solar radiation at the surface. Note that negative values imply a loss of energy at the ocean surface.

Associated with the changes in precipitation and evaporation there are changes in salinity (Fig. 10a, b). In general, salinity decreases at higher latitudes (north of 50°N) and increases at lower latitudes, often following the changes in fresh-water input, primarily precipitation. However, evaporation generally increases with the warmer sea surface temperatures, and this has an effect as well.

The changes in ocean temperature and salinity affect the density distribution in the ocean, which could potentially alter its thermohaline circulation. The changes in stream function for the Atlantic Ocean are shown in Fig. 11a (positive values imply clockwise flow in the plane of the figure). The negative values

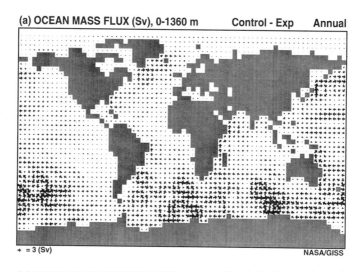

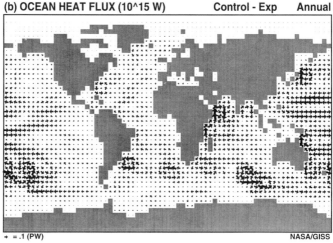

FIGURE 9. Annual average changes in: (a) the mass flux of ocean currents above the thermocline (0–1360 m) and (b) in the associated heat flux.

between 30°–60°N in the Atlantic at a depth of about 1000 m indicate a decrease in North Atlantic deep water (NADW) production on the order of 5–10%. This effect, which has been consistent through the course of the experiment, is generated by the reduction in density owing to the general warming in the North Atlantic along with some small reductions in salinity. It is associated with decreased northward mass flux above the thermocline; as such, it reenforces the surface wind forcing at certain longitudes (i.e., in the western North Atlantic from 30°–45°N), while opposing it at others (i.e., in the western North Atlantic from 45°–60°N). Where the two effects are working together, a net decrease in mass

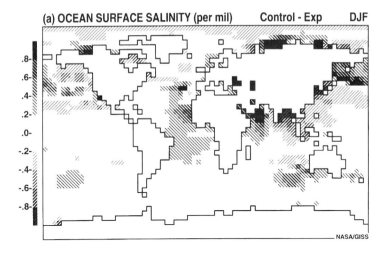

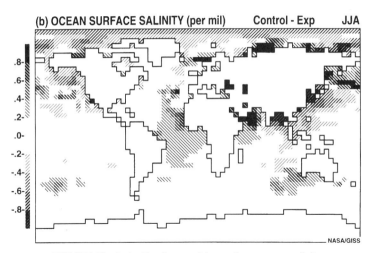

FIGURE 10. As in Fig. 2 except for surface ocean salinity.

flux is apparent (Fig. 9; note the reduction in the southern portion of the Gulf Stream mass flow). Where they are in opposition, the surface wind forcing often predominates, with an increased poleward mass flux despite the decrease in thermohaline circulation intensity.

4.2.2. North Pacific

The ocean surface temperature in this basin generally cools poleward of 50°N, and warms equatorward of that latitude, except for continued cooling in the eastern Pacific (Fig. 7). With intensified cyclonic air flow owing to the stronger

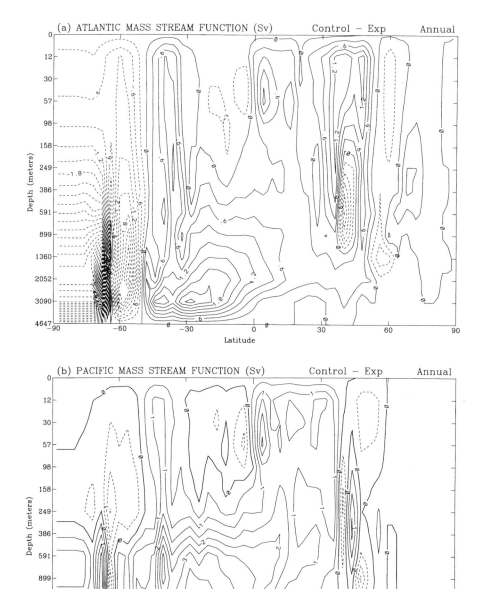

FIGURE 11. Change in annual oceanic streamfunction: (a) Atlantic; (b) Pacific.

Aleutian low, cloudiness generally increases, reducing the absorbed solar radiation (Fig. 8b). Sensible and latent heat fluxes also generally increase (Fig. 8a). Both of these would promote ocean cooling; the only warming influence is the increased poleward heat flux (Fig. 9b) driven by the increased south-to-north wind in the western portion of the basin at midlatitudes. The circulation changes again represent about a 10% effect.

Poleward of 45°N, salinity generally increases in the western North Pacific and decreases in the eastern North Pacific (Fig. 10). Precipitation actually increases over most of this region, so the increase in salinity is due to a decrease in evaporation, associated with the colder sea surface temperatures. With both decreased temperature and increased salinity, the density increases, leading to intermediate water production (at a depth of about 400 m) (Fig. 11b), which enables the currents averaged above the thermocline to deviate from the purely wind-driven effects; note the increased poleward flow in the western North Pacific at upper midlatitudes, which is in opposition to the change in surface wind forcing (Fig. 4).

4.2.3 Global Ocean

The thermohaline circulation changes are not limited to the Northern Hemisphere. In both the Atlantic and Pacific oceans, relative sinking motion occurs in the tropics, with upwelling at lower Southern midlatitudes ($\sim 30°-45°$S) and upwelling further poleward. The general warming of the tropical ocean and cooling in Southern Hemisphere midlatitudes (Fig. 7) are consistent with these circulation changes. The effect on the surface current mass flux can be seen in Fig. 9a as a relative flow from midlatitudes to the tropics along the eastern coasts of the continents. There is a corresponding reduction in heat flux in these same regions (Fig. 9b).

The Southern Hemisphere effect appears to have been initiated by the decrease in North Atlantic circulation. Reduction in NADW decreases tropical upwelling, and the relative tropical subsidence then establishes a (relative) circulation cell extending into the Southern Hemisphere. This result illustrates how ocean circulation effects originating in the Northern Hemisphere may have worldwide influences, at least in the model.

4.3. Climate Changes

Shown in Table 1 are the changes in the global mean temperature and radiation budget quantities. For the globe as a whole, and in the Northern Hemisphere, temperatures have warmed slightly. Sea ice has also decreased somewhat, especially in the Northern Hemisphere. However, the decrease in sea ice did not lead to a smaller planetary albedo as cloud cover remained essentially unchanged, hence the warmer temperatures are not associated with reduced reflection of solar energy out to space. Water vapor in the atmosphere did increase, owing to a small increase in evaporation from the warmer sea surface temperatures (including the small tropical warming). This water vapor change

TABLE 1.   Global and NH Parameters Associated with the Net Radiation Balance

Parameter	Control	Experiment	Change
Surface temperature (°C)	13.20	13.11	0.09
Surface Temperature (°C) NH	13.55	13.41	0.14
Planetary albedo (%)	30.99	30.97	0.02
Planetary albedo (%) NH	31.08	31.01	0.07
Water vapor (mm)	28.70	28.43	0.27
Water vapor (mm) NH	29.44	28.86	0.58
Sea ice (%)	4.11	4.27	−0.16
Sea ice (%) NH	5.15	5.52	−0.35
Net radiation at top (Wm^{-2})	3.30	3.60	−0.30
Sea surface temperature (°C)	26.74	26.46	0.26
Sea surface temperature (°C) NH	28.22	27.55	0.67
Net ocean surface heating (Wm^{-2})	4.29	4.37	−0.08

then helped amplify the warming slightly by increasing the atmospheric greenhouse capacity. Overall, the global and hemispheric changes are small, both in absolute and percentage terms.

As shown earlier, the effects at high latitudes are strongly associated with changes in poleward heat transports. The ocean heat transports for the individual basins in the experiment and control are given in Fig. 12a, b. In the Northern Hemisphere, poleward transports are higher in all three ocean basins in the control run. The increase in the North Atlantic, which extends from the equator to about 50°N, is surprising given the decrease in NADW production; however, the wind velocity changes associated with the increase in sea-level pressure (a strengthening of the subtropical high in the eastern North Atlantic) increased both the flow of warm water northward in the western portion of the basin and cold water southward in the eastern portion. Poleward heat transports decrease somewhat in the Southern Hemisphere, associated with the ocean circulation changes discussed above.

The changes in northward atmospheric energy transport, ocean heat transport, and the total energy transport are shown in Fig. 13. The increased water vapor leads to an increase in atmospheric northward latent energy transport from the tropics, which is more than balanced by a decrease in atmospheric dry energy transport, associated with the reduced latitudinal temperature gradient. In the Northern Hemisphere extratropics, total transports increase slightly, owing primarily to the change in ocean heat transports. The increased transports help reduce sea ice (Table 1). In the Southern Hemisphere, decreased oceanic and total energy poleward transports occur around 30°S and poleward of 60°S, with increases in between. The Northern Hemisphere oceanic response occurs despite the small decrease in thermohaline circulation, while the Southern Hemisphere changes are consistent with the altered thermohaline circulation.

These results emphasize that ocean heat transports are affected by both the density driven mean circulation cells and the surface wind forcing. Note that the

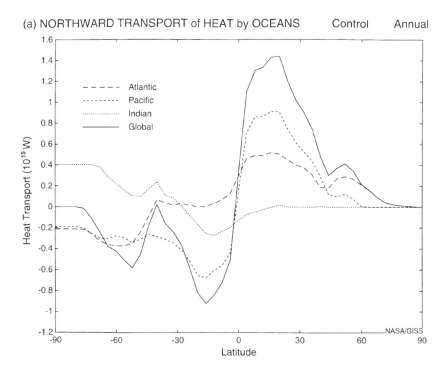

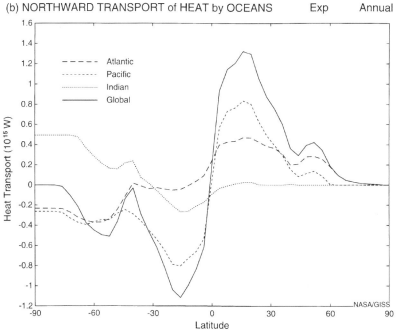

FIGURE 12. Oceanic heat advection in (a) the control and (b) the experiment.

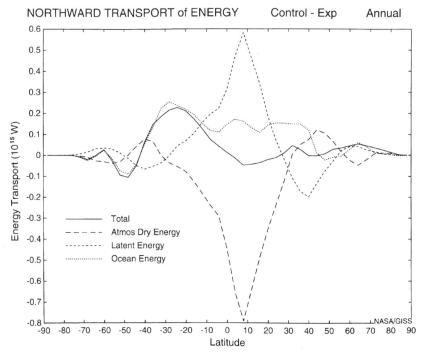

FIGURE 13. Change in northward energy transport by the atmosphere (dry and moist), ocean and total.

change in topography, which increased atmospheric standing eddy energy, did not result in an increase in atmospheric energy transports, as transient eddy effects decreased.

5. DISCUSSION

5.1. Transient versus Equilibrium Response

The change in atmospheric circulation associated with the increased topography produced substantial warming in the North Atlantic, associated with an increase in the subtropical high. In the North Pacific there was general cooling associated with an increase in the Aleutian low. Hence uplift is intensifying these stationary wave features, which are then altering the ocean circulation to produce sea surface temperature responses. It is interesting to note that only the easternmost portion of the Icelandic low is intensified by Himalayan uplift, and that by only a small amount.

A major question concerns whether the tendencies seen in the experiment as run would have been maintained had the experiment been run to equilibrium, say for 1000 years. The short-term responses of the system are likely to be relevant at all time scales: increased topography should amplify the stationary wave features. While the exact configuration of this effect may be influenced by the ultimate sea surface temperature distributions in the two northern ocean basins, the major atmospheric effects would probably prevail on all time scales, as would the wind-driven ocean circulation forced by the resulting changes in wind stress.

From the radiative point of view, the model is already close to being in equilibrium. The change in net radiation at the top of the atmosphere, an indication of approach to equilibrium, is already quite small; with the model's sensitivity of close to $1°C/Wm^{-2}$, the imbalance shown in Table 1 would lead to a small subsequent cooling within the noise level of the model. The net heating of the ocean surface as shown in Table 1 is similarly quite small, again within the noise level.

The major uncertainty concerns the longer-time-scale responses, primarily the thermohaline circulation effects. As was indicated in Fig. 11, there were small changes in the thermohaline circulation in both hemispheres. In the North Atlantic, the thermohaline and wind-driven circulation changes were having opposite effects on ocean heat fluxes, with the wind-driven circulation dominating for the duration of the experiment. In the Southern Hemisphere, the only effect was the change in thermohaline circulation, which then provided a small temperature signal. Simulations of hundreds or thousands of years would be necessary to determine whether the large-scale circulation changes could eventually alter the climate in very different ways from those produced on these short-time scales.

5.2. Comparison with Previous Modeling Studies

These results can be compared with those from previous modeling studies, although there are two major differences. First, the published work so far has emphasized results only for topographic changes over both Eurasia and North America simultaneously (Kutzbach et al.[12] did not discuss their separate uplift experiments in any detail). While North American topographic changes may not have a large effect in Asia, they will likely affect other areas, including the North Atlantic. Secondly, this experiment allowed the ocean temperatures to change owing to dynamical influences, such as the altered surface wind and thermohaline forcing. Changing the ocean temperatures alters land–ocean thermal contrasts, and can affect pressure and precipitation changes.

With those caveats, comparison with the results of Kutzbach et al.[11] show similar responses over Asia and nearby areas in almost all respects: the sea-level pressure, surface wind, and precipitation changes are all at least qualitatively similar. The more anticyclonic circulation in winter and cyclonic circulation in summer has already been discussed. The wetter summer conditions with increased topography in southern Asia and drier conditions in northern Europe and northern Asia are in general agreement, although the patterns are not identical. In other regions the results are somewhat different; there is greater cooling over the Aleutian area with increased topography in this experiment, for example, which is more consistent with observations.[10] In general the temperature differences over the Northern Hemisphere oceans are much larger in this experiment, owing to allowance of dynamical ocean changes.

5.3. Relevance for the Cenozoic

Concerning the relevance of this experiment to uplift during the Cenozoic, a full evaluation of the results depends upon the influence of other conditions that were not changed. Chief among these was using the topography for western North America in its current configuration. Chase *et al.*,[4] with a compilation of paleoelevation proxy data, suggest that most of the Western Cordillera of North America within the United States has been elevated since at least the late Cretaceous. They find some evidence of Oligocene–Miocene collapse of the basin and range province, and some estimates of early Eocene lowering in Wyoming. The Sierra Nevada uplift probably occurred more than 15 million years ago, and uplift of the Colorado Plateau more than 6 million years ago.

This area is important owing to its influence on the planetary long-wave structure in the atmosphere. As the primary effects shown in this study are associated with nonlocal wave forcing from southern Asia, a competing influence would arise from changes in the other major topographical province in the Northern Hemisphere. Rind and Chandler[8] found that uplift of western North American topography, in addition to Asian uplift, alters the long-wave pattern over the North Atlantic by strengthening the western portion of the Icelandic low, with increased northerly flow in the northwestern North Atlantic. As shown here, only the easternmost portion of the Icelandic low was (slightly) affected with Asian uplift alone; increased northerly flow in the western North Atlantic would induce a cooling tendency, opposite to that found in this study, although focused on the other side of the basin. Therefore, whether Cenozoic uplift warms or cools the North Atlantic may depend on which topographical features are being considered. It is obviously crucial to determine with confidence the timing of orography changes (including the Andes); GCM experiments of this nature might in fact add another piece to the puzzle by illustrating the tendencies that resulted from various regions of uplift. The experiment as it stands implies that uplift warmed the North Atlantic by increasing ocean heat transports, somewhat in opposition to the suggested results of Rind and Chandler[8] and Dowsett *et al.*[15]

Other boundary conditions may also be important. The precise bathymetry of the North Atlantic during the Cenozoic, and features such as the closing of the Isthmus of Panama, would have affected the North Atlantic thermohaline circulation. Maier-Reimer *et al.*[16] using an ocean GCM found that closing of the Isthmus of Panama (about 3–4 million years ago, but beginning 7–10 million years ago) would have increased deep water production and poleward ocean heat transports in the North Atlantic. Other influences would therefore have been changing in addition to uplift, with the potential ability to either amplify or completely reverse its effects.

Similar comments might be made concerning the atmospheric CO_2 levels. Raymo *et al.*[17] used marine $\delta^{13}C$ to estimate that Pliocene values were not much larger than those of today. Ruddiman *et al.* (Chapter 9), based on evidence of C_3/C_4 vegetation shifts about 7 million years ago [see Cerling (Chapter 13)], suggests that CO_2 reductions through the 500-ppm threshold could explain observed late Cenozoic circum-Arctic cooling and vegetation changes, without ocean heat transport changes. Varying CO_2 levels could obviously alter regional climate changes induced by topographical uplift.

Finally, the use of full land ice over Greenland could have several consequences: it could affect the temperature of the east Greenland current, and hence the subsequent ocean temperature change, and could affect to some degree the long-wave structure of the atmosphere. Greenland ice apparently appeared very late in the Cenozoic, no earlier than 7 million years ago, and probably later in northern Greenland where forests grew until 3 million years ago.[18] There was a positive mass balance over Greenland in the control run relative to the experiment; warmer ocean temperatures increased the evaporation in the vicinity of Greenland, contributing to this surplus, but again it was occurring with the Greenland ice sheet already in place (and hence with very cold temperatures). From this perspective, perhaps uplift contributed to the growth of the Greenland ice sheet once temperatures became sufficiently cold.

5.4. Relationship to the Cenozoic Climatic Decline

Were the uplift in topography to have produced a much cooler climate, then these results would have implied that uplift was a direct part of the process responsible for the overall climatic cooling during the Cenozoic (as opposed to being an indirect influence through potential weathering effects on atmospheric CO_2 concentrations). However, as indicated earlier in Table 1, the control run was actually slightly warmer than the experiment, driven by the warmer temperatures in the North Atlantic and slight warming in the tropics. Thus the simulations as formulated do not support the notion that uplift by itself cooled the global climate. Again it is important to note that the topography of the American West was not altered, and this could have had an impact on the ocean circulation. Furthermore, if this experiment had been run to equilibrium, it is conceivable that additional changes associated with the North Atlantic and North Pacific thermohaline circulations would have ensued to alter this conclusion. As it stands, the increase in oceanic poleward heat transport in the Northern Hemisphere associated with the altered atmospheric dynamics and surface wind forcing acted to warm high latitudes with uplift, not to provide the extreme cooling noted in the geologic record.

6. CONCLUSIONS

Following are the primary conclusions from this study:

1. Uplift in southern Asia results in winter in increasing geopotential heights across most of Eurasia, the North Atlantic, and the eastern subtropical

Pacific, with decreasing geopotential height over the northern North Pacific. In effect, the North Atlantic and North Pacific subtropical highs are intensified, as is the Aleutian low. In summer uplift leads to more cyclonic flow over southern Asia and northeastern Africa.

2. The circulation change produces an anticyclonic wind-flow tendency over the North Atlantic, with a cyclonic flow over the northwestern North Pacific. These results are generally consistent with those from previous GCM topography-change experiments.

3. Consistent with the wind-flow changes, surface currents change, and temperatures warm in the North Atlantic and cool in the northern Pacific.

4. Precipitation and evaporation changes lead to decreases in salinity in the northern North Atlantic and increases in the northwestern North Pacific.

5. Changes in ocean salinity and temperature produce decreases in North Atlantic deep-water flow, and some increase in North Pacific intermediate water.

6. The ocean circulation changes are not limited to the Northern Hemisphere, as consistent changes also arise in the two major Southern Hemisphere ocean basins, with relative downwelling in the tropics and upwelling at midlatitudes.

7. Poleward ocean transports generally increased in the Northern Hemisphere and decreased in the Southern Hemisphere. The Northern Hemispheric effect was the result of changes in the surface-wind-driven circulation (and was opposed by the thermohaline circulation change), while the Southern Hemisphere effect was consistent with the thermohaline circulation change.

8. A small global warming arose in response to increased atmospheric water vapor, associated with regions of warmer ocean temperatures (especially in the tropics).

9. Total energy transports generally mimicked ocean transport changes, as atmospheric standing/transient eddy and dry/moist energy transports showed a good degree of compensation.

Owing to the short duration of these simulations, this work represents only a first step in defining the effect of southern Asian uplift on the climate of the Cenozoic. Additional simulations should include other variations in boundary conditions and atmospheric composition, which for certain features are not well known. They should also be run for as long as is practical in a coupled mode; in addition to the computation burden such equilibrium simulations would require, coupled ocean–atmosphere models cannot yet produce good reproductions of the present-day climate without the use of flux corrections, which artificially limit the system's degrees of freedom. Therefore, we cannot yet set up or run fully realistic simulations of the effect of uplift on the climate of the Cenozoic.

Nevertheless, the experiment does indicate that ocean circulation changes must be factored into any evaluation of the climate influence of uplift. In addition, the experiment also reveals that rapid changes in the system, over a period of time as short as several decades, can arise in response to topographic changes, which may be relevant to rapid climate variations such as the Allerod/Younger Dryas, which was associated with ice-sheet decay (and thus topographic change). In this respect, the ocean circulation effects, while driven from the Northern Hemisphere, did produce worldwide responses. Not only was the ocean circulation in the Southern Hemisphere directly affected, but tropical warming initiated by the ocean circulation change resulted in increased water vapor, which helped warm the global climate. Mechanisms such as this may ultimately explain synchronous climate oscillations in the two hemispheres.

Acknowledgments

Climate modeling at GISS is supported by the NASA Climate Program.

REFERENCES

1. Hartmann, D. L. (1984). In: *Climate Processes and Climate Sensitivity* (J. E. Hansen and T. Takahashi, eds.), Geophysical Monograph 29, Maurice Ewing Volume 5, pp. 18–28. American Geophysical Union, Washington, D.C.
2. Coney, P. J., and Harms, T. A. (1984). *Geology* **12**, p. 550.
3. Molnar, P., and England, P. (1990). *Nature* **346**, p. 29.
4. Chase, C. G., Gregory, K. M., Parrish, J. T., and DeCelles, P. G. (1996). In: *Tectonic Boundary Conditions for Climate Reconstructions* (T. J. Crowley and K. Burke eds.) (in press) Oxford University Press.
5. DNAG (1989). *Geomorphic Systems of North America.* Geological Society of America, Boulder, Colorado.
6. Manabe, S., and Terpstra, T. B. (1974). *J. Atm. Sci.* **31**, p. 3.
7. Hay, W. W., Barron, E. J., and Thompson, S. (1990). *J. Geol. Soc. London* **147**, p. 385.
8. Rind, D., and Chandler, M. (1991). *J. Geophys. Res.* **96**, p. 7437.
9. Ruddiman, W. F., Prell, W. L., and Raymo, M. L. (1989). *J. Geophys. Res.* **94**, p. 18379.
10. Ruddiman. W. F. and Kutzbach, J. E. (1989). *J. Geophys. Res.* **94**, p. 18,409.
11. Kutzbach, J. E., Prell, W. L., and Ruddiman, W. F. (1993). *J. Geology* **101**, p. 177.
12. Kutzbach, J. E., Guetter, P. J., Ruddiman, W. F., and Prell, W. L. (1989). *J. Geophys. Res.*, **94**, p. 18393.
13. Russell. G. L., Miller, J. R., and Rind, D. H. (1995). *Atmos. Ocean* **33**, p. 683.
14. Ruddiman, W. F., and J. E. Kutzbach (1991). *Sci. Amer.* **264**, p. 66.
15. Dowsett, H., Barron, J., and Poore, R. Z. (1996). *Mar. Micropaleo.* **27**, p. 13.
16. Maier-Reimer, E., Mikolajewicz, U., and Crowley, T. (1990). *Paleooceanography* **5**, p. 349.
17. Raymo, M., Grant, B., Horowitz M., and Rau, G. (1996). *Mar. Micropaleo.* **27**, p. 313.
18. Jansen, E., and Sjoholm, J. (1991). *Nature* **349**, p. 600.

7

Possible Effects of Cenozoic Uplift and CO_2 Lowering on Global and Regional Hydrology

John E. Kutzbach, William F. Ruddiman, and Warren L. Prell

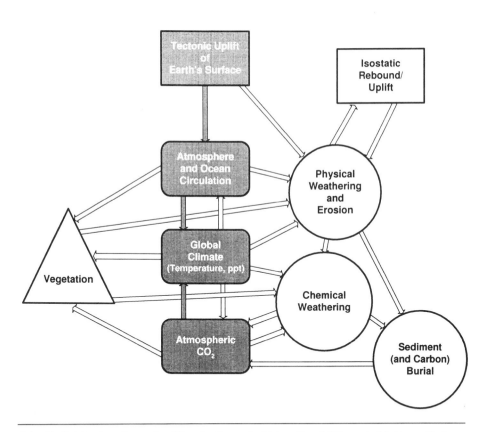

John E. Kutzbach • Center for Climatic Research, University of Wisconsin-Madison, Madison, Wisconsin 53706.   *William F. Ruddiman* • Department of Environmental Sciences, University of Virginia, Charlottesville, Virginia 22903.   *Warren L. Prell* • Department of Geological Sciences, Brown University, Providence, Rhode Island 02912.

Tectonic Uplift and Climate Change, edited by William F. Ruddiman. Plenum Press, New York, 1997.

1. INTRODUCTION

As discussed in other chapters in this volume, much of the highest topography on Earth has been created since the mid-Cretaceous. This includes the dominant mass of the Tibetan–Himalayan complex in southeast Asia (Chapter 2), as well as mountainous extensions westward to the European Alps. Although local mountain ranges existed in the American southwest during the Cretaceous, the broad dome of high topography from the High Plains to the Colorado Plateau did not exist 100 million years ago. Similarly, the high plateau of East Africa is largely a Cenozoic volcanic construction (Chapter 4). Uplift of the central Andean Bolivian Altiplano and much of the northern Andes also dates to the middle and late Cenozoic (Chapter 3). A very substantial portion of this post-Cretaceous uplift has occurred during the last 30–20 million years, including southeast Asia, east Africa, and parts of the South American Cordillera. The high ice domes on Antarctica and Greenland also came into existence in this geologically recent interval.

In this paper we use numerical models to assess the possible effects of Northern Hemisphere uplift on hydrologic processes on regional and global scales. In particular, we describe the effect of uplift on precipitation and runoff. Changes in the averages, the extremes, and the seasonal distribution of these hydrologic variables can influence land surface hydrology, vegetation (Chapter 9), the freshwater flux to the ocean, and erosion, weathering, and sediment transfer to the oceans by rivers.

Because this paper focuses only on the sensitivity of the hydrologic cycle to uplift, the reader is referred to Chapter 5 in this volume and to other studies for a more complete analysis of the climatic response to uplift.[1-9]

2. SIMULATIONS

We report the results of two sets of climate model simulations. The first set of three experiments involves changes of orography alone. The second set of two experiments combines changes of orography and changes in the atmospheric concentration of CO_2, a greenhouse gas.

2.1. Orography Changes

The sensitivity of the simulated hydrologic cycle to surface uplift is estimated from three experiments with specified differences in orography: no mountain (NM), half mountain (HM), and (full) mountain (M). The NM experiment specifies all land at a small, uniform height above sea level (238 m, the average land elevation in the model control, M). This experiment has no analog in geologic time, but, it does provide a useful comparison with the NM experiments made with other climate models.[1-3] The HM experiment specifies the elevation of all mountains and plateaus, and the Greenland and Antarctic ice sheets, at one-half their values in the M experiment (Fig. 1). This experiment is also highly idealized. It captures the general pattern of lower topography that existed in many regions prior to 20 million years ago (Chapter 20) but ignores changes in the positions of the continents and in the atmospheric concentration of CO_2. The M experiment specifies modern orography.

The topography depicted in climate models is a function of the model's spatial resolution. The model used in these experiments (Section 2.3) has a relatively coarse spatial grid (~ 400 km north–south by 750 km east–west). The topography is set to approximate the average elevation within each cell. For example, although the highest elevations in the Himalayas exceed 8800 m, the highest elevation is only 5650 m when averaged over a $1° \times 1°$ grid, roughly 100 by 100 km, and only 4200 m at the resolution used here. Thus the relatively coarse resolution model emphasizes the broad plateaus more than the narrow and high mountain ranges, and produces orographic features that are broader in horizontal extent than observed features. The highest elevations in the Tibetan–Himalayan region are 238 m in NM, 2100 m in HM, and 4200 m in M (Fig. 1).

2.2. Orography and CO_2 Changes

The simulations designed to test sensitivity of climate to uplift (NM, HM, M) have at least two significant limitations in historical accuracy. First, the assumption of uniform uplift (i.e., no mountains everywhere, half-mountains everywhere) is oversimplified; uplift histories on the various continents have not been uniform (Chapter 20). Second, the period of late Cenozoic uplift has also been a period of decreasing concentration of atmospheric CO_2.[10-12] Since both uplift and CO_2 concentration exert a strong influence on climate, we designed an experiment that combined a more accurate specification of regional uplift status for the period around 20–15 million years ago with a corresponding raised concentration of CO_2. We used the following specified lower-than-present orography: for the Tibetan Plateau, the Himalayas, and the highlands of East Africa, elevations were set to one quarter of present; for the North American Rockies, elevations were set to three quarters of present; for Greenland, the ice sheet was replaced by low-elevation land (238 m); all other mountains and plateaus and the Antarctic ice sheet were kept at present elevations. The raised concentration of CO_2 was specified at twice present, or 660 ppmv. The evidence supporting this particular choice of lowered orography and raised concentration of CO_2 is summarized in Chapter 9. Recent evidence indicates that the Bolivian Altiplano and the northern Andes should also have been set significantly lower than present (Chapters 3 and 20). The experiment with lowered mountains and $2 \times CO_2$ is referred to as $LM/2 \times CO_2$. The corresponding control is $M/1 \times CO_2$.

Although $LM/2 \times CO_2$ includes two important changes in boundary conditions appropriate for 20–15 million years ago, other changes are ignored. The

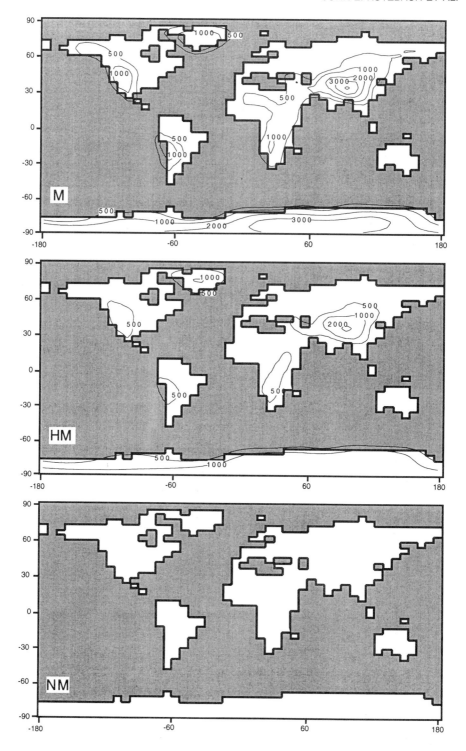

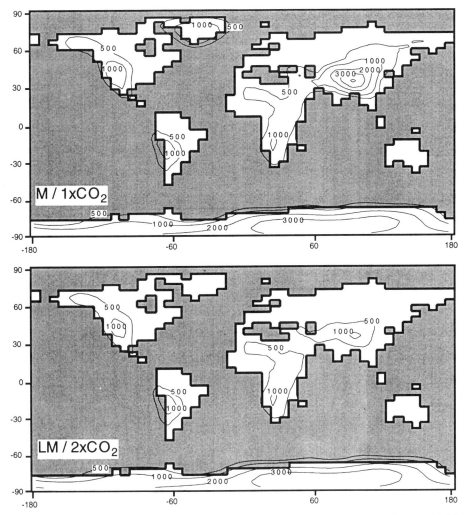

FIGURE 1.   Orography (m) for: (left) no mountain (NM), half mountain (HM), and mountain (M) experiments, and (right) lowered mountain/raised CO$_2$ (LM/12 × CO$_2$) and mountain/1modern CO$_2$ (M/1 × CO$_2$).

relatively small but not necessarily unimportant changes in land/ocean distribution associated with plate movements have been omitted in order to focus attention on only two climate forcing mechanisms: uplift and greenhouse-gas concentration. Changes in the size of ocean gateways would have been an important consideration if our climate model were coupled to an ocean that simulates currents, but are not a factor with the present model, which is coupled to a static mixed-layer ocean (Section 2.3).

2.3. Models

The numerical experiments were made with version 1 of the Community Climate Model (CCM1) of the National Center for Atmospheric Research (NCAR).[13] The model incorporates atmospheric dynamics based upon the equations of fluid motion; it includes radiative and convective processes, condensation, precipitation, and evaporation. Surface energy and hydrologic budget equations permit the calculation of surface temperature, soil moisture, snow melt, and runoff. However, the parameterization of soil hydrology and land surface properties in CCM1 is relatively crude compared with parameterizations now being introduced in climate models.[14,15] The version of the CCM1 used here includes a 50-m-thick mixed-layer ocean for calculating ocean temperature and the location and thickness of sea ice.[16] The ocean has no explicit currents but prescribes a poleward transport of heat. The model has 12 vertical levels in the atmosphere and 1 level in the ocean, and uses a spectral waveform representation, to wave number 15, of the horizontal fields of wind, temperature, pressure, and moisture. When needed for certain calculations, the spectral representation is converted to a grid of 4.4° latitude by 7.5° longitude. The model simulates the entire seasonal cycle, January through December, with a time-step of 30 min.

The experiments were started with all model variables set at values for a particular day of an NCAR CCM1 control simulation (full mountain, CO_2 concentration of 330 ppm) but with orography and CO_2 concentration changed from control conditions as appropriate for each experiment. The model was run for 25 years. During the first 20 years the mixed-layer ocean, sea ice, and soil moisture reached quasi-equilibrium values. The last 5 years were then used for the analyses presented here.

The control case simulations of the CCM1 have known shortcomings in comparison to observations,[13] and these deficiencies may also have an effect on the model's sensitivity to surface uplift. For example, the model's troposphere is colder than the observed atmosphere by several °C and has weaker midlatitude westerlies and weaker subtropical easterlies than the observed atmosphere. Transient disturbances, such as midlatitude storms, are weaker than observed. On the other hand, the general patterns of large-scale circulation, temperature, and precipitation are simulated rather realistically.[17] Because of the low spatial resolution of the model, the simulated boundaries separating warm/cold and wet/dry regions are not as sharply defined as the observed boundaries. The global and annual average precipitation in CCM1 is about 1200 mm, which is perhaps 10–20% higher than observed.[18]

The two control experiments, M and M/1 × CO_2, were run with two slightly different versions of CCM1 and produced slightly different climatologies. The global average precipitation is the same for both models (Table 1a) but, for example, the average precipitation over land is about 2.5% higher in M than in M/1 × CO_2 (Table 1b). However, differences between the two model control experiments are very small compared to the differences produced by uplift and CO_2 lowering (Table 1b).

TABLE 1a. Global and Annual Average Surface
Temperature (T) and Precipitation (P)[a]

Experiment	Global	
	$T(^\circ C)$	P(mm)
M	14.6	1220 (1.0)
HM	15.5	1220 (1.0)
NM	16.0	1220 (1.0)
M/1 × CO$_2$	14.8	1220 (0.92)
LM/2 × CO$_2$	20.0	1330 (1.0)

[a]The fractional change in precipitation, relative to NM (top) and
LM/2 × CO$_2$ (bottom) is included in parentheses.

3. RESULTS

We will present results for HM and M relative to NM (and for M/1 × CO$_2$
relative to LM/2 × CO$_2$) in order to describe the climate trends as a function of
uplift (and uplift/CO$_2$ lowering).

3.1. Global and Latitudinal-Average Changes

Land surface temperature, averaged for all continents, decreases 5.3°K with
uplift (M − NM) (Table 1). The decrease in land surface temperature is explained
in large part by the lowered temperatures in the uplifted regions, but also by
lowered temperature downstream of uplifted regions (Chapter 9). Global average
surface temperature decreases 1.4°K between NM and M (Table 1). Precipitation
over land increases 14% (M − NM) and surface runoff increases 45%. The ratio
of runoff to precipitation increases from 0.30 (NM) to 0.38 (M). The average soil
moisture increases 10% and the annual average snow depth almost doubles
between the NM and M experiments (Table 1). The response to uplift is not
uniform spatially; precipitation increased significantly in some areas and decreased significantly in others. Because global average precipitation is the same
for NM, HM, and M (Table 1), the increase in precipitation over land with uplift
implies a corresponding decrease in precipitation over the ocean.

The combination of uplift and lowered CO$_2$ produces results that are
significantly different from those produced by uplift alone. The decrease of
temperature is larger because of the reinforcing effects of uplift and lowered
CO$_2$ on temperature. The global average surface temperature decreases 5°K
(Table 1). Global average precipitation decreases 8% (M/1 × CO$_2$ − LM/
2 × CO$_2$) (Table 1) because the decreased intensity of the hydrologic cycle
associated with lowered CO$_2$ exceeds the increase of precipitation associated with
uplift. The sensitivity of surface temperature to CO$_2$ doubling for CCM1 is
somewhat larger than that of some other models; the sensitivity of precipitation

TABLE 1b. Mean Elevation, Global and Annual Average Temperature (T), Precipitation (P), Runoff (N), and Soil Moisture (SM), for Land Only[a]

Experiment	Elevation (m)	T(°C)	P(mm)	N(mm)	SM(cm)	N/P	S(fraction)
			Global, land only				
M	722	6.1	1190 (1.14)	455 (1.45)	8.0 (1.10)	0.38	1.93
HM	480	9.0	1130 (1.08)	379 (1.21)	7.7 (1.05)	0.34	1.56
NM	238	11.4	1040 (1.00)	314 (1.00)	7.3 (1.00)	0.30	1.00
M/$1 \times CO_2$	722	7.0	1160 (0.94)	402 (1.0)	7.8 (1.03)	0.35	1.10
LM/$2 \times CO_2$	576	14.0	1240 (1.00)	403 (1.0)	7.6 (1.00)	0.32	1.10

[a]The fractional change in precipitation, runoff, and soil moisture, relative to NM (top) and LM/$2 \times CO_2$ (bottom) is indicated in parentheses. Far right: Runoff ratio N/P (runoff ÷ precipitation); fractional change in snow depth (S), relative to NM (top) and LM/$2 \times CO_2$ (bottom).

is within the range of sensitivities reported for other models. For example, Thompson and Pollard[19] found that global average temperature increased 2°K and precipitation increased 3.3% for CO_2 doubling; using a different model, Manabe and Bryan[20] found that temperature increased 3.2°K and precipitation increased 8%. The corresponding increases for CCM1 are 4°K and 6.7% (Oglesby[21] and personal communication). Averaged over continents only, surface temperature decreases 7°K and precipitation decreases 6% with the combination of uplift and lowered $CO_2(M/1 \times CO_2 - LM/2 \times CO_2)$.

The land surface temperature decreases with uplift (M − NM and M − HM) at all latitudes (Fig. 2). The largest decreases occur at the latitudes of the Tibetan and Colorado plateaus and the Bolivian and South African plateaus. The combined effects of uplift and lowered CO_2 cause much greater cooling at high northern latitudes, associated with increased sea ice and snow cover, and some additional cooling at all latitudes (Fig. 2).

The increased precipitation associated with uplift (M − NM, M − HM) occurs primarily between 40°N and 40°S with the largest increases at the latitudes of the Tibetan and Colorado plateaus and the Bolivian and South African plateaus (Fig. 2). The combination of uplift and lowered CO_2 produces a general decrease in precipitation at all latitudes except 20°–40°N, the region of the Tibetan and Colorado plateaus (Fig. 2). The increase in precipitation at 10°–30°S, notable in M − NM and M − HM, is absent in $M/1 \times CO_2 - LM/2 \times CO_2$ (Fig. 2) because the Bolivian and the South African plateaus were not lowered in $LM/2 \times CO_2$ (see Section 2.2).

3.2. Regional Changes

The increases of precipitation with uplift (M − NM) exceed 50–100% in the vicinity of the Tibetan, South African, Colorado, and Bolivian plateaus and mountains; elsewhere, precipitation decreases (Fig. 3). When the more realistic amounts of regional uplift are combined with lowered CO_2 ($M/1 \times CO_2 - LM/2 \times CO_2$) the areas and extremes of increased precipitation are reduced. Sizable increases in runoff, exceeding 200–500% in limited areas, occur in the regions of increased precipitation (Fig. 3).

Regional precipitation and runoff statistics from CCM1 are summarized for 18 major continental drainage basins, based upon a partitioning of drainage on the CCM1 grid[18] (Figs. 4 and 5). As described in Section 2.3, the CCM1 controls (M, $M/1 \times CO_2$) simulate more precipitation and runoff than is observed; this bias toward excessive precipitation and runoff occurs in most regions. The simulated excesses are particularly large in Africa, central Asia, Australia, and Antarctica; only in southern Asia does the model simulate less precipitation and runoff than indicated by observations.[18] Given that the hydrologic cycle of the model control is biased, we can expect that the uplift experiments are also biased toward excessive precipitation and runoff; moreover, model sensitivity may also be affected.

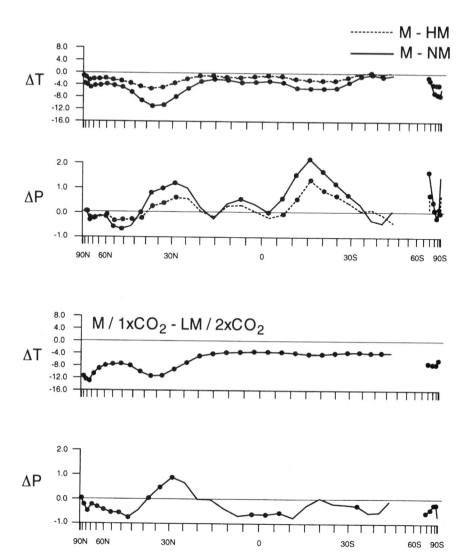

• 95% significance

FIGURE 2.   Differences in annual-average temperature (ΔT, °C) and precipitation (ΔP, mm/day) over land as a function of latitude for: (top) M − HM and M − NM; and (bottom) M/1 × CO$_2$ − LM/ 2 × CO$_2$. Differences that are statistically significant at the 95% level, compared to the model's inherent variability, are indicated with solid circles.

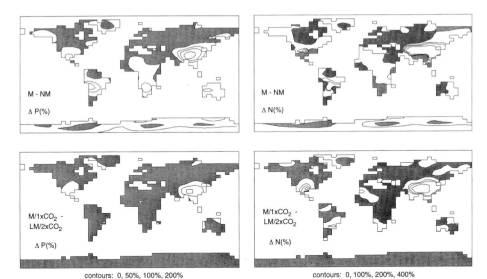

contours: 0, 50%, 100%, 200%   contours: 0, 100%, 200%, 400%

FIGURE 3.   Percentage changes in annual-average precipitation (ΔP, contours at intervals of 50, 100, and 200%) and runoff (ΔN, contours at intervals of 100, 200, and 400%): (top) for $(M - NM)/(NM)$; (bottom) for $(M/1 \times CO_2 - LM/2 \times CO_2)/(LM/2 \times CO_2)$. Areas of decreasing precipitation and runoff are shaded. Percentage changes in runoff exceeding 400% in Asia are not contoured; contours depicting large percentage changes of runoff in areas where runoff approaches zero are also omitted.

Annual precipitation increases with uplift in 13 of the 18 basins (Fig. 4). These increases are caused by significant enhancement of summer monsoon precipitation in the vicinity of high plateaus and mountains, and by enhanced orographic (upslope) precipitation in the vicinity of most mountain ranges.[5,8] Some of the largest changes in precipitation are in southern Asia, where basin-average precipitation increases from 580 mm (NM) to 800 mm (HM) to 985 mm (M) — an overall increase of about 65%, and in eastern Asia, where precipitation increases from 1100 mm (NM) to 1350 mm (HM) to 1640 mm (M) — an overall increase of about 50% (Table 2). Precipitation is unchanged or decreases with uplift across northern and central Eurasia and northern North America (Chapter 5).

Changes in runoff to the ocean (or to interior basins) are even more pronounced (Fig. 5). Runoff (km³/year) is calculated by integrating the local runoff rate per grid cell over all grid cells of each basin. The largest percentage increases occur in southern Asia, where runoff increases by a factor of about 10 $(M - NM)$, and eastern Asia, where runoff increases by a factor of 2.5 $(M - NM)$ (Table 2). Runoff also increases significantly in northern South America, western Africa, and western North America, while decreases occur in northern Eurasia and northern North America.

The combination of more realistic regional uplift and lowered CO_2 $(M/1 \times CO_2 - LM/2 \times CO_2)$ reduces the intensity of the hydrologic cycle in most

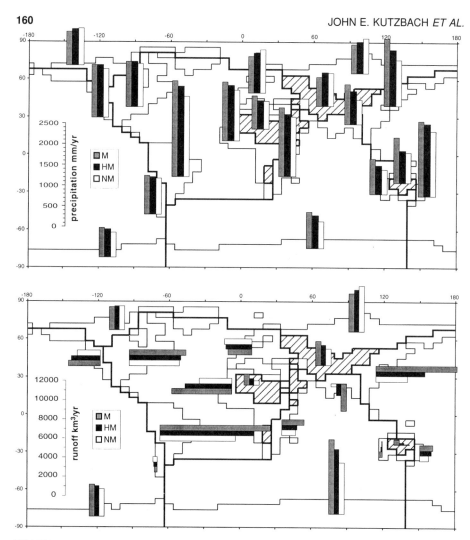

FIGURE 4. Precipitation (mm/year) and runoff (km³/year) for M, HM, and NM simulations for 18 drainage basins. Interior basins are indicated by hatching. The histograms depicting runoff toward the ocean are oriented with their tops pointing in the direction of drainage.

regions (Fig. 5). In contrast to the uplift-only experiments where 13 of 18 regions experienced increases of precipitation, only 2 of 18 regions show increased precipitation: southern Asia (730 to 1020 mm, a 40% increase) and eastern Asia (1400 to 1600 mm, a 15% increase). These same two regions experience large increases in runoff, indicating that in these two areas the effects of uplift on runoff outweigh the effects of lowered CO_2 (Fig. 5). In southern Asia runoff increases by a factor of about 5 and in eastern Asia it increases by a factor of about 2 (Table 2).

These increases of annual precipitation and runoff result primarily from increases in the amplitude of the seasonal cycle of these variables. In southern and eastern Asia the increases occur primarily in the period of the summer monsoon

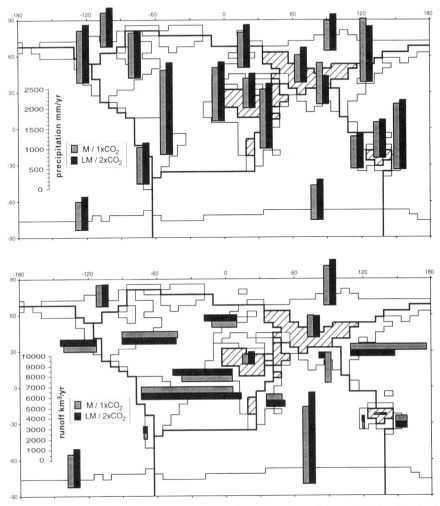

FIGURE 5.   Precipitation (mm/year) and runoff (km^3/year) for M/1 $\times$ CO$_2$ and LM/2 $\times$ CO$_2$ simulations for 18 drainage basins. Interior basins are indicated by hatching. The histograms depicting runoff toward the ocean are oriented with thieir tops pointing in the direction of drainage.

(Fig . 6). In western North America the increases occur primarily in winter when there is maximum orographic precipitation associated with westerly flow from the North Pacific ocean (not shown).

3.3. Changes in Frequency of Extreme Precipitation and Runoff Events

Some erosional processes may depend upon extremes of precipitation and runoff, rather than on mean values.[22,23] To see if the frequency of extreme precipitation and runoff events increases with uplift, we examined daily precipita-

TABLE 2.   Mean Elevation, Annual Average Precipitation
and Runoff for Southern Asia and Eastern Asia[a]

Experiment	Elevation (m)	P (mm)	N (mm)
Southern Asia			
M	1024	985	365
HM	631	803	146
NM	238	584	36
M/$1 \times CO_2$	1024	1022	401
LM/$2 \times CO_2$	408	730	73
Eastern Asia			
M	755	1642	730
HM	496	1350	438
NM	238	1095	292
M/$1 \times CO_2$	755	1606	657
LM/$2 \times CO_2$	286	1387	365

[a]The areas occupied by the southern Asia and eastern Asia basins are shown in
Figs. 4 and 5.

tion and runoff events for June–July–August (the season of maximum rainfall)
for southern and eastern Asia (Table 3). For southern Asia, 2.3% of daily rainfall
rates exceed 24 mm/day for NM. This percentage nearly triples (6.0%) for M. The
percentage of runoff rates exceeding the same threshold increases from 1.6%
(NM) to 5.0% (M). In both southern and eastern Asia, the frequency of
occurrence of extreme rainfall and runoff events increases by factors of 2 to 3
(Table 3), i.e., by factors that are comparable to the increases in the average
values.

3.4. Changes in Freshwater Flux to the Oceans

Uplift could have altered the net freshwater flux to the oceans. To address
this possibility, we examined the budget of continental runoff (N) plus precipita-
tion-minus-evaporation over the oceans [$(P - E)_{ocean}$]. The two CCM1 control
experiments both have small positive biases in freshwater flux (M: 0.15 Sv;
M/$1 \times CO_2$:0.04 Sv). The observations[24] are adjusted to no net flux (Table 4).
The CCM1 simulations of freshwater flux are biased regionally. The flux to the
Pacific is negative, the reverse of the observations. The positive flux to the Arctic
is greater than observed and the negative flux to the Atlantic is less than observed.
Given these biases in the control, only the trends in freshwater flux with uplift are
examined. The largest change occurs in the Indian Ocean where the increased
runoff leads to a net increase in freshwater flux and hence a tendency for
decreased salinity. In the Atlantic, decreased freshwater flux, implying slightly
increased salinity, occurs because $(P-E)_{ocean}$ decreases more than runoff increases.

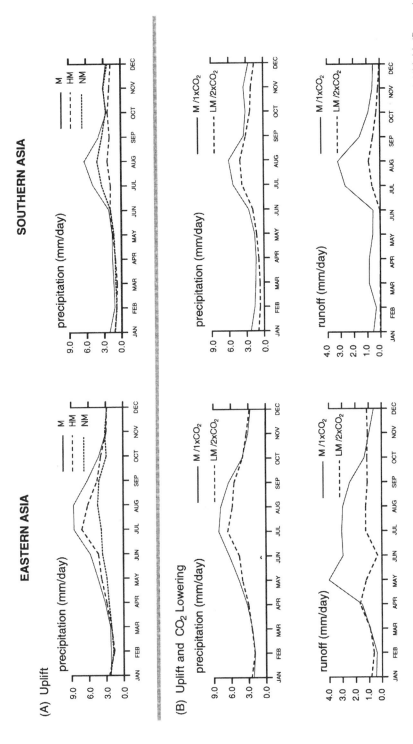

FIGURE 6. (Top) The seasonal distribution of precipitation (mm/day) for eastern Asia and southern Asia: uplift-only simulations — NM, HM, M. (Bottom) The seasonal distribution of precipitation (mm/day) and runoff (mm/day) for eastern Asia and southern Asia: regional uplift and changed CO$_2$ — LM/2×CO$_2$, M/1×CO$_2$. The areas occupied by the eastern Asian and southern Asian basins are shown in Figs. 4 and 5.

TABLE 3. Mean Elevation (m), Average Precipitation Rate (P), Runoff rate (N), and Frequency (%) of Extreme Precipitation and Runoff Greater than 24 mm/day, for June–July–August for Southern Asia and Eastern Asia[a]

Experiment	Elevation (m)	P (mm/day)	$P > 24$ mm/day (%)	N (mm/day)	$N > 24$ mm/day (%)
Southern Asia					
M	1024	4.4	6.0	2.3	5.0
HM	631	3.2	3.7	1.0	3.1
NM	238	2.1	2.3	0.3	1.6
M/1 × CO$_2$	1024	4.6	6.1	2.4	4.8
LM/2 × CO$_2$	408	3.3	3.5	0.8	0.6
Eastern Asia					
M	755	7.6	9.1	3.0	7.3
HM	496	5.9	6.7	1.3	5.1
NM	238	4.0	4.2	0.4	3.2
M/1 × CO$_2$	755	7.4	8.5	2.7	6.2
LM/2 × CO$_2$	286	5.6	6.5	0.9	0.9

[a]The areas occupied by the southern Asia and eastern Asia basins are shown in Figs. 4 and 5.

Freshwater flux to the Arctic decreases slightly with uplift, implying a tendency for the Arctic to become more saline. The Pacific Ocean shows little change. The combined effects of uplift and CO_2 lowering on net freshwater flux show the same trends as uplift alone in the Arctic and Atlantic, but altered trends in the Pacific and Indian oceans (Table 4).

In general, however, the simulated changes in net freshwater flux are relatively small and lead to no clear conclusions concerning possible effects on ocean circulation. Since net freshwater flux is the sum of several processes involving large amounts of water (runoff, precipitation, evaporation), small errors in simulation can have large consequences for even the sign of the flux. Therefore it will be important to calculate these budgets with improved, higher-resolution models.

4. DISCUSSION

High orography has well-documented effects on climate at regional to hemispheric scales, and these in turn play a major role in controlling global weathering processes. The uplifted plateaus are cooled in both seasons by lapse-rate effects on the raised topography, and regions downwind of the plateaus are cooled in winter owing to diversion of lower-level winds. This increases the likelihood of glacial and periglacial processes that cause strong mechanical weathering. In addition, precipitation is enhanced along steep upwind slopes on

TABLE 4. Net Freshwater Flux to Ocean Basins, $(P - E)_{ocean} + N$ (Precipitation minus Evaporation over the Ocean plus Runoff from the Continents[a]

Experiment	Arctic (Sv)	Atlantic (Sv)	Indian (Sv)	Pacific (Sv)	Total (Sv)
M	0.38	−0.24	0.23	−0.22	0.15
HM	0.40	−0.23	0.14	−0.18	0.13
NM	0.42	−0.20	0.12	−0.22	0.12
M/1 × CO$_2$	0.36	−0.26	0.20	−0.26	0.04
LM/2 × CO$_2$	0.39	−0.24	0.24	−0.36	0.03
UNESCO (*observations*)	0.23*	−0.72*	0.33	0.16	0.00

[a]Results (in Sverdrups, 10^6 m^3/sec) are for five simulations with CCM1 and for observational estimates.[24] The asterisk (*) indicates that the UNESCO estimates for the Arctic and the Atlantic cannot be compared directly with the simulated values because UNESCO assigns the drainage basin of Hudson Bay to the Arctic Ocean whereas the simulations assign this drainage to the Atlantic.[18] If the model-simulated runoff from the Hudson Bay region had been assigned to the Arctic rather than the Atlantic, as was the case with the observed estimates, then the net freshwater flux to the Arctic would have been even more positive and the net freshwater flux to the Atlantic would have been even more negative. The areas occupied by the four ocean basins are shown in Figs. 4 and 5.

the margins of high topography, either by increased orographic interception of zonal westerlies and easterlies, or by self-generated monsoonal flows created by the high topography. As a result, global runoff patterns correlate closely with the margins of high-elevation regions that intercept moisture-laden winds.[24] Increased precipitation and runoff along these steep, high-relief slopes bordering high-elevation regions produce maximum global rates of mechanical denudation[24,25] and maximum suspended sediment yield.[26–30]

The positive correlation between topographic relief and precipitation and runoff noted in observations[30] occurs in these uplift experiments (Table 2). Summerfield and Hulton[30] report a large positive correlation between basin relief (the elevation differential between the highest and lowest areas of a drainage basin) and basin-average precipitation and runoff. The ratio of annual runoff to basin relief, as determined by linear regression from observations for 16 large drainage basins, is similar to the ratio calculated from the CCM1 simulations (70 mm/1000 m for the observations; 85 mm/1000 m for the simulations) (Table 5). This close agreement is no doubt partly fortuitous. The corresponding ratios for precipitation and basin relief do not agree as closely (Table 5), perhaps because basin relief itself is poorly represented in the coarse resolution climate model (Table 5). Nevertheless, it is encouraging that the model control (M) simulates the relation between runoff and basin relief with reasonable accuracy, compared to observations. This agreement provides some basis for expecting that the uplift sensitivity estimates, i.e., the increases in runoff (and precipitation) caused by increases in basin relief, may also be realistic. Averaged for 11 drainage basins in the model (see Table 5), runoff and precipitation both increase about 150 mm for

TABLE 5a. Dependence of Annual Precipitation (*P*) and Runoff (*N*) on Basin Relief (BR)[a]

	Slope, P/BR	Slope, N/BR	Range of BR
Observations	208 mm/1000 m	70 mm/1000 m	7010–1066 m
M	64 mm/1000 m	85 mm/1000 m	3540–628 m

[a]As obtained from linear regression analysis of observations for 16 drainage basins ($>10^6$ km^2) summarized by Summerfield and Hulton[30] and of simulations summarized for 11 drainage basins in CCM1 (see Figs. 4 and 5). Interior basins, Australia, and Antarctica were excluded from the analyses. BR is defined as the difference in elevation between the highest and lowest regions of a drainage basin. The observed range of BR for the 16 basins, and the prescribed range of BR for the 11 basins used from CCM1, are summarized.

every increase of 1000 m in basin relief. Whereas precipitation and runoff are shown to increase in a quasi-linear fashion as a function of basin relief, Summerfield and Hulton[30] show that denudation rates increase nonlinearly as a function of basin relief. Therefore the large increases in precipitation and runoff associated with uplift, as simulated by the model, should translate into even larger increases in denudation rates.

Other factors, most of them brought about by uplift-related climatic changes, play a role in mechanical denudation, although secondary to the dominant effect of uplift-induced relief.[30,31] Mechanical denudation rates appear to reach maximum values in two annual-precipitation regimes.[32–34] One is semiarid regions of low annual (200–400 mm) but highly seasonal precipitation. In a study of 33 large externally drained basins, however, Summerfield and Hulton[30] found no clear relationship between runoff variability and denudation. Denudation rates decrease somewhat for mean annual precipitation in the range 400–1000 mm, apparently because of the stabilizing effects of vegetation on the landscape. The second denudation maximum occurs in regions of very high mean annual precipitation (>1000 mm), mainly high-relief borders of orogenic belts. Summerfield and Hulton[30] identified mean annual precipitation and runoff as a significant factor, probably in part because of the high correlation between these factors and

TABLE 5b. Dependence of Changes of Precipitation (Δ*P*) and Changes of Runoff (Δ*N*) on Uplift-Caused Changes in Basin Relief (ΔBR)[a]

	Slope, $\Delta P/\Delta BR$	Slope, $\Delta N/\Delta BR$
M − HM	157 mm/1000 m	180 mm/1000 m
N − NM	176 mm/1000 m	139 mm/1000 m

[a]As obtained from linear regression analysis of simulations NM, HM, and M for the same 11 drainage basins.

relief. A second reason for these high denudation rates could be the increased likelihood of major floods that can overwhelm vegetative cover and may have a disproportionately large impact on long-term erosion.[22,23] Global patterns of flood frequency[35] correlate closely with the high-precipitation margins of uplifted high-elevation regions in southern Asia, the Himalayan-Alpine mountain chain, and the American Cordillera. Finally, varying lithologies within different drainage basins lead to different degrees of rock erodibility and sediment yield.

The results of the simulations also support the expectation of increased denudation rates attributable to the secondary factors summarized above. Uplift enhances the seasonality of precipitation (Fig. 6) and may thus enhance erosion. Uplift also enhances the frequency of extreme event precipitation and runoff (Table 3) and pushes precipitation significantly above 1000 mm, a possible threshold level for high denudation rates in those basins experiencing large uplift and large increases in basin relief. Other factors not considered in these simulations include the possible effect of altered vegetation cover on precipitation and runoff (Chapter 9) and the role of uplift in enhancing the response of monsoon precipitation to orbitally-forced changes in solar radiation (Chapter 8).

Another very important geologic consequence of high topography is increased chemical weathering, which has been proposed as a critical factor in lowering atmospheric CO$_2$ levels and causing glaciation.[36] Increased chemical weathering identified in high terrain[37] results from a number of factors: increased fracturing, faulting and exposure of unweathered rock owing to tectonic compression and uplift[36]; increased mechanical pulverizing of rock at high elevations by glacial and periglacial processes[38]; and increased precipitation and runoff along high-relief slopes of elevated terrain.[36,39] The mechanical fragmentation processes expose much greater rock surface area to chemical weathering processes than is the case in low-relief areas, where bedrock is mantled by a thick cover of soil.[37] Finally, lithologic variability in drainage basins is a very important factor in chemical weathering (Chapter 14).

Richter *et al.*[40] investigated the evolution of the oceanwide ^{87}Sr/^{86}Sr ratio over the last 100 million years and quantitatively evaluated several factors that might control its variations. They showed that changes in dissolution and diagenesis of seafloor carbonates have had negligible effect, and that changes in hydrothermal alteration of seafloor basalt were both small and poorly timed to explain the rise in the Sr isotope curve over the last 40 million years. This analysis left standing only one major explanatory factor: increases in dissolved Sr input by rivers, including both increases in the river fluxes as well as increases in the ^{87}Sr/^{86}Sr ratio of the Sr delivered by rivers. Further analysis targeted on high Sr rivers draining the Himalaya and Tibet showed that this region can explain the entire late Cenozoic rise in ^{87}Sr/^{86}Sr, provided that virtually the entire dissolved river flux in that region today has come into existence within the last 40 million years, and assuming some increase in the Sr ratio delivered from this region due to erosional unroofing of radiogenic granites in the Himalaya. The required increase in dissolved river fluxes within the last 20–15 million years represents a factor of between 35 and 800%, with the large range in these values reflecting the fact that the period of 20–15 million years ago was the time of fastest rise in Sr ratio in the entire record.

In broad terms, these estimated increases in dissolved river fluxes match well with the simulated changes in runoff between $LM/2 \times CO_2$ and $M/1 \times CO_2$, the experiments that were designed to capture important elements in the evolution of climate from 20–15 million years ago to the present (Table 2). The simulated runoff increased by a factor of 2 (100%) in eastern Asia and by a factor of 5 (400%) in southern Asia. These model results are in the middle of the very broad possible range of dissolved river flux increases that can be inferred from the Sr isotope analyses of Richter *et al.*[40] The increase in runoff from the no-orography (NM) to the modern orography (M) case ranges from a factor of 2.5 (150%) to 10 (900%) in the same two regions, and this is consistent with the results obtained by Richter *et al.*[40] indicating that the high levels of runoff and dissolved river fluxes observed today in the Himalayan–Tibetan region are primarily a creation of uplift during the last 40 million years.

While the simulations support the argument that uplift, by itself, increases chemical weathering, they also indicate that the combination of regional uplift and global CO_2 lowering would tend to diminish chemical weathering in non-uplifting regions, where conditions become colder and drier, while still enhancing chemical weathering in uplifting regions, where conditions become colder but much wetter. Without an explicit model of chemical weathering, our model results don't allow us to conclude which of these competing trends is strongest in controlling global chemical weathering.

In summary, these simulations indicate that uplift has a considerable effect on global and regional hydrology. Uplift alone causes land-average precipitation and runoff to increase by 14 and 45%, respectively. In southern and eastern Asia, basin-average precipitation increases by 50–70% and basin-average runoff increases by factors of 2–10. In these same regions, the frequency of extreme-event precipitation and runoff increases by factors of 2–3. The combination of uplift and orbital forcing further enhances these runoff extremes (Chapter 8). Our simulations also indicate that the combination of uplift and CO_2 lowering produces changes in the hydrologic cycle that are significantly different from those due to uplift alone. In most regions and in the global land average, precipitation and runoff decrease slightly. In southern and eastern Asia, however, precipitation still increases by 15–40% and runoff increases by factors of 2–5 because local orographic effects dominate over the effect of CO_2 lowering. These simulated increases in runoff are within the range required to explain all or most of the observed late Cenozoic rise in the oceanwide Sr isotope ratio. Although the climate model used in these experiments simulates about the same relationship between runoff and basin relief as found in observations, its coarse resolution and known biases limit the overall accuracy of the simulation of precipitation and runoff. Our estimates of the sensitivity of the hydrologic cycle to uplift and CO_2 lowering therefore need to be compared to similar estimates obtained with models having higher resolution and improved parameterizations of land surface processes. In order to address directly the relationships between uplift and weathering, it would be desirable to include explicit parameterizations of physical and chemical weathering based upon model-simulated values of temperature, precipitation, and runoff as well as specified surface morphology and lithology. The changes in vegetation caused by uplift and CO$_2$ lowering can influence precipitation and runoff and weathering, and these interactions between vegetation and climate also need to be included in subsequent studies.

Acknowledgments

This research was supported by grants: to the University of Wisconsin-Madison by the National Science Foundation, Climate Dynamics Program, by the Department of Energy, and by computer resources provided by the National Center for Atmospheric Research, which is sponsored by the National Science Foundation; to the University of Virginia by the National Science Foundation, Atmospheric Sciences Division and Ocean Sciences Division; and to Brown University by the National Science Foundation, Ocean Sciences Division, and the Department of Energy.

We thank Pat Behling, Rich Selin, Mike Coe, and JoAnne Kruepke for assistance with model simulations, analyses, and graphics and Mary Kennedy for preparing the manuscript.

REFERENCES

1. Manabe, S., and Tepstra, T. B. (1974). *J. Atmos. Sci.* **31**, p. 3.
2. Hahn, D. G., and Manabe, S. (1975). *J. Atmos. Sci.* **32**, p. 1515.
3. Manabe, S., and Broccoli, A. J. (1990). *Science* **247**, p. 192.
4. Broccoli, A. J., and Manabe, S. (1992). *J. Clim.* **5**, p. 1181.
5. Kutzbach, J. E., Guetter, P. J., Ruddiman, W. F., and Prell, W. L. (1989). *J. Geophys. Res.* **94**, p. 18,393.
6. Ruddiman, W. F., and Kutzbach, J. E. (1989). *J. Geophys. Res.* **94**, p. 18,379.
7. Prell, W. L., and Kutzbach, J. E. (1992). *Nature* **360**, p. 647.
8. Kutzbach, J. E., Prell, W. L., and Ruddiman, W. F. (1993). *J. Geology* **101**, p. 177.
9. deMenocal, P. B., and Rind, D. (1993). *J. Geophys. Res.* **98**, p. 7265.
10. Arthur, M. A. (1991). In: *Advisory Report on Earth System History* (G. S. Mountain and M. E. Katz, eds.), p. 51–74, National Science Foundation, Division of Ocean Studies, Washington, D.C.
11. Freeman, K., and Hayes, J. M. (1992). *Global Biogeochem. Cycles* **6**, p. 185.
12. Cerling, T. E., Wang, Y., and Quade, J. (1993). *Nature* **361**, p. 344.
13. Randel, W. J., and Williamson, D. L. (1990). *J. Clim.* **3**, p. 608.
14. Thompson, S. L., and Pollard, D. (1995). *J. Clim.* **8**, p. 732.
15. Bonan, G. B. (1994). *J. Geophys. Res.* **99**, p. 25803.
16. Covey, C., and Thompson, S. L. (1989). *Palaeogeog, Palaeoclimatol., Palaeoecol.* **75**, p. 331.
17. Pitcher, E. J., Malone, R. C., Ramanathan, V., Blackmon, M. L., Puri, K., and Bourke, W. (1983). *J. Atmos. Sci.* **40**, p. 580.
18. Coe, M. T. (1995). *J. Clim.* **8**, p. 535.
19. Thompson, S. L., and Pollard, D. (1995). *J. Clim.* **8**, p. 732.
20. Manabe, S., and Bryan, K. (1985). *J. Geophys. Res.* **90**, p. 11.

21. Oglesby, R. J., and Saltzman, B. (1992). *J. Clim.* **5**, p. 66.
22. Baker, V. R. (1988). In: *Flood Geomorphology* (V. R. Baker, R. C. Kochel, and P. C. Patton, eds.), pp. 81–95. John Wiley, New York.
23. Komar, P. D. (1988). In: *Flood Geomorphology* (V. R. Baker, R. C. Kochel, and P. C. Patton, eds.), pp. 97–111. John Wiley, New York.
24. UNESCO (1978). *World Water Balance and Water Resources of the Earth*, UNESCO, Paris.
25. Ahnert, F. (1970). *Am. J. Sci.* **268**, p. 243.
26. Gibbs, R. J. (1965). *Geolog. Soc. Am. Bull.* **78**, p. 1203.
27. Milliman, J. D., and Meade, R. H. (1983). *J. Geology* **91**, p. 1.
28. Walling, D. E., and Webb, B. W. (1983). In: *Background to Paleohydrology* (K. J. Gregory, ed.), pp. 69–100. John Wiley, New York.
29. Milliman, J. D., and Syvitski, J. P. M. (1992). *J. Geol.* **100**, p. 525.
30. Summerfield, M. A., and Hulton, N. J. (1994). *J. Geophys. Res.* **99**, p. 13.
31. Summerfield, M. A. (1991). *Global Geomorphology, An Introduction to the Study of Landforms*, Longman, United Kingdom.
32. Langbein, W. B., and Schumm, S. A. (1958). *Trans. Am. Geophys. Union* **39**, p. 1076.
33. Ohmori, H. (1983). *Bull. Dept. of Geogr. Univ. Tokyo* **15**, p. 77.
34. Schmidt, K.-H. (1985). *Earth Surface Processes and Landforms* **10**, p. 497.
35. Hayden, B. C. (1988). In: *Flood Geomorphology* (V. R. Baker, R. C. Kochel, and P. C. Patton, eds.), pp. 13–26. John Wiley, New York.
36. Raymo, M. E., Ruddiman, W. F., and Froelich, P. N. (1988). *Geology* **16**, p. 649.
37. Stallard, R. F., and Edmond, J. M. (1983). *J. Geophys. Res.* **88**, p. 9671.
38. Molnar, P., and England, P. (1990). *Nature* **346**, p. 29.
39. Walling, D. E., and Webb, B. W. (1986). In: *Solute Processes* (S. T. Trudgill, ed.), pp. 251–327. John Wiley, New York.
40. Richter, F. M., Rowley, D. B., and DePaolo, D. J. (1992). *Earth Planet. Sci. Lett.* **109**, p. 11.

8

The Impact of Tibet–Himalayan Elevation on the Sensitivity of the Monsoon Climate System to Changes in Solar Radiation

Warren L. Prell and John E. Kutzbach

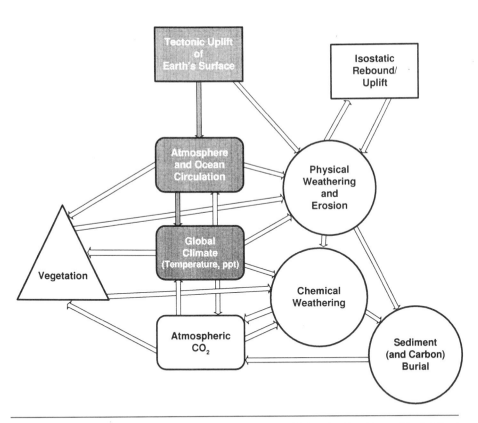

Warren L. Prell • Department of Geological Sciences, Brown University, Providence, Rhode Island 02912.   *John E. Kutzbach* • Center for Climatic Research, University of Wisconsin-Madison, Madison, Wisconsin 53706.

Tectonic Uplift and Climate Change, edited by William F. Ruddiman. Plenum Press, New York, 1997.

1. INTRODUCTION

Two major agents in the evolution of the Earth's climate have been the episodic uplift of mountain ranges and plateaus and changes in the seasonal distribution of solar radiation induced by changes in Earth's orbit around the sun. One approach to identifying the specific regional and global climatic responses that result from changes in orographic and orbital configurations has been the use of atmospheric general circulation models (GCMs). A number of GCM experiments have sought to understand the impact of mountains in general and specifically the uplift of the Tibet–Himalayan complex on the Asian monsoon system.[1-8] Recent "mountain" experiments have included surface boundary conditions with no orographic relief, surface elevations at half of modern values, and several variations with lowered mountains.[3-7] In general, these experiments have shown that the summer monsoon response is highly sensitive to the surface elevation of the Tibet–Himalayan complex and that at least half of the modern elevation of this complex is required to produce the strong southwesterly winds over the Arabian Sea and significant precipitation over India and southern Asia.[5]

Another series of GCM experiments has been performed to understand the periodic climate changes and resulting cyclic sediments that are induced by changes in Earth's orbit (Milankovitch cycles).[9-13] Many sedimentary sequences exhibit periodicities near 41,000 and 23,000 years, which have been attributed to changes in the obliquity and precession of Earth's axis. Other longer cycles (100,000 and 400,000 years) are associated with the eccentricity of Earth's orbit and with nonlinear feedbacks in the climate system, especially with global ice sheets. Many GCM experiments have shown that orbital forcing of the seasonal radiation budget induces changes in the model climate that are consistent with the scale and periodicity of changes observed in the sedimentary record. For example, Kutzbach[9] demonstrated that the orbitally induced increase in solar radiation at 9000 years BP forced a stronger Asian summer monsoon circulation, which was consistent with enhanced upwelling in the Arabian Sea[14-16] and higher lake levels and more humid vegetation types in Africa, the Arabian Peninsula and the Asian subcontinent.[12,13] Although orbitally induced changes in seasonal solar radiation (Milankovitch mechanism) are widely accepted as an important mechanism for climate change, the amplitude of climatic and geological responses is often disproportionate to the original radiation forcing. Hence, the question arises: "What internal processes and feedbacks act to amplify or dampen the amplitude (or sensitivity) of climate and sedimentary responses associated with Milankovitch forcing, and how are they related to changes in major boundary conditions?" Our studies of the Asian summer monsoon response to changes in the elevation of the Tibet–Himalayan complex have led us to the questions: "Is the sensitivity of the monsoon to orbitally induced radiation changes partially dependent on the elevation of the Tibet–Himalayan complex? Or does the amplitude of climate response to Milankovitch forcing change with elevation?"

In previous studies, we defined several linear equilibrium sensitivity coefficients (ESC) to quantify the response of the monsoon to changes in solar radiation (ΔS) and mountain-plateau elevation (ΔE).[5,11] These ESCs were combined with time series of predicted ΔS and hypothetical ΔE to simulate the temporal pattern of monsoon variability and evolution over the past 15 million years.[5] The time-series simulations clearly demonstrated that the monsoon strength in the no-mountain case was distinctly weaker than the modern control and that a relatively high plateau (about half the current elevation) was needed to produce a strong summer monsoon circulation. Comparison with data from the Arabian Sea and Pakistan indicated that the monsoons were exceptionally weak prior to about 8 million years ago.[5,16] Hence, the implication of the ESC modeling was that the plateau must have been less than half the height of its current elevation prior to 8 million years ago.[5]

These time-series models, however, assumed that the responses to ΔS and ΔE were linear and additive and did not include possible interactions between the two major boundary conditions. Here, we use the National Center for Atmospheric Research (NCAR) Community Climate Model-Version 1 (CCM1) (Chapter 7 and see Pitcher *et al.*[17] for a description of CCM1) to examine the sensitivity of the monsoon response, especially the hydrological budget, to *combined* changes in ΔS and ΔE of the Tibet–Himalayan complex. Our objective is to determine if the sensitivity of the monsoon to orbital forcing differs between modern orography (full mountain, FM) and lowered orography (quarter mountain, QM) when ΔS and ΔE are allowed to interact and, if so, to make better estimates of monsoon variability and evolution for comparison to geological and paleoceanographic records.

2. ELEVATION–RADIATION EXPERIMENTS

We used the NCAR CCM1 to estimate the sign and magnitude of response in surface temperature (ΔT), precipitation (ΔP), precipitation minus evaporation ($\Delta P - E$), and runoff (ΔR) over the regions of southern Asia, the Pakistan–Afghanistan region, and northeast Africa (see Fig. 1 for the location of area average regions) to changes in both Milankovitch forcing and elevation of the Tibet–Himalayan complex. We force the model with two sets of orographic boundary conditions (FM and QM) and the maximum range of summer solar radiation expected over the past 15 million years.

The FM experiments use the orography of the CCM1 modern control case, which averages 2688 m over the Tibetan Plateau (Fig. 1a). This modern orography represents the maximum elevation and extent of the Tibet–Himalayan complex. The QM experiments reduce the mountains and plateaus to about one-quarter of their control elevation and average about 850 m over the Tibetan Plateau (Fig. 1b). In all experiments, the actual topography is averaged over the CCM1 grid and smoothed to the model's spectral resolution. Although the

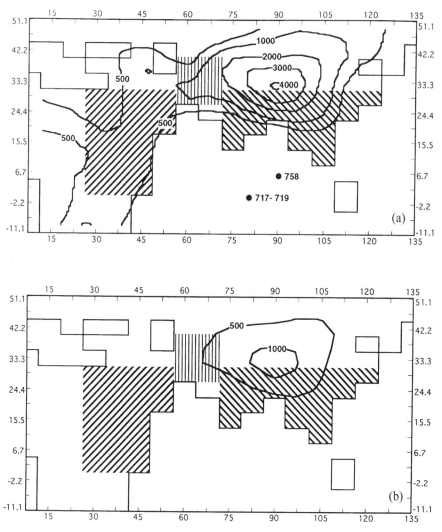

FIGURE 1. The orographic boundary conditions for: (a) full mountain (FM) and (b) quarter mountain (QM) experiments. Elevation contours are 0.5, 1.0, 2.0, 3.0, and 4.0 km. The areas used to calculate climate averages are: southern Asia (diagonal left lines), Pakistan–Afghanistan (vertical lines), and northeast Africa (right diagonal lines). Also shown is the location of ODP Sites 758 on the Ninety East Ridge and Sites 717—719 on the Bengal Fan.

mountains and plateaus are greatly smoothed, they do provide the elevation contrast needed to simulate the major orography-related climate patterns (see Chapter 7). We used QM rather than the more extreme no-mountain case (NM, land elevation at sea level) because not only is QM geologically more appropriate to the early Miocene but the low orography tends to anchor major features of the

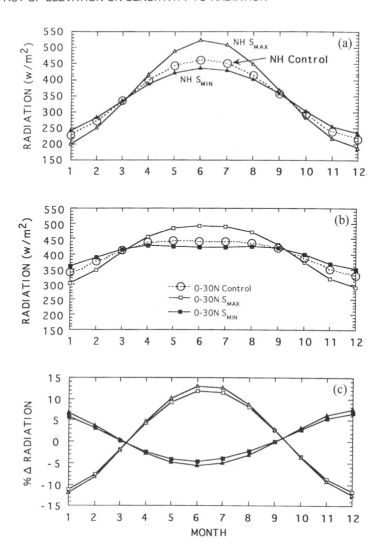

FIGURE 2.   The annual cycle of radiation (W/—m^2) in the control, S_{max}, and S_{min} experiments for: (a) the entire Northern Hemisphere, (b) the low latitude Northern Hemisphere (0°–30°N), and (c) the percent departures from the modern control radiation (data from Berger[18] and Berger and Loutre[19]).

atmospheric circulation in a realistic fashion. The QM case is similar to the lowered mountain (LM) case used in Kutzbach *et al.* (Chapter 7).

   To quantify the maximum range of orbitally induced radiation changes over the Northern Hemisphere (NH), we used June perihelion, an obliquity of 25° and an eccentricity of 0.04 to define the maximum NH summer radiation (ΔS_{max}) and December perihelion, an obliquity of 22° and an eccentricity of 0.04 to define the minimum NH summer radiation (ΔS_{min}). Orbital data are from Berger[18] and Berger and Loutre.[19] The annual cycles of NH and $0°-30°$N solar radiation (W/m^2) for the S_{max} and S_{min} cases are shown in Fig. 2a, b. Compared to the modern control case, this combination of orbital parameters results in NH summer radiation that is 12.5% higher for the ΔS_{max} case and 5.0% lower for the ΔS_{min} case (Fig. 2c). Owing to the low- to mid-latitude pattern of precession-induced radiation changes and the modulation of precession by eccentricity, most of the radiation forcing for the Tibet–Himalayan complex lies in the precession band, which has an average periodicity of about 21,000 years.

The combinations of these four boundary conditions (FM-S_{max}, QM-S_{max}, FM-S_{min}, QM-S_{min}) give experiments that span the maximum range of both orographic and orbital (Milankovitch) forcing that might be expected over the past 15 to 20 million years. Because the explicit path of the Tibet–Himalayan complex elevation versus time is poorly known, these experiments should be considered as sensitivity tests. The model estimates of long-term monsoon trends and of increased/decreased monsoonal variability should provide some constraints on the interpretation of geologic records associated with uplift-induced changes in precipitation and runoff.

3. CLIMATE RESPONSES TO CHANGES IN ELEVATION AND RADIATION

To quantify the climate associated with the evolution of the Tibet–Himalayan complex (especially the summer monsoon intensity), we calculated average annual cycles for the areas of southern Asia (21 model grid points), Pakistan–Afghanistan (9 grid points), and northeast Africa (23 grid points) (Fig. 1). Compared to the annual cycle of the CCM1 control case (FM-S_{mod}), ΔT, ΔP, $\Delta P - E$, and ΔR all show distinct and large responses to the combined changes in ΔE and to ΔS. We first examine the annual patterns of temperature and hydrology in monsoonal southern Asia, Pakistan–Afghanistan and northeast Africa to identify the regional climatic responses to combinations of ΔE and ΔS.

3.1. Surface Temperature

3.1.1. Southern Asia

The annual cycle of surface temperature in southern Asia responds strongly to the combinations of ΔE and ΔS. As elevation increases, temperature decreases uniformly across the entire annual cycle (Fig. 3). The QM cases are consistently warmer than the FM cases, regardless of solar radiation. This systematic offset mostly reflects the FM land surface being higher in the atmosphere (and hence cooler) but also reflects changes in the circulation–climate system (see also Chapters 5–7). All of the orbital–elevation configurations give maximum surface

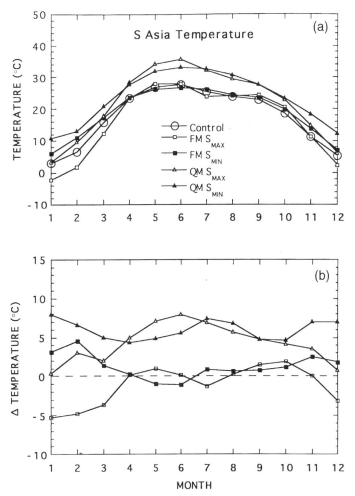

FIGURE 3.   The annual cycle of surface temperature (°C) in the southern Asia area (a) and its departure from the modern control case (b). Open circles are CCM1 control, squares are FM experiments, triangles are QM experiments, open symbols have S_{max} forcing, and filled symbols have S_{min} forcing.

temperatures in May–June and somewhat lower temperatures during the July–August monsoon maximum. These lower monsoon temperatures, especially in the FM-ΔS_{max} experiment, reflect increased cloud cover and evaporative cooling associated with high levels of monsoonal precipitation. This lowering of temperature during the monsoon interval is a distinct feature of the modern regional climatology.[20,21] In general, the summer temperatures reflect the elevation changes, so that even the weak solar forcing at FM produces summer temperatures similar to the control and the FM-ΔS_{max} case. The winter temperatures in the FM-ΔS_{max} are distinctly lower than the control and FM-ΔS_{min} cases.

3.1.2. Pakistan–Afghanistan

In contrast to southern Asia, the annual cycle of surface temperature in the adjacent, but more arid, Pakistan–Afghanistan area tends to have highest/lowest temperatures at max/min ΔS regardless of the Tibet–Himalayan complex elevation (Fig. 4). This region does not experience the monsoon cooling, and thus summer temperatures are higher and tend to peak in July–August. Similarly, the Pakistan–Afghanistan area did not undergo substantial elevation changes in these experiments, and thus does not include the lapse-rate component of cooling as did southern Asia.

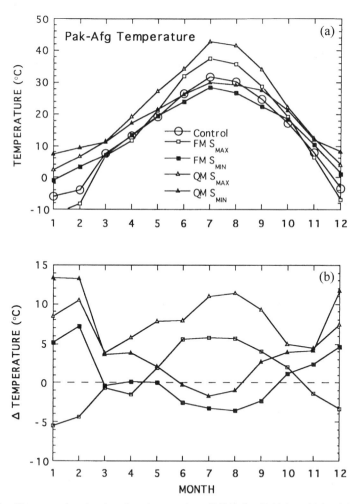

FIGURE 4.   The annual cycle of surface temperature (°C) in the Pakistan–Afghanistan area (a) and its departure from the modern control case (b). Symbols as in Fig. 3.

3.1.3. Northeast Africa

The annual cycle of surface temperature in northeast Africa exhibits a broad, almost bimodal, pattern of high summer temperatures, with the maximum usually in July–August (Fig. 5). Temperatures in this low-latitude region exhibit a lower annual range and seem to respond more to the changes in radiation than to changes in Tibet–Himalayan complex orography. The orography-related responses (difference between FM and QM at the same radiation) seem to occur mostly in the winter season when differences reach 4° to 6°C (Fig. 5). During the summer season, there is relatively little temperature contrast (0–2°C) between the FM and QM cases and the control. These temperature responses indicate that

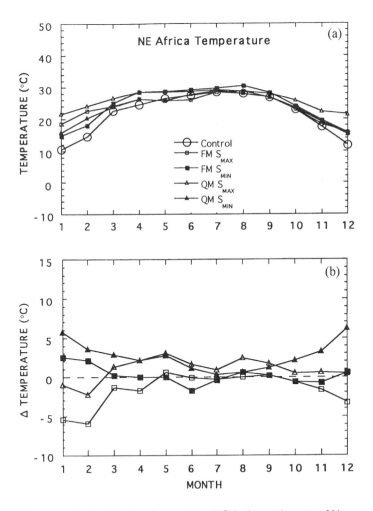

FIGURE 5. The annual cycle of surface temperature (°C) in the northeastern Africa area (a) and its departure from the control case (b). Symbols as in Fig. 3.

climate changes owing to higher/lower Tibet–Himalayan complex elevation do propagate to northeast Africa but that the temperature responses to changes in low-latitude radiation are greater.

3.2. Precipitation

3.2.1. Southern Asia

The annual cycle of precipitation responds dramatically to the combinations of ΔS and ΔE, with the largest response occurring in the summer monsoon season. The FM-S_{max} gives the highest and the QM-S_{min} gives the lowest monsoonal precipitation (Fig. 6). The variable solar forcing produces a larger range of precipitation (and all other hydrologic variables) than observed in Chapter 7.

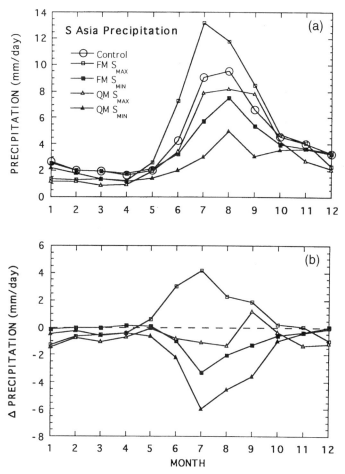

FIGURE 6. The annual cycle of precipitation (mm/day) in the southern Asia area (a) and its departure from the control case (b). Symbols as in Fig. 3.

Even the maximum radiation forcing (ΔS_{max}), when combined with the QM orography, gives 13% less precipitation than the control case. All combinations of radiation with lower mountains give less precipitation than the control. The range of precipitation owing to ΔS forcing is also greater for the FM cases than for the QM cases. Hence, the maximum monsoon precipitation is dependent on both elevation and radiation.

3.2.2. Pakistan–Afghanistan

The annual cycle of precipitation in this area is strikingly different from monsoonal southern Asia in both amplitude and seasonality (Fig. 7). In general, the Pakistan–Afghanistan region tends to get dryer as monsoonal Asia gets

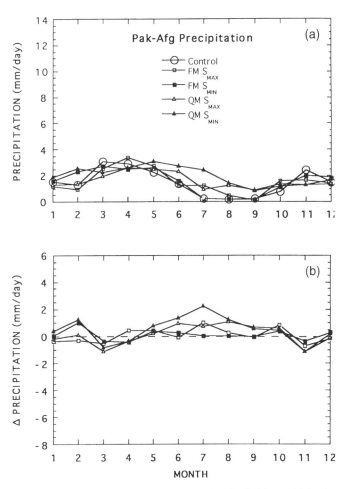

FIGURE 7.   The annual cycle of precipitation (mm/day) in the Pakistan–Afghanistan area (a) and its departure from the control base (b). Symbols as in Fig. 3.

wetter. In addition to the values being lower, especially in the summer months, the seasonal peaks occur in the spring (March–May) rather than during the monsoon season (June–August). Although the changes are small, most radiation–elevation configurations give more summer precipitation than the control case.

3.2.3. Northeast Africa

The pattern of precipitation in this area exhibits a broad maximum that is concentrated in the summer months (June–September) in the ΔS_{max} cases and a weak response in the ΔS_{min} cases (Fig. 8). Both radiation forcing configurations

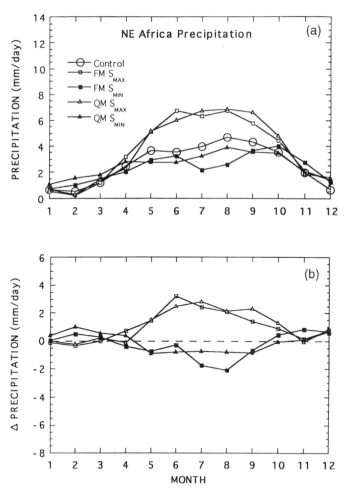

FIGURE 8.   The annual cycle of precipitation (mm/day) in the northeastern Africa area (a) and its departure from the control case (b). Symbols as in Fig. 3.

produce a similar response in the FM and QM cases, indicating that precipitation is dominated by changes in radiation, with the influence of the Tibet–Himalayan complex elevation having relatively little effect.

3.3. Runoff

The land surface component of CCM1 calculates runoff as the amount of precipitation that is not taken up by soil moisture or that is evaporated-transpired (see Chapter 7 for a description). Although difficult to estimate on a small scale, the integration of the runoff over large basins is similar to observed fluxes[22] (Chapter 7 and Coe[22]). In these elevation–radiation experiments, the runoff values should give a sense of potential climate-induced impacts on geologic

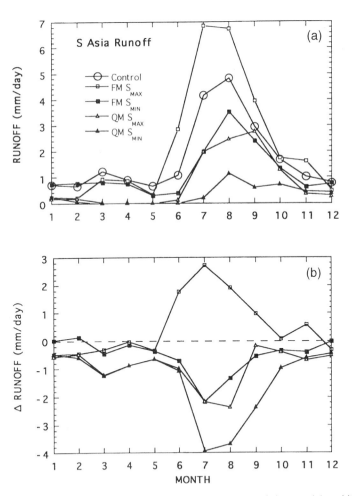

FIGURE 9.   The annual cycle of runoff (mm/day) in the southern Asia area (a) and its departure from the control case (b). Symbols as in Fig. 3.

processes, especially in the summer monsoon region. In addition, geological records, such as the type and accumulation rate of sediments and physical–chemical weathering indices, might be more related to changes in excess water (i.e., runoff) rather than directly to precipitation.

3.3.1. Southern Asia

The estimates of runoff from southern Asia show the largest changes of all the climate parameters. Runoff, like precipitation, is concentrated in the monsoon season and is strongly related to the specific combinations of radiation and elevation. The maximum runoff (6.9 mm/day) occurs in the FM-S_{max} experiment

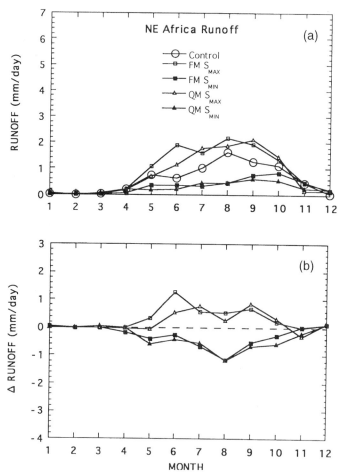

FIGURE 10. The annual cycle of runoff (mm/day) in the northeastern Africa area (a) and its departure from the control case (b). Symbols as in Fig. 3.

(Fig. 9), while the FM-S_{min} and QM-S_{max} cases produce similar but significantly less runoff, and the QM-S_{min} configuration produces almost no runoff (<1 mm/day). The range of runoff response is much greater in the FM cases than in the QM cases. There is virtually no runoff in the Pakistan–Afghanistan region.

3.3.2. Northeast Africa

The annual pattern of runoff in northeast Africa closely follows the radiation maximum and precipitation patterns, with maximum runoff in the ΔS_{max} cases and significantly less runoff in the ΔS_{min} cases (Fig. 10). These patterns indicate that the hydrologic cycle of northeast Africa is only minimally influenced by the elevation changes in the Tibet–Himalayan complex.

4. SENSITIVITY OF THE MONSOON HYDROLOGIC CYCLES TO ELEVATION CHANGE

The above experiments indicate that the climate effects induced by elevation and radiation are greatest during the summer monsoon hydrologic cycle of southern Asia. A summary of the July water budget (Fig. 11) reveals that precipitation (mm/day) changes by a factor of 3 between the QM-S_{min} and FM-S_{max} experiments, while evaporation changes only by a factor of 1 and runoff changes by a factor of 33. The seasonal storage is small and shows a change ($\times 3$) similar to that in precipitation. Thus, much of the variation in the water budget stems from the exceptionally large responses for the FM-S_{max} experiment, es-

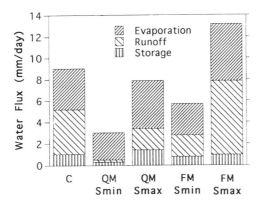

FIGURE 11.   Summary of the summer monsoon (July) water budget for the southern Asia area over the range of elevation–radiation experiments. The height of each column is the total precipitation (mm/day) for the experiment, which is partitioned into evaporation (right diagonal lines), runoff (left diagonal lines), and seasonal storage (vertical lines). The experiment names are as in the text.

TABLE 1.   Response of Water Budget Components Relative to Control and QM-S_{min} Experiment for Southern Asia, July

	Control	QM-S_{min}	QM-S_{max}	FM-S_{min}	FM-S_{max}
Precipitation					
July (mm/day)	9.03	3.02	7.89	5.71	13.17
%ΔP (control)	—	-67	-13	-37	$+46$
%ΔP (QM-S_{min})	199	—	161	89	336
$P - E$					
July (mm/day)	5.19	0.52	3.40	2.77	7.85
%$\Delta P - E$ (control)	—	-90	-34	-47	$+51$
%$\Delta P - E$ (QM-S_{min})	889	—	548	428	1397
Runoff					
July (mm/day)	4.14	0.20	1.97	1.97	6.86
%ΔR (control)	—	-95	-53	-52	$+66$
%ΔR (QM-S_{min})	1930	—	864	867	3264

pecially in the runoff component (Fig. 11). The impact of changes in elevation and radiation on the components of the hydrologic cycle is better illustrated by the relative changes (percent departure from control) among experiments (Table 1, Fig. 12). Using the CCM1 modern control as the reference, the hydrologic components exhibit the largest range of response to radiation forcing ($S_{max} - S_{min}$) in the FM experiments. For precipitation, the response range is 83% for the FM and 54% for the QM experiments (Table 1, Fig. 12). $P - E$ has a range of 98% in the FM and 56% in the QM experiments, while runoff has a range of 118% in the FM and 42% in the QM experiments (Table 1, Fig. 12). These departures are consistent, but are larger than those reported in Chapter 7 that considered only changes to the orography. If the range of these hydrologic responses were directly transformed into geological indices, the cyclicity of that index associated with orbitally induced radiation (Milankovitch cycles) would have much greater amplitude in the FM world. Hence, as the mountain–plateau elevations increase with time, the sedimentary Milankovitch cycles associated with changes in the monsoon hydrologic system are also expected to increase in amplitude.

The relative amount of radiation-forced climate change as mountain–plateau elevations increase can be estimated by comparing the range of climate response in the QM and FM cases with the same radiation forcing. For example, relative to the QM case, the range of runoff in the FM case increases 180% [($\Delta FM - \Delta QM)/\Delta QM = (118\% - 42\%)/42\% = 180\%$]. The range in precipitation increases 54% and the increase in the range of $P - E$ is 75% (see Table 1, Fig. 12).

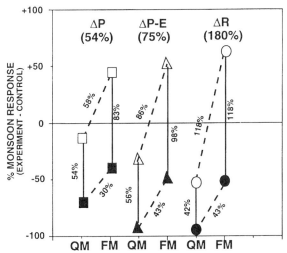

FIGURE 12.   Summary of percent changes (relative to the modern control) of precipitation (ΔP), precipitation minus evaporation ($\Delta P - E$), and runoff (ΔR) in the southern Asia monsoon hydrologic cycle. Experiment orography is indicated by QM or FM. Open symbols are ΔS_{max} and filled symbols are ΔS_{min}. The percent values on the diagonal dashed lines are the change between QM and FM with the same ΔS forcing, and the percent values on the vertical lines are the changes between S_{max} and S_{min} forcing at the same elevation. The percent change in the range of the response of the three climate variables to orbital forcing in FM, relative to QM, is shown in parentheses at the top of the figure. These percent changes, all positive, reflect the increased length of the vertical lines ($\Delta S_{max} - \Delta S_{min}$) in FM, relative to QM.

The change in sensitivity of the hydrologic components can be estimated by comparing the range of the response to the change in forcing (i.e., $\Delta R/\Delta S$) for the QM and FM cases. The difference between the ΔS_{max} and ΔS_{min} forcing is 17.5% (Fig. 2c). Hence, the sensitivity of precipitation ($\Delta P/\Delta S$) increases from 3.0 (54%/17.5%) in the QM case to 4.7 (83%/17.5%) in the FM case, an increase of over 50%. In a similar manner, the sensitivity of $P - E$ increases from 3.2 in the QM case to 5.6 in the FM case, for an increase of about 75%. The response of $\Delta P - E$ in the FM $- \Delta S_{max}$ experiment is 85% higher than the QM-ΔS_{max} value, and the range of $\Delta P - E$ to orbital forcing is almost twice as large in the FM case (98%) as in the QM case (56%). The response of runoff to ΔS_{max} forcing increases 118% from the QM to the FM experiment, and the range of ΔR to solar forcing ($\Delta S_{max} - \Delta S_{min}$) varies by a factor of 3 from the FM case (118%) to the QM case (42%). Hence, the sensitivity coefficient of runoff increases from 2.4 in the QM case to 6.8 in the FM case, an increased sensitivity of 283%. If the QM-S_{min} experiment were used as the reference, the runoff in the FM-S_{max} case would have increased 3300%.

The origin of the increased sensitivity in all hydrologic variables is the large response in the FM-S_{max} case. In fact, the response to maximum radiation forcing at lowered mountains (QM-S_{max}) and to minimum radiation forcing at high mountains (FM-S_{min}) are often similar, especially in runoff (Figs. 11 and 12).

Hence, sedimentary or geochemical indices that reflect runoff (i.e., the amount of excess water in the system) might be expected to correlate positively with the evolution of the S_{max} peaks of runoff.

5. ESTIMATES OF MONSOON VARIABILITY AND EVOLUTION

The QM − FM and S_{max} − S_{min} experiments reveal that the hydrologic cycle in monsoonal southern Asia is highly sensitive to changes in both the elevation of the Tibet–Himalayan complex and solar radiation. However, these two forcings have very different temporal patterns of change, with the surface elevation changing on the scale of 10^6 to 10^7 years and the radiation changing on the scale of 10^4 to 10^5 years.

5.1. Equilibrium Sensitivity Coefficient Model for Monsoon Runoff

To quantify the combined effects of tectonic (elevation) and radiation (Milankovitch) forcing, we have incorporated the dependence of the Milankovitch response upon elevation into the ESC model[5] to estimate runoff for southern Asia. We derived the partial ESC to elevation change by taking the difference between the ESCs for the FM and QM cases (6.8 − 2.4 = 4.4). This approach still assumes that the terms are additive, but now explicitly incorporates the dependence of the radiation-forced runoff response upon elevation.

The July runoff equation is

$$\Delta R_t(\%) = [\beta_s + \beta_{s,e}(\Delta E_t)]\Delta S_t + \beta_e(100^*\Delta E_t) + K \qquad (1)$$

Where ΔR_t is the percent change in runoff; β_s ($= 2.4$) is the sensitivity of percent runoff to percent changes in solar radiation in the QM case (ΔS_t); $\beta_{s,e}$ ($= 4.4$) is the combined sensitivity of runoff to ΔE_t and ΔS_t; β_e ($=0.64$) is the sensitivity of runoff to elevation changes ($100^*\Delta E_t$); and $K = -83$ is a constant derived from the simultaneous solution of runoff departures (see Fig. 12). Hence,

$$\Delta R_t \ (\%) = [2.4 + 4.4(\Delta E_t)]\Delta S_t + 0.64(100^*\Delta E_t) - 83 \qquad (2)$$

The β_s and β_e terms determine the mean value of runoff versus time, while the interactive term, $\beta_{s,e}(\Delta E_t)$, controls the envelope of the variability and is responsible for the increased variability with increased elevation. This latter feature was not included in earlier ESC models of monsoon evolution,[5] which had a constant envelope of variability versus time. The runoff equation (2) requires that ΔS_t time series is in percent departure from average modern Northern Hemisphere summer radiation and that the ΔE_t time series is scaled from 0 (QM elevation) to 1.0 (FM or modern elevation).

5.2. *Radiation and Elevation Forcing Time Series*

Given this new ESC runoff model, we need time series of ΔS and ΔE to calculate the evolution and variability of the monsoon over the past 15 million years. We use the solar radiation forcing from Prell and Kutzbach,[5] which captures the full range of the $\Delta S_{max} - \Delta S_{min}$ experiments (Fig. 13a). This time

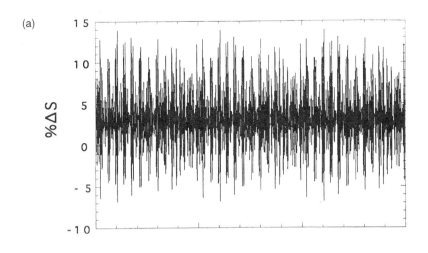

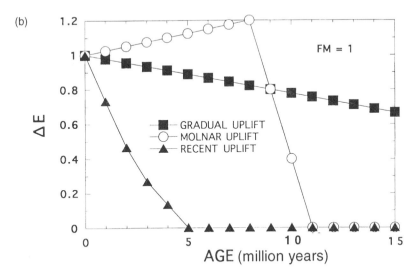

FIGURE 13. Summary of solar radiation (a) and elevation (b) time series used to force the equilibrium sensitivity coefficient runoff model. The radiation changes are in percent departure from the control Northern Hemisphere summer values (data from Berger[18] and Berger and Loutre[19]). The hypothetical elevation histories (gradual, Molnar, and recent) are scaled so that FM = 1 and QM = 0.0. The basis for the elevation curves is described in the text.

series of departures from modern Northern Hemisphere summer radiation is dominated by variability in the precession (23,000-year), obliquity (41,000-year) and precession-eccentricity modulated (100,000- and 400,000-year) periods.[18,19] The actual orbital calculations extend to 5 million years and were replicated to give the form and periodicity of the orbital forcing for the past 15 million years. The ΔS time series does not contain long-term trends (nor do more recent calculations), and the modern solar radiation is about 3% less than the long-term average. Hence, the ΔS_{max} forcing gives relatively large changes from the control climate. This time series of changes in solar radiation should provide a realistic short-term forcing for the monsoon simulations.

We adopt several elevation models (Fig. 13b) to test various hypotheses about the surface uplift history. In our ESC model, surface elevation is scaled so that the FM case is unity (1.0) and the QM case is zero (0.0). A gradual uplift model assumes relatively constant uplift since the hard collision of India with Asia about 45 million years ago and thus exhibits only slight elevation change (scaled as 0.67 to 1.0) over the past 15 million years. A second elevation model follows suggestions by Molnar et al.[23] and incorporates QM elevation from 15 to 11 million years ago, a rapid increase to slightly higher than the FM elevation 8 million years ago and a linear decrease to the FM elevation at the present time. A third uplift model assumes that the major uplift is roughly coincident with the initiation of loess deposits and deserts in Asia. Because previous experiments indicated that significant climate changes would require about half of the current (FM) elevation, the recent uplift model maintains QM elevations from 15 to 5 million years ago, increases to half elevation from 5 to 2.5 million years ago ($\Delta E = 0.33$ on our FM $-$ QM scale), and increases to FM elevations during the past 2.5 million years.

The incorporation of these elevation (ΔE) and radiation (ΔS) change time series into the ESC runoff model enables us to evaluate potential patterns of monsoon variability and evolution over the past 15 million years.

5.3. Simulations of Monsoon Evolution and Variability

The three simulations of monsoon runoff variability and evolution (Fig. 14) are similar in that they all contain the orbital periodicities (23,000, 41,000, 100,000, and 400,000 years). Hence, geologic indices that record runoff-related processes should also be expected to contain these periods. However, the three simulations differ in the long-term trend (mean value) of runoff and its short-term amplitude variability, both of which are directly related to the elevation model.

The gradual uplift simulation (Fig. 14a) exhibits only slight increases in mean value and amplitude over the past 15 million years. These small changes are a consequence of the elevation 15 million years ago being 67% of the FM elevation so that much of the uplift increase occurred earlier than 15 million years ago. In general, this simulated pattern of long-term trend and variability in runoff would be difficult to distinguish from the modern monsoon and would predict little change in the monsoon indices over the past 15 million years, and especially over the past 10 million years.

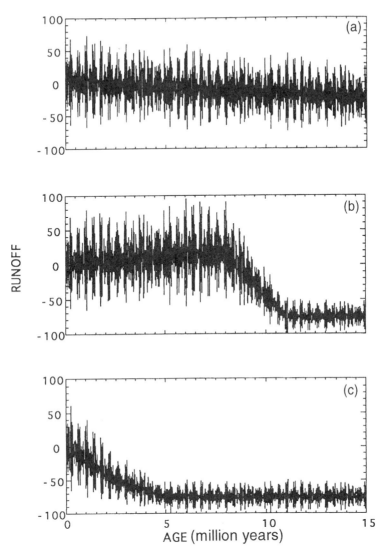

FIGURE 14. Simulated time series of percent change (from the control case) of runoff from southern Asia using the ESC model to combine the different uplift histories with changes in solar radiation: (a) gradual uplift model, (b) Molnar uplift model, and (c) recent uplift model. See text for discussion of the temporal changes in elevation.

The Molnar uplift simulation (Fig. 14b) gives a distinctly different pattern of monsoon runoff evolution and variability. Runoff is extremely low, with low variability from 15 to 11 million years ago, and increases rapidly in both mean value and amplitude from 11 to 8 million years ago. The increase in mean values is directly driven by the elevation change, whereas the increased envelope of variability reflects the increased sensitivity of the solar radiation response to elevation. The highest and most variable runoff occurs about 8 million years ago when the elevation is 20% higher than in the FM case. This simulation predicts a rapid and dramatic increase in mean value and variability with increased elevation. The simulation also predicts that the increased amplitude of runoff should occur early in the uplift history, which lends support to the view that elevation-induced precipitation might cause an erosional feedback that could increase local uplift. If the relatively rapid uplift rates and general timing (i.e., uplift during the interval from 11 to 7 million years ago) of the Molnar elevation model are correct, geologic indicators should exhibit distinctive changes over the past 15 million years.

The recent uplift simulation (Fig. 14c) also gives a distinctive pattern of increased amplitude and mean of monsoon runoff. However, in this case low runoff and low variability extend to 5 million years ago, with the rapid increase in both mean runoff and variability concentrated over the past 2 to 3 million years. This recent uplift runoff response can clearly be distinguished from the Molnar and gradual uplift simulations.

6. SIMULATIONS FOR UPLIFT–MONSOON INTERACTIONS

6.1. Implications for Monsoonal Runoff Patterns

The three end-member uplift scenarios (Fig. 13b) give distinct temporal patterns (Fig. 14) of both mean runoff values and variability that should provide templates for the evaluation of geologic records related to monsoon evolution and variability. The patterns of monsoonal response that might be recorded by geologic indices include the following:

1. *Runoff Sensitivity:* The monsoonal (July) runoff from southern Asia varies greatly with elevation and radiation forcing. Runoff increases by a factor of 3 between the average QM and FM cases $[(S_{min} + S_{max})/2]$ and by a factor more than 300 if the QM-S_{min} case is used as the baseline. In addition to increased mean runoff, the frequency of extreme precipitation–runoff events would also greatly increase in the FM-S_{max} conditions. In short, the summer monsoon hydrologic regime, especially runoff, is significantly different between lowered QM elevations (reflecting early Miocene orography?) and the more recent FM elevations (Fig. 14).

2. *Mean Runoff:* The mean runoff is directly related to the elevation of the Tibet–Himalayan complex and will thus reflect the elevation history model used in the simulation. Although the elevation histories (Fig. 13b) and hence the simulated runoff histories (Fig. 14) reflect somewhat arbitrary end-member scenarios, the difference between runoff values in the lower QM orography (Fig. 14b, c) and modern FM orography (Fig. 14) is large and the transition between the two levels of response should be detectable in geologic indices.

3. *Variability of Runoff:* The large increase in the variability of runoff at the orbital time scale is associated with increased elevation of the Tibet–Himalayan complex and reflects the elevation-dependent portion of the radiation response.

The larger envelope of variability is characteristic of the coupled elevation–radiation system for this monsoonal region and is a potential criterion for the identification of uplift-forced monsoonal climate responses. The envelope of variability is distinctly different among the three simulations (Fig. 14) and is likely to be recorded by geologic indices, especially those related to high runoff. However, to fully measure these responses, the sediment record must accumulate rapidly enough and be sampled frequently enough to resolve the dominant precessional (21,000-year) variability.

4. *Peak Runoff Events:* The large elevation dependence of the radiation-forced runoff [Eqs. (1) and (2)] implies that the increase in orbital-scale high-runoff events would occur early in the transition between QM and FM conditions. This early response and the more frequent extreme precipitation–runoff events would enhance erosional feedbacks (valley incision and local isostatic uplift) that would increase the rate of local uplift.

Hence, sedimentary cycles, especially those related to runoff and sediment transport at the orbital timescales, might be expected to increase in amplitude by a factor of 2 to 3 as mountain-plateau elevation increases from QM to FM. If sediment transport is assumed to be linearly related to the runoff, terrigenous fluxes from the continent to the ocean should increase with time and uplift. Thus, the long-term terrigenous flux as well as the amplitude of orbital scale terrigenous-runoff cycles should increase (Fig. 14). All else being equal, the continuous record of terrigenous sediment flux might be the most reliable record of when uplift-induced responses occurred in the Miocene.

6.2. Associations among Surface Uplift, Monsoonal Runoff, Weathering, Sediment Transport, and Sea-Level Changes

A number of papers [5,16,23] have noted the coincidence of a wide variety of uplift-related climatic and geologic responses during the interval from about 10 to 7 million years ago. These responses include the intensification of monsoonal winds and upwelling in the Arabian Sea,[16,24] a transition from C_3 to C_4 vegetation and to more seasonal environments as indicated by stable isotopes in soil carbonates (see Chapter 13 and Quade *et al.*[25]), increased terrigenous flux to the Indian Ocean,[26–29] and thrust faulting in the Indian Ocean crust.[23,26,30] One explanation for the near synchroneity of these disparate events and processes is that relatively rapid surface uplift occurred around 8 million years ago. This hypothesis is, in part, the basis for the Molnar elevation model[23] (Figs. 13 and 14). GCM experiments indicate that a weak monsoon response implies that the elevation of the Tibet–Himalayan complex was less than the half-mountain orography prior to 10 million to 8 million years ago and increased rapidly to modern FM orography between 10 and 7 million years ago[5] or that an orographic–climatic threshold was exceeded during this time.

Other papers[23,31] and Chapters 2, 12, and 13 in this volume have focused on the apparent discrepancies among different indicators of uplift, monsoonal activity, and sedimentary records, especially from the Bengal Fan and foreland basins of the Himalayas. The turbidite sequences of the Bengal Fan have been interpreted to show a different history of sediment transport/weathering over the past 15 million years (see Chapter 12, Cochran et al., [26] Burbank et al., [31]). From about 15 to 7 million years ago, the fan sediments are characterized as having high accumulation rates, larger grain size, more clastic clays (chlorite-illite) and C_3 plant carbon. The physical character and chemistry of the sediments are interpreted to reflect low weathering intensity and low residence time in the coastal flood plains (Chapter 12). However, for the late Miocene–Quaternary interval from about 7 to 1 million years ago, the Bengal Fan sediments are characterized as having lower accumulation rates, smaller grain size, chemically weathered clays (kaolinite-smectite) and mixed C_3 and C_4 plant carbon. These sediments are interpreted as indicating lower sediment transport, increased weathering, and increased residence in the flood-plain environment. About 1 million years ago, the sediments revert to the more clastic pattern (Fig. 15c).

6.2.1. Terrigenous Flux as an Indicator of Uplift

The observations noted above lead to the question: "If monsoon runoff increases with uplift at about 8 million years ago, why do Bengal Fan sediments decrease in accumulation rate and grain size, and reflect more weathering?" This question tacitly assumes that increased uplift and the associated monsoonal precipitation and runoff will directly lead to increased erosion, low residence time in flood plains, little time for chemical weathering, and increased clastic sediment transport to the deep-sea fan. However, sediment transport to deep-sea fans is also highly dependent on changes in sea level.[24,27,30,32–34] Given the prominence of low-lying flood plains and deltas in the weathering cycle, the interaction of monsoonal runoff and sediment transport with changes in sea level becomes a critical factor. As discussed by Stow et al.,[32] Rea,[27] and Flood et al.,[34] sea level controls the input of sediment to the deep sea by trapping sediment in deltas and flood plains during intervals of rising sea level and thus reducing the input to the deep sea, especially to the turbidite deposition in deep-sea fans. Conversely, during intervals of low sea level, falling sea level, or rapidly fluctuating sea level, sediment flux to the deep sea is higher, as the base sea level is lower and coastal deltas and flood plains are eroded and bypassed. Hence, the trend of the high-stands of sea level and the amplitude of changes in sea level are the important factors that will determine if sediments, especially coarse clastics, are deposited in shelf deltas or are transported directly to the deep sea fans. In short, the accumulation rates, the facies, and even the composition of terrigenous deep sea sediments are a complex function of both supply (uplift and erosion) and transport/storage mechanisms (runoff and sea level).

6.2.2. ODP Site 758, Ninety East Ridge

To examine the history of terrigenous flux from the Ganges–Brahmaputra system from another perspective, we use the continuous sediment record at ODP Site 758 as an alternative to the Leg 116 Bengal Fan sites. Site 758 is located in the Bay of Bengal at about 5°N, 90°E on the Ninety East Ridge at 2920 m water depth[35] (Fig. 1). At this location, the site lies about 2000 km south of the Ganges–Brahmaputra delta complex and about 800–1000 m above the adjacent Bengal Fan.[35] As the crustal age at Site 758 is about 80 million years, the thermal subsidence of the Ninety East Ridge during the past 10 million years is relatively small and not likely to cause the observed trends of decreased carbonate content and increased terrigenous sediments.[28,29,35]

The sediments at Site 758 are primarily pelagic carbonates with admixtures of terrigenous silts and clays and volcanic ash.[35] We have combined the age models from Farrell and Janecek[29] and Hovan and Rea[28] to examine the sedimentation rates for ODP 758 (Fig. 15). During the interval from 15 to about 9 million years, the sediments are predominantly pelagic carbonates with some volcanic ash and exhibit quite low accumulation rates, averaging less than 5 m/million years.[28,35] (Figs. 15 and 16). About 9 million years ago, the carbonate content begins to decrease systematically (Fig. 16a) and the sedimentation rates increase threefold to 15–20 m/million years (Fig. 15). These rates remain high but variable until the present[28,29,35] (Fig. 15). The higher accumulation rates and lower $CaCO_3$ contents are attributed to increased terrigenous flux, which dilutes the pelagic sediments, rather than to variations in pelagic productivity and carbonate dissolution.[28,29] The source of the terrigenous flux at this site has been interpreted to reflect suspended load from the Ganges–Bhahmaputra system and/or fine sediments suspended by Bengal Fan turbidites.[28,29,35]

Estimates of the terrigenous flux at Site 758 have been made by Farrell and Janecek[29] on the basis of the noncarbonate component and by Hovan and Rea,[28] who extracted and measured the mineral content (Fig. 16b). They found that the terrigenous flux began a systematic increase about 9 million years ago, and is characterized by several pulses during the late Miocene and Pliocene (Fig. 16b). Another indicator of terrigenous material is the magnetic susceptibility of the sediment. A potential advantage of magnetic susceptibility is its rapid, closely spaced (5–10 cm) measurement, which provides a relatively continuous, high-resolution record of the terrigenous content that is more comparable with the model simulations of runoff (Fig. 14). The magnetic susceptibility record for Site 758 has been shown to approximate the terrigenous fraction but is complicated by occasional ash deposits, which produce spikes in the magnetic susceptibility record.[29,35] Following the shipboard core descriptions,[35] we removed obvious ash-related spikes in the magnetic susceptibility data and have calculated the flux of magnetic susceptibility over the past 15 million years. We used the age models, linear sedimentation rates, and dry bulk density values from Farrell and Janecek[29] (ranging from the present to 7 million years ago) and Hovan and Rea[28] (7 to 15 million years ago) and the magnetic susceptibility values from Peirce and Weissel.[35] The flux of magnetic susceptibility is similar to the estimates of terrigenous flux (Fig. 16b, c). The values are low from 15 to 10 million years ago and from 8 to 6 million years ago, with an interval of higher values between 10 and 8 million years ago (Fig. 16c). A trend toward higher and more variable values begins at about 6 million years ago. Although the effects of dispersed ash

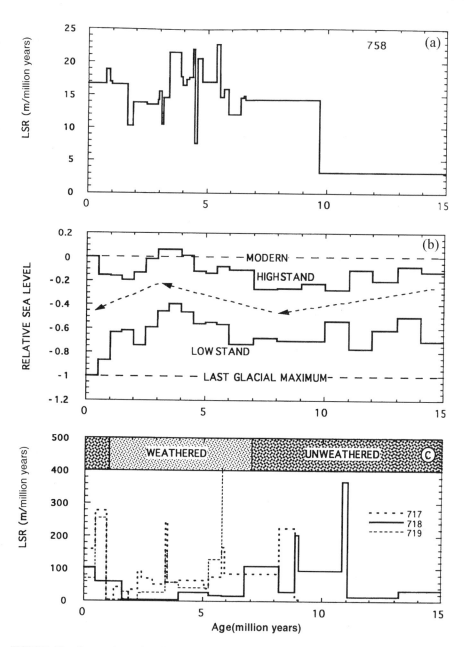

FIGURE 15. Comparison of sedimentation rates in (a) Site 758 and (c) Sites 717–719 and (b) sea-level changes over the past 15 million years. Sedimentation rates for ODP 758 are from Hovan and Rea[28] and Farrell and Janecek[29] and rates for ODP 717–719 are from Gartner.[36] The upper and lower envelope of eustatic sea level change for the past 15 million years is from Prentice and Matthews.[37] Intervals are 0.5 or 1.0 million years depending on data density. Also shown in (c) is the weathered/unweathered facies of Derry (Chapter 12).

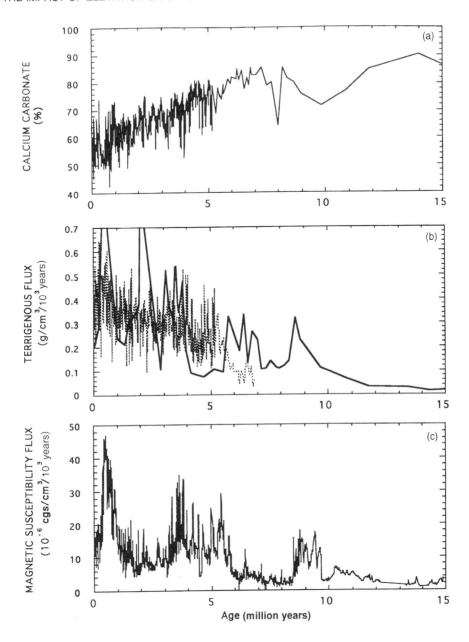

FIGURE 16. Carbonate and terrigenous flux in ODP Site 758 over the past 15 million years: (a) Carbonate percent,[28,29,35] (b) Terrigenous flux[28] (solid line) and noncarbonate flux[29] (dashed line), (c) Magnetic susceptibility flux calculated from Farrell and Janecek[29] and Peirce and Weissel.[35]

complicate the interpretation, the broad pattern and trend of the magnetic susceptibility data over the past 15 million years are consistent with the pattern of decreasing carbonate content, increasing sedimentation rates, and increase in terrigenous accumulation. (Figs. 15 and 16).

6.2.3. ODP Sites 717–719, Bengal Fan

To examine the variability of sedimentation rates at the Bengal Fan sites, we used the calcareous nannofossil datums of Gartner[36] to calculate the linear sedimentation rates for Sites 717–719 (Fig. 15c). These fan locations exhibit extremely variable sedimentation both within sites and between sites (Fig. 15c). This variability is expected because, in addition to the tectonic and eustatic sea-level controls on fan sedimentation, Stow et al.[32] noted that normal processes of fan growth such as lobe switching, channel avulsion, supply direction, and intraplate deformation of the fan could greatly affect the local sedimentation rates. Although variable, the sedimentation pattern of the Bengal Fan sites can be characterized as having low rates from 15 to about 11 million years ago (718), relatively high but variable rates from 11 to 7 million years ago (718) or to 5 million years ago in Site 717. An interval of moderate or low rates extends from 7–5 million to about 1 million years ago, when rates at all sites increase toward the present. This pattern of sedimentation is similar to that found by Rea[27] and is more complex than the weathered/unweathered facies description (see Burbank et al.[31] and Chapter 12) (Fig. 15c).

6.2.4. Sea-Level Effects

To compare these sedimentation rate patterns with changes in sea level, we calculated an upper and lower envelope of sea-level variation over the past 15 million years (Fig. 15b). The envelope of sea-level variation is approximated for intervals of 0.5 or 1 million years, depending on data density, by taking the mean plus and minus two standard deviations of the temperature-corrected $\delta^{18}O$ of planktonic foraminifera.[37] The relative sea-level changes are scaled so that modern sea level is zero and the sea level at the last glacial maximum is one. The summary shows that sea-level highstands are falling and sea-level variability is relatively low from 15 to about 10 million years ago, a trend that is consistent with interpretation from sequence stratigraphy.[32,38] Sea-level highstands are relatively stable until about 7 million years ago, when they begin to rise, until about 3 million years ago. Starting 3 million years ago, lowstands fall continuously, while highstands initially drop and remain constant until 0.5 million years ago. The range of sea-level variability is relatively low until about 1 million years ago. This pattern of long-term trend and variability of sea-level change provides a framework with which to interpret the terrigenous accumulation on both the Bengal Fan (Sites 717–719) and the Ninety East Ridge (Site 758).

On the basis of sea-level trends, one would expect the interval from 15 to 10–9 million years ago to have high accumulation rates and coarser, less-weathered sediment because the falling highstands would increase the hydrologic gradients in coastal rivers, decrease the stability of shelf and deltaic deposits, increase the direct sediment transport to the deep-sea fans, and bypass the deltaic weathering environments. From about 10–9 million to about 3 million years ago, one would expect lower accumulation rates and finer, more weathered sediment because the rising highstands and low sea level variability tend to flood the shelf and deltaic environments, decrease the coastal hydrologic gradients, and sequester sediments in coastal weathering environments, which decreases the episodic turbidite deposition. During the last 1 million years, one would expect a return to the high accumulation regime with the onset of the extremely high sea level (>100 m) variability associated with the Quaternary Ice Ages. These glacial lowstands would again downcut existing coastal deposits, increase sediment pumping to the deep sea, and bypass the deltas to provide direct input to the deep sea fans. This pattern of rapid turbidite deposition during glacial lowstands and low hemipelagic accumulation during interglacial highstands has been widely observed in the Indus Fan[24,33] and Amazon Fan.[34]

6.2.5. Uplift, Monsoonal Runoff, and Terrigenous Flux

Although the Ganges–Brahmaputra source of the terrigenous sediments is identical for the Ninety East Ridge and Bengal Fan sites, the physiographic location of the sites and depositional processes may produce different temporal records of terrigenous flux. Deposition at Site 758 is characterized by relatively continuous, fine-grained suspended terrigenous material, whereas deposition at Sites 717–719 is characterized by episodic, turbidite bottom transport of coarser terrigenous material. Because the two major controls on terrigenous flux are the uplift-related erosional supply and eustatic sea level changes, we can use the sea level envelope (Fig. 15b) to help identify the trends that may reflect the uplift of the Tibet–Himalayan complex.

The record at Site 758 shows high carbonate and low accumulation rates from 15 million to about 9 million years ago and increasing terrigenous flux starting 9–7 million years ago. The trend of decreasing carbonate (Fig. 16a), and increasing sedimentation rate (Fig. 15a) and terrigenous flux (Fig. 16c) occurs during an interval of rising sea-level highstands. Hence, this trend is probably related to the uplift-erosional history rather than to sea level change. This pattern is broadly consistent with the other indicators of monsoon intensification and has the general timing of the Molnar uplift model (Figs. 13 and 14).

The Bengal Fan sediments show a more complex history. The sedimentation rates are low from 15 to 11 million years ago during an interval of falling sea-level highstands. The interval of higher rates (11 to about 7–5 million years ago) coincides with higher rates at Site 758 and a period of relatively constant sea level. However, the lower rates from about 7–5 million to 1 million years ago do not coincide with Site 758 trends but occur largely during a regime of rising sea level. Similarly, the rapid increase in rates 1 million years ago has no counterpart at Site 758 but does coincide with higher variability of sea level. A tentative interpretation of these Bengal Fan trends is that the high accumulation interval (11 to 7–5 million years ago), which coincides with increased rates in Site 758 and relatively constant sea level, is an increase in uplift-related terrigenous flux. However, the low rates that occur during rising sea level and the high rates of the last 1 million years are probably related to sea level control of sedimentation. Hence the distribution of facies (i.e., highly weathered versus less weathered) in the Bengal Fan sediments is interpreted to reflect sea-level control rather than changes in tectonic uplift-related sediment supply.

The common trends between the Bengal Fan and Ninety East Ridge sites point to low terrigenous fluxes from 15 to about 11–10 million years ago and then an interval of high fluxes (11–10 million to 7–5 million years ago) (Figs. 15 and 16). These trends seem unlikely to be dominated by sea-level changes and therefore should be interpreted as uplift and possibly monsoon-related fluxes. If correct, this timing is still consistent with the previous studies[5,16,23,25–27] that proposed a late Miocene (10–7 million years ago) enhancement of uplift and the uplift-related climatic responses. Such a scenario is most like our Molnar uplift model (Figs. 13 and 14) although the exact timing of the uplift is not highly constrained. However, these sedimentation rate data suggest that an event around 10 million years ago is likely to be tectonic in origin. Further studies of both fan and ridge sites and their synthesis should lead to better temporal constraints and a better understanding of how uplift and increased monsoonal runoff is related to sediment flux and weathering.

7. SUMMARY

Our review of terrigenous sedimentation rates on both the Bengal Fan and Ninety East Ridge shows that they are most consistent with our ESC runoff simulations (Fig. 14) that use the Molnar uplift model (Figs. 13 and 14) or variations of it. The terrigenous flux data do not seem to support the gradual uplift model or the recent uplift model. Although some records hint at the amplified variability indicated by the GCM experiments, most of the records do not have sufficient resolution to test the amplification hypothesis.

REFERENCES

1. Manabe, S., and Terpstra, T. B. (1974). *J. Atmos. Sci.* **31**, p. 3.
2. Hahn, D. G., and Manabe, S. (1975). *J. Atmos. Sci.* **32**, p. 1515.
3. Kutzbach, J. E., Guetter, P. J., Ruddiman, W. F., and Prell, W. L. (1989). *J. Geophys. Res* **94**, p. 18,393.
4. Ruddiman, W. F., and Kutzbach, J. E. (1989). *J. Geophys. Res.* **93**, p. 18,409.
5. Prell, W. L., and Kutzbach, J. E. (1992). *Nature* **360**, p. 647.
6. Kutzbach, J. E., Prell, W. L., and Ruddiman, W. F. (1993). *J. Geol.* **101**, p. 177.
7. DeMenocal, P. B., and Rind, D. (1993). *J. Geophys. Res.* **98**, p. 7265.
8. Dong, B., Valdes, P. J., and Hall, N. M. J. (1996). *Paleoclimates: Data and Modelling* **1**, p. 203.
9. Kutzbach, J. E. (1981). *Science* **214**, p. 59.
10. Kutzbach, J. E., and Guetter, P. J. (1986). *J. Atmos. Sci.* **43**, p. 1726.

11. Prell, W. L., and Kutzbach, J. E. (1987). *J. Geophys. Res.* **92**, p. 8411.
12. Cohmap Members (1988). *Science* **241**, p. 1043.
13. Wright, Jr., H. E., Kutzbach, J. E., Webb, III, T., Ruddiman, W. F., Street-Perrott, F. A., and Bartlein, P. J. (1993). *Global Climates Since the Last Glacial Maximum*, University of Minnesota Press, Minneapolis.
14. Prell, W. L. (1983). In: *Climatic Processes and Climate Sensitivity.* Geophysics Monograph Series Vol. 29 (J. E. Hansen and T. Takahashi, eds.), pp. 48–57. American Geophysical Union, Maurice Ewing Series, Washington, D.C.
15. Prell, W. L. (1984). In: *Milankovitch and Climate, Part 1* (A. Berger, J. Imbrie, J. Hays, G. Kukla, and B. Saltzman, eds.), pp. 349–366, Reidel, Hingham, MA.
16. Prell, W. L., Murray, D. W., Clemens, S. C., and Anderson, D. M. (1992). In: *Synthesis of Results from Scientific Drilling in the Indian Ocean*, Geophysical Monograph Vol. 70, (R. A. Duncan, D. K. Rea, R. B. Kidd, U. von Rad, and J. K. Weissel, eds.), pp. 447–469. American Geophysical Union, Washington, D.C.
17. Pitcher, E. J., Malone, R. C., Ramanathan, V., Blackmon, M. L., and Bourke, W. (1983). *J. Atmos. Sci.* **40**, p. 580.
18. Berger, A. L. (1978). *Quat. Res.* **9**, p. 139.
19. Berger, A. L., and Loutre, M. F. (1991). *Quat. Sci. Rev.* **10**, p. 297.
20. Hastenrath, S. (1985). *Climate and Circulation of the Tropics.* Reidel, Boston.
21. Fein, J. S., and Stephens, P. L. (1987). *Monsoons*, John Wiley, New York.
22. Coe, M. T. (1995). *J. Clim.* **8**, p. 535.
23. Molnar, P., England, P., and Martinod, J. (1993). *Rev. Geophys.* **31**, p. 357.
24. Prell, W. L., Niitsuma, N., *et al.* (1989). *Proc. ODP, Init. Repts. Vol. 155*, Ocean Drilling Program, College Station, Texas.
25. Quade, J., Cerling, T. E., and Bowman, J. R. (1989). *Nature* **342**, p. 163.
26. Cochran, J. R., and Stow, D. A. V., *et al.* (1989). *Proc. ODP, Init. Repts. Vol. 116*, Ocean Drilling Program, College Station, Texas.
27. Rea, D. K. (1992). In: *Synthesis of Results from Scientific Drilling in the Indian Ocean*, Geophysical Monograph, Vol. 70, (R. A. Duncan, D. K. Rea, R. B. Kidd, U. von Rad, and J. K. Weissel, eds.), pp. 387–402. American Geophysical Union, Washington, D.C.
28. Hovan, S. A., and Rea, D. K. (1992). *Paleoceanography* **7**, p. 833.
29. Farrell, J. W., and Janecek, T. R. (1991). In: *Proc. ODP, Sci. Results, Vol. 121*, (J. Peirce and J. Weissel, eds.), pp. 297–355. Ocean Drilling Program, College Station, Texas.
30. Cochran, J. R. (1990). In: *Proc. ODP, Sci. Results, Vol. 116*, (J. R. Cochran and D. A. V. Stow, eds.), pp. 397–414. Ocean Drilling Program, College Station, Texas.
31. Burbank, D. W., Derry, L. A., and France-Lanord, D. (1993). *Nature* **364**, p. 48.
32. Stow, D. A.V., Amano, K., Balson, P. S., Brass, G. W., Corrigan, J., Raman, C. V., Tiercelin, J.-J., Townsend, M., and Wijayananda, N. P. (1990). In: *Proc. ODP, Sci. Results., Vol. 116*, (J. R. Cochran and D. A. V. Stow, eds.), pp. 377–396. Ocean Drilling Program, College Station, Texas.
33. Kolla, V., and Coumes, F. (1987). *AAPG Bull.* **71**, p. 650.
34. Flood, R. D., Piper, D. J. W., Klaus, A. *et al.* (1995). *Proc. ODP, Init. Repts., Vol. 117*, Ocean Drilling Program, College Station, Texas.
35. Peirce, J., and Weissel, J. (1989). *Proc. ODP, Init. Repts, Vol. 121*, Ocean Drilling Program, College Station, Texas.
36. Gartner, S. (1990). In: *Proc. ODP, Sci. Results, Vol.116* (J. R. Cochran and D. A. V. Stow, eds.), pp. 165–187. Ocean Drilling Program, College Station, Texas.
37. Prentice, M. L., and Matthews, R. K. (1991). *J. Geophys. Res.* **96**, p. 6811.
38. Haq, B. U., Hardenbol, J., and Vail, P. R. (1987). *Science* **235**, p. 1156.

Testing the Climatic Effects of Orography and CO₂ with General Circulation and Biome Models

William F. Ruddiman, John E. Kutzbach, and I. Colin Prentice

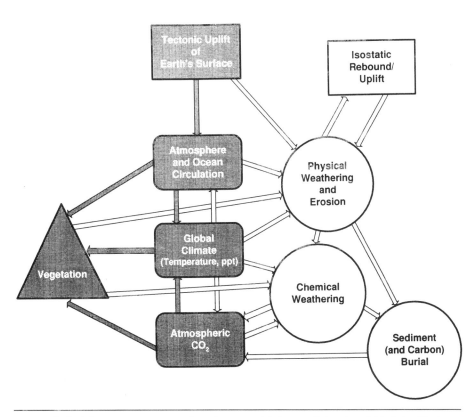

William F. Ruddiman • Department of Environmental Sciences, University of Virginia, Charlottesville, Virginia 22903.   *John E. Kutzbach* • Center for Climatic Research, University of Wisconsin-Madison, Madison, Wisconsin 53706.   *I. Colin Prentice* • School of Ecology, Lund University, 223 62 Lund, Sweden.

Tectonic Uplift and Climate Change, edited by William F. Ruddiman. Plenum Press, New York, 1997.

1. INTRODUCTION

Uplift of large-scale plateau orography has been proposed as a forcing function of long-term climate, both because it can alter the large-scale circulation of the atmosphere,[1-2] and because it may influence chemical weathering and thus atmospheric CO_2 levels.[3] These hypotheses suggest that much of the major climatic change of the last 40 million years could result from uplift in southern Asia and elsewhere.

The proposed physical effects of uplift on atmospheric circulation involve mainly enhancement of the amplitude of meanders in the midlatitude westerlies and of seasonally reversing monsoonal circulations at lower midlatitudes. These changes cause a greater geographic differentiation of wetter and drier regions at lower–middle latitudes, and, in the case of monsoons, increased seasonality of precipitation. The inferred increase in chemical weathering results largely from heavier summer monsoon rains on the steep slopes of the uplifted areas. This may reduce atmospheric CO_2 and cause global cooling.

One way to assess the importance of the hypothesized effects of uplift on climate is to run sensitivity experiments with general circulation models (GCMs) and compare the model output with geologic evidence. Ruddiman and Kutzbach[1] ran a series of GCM experiments with varying orography and compared the climate data output against evidence available from the paleobotanical literature. The model and data agreed fairly well on the long-term evolution of moisture–balance trends at middle latitudes, but the model results did not reproduce enough high-latitude cooling to match paleobotanical (and other) evidence from the geologic record. This mismatch was attributed to the fact that atmospheric CO_2 was not altered in the model runs.

This initial data-model vegetation comparison was qualitative and somewhat crude. Sloan[4,5] subsequently reported a quantitative comparison of GCM output and vegetation change in western North America since Eocene time for several hypothesized climatic forcings. The PRISM Group[6,7] used evidence of mid-Pliocene vegetation types as boundary-condition input for GCM simulations.

Prentice *et al.*[8] developed a biome model that quantitatively links the distribution of functional vegetation types to limiting climatic variables. Biomes arise in this model as combinations of potentially dominant plant functional types. Using output from GCM experiments as input to the biome model makes it possible to implement data–model comparisons of past vegetation change in a quantitative way and on a global basis.

Here, we first assess the possible effects of uplift on climate with three GCM experiments: one designed to isolate the physical effect of uplift; a second that isolates the effect of CO_2; and a third that combines the effects of uplift and CO_2. All of these experiments are sensitivity tests: only orography and/or CO_2 are altered, while modern land–sea distributions are maintained. We then use the climate data output from the orography–CO_2 experiment to simulate biome changes since the early Miocene 20 million years ago, and we compare these changes to paleobotanical data.

2. REVIEW OF MODELS

2.1. Biome Model

The biome modeling approach of Prentice et al.[8] is based on plant physiological constraints to the distribution of the major plant functional types. These constraints are quantified in terms of bioclimatic variables, principally mean temperature of the coldest month, accumulated temperature above 5°C, and an index of plant-available moisture that integrates the effects of precipitation amount and seasonality and soil water-holding capacity. Based on these constraints to plant growth, reproduction, and survival, the model determines the suite of plant functional types that potentially are present, and then uses simple rules to determine which one or more of these types may be dominant. Biomes are defined as unique combinations of dominant types.

The biome modeling approach builds on earlier work on plant functional relationships[9,10] and it avoids two problems inherent in previous attempts to employ climate–vegetation classification schemes[11,12] for use in the geologic past: (1) the restriction to two climate variables that do not adequately express the functional dependence of plant performance on climate; and (2) the implicit assumption in earlier schemes that past vegetation units contained the same groupings of plants as present vegetation units, allowing biome distributions to be determined solely on the basis of empirical correlations with contemporary climate. In contrast, the physiologically based approach uses a larger array of climate variables, designed to approximate the environment as "sensed" by the plants, and it permits plant types to respond individualistically to climate. This is consistent with trends observed in late Pleistocene vegetation, which repeatedly fragmented into different plant groupings during glacial–interglacial climatic cycles.[13]

The model of Prentice et al.[8] is, however, restricted in its ability to simulate realistically the nature of competition between trees and grasses (treated as a simple dominance relationship), and it does not take into account the physiological effects of changing atmospheric CO_2 concentrations on plants, which include changes in photosynthesis and transpiration rates and altered competitive relationships between plants using the C_3 and C_4 photosynthetic pathways. Here we use the Biome 3 model of Haxeltine and Prentice,[14] which incorporates these processes explicitly. For example, Biome 3 allows discrimination between forests and savannas in the tropics, and allows quantitative assessment of the potential physiological effects of changing CO_2 concentration on the locations of these and other boundaries.

Biome models can be directly linked to climatic-variable output from GCMs. For example, Prentice et al.[15] used GCM-generated climatic output to derive anomaly values (differences between last glacial maximum simulations and the present), which were then used to modify the present-day climatology and simulate global biome distributions 20,000 years ago at the last glacial maximum, using the model of Prentice et al.[8] We use the same approach here with the Biome 3 model.

2.2. *General Circulation Model (CCM1)*

For the GCM portion of our experiments, we use CCM1 (version 1) of the Community Climate Model of the National Center for Atmospheric Research, described in Randel and Williamson.[16] It incorporates atmospheric dynamics (equations of fluid motion) and includes radiative–convective processes, condensation, precipitation, and evaporation. Surface temperature, soil moisture, snow melt, and runoff are all calculated from surface energy and hydrologic budget equations. The version of CCM1 employed here uses modifications introduced by Covey and Thompson,[17] including a thermodynamic mixed-layer (50 m) "slab" ocean for calculating ocean temperature and the extent and thickness of sea ice. Ocean currents are not simulated explicitly; instead, there is a prescribed oceanic poleward transport of heat that is longitudinally uniform and that exports heat from the tropical ocean and releases it in middle and high latitudes. The model has 12 vertical levels in the atmosphere and one in the ocean. Horizontal fields of temperature, pressure, and moisture are resolved spectrally to wave number 15, equivalent to a grid of 4.4° latitude by 7.5° longitude. The model simulates the full seasonal cycle with a time-step of 30 min. Cloud cover, soil moisture, and snow cover are computed interactively. Years 15–20 of a 20-year model integration are used, to allow the slab-ocean and sea-ice models to equilibrate.

Overall, the CCM1 control case realistically simulates many of the large-scale patterns of atmospheric circulation, but with several shortcomings.[16] The model troposphere is colder than observed values by several °C. Both midlatitude westerlies and subtropical easterlies are weaker than in the modern atmosphere. Transient disturbances, particularly midlatitude storms, are also weaker in the model than the observations. Model-simulated temperature and moisture boundaries between sharply contrasting regions are not as abrupt as in the modern world because of the low spatial resolution of the model. This results in topographic smoothing that flattens high mountain peaks and fills in valleys, reducing orography to broad, rounded domes. This kind of smoothing is inherent in even the highest-resolution GCM models currently available. Modeled sea-ice limits in the Northern Hemisphere are considerably more extensive than observed, making the Northern Hemisphere continents cooler at high latitudes.

3. EXPERIMENTAL DESIGN

We describe here three aspects of the experimental design: (1) the method by which the CCM1 model output and the biome model are linked; (2) the altered boundary conditions in the CCM1 experiments; and (3) the use of paleobotanical data to test results obtained from the biome model simulations.

3.1. *Combined Use of Biome and CCM1 Models*

Present-day biome distributions[8] are based on modern observations. Boundaries to the present-day distributions reflect limits imposed by environmental variables in the modern climate.

TABLE 1.   CCM1 and Biome Model Simulations and Experiments

CCM1 simulation	CCM1 experiment	Biome simulation
Full orography		
	Orography Experiment	°
	(M−NM)	
No orography (NM)		
Modern CO_2 ($1 \times CO_2$)		
	CO_2 Experiment	
	($1 \times CO_2 - 2 \times CO_2$)	
Double CO_2 ($2 \times CO_2$)		
Modern control		[Modern biomes][a]
(M/$1 \times CO_2$)		
	Orography/CO_2 Experiment	
	(M/$1 \times CO_2$ − LM/$2 \times CO_2$)	
Early miocene		Early Miocene
(LM/$2 \times CO_2$)		Biomes

[a] Modern biomes based on modern vegetation.

In contrast, the paleobiome distributions reported here are based on output from the CCM1 (Table 1). The biome model is driven by using the differences in climatic variables (temperature, precipitation, etc.) derived from the control-case CCM1 run versus the sensitivity-test simulation. These differences are added to (or subtracted from) the modern observed climate data, and these revised climate variables then determine the revised biome distributions. This approach is commonly used in modeling to try to minimize shortcomings in the control case, because model biases are expected to repeat in both the control-case and sensitivity-test simulations and thus largely cancel out in the differences between simulations.

One inadvertent consequence of this approach is that much of the "texture" of the modern biomes carries over into the paleobiome patterns. This occurs for two reasons: (1) modern climate data contain high-resolution information that fixes the modern biome distributions, and (2) all GCM output data (for both the control-case and the sensitivity-test simulations) are smoothed as noted above. In effect, we are thus differencing low-resolution (smoothed) changes in climate variables against a baseline climate-variable data set that has high-resolution topographic texture. As a result, the high resolution produced by modern topographic relief carries over into the patterns of altered biome distributions. For this reason, we restrict our attention in this paper mainly to biome changes affecting large regions away from high topography.

3.2. Boundary Conditions for Sensitivity Tests

The modern control-base simulation was used as the standard of reference for the experiments. For the control run, the continents and oceans are in their modern locations, there is full plateau and mountain orography (although smoothed by the coarse grid-box resolution and the spectral representation), and atmospheric pCO_2 values are at mid-20th-century levels of 330 ppm (above the preindustrial value of 270–280 ppm). We use the control case as a standard against which to calculate changes in climate variables caused by the altered boundary conditions in the sensitivity-test simulations (Table 1). Slightly different climatologies are used here for the control-case runs referred to as M and $M/1 \times CO_2$, because these were run with slightly different versions of the CCM1 model. The differences between the two control experiments are small compared to the differences produced by the imposed changes in orography and CO_2.

3.2.1. Orography Experiment (M–NM)

The purpose of the first experiment was to define the maximum possible effects of orography on climate in a world otherwise having today's geographic configuration. We ran a simulation called NM with all orography lowered to sea level. We will refer to the difference between this NM simulation and the full-orography (M) simulation as the "orography experiment." Other geographic variables and CO_2 levels were left identical to the control case. These boundary conditions repeat the standard "mountain/no-mountain" simulations summarized in Chapter 1. This experiment provides some insight into the impact of widespread Cenozoic tectonic uplift[18,19] although it overstates the impact of uplift in some regions and does not take account of lateral plate motions or sea-level changes.

3.2.2. CO_2 Experiment ($1 \times CO_2$–$2 \times CO_2$)

Here, we utilize previous simulations by Ogelsby and Saltzman[20] using the CCM1 model to isolate the impact of CO_2 on climate. These simulations spanned a range of CO_2 values from 100–1000 ppm, including modern (330 ppm) and twice-modern (660 ppm) levels. We will refer to the difference between the modern ($1 \times CO_2$) and the twice-modern ($2 \times CO_2$) simulations as the "CO_2 experiment."

3.2.3. Orography–CO_2 Experiment ($M/1 \times CO_2$–$LM/2 \times CO_2$)

The purpose of this experiment is to evaluate the combined climatic effects of plausible changes in Northern Hemisphere orography and CO_2 since the early Miocene (20 million years ago). Because orography was selectively reduced to lower (LM) levels and atmospheric CO_2 was doubled, we refer to the early Miocene simulation as "$LM/2 \times CO_2$." We call the difference between it and the modern control case ($M/1 \times CO_2$) the "orography–CO_2 experiment" ($M/1 \times CO_2$–$LM/2 \times CO_2$).

Northern Hemisphere elevations were reduced on a region-by-region basis, although estimating paleoelevations is difficult.[21] Orography in Asia was set at one-quarter of that today, consistent with the lack of terrigenous sediments on the Indus Fan until the early Miocene,[22] the lack of monsoonal upwelling indicators off the Somali coast until later in the Miocene,[23] and conclusions reached in

Chapters 2, 8, and 12. African orography was also set at one-quarter of modern elevations, because construction of the high-standing volcanic edifice of East Africa began around 30–25 million years and has continued into recent times[24] (Chapter 4).

North America was placed at three-quarters of its modern elevation. Molnar and England[21] argued that much of western North America has been near or above modern elevations since early in the Cenozoic when convergent-margin crustal thickening and mountain building stopped.[25] Moderate-scale (1 km) uplift or magmatic construction appears to have occurred owing to mantle-plume "hotspot" activity in the northwestern United States during the last 17 million years,[26] and possibly in the Sierra Range during the last 10 million years.[27] The age of uplift of the Colorado Plateau is not known and may have occurred as early as the late Cretaceous[25] (Chapter 3) or as late as the late Cenozoic.[28]

Ice-sheet orography on Greenland was eliminated, consistent with oceanic evidence that ice did not form there until after 7 million years ago.[29] The Greenland ice sheet forms a mass of high topography comparable to many mountain ranges, and its presence in this simulation would have unrealistically perturbed Northern Hemisphere circulation trends.

Elevations in the Southern Hemisphere were left at modern levels for several reasons, including the poor performance of all GCMs in reproducing modern circum-Antarctic westerly flow, and the extreme model smoothing of youthful topography in the Central Andes[30] (Chapter 3). The region of moderately high (1 km) terrain in southern Africa largely dates from the Cretaceous breakup of Panaea.[31]

Early Miocene CO$_2$ values are also only partially constrained by geologic evidence. Analyses of δ^{13}C of total marine organic carbon[32] and of specific organic carbon biomarkers[33] have been used to estimate past changes in CO$_2$, but recent evidence suggests that CO$_2$ estimates derived from marine organic carbon may be overridden by effects of phytoplankton growth rates.[34]

Cerling *et al.*[35] found δ^{13}C evidence from late Miocene soils and mammal teeth of a simultaneous vegetation shift from the C$_3$ to C$_4$ metabolic–photosynthetic pathway around 7 million years ago in Asia, Africa, and North and South America (Chapter 13). This widespread change indicates a global-scale explanation, and Cerling *et al.*[35] invoked a late Miocene decline in atmospheric pCO$_2$ values through the 500-ppm-threshold region that determines C$_3$ versus C$_4$ pathway utilization in greenhouse experiments. This evidence is broadly consistent with our choice of twice the modern pCO$_2$ values earlier in the Miocene around 20 million years ago, although simple linear extrapolation back in time from 7 million years ago would suggest a significantly higher early Miocene CO$_2$ level.

3.3. Paleobotanical Data and Biome Simulations

The GCM experiments run here are designed to test uplift and CO$_2$ effects during the middle and late Cenozoic, an interval for which no published paleobotanical database exists. Scott Wing of the Department of Paleobiology in the Smithsonian Institute allowed us to use an in-preparation database of some 125 sites, mainly from North America, and we supplemented this compilation with other published data and summary articles.[36,37] This compilation falls far short of the kind of database actually needed for meaningful large-scale data–model comparisons.

We restricted the data–model comparisons to regions meeting the following criteria: (1) GCM-simulated climatic changes in these regions were significant relative to model variability, broad in extent, and caused by well-understood physical processes linked to changes in orography and/or CO_2; (2) the simulated biome changes were also broad in extent and clearly linked to the climatic changes. Given the sparse paleobotanical data, only a very broad-scale evaluation of the biome simulation was possible.

4. GCM RESULTS: SIMULATED CLIMATE CHANGES

In this section, we summarize results from the three CCM1 experiments exploring both the separate and combined climatic effects of orographic uplift and declining CO_2. We portray results from all experiments as differences between the modern control case (full orography and/or modern CO_2) minus the sensitivity-test experiment (lower or no orography and/or higher CO_2). This sense of difference is used because it follows "time's arrow," thus matching the geologic evolution from earlier Cenozoic to modern conditions.

Although the CCM1 simulates changes in climate variables across the globe, we focus on two changes: (1) cooling at high and middle latitudes of the Northern Hemisphere, and (2) intensified moisture contrasts in the midlatitudes. These changes are broad in areal extent, statistically significant, and caused by clearly identifiable physical processes.

4.1. Temperature Changes

Our main findings here are as follows: CO_2 is more important than orography in cooling circum-Arctic Asia and North America, orography is more important than CO_2 in cooling uplifted plateaus, and both are significant in cooling late Cenozoic Northern Hemisphere land masses.

The largest cooling effects due to plateau orography are concentrated over and around the uplifted areas. Compared to an earlier CCM0 orography experiment[38] lacking interactive snow cover and sea ice and thus albedo-temperature feedback, the CCM1 model significantly increases the estimated cooling owing to orography (Figs. 1a, 2a, and 3a). Estimated seasonal and annual temperatures decrease by 16–24°C over Tibet and by 8°C over the high terrain of the American Southwest, with similar downwind temperature decreases in winter owing to dynamical effects associated with strong winter westerlies. In addition, winter snow cover moves southward over the Tibetan Plateau.

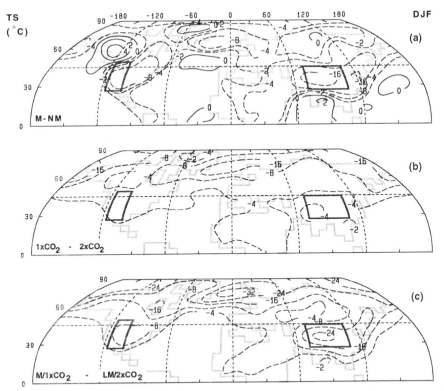

FIGURE 1. Changes in simulated Northern Hemisphere winter temperature for three experiments described in the text: (a) orography experiment, showing changes owing to uplift; (b) CO$_2$ experiment, showing changes owing to falling CO$_2$; and (c) orography–CO$_2$ experiment, showing changes owing to the combined effects of plausible changes in uplift and CO$_2$.

Temperatures in high northern latitudes change modestly in the orography experiment. Circum-Arctic winter and summer temperatures over land generally cool by 2°–4°C. The largest summer cooling (>8°C) is over Greenland, owing to formation of the ice sheet. Winter temperature changes range from a 4°–6°C cooling over western Europe (owing to increased sea ice in the North Atlantic) to a 2°–4°C warming over Alaska (owing to greater advection of warm air from the south).

The CO$_2$ experiment (Figs. 1b, 2b, and 3b) simulates very great winter cooling at high latitudes near the seasonal sea-ice limits. North of 50°N, cooling of the land surface ranges up to 16°C in winter, with a similar cooling over the Greenland–Norwegian Sea and in the Bering Sea (Fig. 1b). This cooling is linked to increased sea ice around the Arctic margins. The amount of winter cooling in Fig. 1b is comparable to that simulated in a sensitivity experiment designed to isolate climatic effects of changes in Arctic sea ice.[39] The cooling occurs because sea-ice cover allows overlying air temperatures to cool far below 0°C, whereas

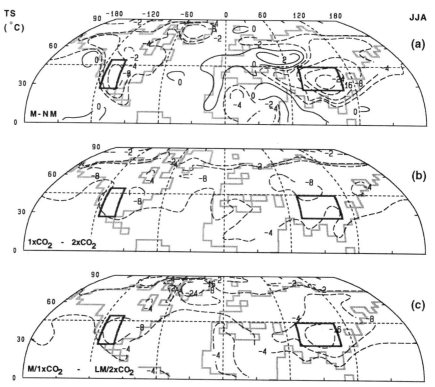

FIGURE 2. Changes in simulated Northern Hemisphere summer temperature for three experiments described in the text: (a) orography experiment, showing changes owing to uplift; (b) CO_2 experiment, showing changes owing to falling CO_2; and (c) orography–CO_2 experiment, showing changes owing to the combined effects of plausible changes in uplift and CO_2.

open water keeps winter air temperatures near $-1.8°C$. The winter cooling of the Atlantic also affects maritime northwest Europe. The high-latitude cooling in summer in the CO_2 experiment (Fig. 2b) is smaller, generally $2°–4°C$, because air temperatures over melting sea ice and open water are similar in summer (slightly above freezing). Significant summer cooling occurs in the interiors of the northern continents, particularly Asia (Fig. 2b).

In the combined orography–CO_2 experiment (Figs. 1c, 2c, and 3c), large decreases in estimated temperature occur at high and middle latitudes. CO_2 effects dominate at higher latitudes, especially in winter in the circum-Arctic region. Cooling over the broad area of uplifted terrain in Asia exceeds $16°C$ in both seasons, with most of it due to uplift/lapse-rate effects. Cooling over the region of lesser uplift in the American West is about $8°C$, with CO_2 accounting for a significant fraction of his change.

Simulated Northern Hemisphere zonal mean changes in temperature over land (winter, summer, and annual) are shown in Fig. 4, with latitude plotted as a

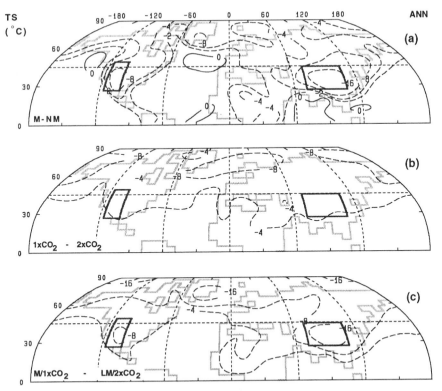

FIGURE 3. Changes in simulated Northern Hemisphere annual temperature for three experiments described in the text: (a) orography experiment, showing changes owing to uplift; (b) CO$_2$ experiment, showing changes owing to falling CO$_2$; and (c) orography–CO$_2$ experiment, showing changes owing to the combined effects of plausible changes in uplift and CO$_2$.

cosine function that compresses higher latitudes into the reduced land fraction of the Earth that they actually encompass. For the orography experiment (Fig. 4a), cooling associated with lapse-rate effects on uplifted terrain at middle latitudes strongly influences zonal mean land temperatures. The strongest zonal mean cooling (12°C in winter, 10°C in summer) occurs at the latitudes of the high plateau terrain (30°–40°N).

In the CO$_2$ sensitivity experiment (Fig. 4b), cooling associated with falling CO$_2$ levels is strongest at higher latitudes ($>50°$N), especially in winter. This reflects the impact of sea ice and snow albedo feedback, compared to the smaller CO$_2$-induced cooling at lower latitudes owing to water vapor feedback.

Temperature changes for the combined orography–CO$_2$ experiment (Fig. 4c) are similar to those from the orography experiment at latitudes 0°–45°N, but are larger north of this latitude. The cooling north of 45°N is due mainly to CO$_2$, while the cooling below 45°N is due to both CO$_2$ and uplift, with uplift predominant between 25° and 45°N.

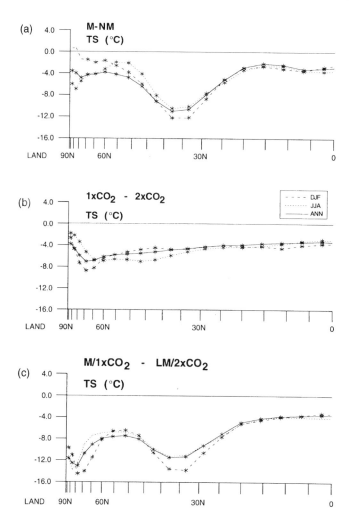

FIGURE 4. Changes in simulated Northern Hemisphere zonal mean temperature over land for experiments mapped in Figs. 1–3: (a) orography experiment; (b) CO_2 experiment; (c) orography–CO_2 experiment. Asterisks show changes significant at the 95% level.

The simulated hemispheric mean coolings over land for the three CCM1 experiments are shown in Table 2. Both CO_2 and orography are potentially very substantial forcing factors of late Cenozoic cooling, accounting for as much as a 7.6°–8.7°C mean cooling over land.

4.2. Precipitation and Surface Wetness Changes

Our findings here are as follows: strongly enhanced summer monsoonal precipitation is focused on the margins of rising plateaus in lower midlatitudes;

TABLE 2.   Mean Cooling over North Hemisphere Land Simulated from Climate Model (CCM1) Experiments

	Orography experiment °C	CO_2 experiment °C	Combined orography and CO_2 experiment °C
Winter	5.1	5.4	8.7
Summer	5.0	5.3	7.6

precipitation declines with uplift in midlatitude continental interiors, particularly in winter; and declining CO_2 causes additional drying at midlatitudes in winter.

The orography experiment confirms a previously identified trend[1,19,23,38]: intensified midlatitude precipitation contrasts in summer in the vicinity of the rising plateaus (Figs. 5a, 6a, 7a, 8a, 9a, and 10a). South and east of the plateaus,

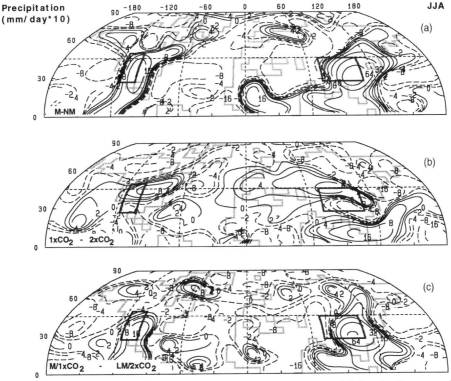

FIGURE 5.   Changes in simulated Northern Hemisphere summer precipitation for three experiments described in the text: (a) orography experiment, showing changes owing to uplift; (b) CO_2 experiment, showing changes owing to falling CO_2; and (c) orography–CO_2 experiment, showing changes owing to the combined effects of plausible changes in uplift and CO_2.

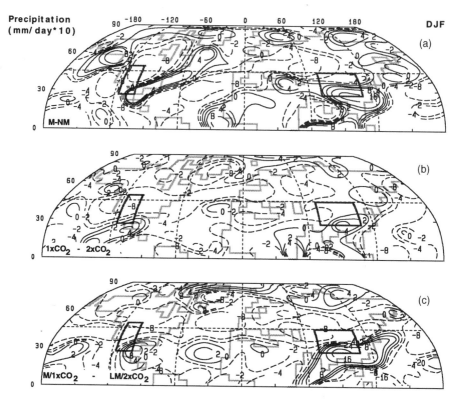

FIGURE 6.   Changes in simulated Northern Hemisphere winter precipitation for three experiments described in the text: (a) orography experiment, showing changes owing to uplift; (b) CO_2 experiment, showing changes owing to falling CO_2; and (c) orography–CO_2 experiment, showing changes owing to the combined effects of plausible changes in uplift and CO_2.

very strong increases in summer and annual precipitation (Figs. 5a and 7a) and surface wetness (Figs. 8a and 10a) occur owing to uplift. This happens because moist ocean air is drawn in toward and intensified low-pressure cell over the plateau in summer. Increased cloud cover also produces cooling that suppresses evaporative heating.

In continental interiors north of the plateaus, the orography experiment simulates decreased annual precipitation and surface wetness (Figs. 7a and 10a), along with increased annual evaporation and decreased $P - E$ (not shown). Several factors contribute to this drying trend. In Asia during summer, it partly reflects rainshadow effects behind the rising plateau (Fig. 5a), as well as outflow of dry air from the Asian interior as part of the circulation around the intensified low-pressure cell. This drying impacts regions of Asia west of Tibet and east of the Caspian Sea, and it extends southeastward into Arabia and northeastern Africa. Summer drying in North America is largely restricted to the west side of the high plateau region, in association with an intensified subtropical high over the North Pacific.

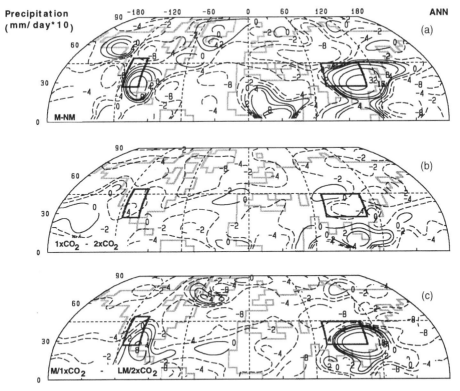

FIGURE 7. Changes in simulated Northern Hemisphere annual precipitation for three experiments described in the text: (a) orography experiment, showing changes owing to uplift; (b) CO$_2$ experiment, showing changes owing to falling CO$_2$; and (c) orography–CO$_2$ experiment, showing changes owing to the combined effects of plausible changes in uplift and CO$_2$.

In the midlatitudes of both Asia and North America, drying is more prevalent during the winter season (Fig. 6a). This reflects factors discussed by Broccoli and Manabe in Chapter 5: rainshadow effects, anchoring of the stationary waves in positions unfavorable to moisture penetration into continental interiors, and strengthening of winter high-pressure cells over land.

In the CO$_2$ experiment, changes in precipitation (Figs. 5b, 6b, and 7b) and surface wetness (Figs. 8b, 9b, and 10b) are generally smaller than those in the orography experiment. In summer, the Asian monsoon weakens as CO$_2$ decreases, confirming an earlier observation that the long-term effect of decreasing CO$_2$ opposes the long-term effect of higher orography.[23] Over high-latitude land, both summer and winter precipitation decrease with falling CO$_2$ (Figs. 5b and 6b), in agreement with earlier $2 \times$ CO$_2$ experiments summarized by Schlesinger and Mitchell.[40]

Precipitation and soil moisture changes in the orography–CO$_2$ experiment (Figs. 5c–10c) generally resemble those from the orography experiment (Figs. 5a–10a), indicating the dominant impact of uplift. To the south and east of the

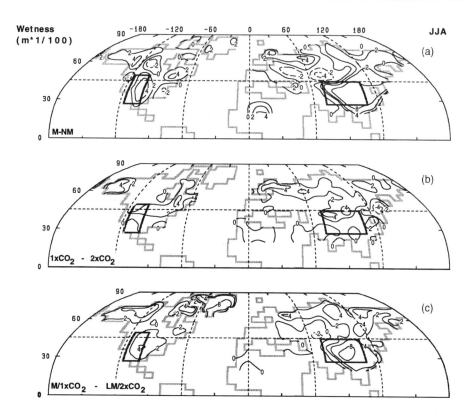

FIGURE 8. Changes in simulated Northern Hemisphere summer soil moisture for three experiments described in the text: (a) orography experiment, showing changes owing to uplift; (b) CO_2 experiment, showing changes owing to falling CO_2; and (c) orography–CO_2 experiment, showing changes owing to the combined effects of plausible changes in uplift and CO_2.

plateaus, summer and annual precipitation and surface wetness increase owing to the strengthened monsoons (Figs. 5c, 7c, 8c, and 10c), with the changes in Asia far larger than those in North America. The summer drying trend north and west of Tibet is still evident (Figs. 5c and 8c), but the reduced summer precipitation associated with the stronger subtropical high along the west coast of North America has shifted offshore (Fig. 5c). In winter, a broad, nearly zonal band of reduced precipitation spans the entire Northern hemisphere at latitudes $35°–65°N$ (Fig. 6c), with a somewhat weaker version evident in the annual average (Fig. 7c). Surface wetness is significantly decreased in both continental interiors in winter (Fig. 9c) and in the annual average (Fig. 10c). This winter drying trend reflects factors noted above.

Simulated zonal mean changes in Northern Hemisphere precipitation over land are shown in Fig. 11. The main feature of the orography experiment (Fig.

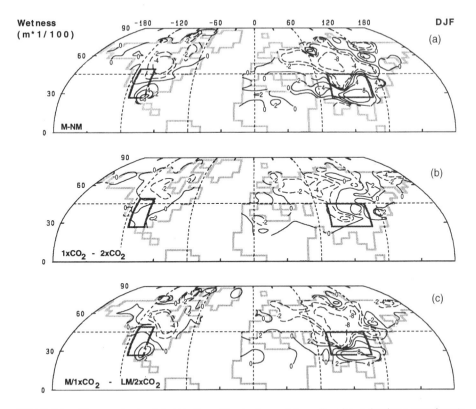

FIGURE 9. Changes in simulated Northern Hemisphere winter soil moisture for three experiments described in the text: (a) orography experiment, showing changes owing to uplift; (b) CO$_2$ experiment, showing changes owing to falling CO$_2$; and (c) orography–CO$_2$ experiment, showing changes owing to the combined effects of plausible changes in uplift and CO$_2$.

11a) is a 1.5 mm/day (or larger) increase in summer precipitation across latitudes (25°–45°N as a result of the strengthened summer monsoon circulation. There is a small (<1 mm/day) but statistically significant decrease in precipitation in the winter and annual averages at latitudes above 45°N.

In the CO$_2$ experiment (Fig. 11b), zonal mean precipitation decreases in both seasons at most latitudes, with most of the statistically significant decreases occurring north of 40°N. Increased precipitation occurs in the northeast portions of the two plateau regions.

Simulated changes in zonal mean precipitation in the combined orography–CO$_2$ experiment (Fig. 11c) resemble those in the orography experiment, slightly modified by the effects of CO$_2$. The band of increased summer monsoon precipitation is somewhat weakened and shifted southward, allowing the band of decreased midlatitude winter precipitation to extend to lower latitudes.

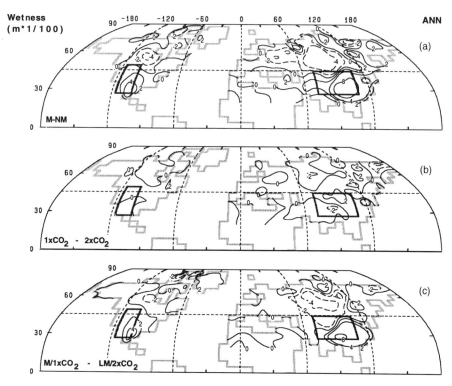

FIGURE 10. Changes in simulated Northern Hemisphere annual soil moisture for three experiments described in the text: (a) orography experiment, showing changes owing to uplift; (b) CO_2 experiment, showing changes owing to falling CO_2; and (c) orography–CO_2 experiment, showing changes owing to the combined effects of plausible changes in uplift and CO_2.

5. BIOME MODEL RESULTS: SIMULATED VEGETATION CHANGES

In this section, we examine the three Northern Hemisphere biome-distribution maps shown in Fig. 12. The modern biome distributions are shown in Fig. 12a. Figures 12b and 12c show two paleobiome reconstructions for the early Miocene derived from the Biome 3 model, both based on the $LM/2 \times CO_2$ simulation with the CCM1 model (Table 1). The simulation shown in Fig. 12c includes the physiological effects of higher CO_2 (fertilization effects and its impact on the choice of C_3 versus C_4 photosynthetic pathways), whereas the simulation in Fig. 12b omits these effects. As before, we structure the discussion to follow the sense of geologic evolution from lower orography and higher CO_2 of the early Miocene to the higher orography and lower CO_2 of the modern world.

We focus on simulated biome changes in three regions: (1) southward biome shifts across high northern latitudes ($>45°N$); (2) development of biomes indicating intensified moisture contrasts in southern Asia, Arabia, and eastern North Africa; and (3) development of drier biomes in the interior of North America and

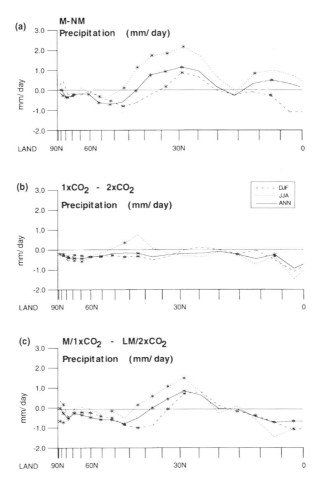

FIGURE 11. Changes in simulated Northerm Hemisphere zonal mean precipitation for experiments mapped in Figs. 5–7: (a) orography experiment; (b) CO$_2$ experiment; (c) orography–CO$_2$ experiment. Asterisks show changes significant at the 95% level.

Asia. For gridbox areas in several of these regions (Fig. 13), we also show selected seasonal temperature and precipitation changes from the CCM1 model output for the combined "orography–CO$_2$ experiment" (Figs. 14 and 15).

5.1. Southward Biome Shifts in High Northern Latitudes

Major southward biome shifts across northern Eurasia and North America at latitudes north of 45°N are evident from comparing the early Miocene biomes (Figs. 12b, c) with the modern biomes (Fig. 12a). The climate modeling simulations discussed in Section 4.1 indicate that decreasing CO$_2$ outweighs changes in orography as the main control on these vegetation changes.

Modern

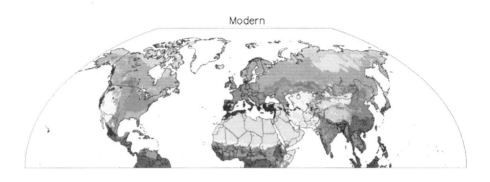

Early Miocene (Lm/2xCO2)

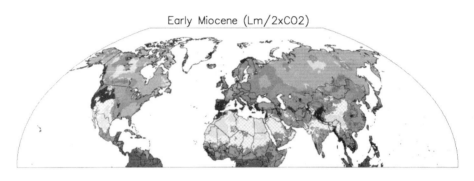

Early Miocene (Lm/2xCO2 plus physiological effects)

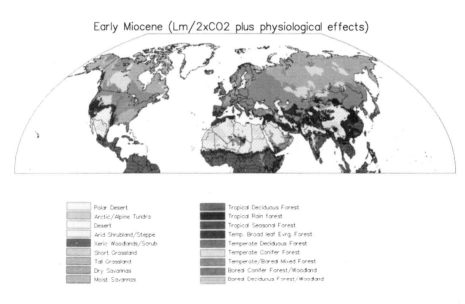

Polar Desert	Tropical Deciduous Forest
Arctic/Alpine Tundra	Tropical Rain forest
Desert	Tropical Seasonal Forest
Arid Shrubland/Steppe	Temp. Broad leaf Evrg. Forest
Xeric Woodlands/Scrub	Temperate Deciduous Forest
Short Grassland	Temperate Conifer Forest
Tall Grassland	Temperate/Boreal Mixed Forest
Dry Savannas	Boreal Conifer Forest/Woodland
Moist Savannas	Boreal Deciduous Forest/Woodland

FIGURE 12. Vegetation biomes for three cases: (top) modern biomes[8]; (middle) Biome 3 simulation of early Miocene vegetation based on climate-variable output from CCM1 simulation run with regionally reduced orography and doubled CO_2 levels (LM/2 × CO_2); (bottom), same as center, but with physiological effects of higher CO_2 levels included. (See color reproduction on facing page.)

Modern

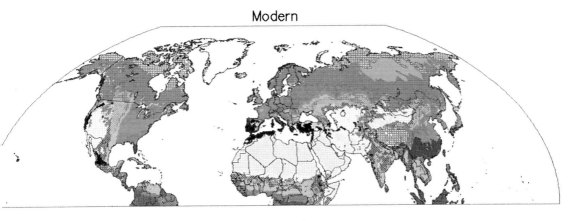

Early Miocene (Lm/2xCO2)

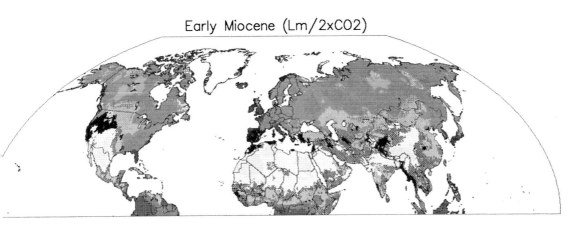

Early Miocene (Lm/2xCO2 plus physiological effects)

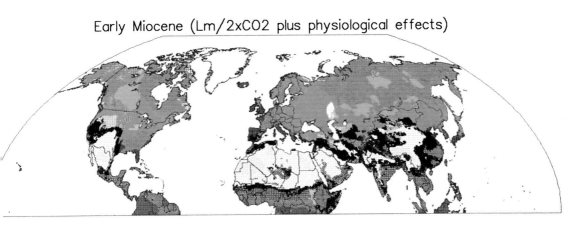

Polar Desert
Arctic/Alpine Tundra
Desert
Arid Shrubland/Steppe
Xeric Woodlands/Scrub
Short Grassland
Tall Grassland
Dry Savannas
Moist Savannas
Tropical Deciduous Forest
Tropical Rain forest
Tropical Seasonal Forest
Temp. Broad leaf Evrg. Forest
Temperate Deciduous Forest
Temperate Conifer Forest
Temperate/Boreal Mixed Forest
Boreal Conifer Forest/Woodland
Boreal Deciduous Forest/Woodland

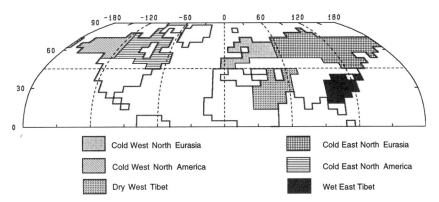

FIGURE 13.   Map of gridbox areas for which seasonal temperature and/or precipitation changes are shown in Figs. 14 and 15.

The following biomes move southward and successively displace each other: tundra, boreal conifer forest/woodland, boreal deciduous forest/woodland, temperate/boreal mixed forest, and temperate deciduous forest. The early Miocene simulations (Figs. 12b, c) have tundra largely confined to the northernmost Arctic archipelago in North America, present only in a thin band along the northernmost margin of central Asia, and nonexistent in northern Europe. By comparison, the modern biome map shows tundra on average some 10° (1000 km) farther south, reaching latitudes 55°–65°N in North America and 60°–70°N in Eurasia.

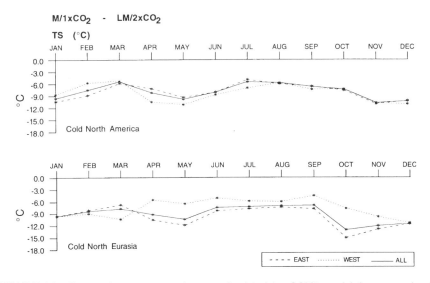

FIGURE 14.   Seasonal temperature changes simulated by CCM1 model for orography–CO$_2$ experiment in four high-latitude regions encompassing gridboxes shown in Fig. 13: (top) North America and (bottom) Eurasia, both subdivided into eastern and western subregions. Asterisks show changes significant at the 95% level.

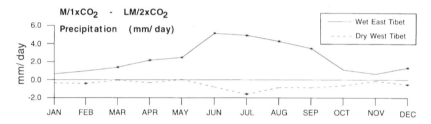

FIGURE 15. Seasonal precipitation changes simulated by CCM1 model for orography–CO_2 experiment in two midlatitude regions of Asia encompassing gridboxes shown in Fig. 13: region east of Tibet, which becomes wetter with uplift, and region west of Tibet, which becomes drier with uplift. Asterisks show changes significant at the 95% level.

In North America, the southward biome shifts are comparable at all longitudes. The magnitude of displacement of successively more temperate biomes diminishes toward middle latitudes, with the southern boundary of the temperate deciduous forest biome moving southward by only 3°–4° of latitude. This decreasing offset underscores the importance of Arctic-centered temperature changes in altering high-latitude biomes. It particularly underscores the importance of winter cooling (Fig. 1c) owing to increased Arctic sea ice.

Although there are also latitudinal shifts in Eurasia, the trends are more complex. Several biomes (e.g., temperate/boreal mixed forest) shift 15–20° of longitude toward the west into eastern Europe. This reflects increased continentality associated with major winter cooling (Fig. 1c) owing to increased sea-ice extent along the northwestern margin of Europe.

Very large southward biome shifts occur along the northeast coast of Asia. The tundra biome, which is absent in the maps based on the early Miocene simulation (Fig. 12b, c), moves more than 10° of latitude onto the continent (Fig. 12a). And the boreal conifer forest/woodland biome shifts southward by 15° of latitude along the east coast of Asia. These changes reflect not just CO_2-induced winter cooling and sea-ice advance in the Arctic, but also the effects of midlatitude uplift of the Tibetan Plateau in strengthening the Siberian High and increasing the winter outflow of cold air toward the southeast.

The simulated seasonal patterns of cooling over high-latitude North America and Asia owing to combined orographic and CO_2 effects are shown in Fig. 14. For much of the circum-Arctic, the simulated cooling averages near 9°C. In North America, both east and west of the Rocky Mountains, the largest seasonal temperature decreases occur in April–May and November–January, with smaller cooling simulated for March and July–October (Fig. 14a). In northern Eurasia, the simulated cooling is larger in Asia (east of the Urals) and reaches maximum values in October–November (Fig. 14b). The simulated cooling in Europe and in Asia west of the Urals is smaller and mainly in November–March.

There is also a biome shift toward colder vegetation on top of the Tibetan Plateau. In effect, the choice of major uplift in this region as a boundary condition inevitably imposes this climatic and vegetation response. The simulated biome shift over the central and northern Tibetan Plateau is from an Early Miocene combination of hot desert, arid shrubland/steppe, xeric woodlands/scrub and dry savanna (Fig. 12b, c) to mainly Alpine tundra today (Fig. 12a). These changes mainly reflect major cooling owing to uplift and lapse-rate effects,[41] along with some cooling from reduced CO$_2$.

5.2. Intensified Moisture Contrasts in Southern Asia

Major changes in midlatitude biomes in the vicinity of the Tibetan Plateau are evident from comparisons of the early Miocene biomes (Figs. 12b, c) against the modern biomes (Fig. 12a). The climate model (CCM1) results discussed in Section 4.2 indicate that uplifted orography is the main cause of the simulated biome changes in this region, by virtue of its impact on precipitation and surface wetness patterns.

East and southeast of Tibet, the early Miocene biomes in Fig. 12b, c are dry-adapted vegetation: arid shrubland/steppe, xeric woodland/scrub, dry and moist savannas, and some seasonal tropical forest. By comparison, the same regions of southeast Asia today are predominantly temperate broadleaf and temperate deciduous forest biomes (Fig. 12a).

In the absence of the Tibetan Plateau or Himalayan Mountains, summer circulation in this region would be dominated by the large-scale Hadley Cell circulation, which promotes subsidence and drying in the subtropics.[1,2] With one-quarter of the plateau orography present in the early Miocene, these drying tendencies produce savannas and xeric woodlands (Figs. 12b, c). With full modern orography (Fig. 12a), a strong low-pressure cell develops over Tibet in summer and draws moist ocean air into southeast Asia. This causes a regional shift from relatively arid biomes to wetter ones. The simulated seasonal precipitation changes from the CCM1 east of Tibet (Fig. 15a) show increased precipitation in all months, with the largest and most significant increases during the summer monsoon months (June–September).

A broad region west of Tibet shows an opposite sense of biome change toward dry-adapted types. This region is bordered on the northwest by the Caspian Sea and extends southward across Iran and Iraq and southwestward into Arabia. Today, this region is largely arid shrubland/steppe (Fig. 12a) southward to the Persian Gulf, with local desert biomes in Iran and Iraq. Farther south, Arabia and eastern North Africa are almost entirely desert today. In contrast, the biome simulations for the early Miocene (Fig. 12b, c) show significantly less arid biome types in these areas. Dry savanna and arid shrubland/steppe biomes are simulated in areas of Arabia and east-central North Africa that are now desert. The broad area from the Persian Gulf north to the Caspian and Ural Seas that is now arid shrubland/steppe (Fig. 12a) appears in the early Miocene simulations mainly as dry savanna (Fig. 12b) and xeric woodlands/steppe or even temperate deciduous forest (Fig. 12c).

This large-scale simulated biome shift toward dry-adapted types in central Asia, the Middle East, Arabia, and northeast Africa in part reflects the uplift of Tibet. The counterclockwise flow of air created by the summer low-pressure cell over Tibet sends dry, hot air from the interior of Asia southwestward over Arabia and northeast Africa. This reverses a more persistent westerly flow prior to uplift that had provided some moisture from the Atlantic and Mediterranean to these regions, thus causing summer drying over a very large region (Figs. 5a,c and 8a,c). The simulated seasonal precipitation from the CCM1 for the region west of Tibet (Fig. 15) shows decreases in precipitation in all months. Although only the February and July changes are significant at the 95% level, all of the seasonal precipitation decreases are in the range of 25–40% below already-low levels.

The shift toward dry-adapted vegetation types in this region is also related to physiological effects associated with the long-term decline of atmospheric CO_2. The early Miocene biome simulation in Figure 12c incorporates additional physiological effects not included in the biome simulation shown in Figure 12b. In general, grass vegetation using the C_4 photosynthetic pathway and adapted to low modern CO_2 levels replaces tree and shrub C_3 vegetation adapted to the higher levels of the early Miocene. Addition of these physiological effects helps to transform the region between the Caspian Sea and Tibet that sustained C_3 woodland vegetation in the early Miocene simulation (Fig. 12c) into the arid shrubland and steppe of the present world (Fig. 12a).

5.3. Drier Biomes in Interior North America and Asia

A third biome shift shown in Fig. 12 is a shift from forest to grassland biomes in the continental interiors of North America and Asia. This change shows up most clearly in the contrast between the modern biomes (Fig. 12a) and the early Miocene simulation that incorporates the physiological effects of changing CO_2 (Fig. 12c). Large areas of temperate deciduous forest give way to tall and short grasslands in the interiors of North America and Asia.

This change is not apparent in a comparison of modern biomes (Fig. 12a) with the early Miocene simulation lacking the physiological effects of CO_2 (Fig. 12b). This is somewhat surprising, given the prominence of the winter-season drying of the midlatitude interiors of Asia and North America in the GCM simulations (Figs. 6c and 9c). This suggests that neither orography nor decreasing CO_2 causes enough drying to alter biomes in the continental interiors (except in the region just discussed in southwest Asia influenced by summer drying). This in turn suggests that physiological effects involved in the change from C_3 to C_4 vegetation exceed climatic effects in determining modern locations of grasslands in the continental interiors. A contributing factor may be the fact that a reduction of precipitation in winter (Fig. 6) is not as critical to vegetation as it would have been during the growing season.

6. COMPARISON OF BIOMES WITH PALEOBOTANICAL DATA

6.1. High Northern Latitudes

The Cenozoic paleobotanical record of high northern latitudes in North America and Eurasia during the Cenozoic[36,37] is fundamentally a progression from the warm-adapted evergreen and warm-temperate vegetation of the Paleocene and Eocene to the tundra, cold deciduous forest, and cool mixed forest of the Pliocene and Pleistocene. Although there are intervals of more abrupt cooling, as well as temporary reversals or pauses in this trend, the overall pattern is one of ongoing cooling. The biome comparisons in Fig. 12 simulate the impact of the last 20 million years of this ongoing cooling, the portion from the relative warmth of the early Miocene to the cold of the present day.

Comparison of the early Miocene biome simulations (Fig. 12b, c) with the modern biomes (Fig. 12a) generally shows boreal conifer forest/woodland being replaced by tundra. In more limited regions to the south, tundra replaces temperate/boreal mixed forest or boreal deciduous forest/woodland. Farther south, temperate/boreal mixed forest is replaced by boreal conifer forest/woodland.

The simulated early Miocene pattern showing boreal conifer forest/woodland ringing the Arctic Ocean agrees with Wolfe[36], who showed mixed coniferous forest in a broad circum-Arctic band for the interval from 22 to 18 million years. Adequately dated paleobotanical sites in the Arctic are, however, much too sparse to provide a sufficient test of the biome simulations. Most well-dated Miocene sites are restricted to western North America at latitudes south of 50°N, outside the region of largest simulated biome changes. Here, we cite the few well-dated high-latitude sites.

The early Miocene Haughton Formation on Devon Island (75°22'N, 89°40' W) in the high Canadian Arctic occupies an impact crater that is 23 million years old. The site today lies well within the present tundra belt. Pollen evidence[42] suggests that early Miocene vegetation in this region was a mixed conifer–hardwood forest, compared to the biome simulations (Figs. 12b, c) of boreal conifer forest/woodland.

The warmth of the early Miocene appears to have lasted through the middle Miocene 14 million years ago. The lower portion of the Kenai Group of Cook Inlet on the south coast of Alaska is 15 million years old. Pollen evidence[43] suggests that the Miocene vegetation in this region was mixed northern hardwood forest, while the biome simulations (Fig. 12b, c) indicate temperate/boreal mixed forest. Wolfe[43] inferred a mid-Miocene (13–12 million years ago) cooling of 9°C in summer and 5°C in winter, with subsequent further winter cooling. This compares with the annual and winter cooling of 8°C in the orography–CO_2 experiment (Figs. 1c, 2c and 3c).

Several sites in Alaska and Siberia with poorly constrained ages appear to have had vegetation similar to that at Devon Island and Cook Inlet,[43] implying a very broad distribution of this early–middle Miocene Arctic vegetation. Pollen in the Nenana coal fields in interior Alaska[44] suggests a mixed northern hardwood forest with coniferous elements, in basic agreement with the biome simulation of temperate/boreal mixed forest (Fig. 12b, c). Cooling since the middle Miocene has lowered mean annual temperature by an amount estimated at 5°–7°C,[44] close to the GCM-simulated cooling of 8°C (Fig. 3c). The very sparse early–middle Miocene paleobotanical data thus match the climate model and biome simulations reasonably well.

Circum-Arctic paleobotanical data are also scarce for subsequent intervals,[37] but they suggest that a substantial portion of the post-early Miocene change in high-latitude vegetation may have occurred late in the Cenozoic, and especially during the last 5 million years. In Siberia, Taiga first appeared north of 65°–70°N about 5 million years ago.[36] On Meighen Island in the high Canadian Arctic (80°N), sediments dated to the interval from 5 to 2.5 million years ago contain abundant pine and birch pollen[37] in an area now located deep in the tundra belt. Tundra appears to have first arrived during the late Pliocene both in the Alaskan coastal plain[45] and in the lowlands of northern Siberia.[46] Arctic tundra thus appears to be mainly Pliocene or later in origin,[36,47,48] and the subpolar ocean has cooled substantially since this time as well.[49]

The large vegetation changes during the last 5 million years have raised the issue of whether or not there has been sufficient CO_2 forcing to explain the observed cooling. The late Miocene C_3–C_4 vegetation shift (Chapter 13) implies a major reduction in CO_2 during the last 7 million years, from 500 ppm or higher to the preindustrial level of 270–280 ppm. This means that a large fraction of the CO_2 difference between the value we used as a boundary condition for the early Miocene simulation and the modern value occurred subsequent to 7 million years ago. A substantial CO_2 reduction during this interval is also at least broadly supported by evidence of widespread late Miocene initiation of mountain glaciation in Alaska,[50] South America,[51] and around the Norwegian Sea.[29]

The GCM experiments reported in Section 4 show that much of the high-latitude Northern Hemisphere cooling is related to sea-ice expansion driven by falling CO_2 levels. We thus infer that the ongoing Cenozoic decrease in CO_2 reached levels sufficient to develop winter sea ice along the Arctic margins of Eurasia and North America and in the Canadian archipelago during the last several million years. This in turn produced a major expansion of tundra and other cold northern biomes. Earlier sea-ice expansion that had not yet reached the coastlines presumably had a lesser impact on climate and vegetation over Northern Hemisphere continents.

Several uncertainties preclude more detailed analyses of data–model agreements and disagreements in the Arctic region. In the first place, the paleobotanical record is still far too patchy for truly definitive comparisons. In addition, there are uncertainties in the climate models. On the one hand, the true sensitivity of the climate system to CO_2 forcing may be lower than that in the CCM1 model,[40] which would lessen the amount of simulated climatic cooling in the last 20 million years. Circum-Arctic sea-ice limits in the modern CCM1 control case are larger than those actually observed, potentially making this model oversensitive to climatic forcing because of excessive ice-albedo feedback. Also, the CCM1 sea-ice model does not allow for leads in the ice or for ocean drift of ice, both of which might have significant effects on sea-ice limits.

On the other hand, the doubled CO_2 level (660 ppm) used as an approximation for values 20 million years ago may be too low, based on the value obtained by linearly extrapolating from the inferred C_3–C_4 threshold value of 500 ppm (or higher) 7 million years ago. Using a higher early Miocene CO_2 value would act to offset model oversensitivity. Further, the CCM1 model lacks feedbacks that may be important at high latitudes. The very large increase in albedo feedback resulting from the change from (dark) boreal forest to (light) tundra[52] can almost double regional temperature responses at high latitudes to climate forcing.[53,54]

Given these caveats, we simply conclude that the basic sense of data–model agreement here is sufficiently good to support the hypothesis that CO_2-driven cooling provides a good first-order explanation of the pattern of late Cenozoic vegetation change at high latitudes. This agreement warrants future comparisons using models that incorporate in a more realistic way more of the key feedbacks of the climate system, as well as attempts to obtain much better paleobotanical coverage.

We omitted from this comparison a discussion of climatic or biome changes directly over the Tibetan Plateau or the high topography of the American West, partly because the elevation changes are prescribed for the experiment (and thus not independently available for verification), and partly because the model smooths the plateau orography. This eliminates critical mountain–valley topographic texture, particularly in the Himalayan region, which is the site for much of the paleobotanical data. It also reduces Himalayan rainshadow effects northward over Tibet. Nevertheless, the simulated biome changes in Fig. 12 over the Tibetan Plateau are in a direction crudely consistent with changes summarized by Mercier et al.[55] for southern Tibet. In western North America, where we assumed little late Cenozoic uplift for the boundary-condition changes, the simulated biome shifts in Fig. 12 are accordingly small.

6.2. Southern Asia

Across a broad region of southwest Asia, Arabia, and northeast Africa, where the GCM simulations indicate summer drying (Fig. 5a, c), there is a similarly large-scale shift toward drier biomes (Fig. 12). Unfortunately, there is almost no paleobotanical data with which to test this simulated trend. Traverse[56] found that forest/steppe pollen ratios in a Black Sea core defined a progressive late Cenozoic change from deciduous forest to steppe vegetation, although this region lies somewhat west of the area of main summer drying and may be influenced by changes in steppe vegetation to the north.

In northeast Africa, Axelrod and Raven[57] inferred that savanna-woodland and thorn scrub occupied portions of what is now the Saharan–Libyan Desert in the late Oligocene (30–25 million years ago).This matches fairly well the biome simulations of a change from early Miocene arid shrubland/steppe or dry savanna (Fig. 12b, c) to the modern hyperarid desert (Fig. 12a). Evidence for trends toward drier vegetation in North Africa much later in the Cenozoic[58,59] are, however, likely to reflect the overprint of orbital-scale changes in Northern Hemisphere ice sheets, and possibly of local tectonics in the East African plateau.

Carbon isotopic evidence of C_3–C_4 vegetation changes also bear on this regional biome shift. Data from Pakistan[60] show an abrupt shift from C_3 to C_4

vegetation 8–6 million years ago. This points to the likelihood of a major expansion of summer-wet C_4 grassland and desert scrub in the region west of Tibet owing to physiological effects rather than to precipitation changes. This agrees with the simulated biome changes from Fig. 12c to Fig. 12a.

In East Africa, $\delta^{13}C$ data suggest that C_4 vegetation replaced C_3 vegetation relatively gradually over the very late Cenozoic,[61] rather than in a relatively abrupt shift some 8–6 million years ago. This may indicate that a combination of climatically induced drying and physiologically effects of decreasing CO_2 have contributed to shifts toward "drier" biomes in southwest Asia, Arabia, and northeast Africa since the early Miocene.

In a broad area east of Tibet, CCM1 results simulate a major intensification of summer monsoonal rainfall owing to plateau uplift (Fig. 5a, c). This results in a simulated shift toward wetter biomes in this region (Fig. 12). Paleobotanical data give at best mixed support to this simulated result.

In Japan at 38°N, broad-leafed evergreen vegetation now occupies an area dominated by broad-leafed deciduous forest in the early Miocene.[62] On Taiwan (30°N), paratropical rain forest now occupies areas in which early Miocene (notophyllous) vegetation indicates somewhat lower moisture (and temperature) levels than today.[63] From this evidence, Wolfe[36] inferred that evergreen vegetation during the Miocene was restricted to lower latitudes than at present in easternmost south Asia. The subsequent northward movement of warm- and wet-adapted biomes is in direct opposition to the general global trend toward cooler, drier vegetation during the late Cenozoic. These shifts are, however, consistent with orographic forcing toward warmer and slightly wetter climates in this region.

Closer to the east side of Tibet, the simulation of a change from drier to wetter biomes does not agree with the paleobotanical data. Relatively wet-adapted vegetation (tropical and subtropical temperate broad-leafed evergreen and deciduous forests) prevailed at numerous sites east of Tibet as far back as the Oligocene, and the only evidence of slightly more dry-adapted subtropical vegetation is from the Eocene.[55,62] The indication is clear that wet-adapted vegetation existed in this region by the Oligocene. The reason for this data–model disagreement is not clear. Part of the explanation may be that mountainous topography capable of trapping orographic precipitation existed in this region by the Oligocene, either inherited from the precollision subduction of Tethys ocean crust under Asia, or created by "tectonic escape" of Asian continental crust to the east during the Eocene and Oligocene collision (Chapter 2).

6.3. Interior North America and Asia

There is good paleobotanical evidence in the continental interior of North America to match the biome simulations of a shift from early Miocene temperate deciduous forest (Fig. 12c) to modern grassland (Fig. 12a). Wolfe[36] mapped hardwood forests in the interor of North America during the early Miocene,

22–18 million years ago. Several studies[64,65] show that grasslands expanded in the latest Cenozoic. Although this has been interpreted as a consequence of rainshadow drying owing to late Cenzoic uplift of western North America,[1,66] evidence for earlier uplift in North America[21,25] argues against this explanation.

An abrupt change from C_3 to C_4 plants occurred on the plains of the western United States between 8 and 6 million years ago[67] (Chapter 13), indicating the spread of C_4 grasses. As noted previously, this appears to be best explained by the decline of atmospheric CO_2 through a critical threshold.[35] This explanation for the development of grassland is supported by the simulated changes in biomes between Fig. 12c (early Miocene simulation with physiological factors included) and Fig. 12a (modern biomes).

Indications of a similar change in the northern interior of Asia are only faintly evident in the biome simulations (Fig. 12). In a small area north and east of the Ural Sea, grass/shrub moves into regions previously occupied by temperate deciduous forest and cool grass forest. These simulated midcontinent biome shifts may reflect a general midlatitude winter drying owing partly to orography (Fig. 6a) and to lower CO_2 (Fig. 6b), but may also be affected by the C_3 to C_4 vegetation shift.

We have not analyzed here the high-elevation westernmost portion of the North American continental interior, because it is within the region in which we have specified orography as a boundary condition. There are major disagreements over paleoelevations in the American West during the Cenozoic,[21,25−28] as well as over the degree of climatic equability in the early Cenozoic based on floral and faunal data[68] versus GCM modeling results.[4,5]

7. SUMMARY

These experiments in large part confirm previous claims that uplift has a first-order impact on late Cenozoic changes in climate and vegetation via changes in orography[1,2] and via changes in chemical weathering and its effect on CO_2 in the atmosphere.[3]

GCM experiments indicate that CO_2 has a major effect on high-latitude temperature (Figs. 1b, 2b, and 3b), and that orography strongly affects a comparably large area in the middle latitudes (Figs. 1a, 2a, and 3a). Plausible changes in both factors during the last 20–15 million years (Figs. 1c, 2c, and 3c) caused a simulated cooling over Northern Hemisphere land of more than $7°–9°C$. These uplift-related climate changes produce simulated southward shifts in high-latitude biomes (Fig. 12) that agree with observed trends from a very sparse paleobotanical database.

As noted above, these simulations have numerous limitations: uncertainties in model sensitivity to CO_2 forcing; past CO_2 values; simplifications in the slab-ocean and sea-ice model; and important feedbacks that are not included, such as vegetation–albedo feedback along the tundra–boreal forest boundary.

TABLE 3.   Fractional Sea-Ice Cover (Percent Annual) for
Two GCM Experiments

Latitude (°N)	$M/1 \times CO_2$	$LM/2 \times CO_2$
86	100	68
82	98	50
78	84	43
73	73	42
69	73	27
64	65	20
60	38	10
56	10	3
51	2	0

Given these limitations, we conclude only that the basic sense of data–model agreement is sufficiently good to support the hypothesis that CO_2 forcing explains the basic pattern of late Cenozoic vegetation change at high latitudes.

The argument has been put forward that CO_2 changes are insufficient to explain Arctic cooling since the middle Pliocene 3 million years ago, and that it is also necessary to call on decreased ocean heat transport.[69,70] This conclusion was largely based on estimates of past CO_2 from $\delta^{13}C$ of marine organic carbon, but it has been shown subsequently that vital effects can invalidate such estimates.[34]

We infer that changes in ocean heat transport may not be necessary to explain cooling of the Arctic. The widespread change from C_3 to C_4 vegetation between 8 and 6 million years[35] indicates a large drop in CO_2 since that time. Table 3 indicates that CO_2 reductions of from twice-modern to modern values would have had a major impact on the southern limit of sea ice (note the changes in the 50% limit of ice cover between experiments). This sea-ice advance appears to have brought winter sea ice to the northern shores of Eurasia and North America and into the Canadian archipelago very late in the Cenozoic, thus promoting the observed spread of tundra during the Pliocene and Pleistocene. In addition, simulations in Chapter 6 with a coupled ocean–atmosphere GCM suggest that northward heat transport in the Atlantic Ocean during the late Cenozoic owing to uplift in southern Asia increased, rather than decreased.

GCM experiments also confirm that orography has had a major effect on moisture balances in Asia, where a pattern of increased contrast east and west of Tibet developed owing to the circulation around an intensified summer monsoon low-pressure cell (Figs. 5–10). There are almost no paleobotanical data available to test the simulated change to drier biomes west of Tibet and southwestward into Arabia and northeast Africa (Fig. 12). The data that do exist suggest that the late Miocene C_3 to C_4 shift was the most prominent large-scale change in this area. The simulated change toward wetter biomes east of Tibet tends to be contradicted by the data, which show wet-adapted vegetation in that area since the Oligocene.

This disagreement may be due to preexisting topography east of Tibet, formed before collision or early in the collision process.

GCM experiments show a significant winter (and annual) drying of the continental interiors of North America and Asia (Figs. 6c, 7c, 10c, and 11c). Although orographically induced aridity has been invoked to explain the development of C$_4$ grasslands in the American plains,[1,66] the biome simulations incorporating the physiological effects of CO$_2$ on vegetation indicate that the C$_3$ to C$_4$ shift owing to declining CO$_2$ was a more important factor, consistent with $\delta^{13}C$ evidence.[67]

Finally, the experiments presented here are sensitivity tests focused only on the effects of orography and CO$_2$. Several well-known late Cenozoic tectonic changes are not included in these experiments, and these could affect climate on regional or larger scales. Several isthmuses have opened or closed (or both) in the last 20 million years, including Panama and both ends of the Mediterranean. These obviously have the capability of changing regional climates by altering the transport of sensible and latent heat in the ocean. Also, effects of plate tectonic motions on latitudinal position of continents are not included in these experiments. The largest such effects in the Cenozoic are in the Southern Hemisphere (e.g., Australia); changes in the Northern Hemisphere are predominantly latitude-parallel and thus not across large temperature–precipitation gradients.

Acknowledgments

We thank: P. Behling, R. Selin, and M. Sykes for preparing the figures, A. Haxeltine for prepublication access to the Biome3 model, S. Wing for sharing his paleobotanical database, and T. Anderson for help in researching additional paleobotanical sources. The research was sponsored by U.S. National Science Foundation grants to the University of Virginia and the University of Wisconsin-Madison. The computer calculations were performed at the National Center for Atmospheric Research, which is sponsored by the U.S. NSF.

REFERENCES

1. Ruddiman, W. F., and Kutzbach, J. E. (1989). *J. Geophys. Res.* **94**, p. 18409.
2. Ruddiman, W. F., and Kutzbach, J. E. (1991). *Sci. Amer.* **264**, p. 66.
3. Raymo, M. E., Ruddiman, W. F., and Froelich, P. N. (1988). *Geology* **16**, p. 649.
4. Sloan, L. C., and Barron, E. J. (1992). *Palaeogeogr., Palaeoclim., Palaeoecol.* **93**, p. 183.
5. Sloan, L. C. (1994). *Geology* **22**, p. 881.
6. Thompson, R. S., and Fleming, R. F. (1996). *Mar. Micropal.* **27**, p. 27.
7. Sloan, E. L., Crowley, T. J., and Pollard, D. (1996). *Mar. Micropal.* **27**, p. 51.
8. Prentice, I. C., Cramer, W., Harrison, S. P., Leeman, R., Monserud, R. A., and Solomon, A. M. (1992). *J. Biogeogr.* **19**, p. 117.
9. Box, E. O. (1981). *Macroclimate and Plant Forms: An Introduction to Predictive Modelling in Phytogeography.* Junk, The Hague.

10. Woodward, F. I. (1987). *Climate and Plant Distribution.* Cambridge University Press, Cambridge.
11. Koppen, W. (1936). Das geographisches system der Klimate: In: *Handbuch der Klimatologie I(C)* (W. Koppen and R. Geiger, eds.). Gebruder Borntraeger, Berlin.
12. Holdridge, L. R. (1947). *Science* **105**, p. 367.
13. Davis, M. B. (1989). *Bull. Ecol. Soc. Am.* **70**, p. 222.
14. Haxeltine, A., and Prentice, I. C. (1996). *Global Geochemical Cycles* **10**, p. 693.
15. Prentice, I. C., Sykes, M. T., Lautenschlager, M., Harrison, S. P., Denissenko, O., and Bartlein, P. J. (1993). *Global Ecol. Biogeogr. Lett.* p. 67.
16. Randel, W. J., and Williamson, D. I. (1990). *J. Clim.* **3**, p. 608.
17. Covey, C., and Thompson, S. L. (1989). *Palaeogoegr. Palaeoclimat. Palaeoecol.* **75**, p. 331.
18. Ruddiman, W. F., Prell, W. L., and Raymo, M. E. (1989). *J. Geophys. Res.* **94**, p. 18379.
19. Kutzbach, J. E., Prell, W. L., and Ruddiman, W. F. (1993). *J. Geol.* **101**, p. 177.
20. Ogelsby, R. J., and Saltzman, B. S. (1992). *J. Clim.* **5**, p. 66.
21. Molnar, P., and England, P. (1990). *Nature* **346**, p. 29.
22. Cochran, J., Stow, D. A. V. *et al.* (1989). Initial Reports, Ocean Drilling Program, Leg 116, 388 pp., U.S. Government Printing Office, Washington, D.C.
23. Prell, W. L., and Kutzbach, J. E. (1993). *Nature* **360**, p. 647.
24. Baker, B. H., and Wohlenberg, J. (1971). *Nature* **229**, p. 538.
25. Coney, P. J., and Harms, T. A. (1984). *Geology* **12**, p. 550.
26. Pierce, K. L., and Morgan, L. A. (1992). In: *Regional Geology of Eastern Idaho and Western Wyoming* (P. K. Link, M. A. Kuntz, and L. B. Platt, eds.). Geological Society of America, Boulder, **179**, pp. 1–53.
27. Huber, N. K. (1981). *U.S. Geol Surv. Prof. Paper* **1197**, p. 1.
28. Lucchitta, I. (1979). *Tectonophysics* **61**, p. 63.
29. Jansen, E., and Sjoholm, J. (1991). *Nature* **349**, p. 600.
30. Isacks, B. L. (1988). *J. Geophys. Res.* **93**, p. 3211.
31. Westaway, R. (1993). *Earth Planet Sci. Lett.* **119**, p. 331.
32. Arthur, M. A. (1991). In: *Advisory Report on Earth System History* (G. S. Mountain and M. E. Katz, eds.). Natural Science Foundation Division of Ocean Studies, Washington, D.C., pp. 51–74.
33. Freeman, K. H., and Hayes, J. M. (1992). *Global Biogeochem. Cycles* **6**, p. 185.
34. Laws, E. A., Popp, B. N., Bidigare, R. B., Kennicutt, M. C., and Macko, S. A. (1995). *Geochim. et Cosmochim. Acta* **59**, p. 1131.
35. Cerling, T. E., Wang, Y., and Quade, J. (1993). *Nature* **361**, p. 344.
36. Wolfe, J. A. (1985). In: *The Carbon Cycle and Atmospheric CO_2: Natural Variations Archean to Present* (E. T. Sundquist and W. S. Broecker, eds.), Geophysics Monograph Series., Vol. 32, AGU, Washington, D.C., pp. 357–375.
37. Mathews, J. V., and Ovendon, L. E. (1990). *Arctic* **43**, p. 364.
38. Kutzbach, J. E., Guetter, P. J., Ruddiman, W. F., and Prell, W. L. (1989). *J. Geophys. Res.* **94**, p. 18393.
39. Raymo, M. E., Rind, D., and Ruddiman, W. F. (1990). *Paleoceanography* **5**, p. 367.
40. Schlesinger, M. E., and Mitchell, J. F. B. (1987). *Rev. Geophys.* **25**, p. 760.
41. Birchfield, G. E., Weertman, J., and Lunde, A. T. (1983). *Science* **202**, p. 305.
42. Whitlock, C., and Dawson, M. R. (1990). *Arctic* **43**, p. 324.
43. Wolfe, J. A. (1994). *Palaeogeogr. Palaeoclim. Palaeoecol.* **108**, p. 207.
44. Liu, G., and Leopold, E. B. (1994). *Palaeogeogr. Palaeoclim. Palaeoecol.* **108**, p. 217.
45. Brigham-Grette, J., and Carter, L. D. (1992). *Arctic* **45**, p. 74.
46. Sher, A. V., Kaplina, T. N., Giterman, R. E., Lozhkin, A. V., Arkhangelov, A. A., Kiselyov, S. V., Kouznetsov, Y. V., Virina, E. I., and Zazhigin, V. S. (1979). In: 14th Pacific Scientific Congress Excursion 11 Guidebook. Akad. Nauk. SSR, Moscow. 115 pp.
47. Willard, D. A. (1994). *Rev. Paleobot. Palynol.* **83**, p. 275.
48. Thompson, R. S., and Fleming, R. F. (1996). *Mar. Micropal.* **27**, p. 27.
49. Dowsett, H., Barron, J., and Poore, R. Z. (1996). *Mar. Micropal.* **27**, p. 13.
50. Armentrout, J. M. (1983). In: *Glacial Marine Sedimentation* (B. F. Molnia, ed.). Plenum, New York, pp. 629–666.

51. Mercer, J. H., and Sutter, J. (1981). *Palaeogeogr. Palaeoclim. Palaeoecol.* **38**, p. 185.
52. Bonan, G. B., Pollard, D., and Thompson, S. L. (1992). *Nature* **359**, p. 716.
53. Foley, J. A., Kutzbach, J. E., Coe, M. T., and Levis, S. (1994). *Nature* **371**, p. 52.
54. Gallimore, R. G., and Kutzbach, J. E. (1996). *Nature* **381**, p. 503. *Note added in proof*: Recent work indicates that this effect may have been as large as 5°C across broad regions of Northern Asia and North America [Dutton, J. F., and Barron, E. J. (1997). *Geology* **25**, p. 39].
55. Mercier, J.-L., Armijo, R., Tapponier, P., Covey-Gailhardis, E., and Tonglin, H. (1987). *Tectonics* **6**, p. 275.
56. Traverse, A. (1982). *Alcheringa* **6**, p. 197.
57. Axelrod, D. I., and Raven, P. H. (1978). In: *Biogeography and Ecology of South Africa* (M. J. A. Wager, ed.), pp. 77–130. Junk, The Hague.
58. Bonnefile, R., Vincens, A., and Buchet, G. (1987). *Palaeogeogr. Palaeoclim. Palaeoecol.* **60**, p. 249.
59. Leroy, S., and Dupont, L. (1994). *Palaeogeogr. Palaeoclim. Palaeoecol.* **109**, p. 295.
60. Quade, J., Cerling, T. E., and Bowman, J. R. (1989). *Nature* **342**, p. 163.
61. Cerling, T. E., Bowman, J. R., and O'Neill, J. R. (1988). *Palaeogeogr. Palaeoclim. Palaeoecol.* **63**, p. 335.
62. Tanai, T. (1961). *J. Fac. Sci. Hokkaido Univ.* **11**, p. 119.
63. Chaney, R. W., and Chuang, G. C. (1968). *Geol. Soc. China II*, p. 3.
64. Chaney, R. W., and Elias, M. K. (1936). *Carnegie Inst. Wash. Publ.* **476**, p. 1–72.
65. Thomasson, J. R. (1979). *Kansas Geol. Surv. Bull.* **218**, p. 1–67.
66. Axelrod, D. I. (1985). *Botan. Rev.* **51**, p. 164.
67. Wang, Y., Cerling, T. E., and Mcfadden, B. J. (1994). *Palaeogeogr. Palaeoclim. Palaeoecol.* **107**, p. 269.
68. Wing, S. L., and Greenwood, D. R. (1993). *Phil. Trans. R. Soc., Lond. Ser. B* **341**, p. 243.
69. Dowsett, H. J., Cronin, T. M., Poore, R. Z., Rhompson, R. S., Whatley, R. C., and Wood, A. M. (1992). *Science* **258**, p. 1133.
70. Raymo, M. E., Grant, B., Horowitz, M., and Rau, G. H. (1996). *Mar. Micropal.* **27**, p. 313.

IV

Geological and Geochemical Evidence of Uplift Effects on Weathering and CO_2

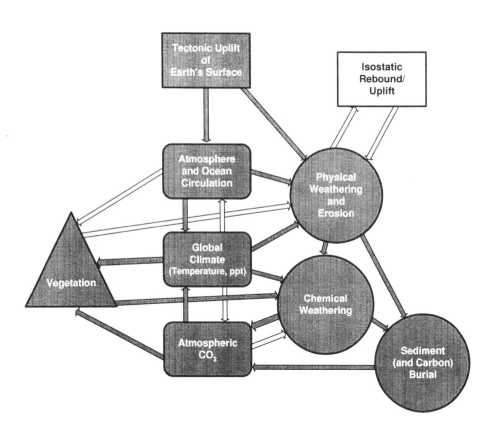

10

Fluvial Sediment Discharge to the Sea and the Importance of Regional Tectonics

John D. Milliman

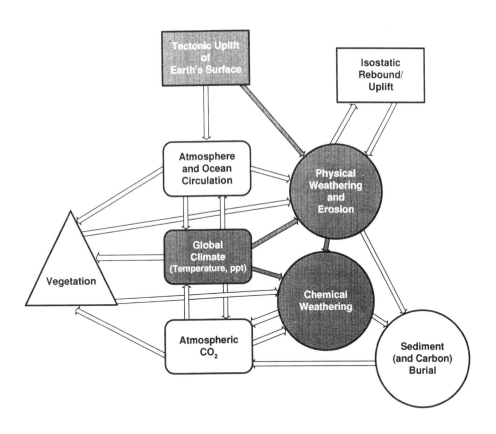

John D. Milliman • School of Marine Science, College of William and Mary, Gloucester Point, Virginia 23062.

Tectonic Uplift and Climate Change, edited by William F. Ruddiman. Plenum Press, New York, 1997.

1. INTRODUCTION

In recent years erosion and sediment discharge to the oceans have become topics of increasing interest among a wide diversity of scientists. Not only can a quantitative understanding of erosion and discharge help us delineate the various processes controlling the erosion, transport, and deposition of fluvially derived materials, but it also can have important implications for environmental management. The discussion in this paper shows that tectonic character (and associated factors) and the area of the drainage basin play the major roles in determining the sediment (both suspended and dissolved) discharge from rivers, but that climate, geology, and human factors also have marked impacts.

Part of the increased interest in fluvial systems comes from the increasing database with which we can work. The classic studies by Fournier[1] and Holeman[2] of global river discharge were based on relatively few data, many of which were of questionable quality. A subsequent study by Milliman and Meade[3] contained data from about 100 rivers, and the overall quality and length of individual river measurements were markedly better. By the early 1990s, we had access to data from nearly 300 rivers,[4] and presently the database exceeds 500,[5,6] although its quality remains uneven (see below).

As the database expands, our ability to parameterize the factors controlling sediment discharge to the oceans increases correspondingly. We have progressed from the rather simplistic models of Fournier and Holeman to a multidimensional model[7] that relates fluvial sediment discharge to a variety of factors, including climate, tectonics, geology, and land-use.

Because sediment loads are strongly affected by topography and therefore tectonics, we may be able to relate sediment accumulation rates noted in sedimentary basins to sediment discharge of specific river systems and therefore estimate the elevations of paleodrainage basins. Changes in rates of sediment accumulation, for example, may give us some indication of when and at what rate uplift (or denudation) of that drainage basin occurred. In this paper I discuss the factors controlling sediment discharge to the ocean, and, for the first time, dissolved sediment load.

It should be emphasized at the outset that fluvial data and their interpretation represent delivery to coastal areas. They do not necessarily represent total basin erosion, as some of the eroded material never reaches the coast but rather is stored within the watershed. Nor do the fluvial data necessarily indicate what is discharged to the ocean, since some of the material may be stored in coastal lowlands downstream of the gauging station. This problem of sediment storage is, in fact, an important determinant in understanding the sediment yield of various sized watersheds.

2. FACTORS AFFECTING SUSPENDED SEDIMENT DISCHARGE

2.1. Topography and Tectonics

As will be discussed in the following section, the sediment load of a river is partly dependent on basin size and runoff. However, if we look at rivers of similar

TABLE 1. Comparison of Water Discharges, Suspended Sediment loads, Sediment Yields, Suspended Sediment Concentrations, and Dissolved Sediment Loads for Rivers with Roughly Similar Drainage Basin Areas

River	Area (10^6 km^2)	Maximum elevation (m)	Q km^3/year	Suspended sediment (10^6 tons/year)	Sediment yield (tons/km^2/year)	Suspended sediment (mg/liter)	Dissolved sediment (10^6 tons/year)
Rappahanock (USA)	16	290	2.1	0.087	56	41	
Tano (GNA)	16	<500	4.5	0.35	22	78	0.32
Pearl (USA)	17	<500	6.6	1.6	46	242	0.94
Delaware (USA)	17	570	10.0	1.0	39	100	5.9
Tiber (Italy)	16	1300	7.4	7.5	350	1010	
Solo (Indonesia)	16	2600	15.0	19.0	1200	1270	
Ardour (France)	16	2800	11.0	0.24	18	22	0.75
Sous (MOR)	16	3300	0.31	1.6	260	5160	0.04

[a]Discharge has little control on sediment load, the two rivers with the greatest loads are the Tiber and Solo rivers, having the lowest and highest water discharges, respectively. What is clear is that mountainous rivers have greater sediment loads than do rivers draining low-lying terrain, and rivers draining higher mountains have the highest sediment yields. The Ardour (France) is the exception, with a load more like that of a lowland river; presumably this reflects the nonerosional nature of the older rocks within the Ardour's drainage basin. After Milliman et al.[5] and Meybeck and Ragu.[6]

drainage areas and water discharges, we see a great range of sediment loads and mean suspended matter concentrations (Table 1). Much of this variation can be explained by the morphology of the drainage basin. Rivers draining mountains have greater sediment loads (and sediment yields, i.e., load normalized for basin area) than do rivers whose headwaters drain uplands or lowlands. Determining mean basin topography or morphology, however, is laborious, and even then may not be as accurate an indicator of the topographic effect on sediment load as maximum elevation. Africa, for instance, has the highest mean elevation of any continent because of its many high plateaus, but with the exception of the Rift Valley and related mountains in the east and the mountains in the northwest, the continent generally lacks mountainous terrain; as a result, sediment discharge from Africa is lower than for any other large continent.[3]

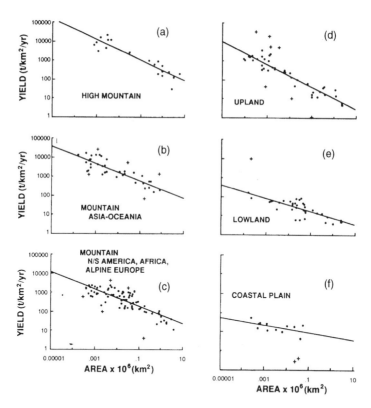

FIGURE 1. Sediment yields of rivers whose headwaters drain (a) high mountains (>3000 m maximum elevations); (b) Asian mountains (maximum elevations between 1000 and 3000 m); (c) other mountainous areas of the world; (d) uplands (500–1000 m maximum elevations); (e) lowlands (100–500 m maximum elevations); and (f) coastal plains (maximum elevations less than 100 m). Although these log–log plots tend to minimize the amount of scatter, fewer than 10% of the data points deviate by more than one standard deviation from the trend lines in each plot (after Milliman and Syvitski[4]).

A far easier topographic parameter to delineate is maximum elevation of a river basin, as that can be estimated from a regional topographic map. A direct relation between the maximum elevation of a river's headwaters and its sediment load was shown by Milliman and Syvitski.[4] High mountain rivers, whose maximum headwaters drain terrain above 3000 m, have sediment loads two to three orders of magnitude greater than similar size lowland rivers (maximum elevations between 100 and 500 m) (Fig. 1). Maximum elevation, of course, is no more than a simple indication of drainage basin topography, which in turn is a surrogate for basin tectonics and geology. Mountainous areas are more likely to experience earthquakes and mass failures as well as, e.g., heavy rains and snow melts, all of which facilitate erosion.

The effect of terrain upon the sediment load of a river can best been seen by the fact that all the major rivers in the world (sediment loads greater than 50×10^6 tons/year), drain mountains. Interestingly, the best documented rivers are those with extensive drainage basins that empty into trailing-edge margins (e.g., Amazon, Nile, Mississippi, Danube, Yangtze, Ganges–Brahmaputra, and Indus). These rivers often are well-documented because of their importance to transportation, irrigation, flooding and hydroelectric power as well as for their hydrocarbon potential. Smaller rivers, often discharging to the leading edge of the plates, tend to be less well-documented, but their small basin sizes suggest that they may discharge a much larger volume of sediment than previously envisaged, a concept discussed in the following paragraphs.

2.2. Basin Area

No reader would be surprised to learn that large rivers often have larger sediment loads than small rivers. However, if we normalize sediment loads with respect to drainage basin area (i.e., sediment yield), we see that smaller basins generally have higher sediment yields than do larger basins (Fig. 1). For instance, the Amazon River has a much greater sediment load (1.2×10^6 tons/year) than the Fly River in New Guinea (0.085×10^6 tons/year); but the Amazon's drainage basin is 100 times larger than that of the Fly (6.1 versus 0.06×106 km^2), meaning that the Fly's sediment yield is more than six times greater than the Amazon's (1300 versus 200 tons/km^2/year).

The increased sediment yield with decreasing basin area results from three prime factors:

1. Less sediment tends to be stored in smaller basins. Only a small portion of the sediment eroded from higher terrain within a large drainage basin may actually reach the sea, much of it being stored at lower elevations. In contrast, most of the sediment eroded from very small basins reaches the sea simply because there is no available storage space.
2. Smaller basins have greater mean topographic gradients than do large basins with similar elevations, a factor that favors erosion and sediment removal rather than deposition.

3. Smaller drainage basins are more susceptible to episodic events, whereas larger basins tend to modulate such events. Because of its very large basin size and generally rainy climate, for instance, discharge from the Amazon tends to be remarkably uniform both seasonally and interannually; half a dozen spot-measurements at Obidos can give a fairly accurate measure of the river's annual and longer-term discharge.[8] Similarly, in the more than 30 years that the undammed Indus was monitored at Kotri, its mean annual discharge varied only by about a factor of two,[9] and similar trends have been noted for the Yangtze River.[10] Even in large rivers whose watersheds are subject to dramatic climatic shifts, such as the Yellow River, annual mean discharge seldom varies by more than an order of magnitude.

Compare this to a small, semiarid drainage basin such as the Santa Clara River ($0.4 \times 10^6 \, km^2$) north of Los Angeles. Annual sediment discharge is often less than 0.1×10^6 tons/year, except during El Niño-induced rains, when discharges up to 25×10^6 tons have been recorded during a 24-h period! Annual discharges during El Niño years, in fact, can be four orders of magnitude greater than during non-El Niño years (Fig. 2). For the 14 non-El Niño years that this river was gauged, the average annual sediment load was 0.9×10^6 tons. In contrast, during the 4 El Niño years the average load was 26.8×10^6 tons/year,

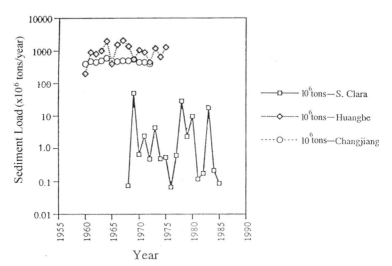

FIGURE 2. Comparison of annual sediment loads of a small mountainous river (Santa Clara, drainage basin area $0.041 \times 10^6 \, km^2$) with the very large Changjiang (Yangtze River, basin area $1.8 \times 10^6 \, km^2$) and Huanghe (Yellow River, basin area $0.75 \times 10^6 \, km^2$). The Huanghe's annual load varies by up to an order of magnitude, largely depending on the amount of rainfall that year, and the Changjiang's annual load rarely varies by more than a factor of two. In contrast, El Niño-caused floods on the Santa Clara can increase sediment discharge by more than four orders of magnitude, much of which escapes directly to the adjacent Santa Barbara Basin.

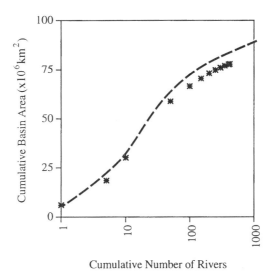

Cumulative Number of Rivers

FIGURE 3. Cumulative basin area of the 420 largest rivers as ranked by Meybeck and Ragu.[6] The Amazon River accounts for $6 \times 10^6 \text{km}^2$; combined with the next four largest rivers, the total basin area is $18.5 \times 10^6 \text{km}^2$. The 100 largest rivers listed by Meybeck and Ragu total $66 \times 10^6 \text{km}^2$, and the total 420 rivers are $78 \times 10^6 \text{km}^2$. Given the shape of this curve, total land area discharging to the sea might be expected to be about $85 \times 10^6 \text{km}^2$, whereas the actual area is closer to $100 \times 10^6 \text{km}^2$.[11] Meybeck and Ragu's list, however, misses some rivers whose basin areas range between 30,000 and 10,000 km². In Mexico alone, for example, Meybeck and Ragu list 8 rivers with areas greater than 10,000 km², whereas this writer has found 25. Assuming a large number of other undocumented rivers, the actual trend of cumulative river basin area (dashed line) must lie above the data in this plot. Clearly, in terms of numbers, small rivers become important; moreover, as discussion in this paper shows, small rivers have significantly greater suspended sediment yields than do larger rivers.

with a maximum annual load (1969) of 50.5×10^6 tons. In fact, 24-hr loads in 2 years each equaled or exceeded the total loads for the 14 non-El Niño years.

The increased sediment yield with decreasing basin size has a marked impact on the sediment discharge to the oceans, as there are many more smaller rivers than larger rivers discharging into the ocean. Meybeck and Ragu[6] have listed 423 rivers with areas greater than 10,000 km², but in reality this number is much greater, since these authors list primarily those rivers for which some measurements are available, and many rivers smaller than 50,000 km² have not been gauged. The Amazon drainage basin alone accounts for more than $6 \times 10^6 \text{km}^2$, or about 6% of the global drainage basin area emptying into the ocean. The 8 largest drainage basins account for more than $25 \times 10^6 \text{km}^2$, the 30 largest for more than $50 \times 10^6 \text{km}^2$, and the 250 largest for more than $75 \times 10^6 \text{km}^2$; but the next 150 largest rivers account for only an additional $3 \times 10^6 \text{km}^2$ (Fig. 3). One clearly needs to include thousands of very small rivers to approach a total of $100 \times 10^6 \text{km}^2$ of land area draining into the global ocean, and these small rivers tend to have much greater sediment yields than larger rivers draining similar elevations.

2.3. Precipitation

Precipitation can have an orographic dependence—rainfall and snowfall tending to increase with elevation. Mountain ranges also create precipitation shadows as well as affect weather patterns (the monsoon climate in southwest Asia being an obvious example), with maximum precipitation falling on windward slopes.

The relationship between sediment erosion and precipitation, however, is not easy to sort out from the other factors affecting soil erosion. It has long been understood that erosion in most areas does not increase linearly with increased precipitation. Rather, the yield is high in low-rainfall environments (the result of poor vegetation cover and periodically heavy rainfall), lower in areas with moderate rainfall (the result of heavy vegetation), and high again in areas with particularly heavy rainfall.[12,13] As pointed out by Walling[14] and Milliman and Syvitski,[4] however, the relationship between average annual rainfall (or runoff) and sediment erosion is actually rather poor. Rather, erosion seems to increase in some nonlinear relationship of peak rainfall to average rainfall.[1] Thus, sediment delivery from the mountainous rivers in northwest Africa is much greater than would be predicted by the average rainfall/funoff, primarily because of the effect of episodically heavy rainfall in a normally arid climate with little protective vegetation.

To appreciate fully the limited impact of precipitation on sediment discharge, let us look at the suspended loads of rivers that drain tropical to subtropical young mountains (less than 250 million years old; most of them in southern Asia): moderate- and high-runoff (250–800 mm/year, and >800 mm/year, respectively) rivers have similar loads and yields; low runoff (<250 mm/year) rivers deviate somewhat, but generally have similar loads (Fig. 4). More locally, however, precipitation can have considerable control, as can be seen from the case in New Zealand described below. Climate and precipitation also have greater importance in determining chemical weathering and therefore the dissolved load, as will be discussed in a subsequent section.

2.4. Geology

The ability of a river to erode and transport depends on the nature of substrate that it drains. One graphic indication of this (as well as of the local role of precipitation) comes from an analysis of New Zealand rivers by Hicks and co-workers.[16] Based simply on topography and basin area, sediment yields of the rivers draining subtropical to temperate mountains in New Zealand lie scattered around the trends for tropical–subtropical and temperate–subarctic mountains (Fig. 5a). This scatter is due to variable precipitation and geology in the river basins: yields increase with increased precipitation, and the sediment

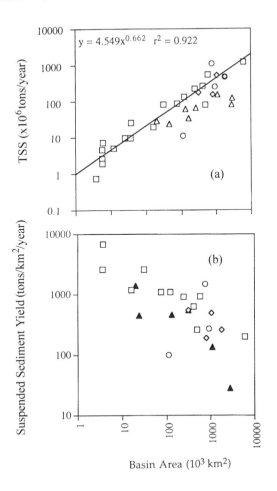

$$y = 4.549x^{0.662} \quad r^2 = 0.922$$

FIGURE 4.   Sediment loads for tropical and subtropical high-mountain and mountain rivers. The correlation for rivers with high runoff (> 800 mm/year) and moderate runoff (250–800 mm/year) is extremely high ($r^2 = 0.922$); rivers with runoff less than 250 mm/year (Indus, Huanghe, and Chao Phyrya) show greater scatter. Rivers draining young mountains (i.e., rocks less than 250 million years old; *sensu*[15]) have considerably higher yields than those draining older mountains (mostly those draining central and southern India). High-runoff mountainous rivers include the Amazon, Brahmaputra, Cimanuk, Citamandy, Fly, Hong (Red), Irrawady, Magdalena, Purari, Solo, and Zhujiang. Moderate-runoff rivers are the Ganges, Godovari, Mekong, and Changjiang (Yangtze). Old mountain rivers with high or moderate runoff include the Da, Damovar, Mahandi, Parana, Uruguay, and Orinoco (part of which drains the Guyana Highlands). Most data are from Milliman et al.,[5] with supplementary data from Meybeck and Ragu.[6]

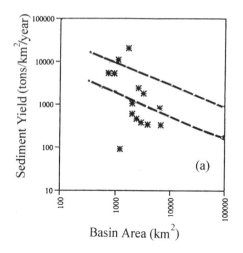

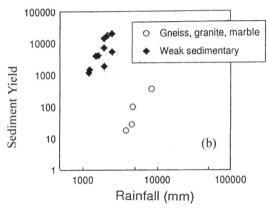

FIGURE 5. (a) Mountainous New Zealand rivers, which were not included in Fig. 3, have yields (unpublished data courtesy of M. Hicks, 1996) that show considerable deviation from the yield curves for rivers draining tropical to subtropical young mountains (T-ST YM; see Fig. 4b) or temperate to subarctic mountainous rivers (T-SA YM; yield-area curve from Fig. 1c). Thus basin area and elevation alone are poor indicators of sediment yield for New Zealand rivers. (b) Plot showing that much of the deviation seen in (a) can be explained by local variations in precipitation and geology (after Hicks et al.[16]).

yield from basins with easily erodible sedimentary rock is about three orders of magnitude greater than for rivers draining predominantly crystalline rocks (Fig. 5b).

Extrapolating this relationship to global rivers, one sees that mountainous rivers draining younger sedimentary rocks have considerably higher yields than those draining harder, older rocks (Fig. 4), perhaps explaining part of the difference between the loads of southern Asian and other mountainous rivers noted by Milliman and Syvitski.[4]

2.5. Human Activity

The impact of runoff and erosion is strongly controlled by the amount and nature of vegetation in the drainage basin. Sediment yields in undisturbed forests in peninsular Malaysia, for instance, range from about 20 to 100 tons/km^2/year, whereas land areas with mixed use have yields ranging from 100 to 300; cultivated land in Java can have yields as high as 7000 tons/km^2/year.[17]

The impact of changing land use upon sediment delivery from modern rivers, however, is less well-documented,[18] although Milliman and Syvitski[4] have speculated that the generally higher yields of Asian rivers may be partly the result of human activity. Sediment loads in the Yellow River, for example, increased about an order of magnitude 2500 years ago, when man began farming the highly erodible loess hills in northern China.[10] Prior to that, in fact, the river was not "Yellow," but was called Big River (Dahe). The impact of human activity also can be seen by the fact that sediment yield increases by a ratio of about 1.7:1 relative to population growth (see Fig. 4 in Walling et al.[18]).

Reforestation, however, can dramatically retard erosion, and dams can partially or completely trap sediment and prevent it from reaching the sea. The Nile and Colorado rivers presently deliver no sediment to the ocean, and the Indus, Mississippi, Rhone, and Zambesi (to name only a few examples) discharge only a fraction of what they did 50 years ago. Thus, ironically, land erosion seems to be increasing (particularly in mountainous developing countries) while sediment delivery to the oceans is decreasing.

2.6. Some Cautions

Before the reader assumes that the above discussion provides all the parameters necessary to understand fluvial discharge, there are several key disclaimers that should be noted. First, we depend heavily on the river data from Asia and the high-standing islands of Oceania, which are generally few, short term, and difficult to judge as to quality. This problem is exacerbated by the fact that few small rivers have been adequately monitored (if at all), even though the importance of episodic events (see above) necessitates a relatively long-term database.

Second, and quite apart from data reliability, is the question of whether sediment load data are accurate indicators of the actual sediment discharge to the ocean. The total of suspended solids (TSS) clearly does not include bed load, although as a rule bed load transport near the mouths of large rivers probably accounts for no more than 10–20% of the total load[3]; thus we probably have not underestimated the sediment discharge from large rivers. More problematic is the fact that rivers usually are gauged upstream of any tidal influence. On the Amazon River, for instance, the last downstream station is at Obidos, 600 km from the river mouth; on the Mississippi it is near Baton Rouge, 200 km from the delta front. The assumption that the sediment passing these upstream stations actually reaches the ocean is, at best, naive, since some of the sediment load is deposited before reaching the ocean, often as channel or floodplain deposits. In the Huanghe, for instance, the sediment load decreases from 1600×10^6 tons/year at Shamenxia (just downstream from the loess hills) to 1100×10^6 tons/year at Lijin, at the head of the delta. Similarly, there is increasing evidence that a substantial portion of the very large documented loads of the Ganges and Brahmaputra rivers, measured upstream in India and Bangladesh, respectively, are deposited along the rivers' lower course; the amount reaching the Bay of Bengal may be substantially lower than the oft-quoted numbers would suggest. This probably is a not serious problem for smaller rivers because of the much smaller area over which sediment can be stored; also, periodic floods presumably flush the system. Because the amount of sediment discharge from large rivers may be higher than commonly cited (see Mead et al.,[20] for a discussion), TSS yield curves may actually be steeper than shown in Figs. 1 and 4, large rivers having somewhat lower yields.

Finally, we must understand that nearly all rivers not only reflect anthropogenic impact (see above) but also the late Holocene highstand conditions. During sea-level regression and lowstands, for example, much of the sediment stored on land was eroded and transported beyond the shelf edge to the deep sea. During these episodes, sediment discharge from larger rivers must increase, in some cases perhaps dramatically.

3. FACTORS AFFECTING DISSOLVED SEDIMENT DISCHARGE

The fluvial discharge of dissolved sediments to the ocean has been discussed by a number of researchers,[11,21-23] but most notably by Meybeck.[24-26] Much of the emphasis of published discussions has been on the nature and flux of organic matter as well as particulate versus dissolved materials (see, e.g., Maybeck[25] and Degens et al.[27]).

Until recently our database was too small to allow a comprehensive discussion of the tectonic control on total dissolved solids (TDS) let alone the concentrations or ratios of concentrations of various dissolved ions. While absolute and relative concentrations of dissolved species depend on a number of interrelated factors, dissolved solids in most rivers come from rock weathering.[23,24] Dissolved (as well as suspended) solid loads increase with increased river flow,[23] but the role of basin area has not been discussed previously.

The topographic factor is somewhat more clear: rivers draining areas with low relief or crystalline shields have low dissolved loads and yields[23,25] because of the protective nature of the overlying sediment (Chapter 15 and Edmond et al.[28]). Meybeck[26] (see his Fig. 4.6) has shown that maximum TDS (also TSS) values occur in rivers draining humid mountainous terrain, and minimum values occur in arid and arctic rivers. Meybeck also has shown that the TSS/TDS ratios for mountainous rivers fall between about 2 and 20, and are much lower for rivers from lower drainage basins. No mention, however, was made of how TDS or the TSS/TDS ratio vary among mountainous rivers draining different terrain or

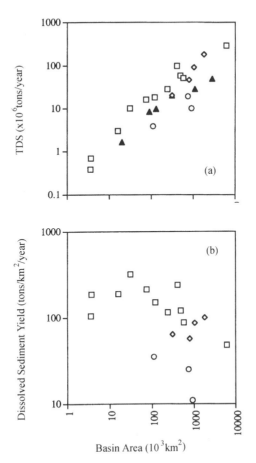

FIGURE 6.   Total dissolved solid (TDS) loads (a) and yields (b) of tropical and subtropical young mountainous rivers. Note that loads and yields of rivers with high (>800 mm/year) and moderate (250–800 mm/year) runoffs are similar, whereas low-runoff rivers have lower TDS values, as do rivers draining older mountains. Most data are from Milliman et al.,[5] with supplementary data from Meybeck and Ragu.[6]

different-sized watersheds, in part because of the paucity of data. The recent collation of available TDS data by Meybeck[5,6] now provides us with a sufficient database to allow us to begin delineating, if not completely understanding, global controls on TDS. In the following paragraphs I present data for the same humid tropical and subtropical mountainous rivers for which I have presented TSS data. This discussion, however, does not deal with ionic concentrations or ratios, which presumably Meybeck will synthesize in the near future.

For the moderate- to high-runoff tropical mountain rivers shown in Fig. 6a, TDS loads range from less than 0.5 to nearly 300×10^6 tons/year, the dissolved loads increasing with increasing basin size; the coefficient of determination (r^2) is,

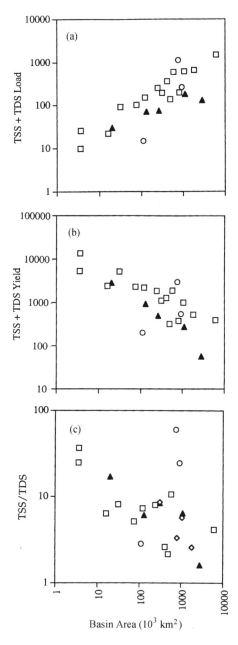

FIGURE 7. Total solid and dissolved sediment loads (a) and yields (b) for moderate- and high-runoff rivers draining young and old mountain rivers (see Figs. 4 and 6). Although the database is small and there is considerable scatter, large rivers tend to have lower TSS/TDS ratios (c) than do rivers draining small basins, where the ratios can exceed 10. The TSS/TDS ratios in rivers with low runoff are considerably higher, whereas ratios of rivers draining older mountains are lower. Most data are from Milliman et al.[5] with supplementary data from Meybeck and Ragu.[6]

in fact, actually marginally better (0.94) than for suspended loads (0.92). TDS values are roughly one-fifth to one-tenth those of TSS, but are perceptibly lower for low-runoff rivers and for rivers draining older mountains. In the first instance low values are a function of low levels of chemical weathering. In the second instance they may reflect the resistant nature of the rock itself. These data signify the importance of climate and lithologic character of the watershed[13,15] in chemical weathering (Chapter 15), although topography (even as a surrogate— see above) plays an equally (if not more important) role (Chapter 14 and Raymo and Ruddiman[29]). Climate may have a greater impact on the composition of the dissolved constituents than on their quantity; again, however, verification of this suggestion awaits further analysis of available data.

TDS yield (Fig. 6b) is more or less the inverse of TDS load, as would be expected, although yields do not increase in smaller basins, as do TSS values, but rather appear to decrease in basins smaller than about 15,000 km^2. The reader is cautioned, however, that this observation is based on very few data, some or all of which may be subject to error.

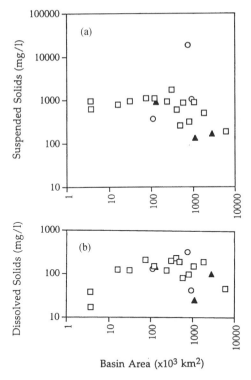

Basin Area (x10^3 km^2)

FIGURE 8.   Suspended (a) and dissolved (b) solid concentrations in the same rivers as in Figs. 4, 6, and 7. In this plot, moderate- and high-runoff rivers have been shown as open squares, low-runoff rivers as open circles, and older mountainous rivers as solid triangles; several rivers have been excluded where discharge is not documented. The high-value outlier in (a) is the Huanghe (Yellow River), whose mean suspended sediment concentration is about 20 grams/liter.

Although there is considerable scatter, the TSS/TDS ratio increases with decreasing drainage basin area (Fig. 7c). Rivers draining basins larger than $10^6 \, km^2$ tend to have TSS/TDS ratios less than 2 to 4, whereas basins smaller than $10^5 \, km^2$ have yields greater than 7; rivers draining basins smaller than $10^4 \, km^2$ have yields greater than 10. These higher ratios suggest that as sediment storage decreases, chemical weathering decreases relative to mechanical weathering, presumably the result of insufficient storage. Said another way, as basin area decreases, the sediment load of mountainus rivers is increasingly dominated by suspended rather than dissolved solids. In contrast, TSS/TDS ratios for lowland rivers appears to remain relatively constant, between 0.5 to 2.0, regardless of basin size (Milliman, unpublished data), again indicating the importance of watershed topography in controlling chemical weathering.

An alternative explanation is that these trends are more related to water discharge than to basin area, large rivers discharging more water (and therefore more suspended and dissolved solids) than small rivers. To some degree this is true, although the admittedly small database suggests that rivers larger than about $2 \times 10^5 \, km^2$ carry proportionately lower concentrations of suspended solids (Fig. 8a), while rivers smaller than about $15 \times 10^3 \, km^2$ carry proportionately lower concentrations of dissolved solids (Fig. 8b). Again, this supports the concept of loss (storage) of sediment in larger rivers and the ineffectiveness of chemical weathering in small drainage basins. A better database, however, may disprove or modify some of these statements.

4. TECTONIC EFFECTS ON THE SEDIMENTARY RECORD

Based on the above discussion, the impact of rivers draining a tectonically active mountain belt is somewhat less straightforward than one might suspect. Sediment erosion (both mechanical and chemical) in mountainous river basins increases with increasing watershed and is greater for young mountains than for older mountains (Fig. 7a). Total yield, however, decreases with increased drainage basin area (Fig. 7b), as does the TSS/TDS ratio, but the Indus and Yellow rivers, which have low runoff, have much greater TSS/TDS ratios than their size would suggest (Fig. 7c), indicating incomplete chemical weathering.

As the mountains are uplifted, sediment yields should increase; for every 1000 m of uplift, the suspended load of a river should increase by a factor of 2 to 5, assuming, of course, that the size of the drainage basin remains constant. However, if deltaic and shoreward accretion enlarges the river basin, the TSS yield might decrease owing to increased sediment storage, and the TSS/TDS ratio would then decrease (Fig. 7).

However, the analyses in this paper also indicate the major roles that climate and geology play in determining the sediment load of a river. Young mountains contain, by definition, younger, softer rocks, which are more easily erodible; therefore their sediment loads, both solid and dissolved, should be greater. Rivers draining the Deccan traps in central and southern India, for example, have loads three- to fivefold smaller than similar-sized rivers draining the younger Himalayas (Fig. 7). In part, of course, younger mountains generally have higher elevations and steeper gradients; but even in the Himalayas, older rocks seem to erode at slower rates than do younger rocks (Chapter 12).

One can thus imagine the following scenario in southern Asia. As the Himalayas achieved significantly high topography about 20 million years ago, erosion rates increased, as seen by increased accumulation rates in coastal and deep-sea basins (Chapters 12 and 13). With the uplift of the Himalayas, orographically controlled precipitation increased; the resulting monsoon climate — intervals of heavy rain and rain-free periods — provided optimal conditions for sediment erosion. Thus one can speculate that rivers draining the young Himalayas were relatively small but had high yields. As these rivers coalesced into larger rivers and as their floodplains grew, the sediment yields presumably decreased, but the TDS would have increased relative to the TSS (Fig. 7). This scenario achieves the same result as the latest Cenozoic lowering of the elevation of the Himalayas suggested by Derry and France-Lanord (Chapter 12).

Given the factors discussed above, it is not surprising that the mountains of southern Asia have the highest sediment yields (for given drainage basin areas) in the world: the mountains are high, precipitation is heavy and seasonal, and the strata are relatively young and erodible. Human impact has only accentuated what must have been large-scale sediment shedding of the uplifted Himalayas.

One might ask when the Himalayas reached a critical elevation to accelerate erosion and sediment delivery? Even a casual reading of the literature gives a tentative answer. During the late Oligocene to early Miocene, once flourishing reefs were drowned by terrigenous sediments in the South and East China seas (Fullthorpe[30] and Milliman and Scott, unpublished data), and there was a major influx of terrigenous sediments in the Bengal[31] and Indus basins as well as the adjacent deep-sea fans (see Whiting et al.[32] and references therein). The tectonic control on this major increase in sediment discharge cannot be ignored.

5. SUMMARY

The geomorphic–tectonic character of the drainage basin and the actual basin area are first-order controls on the sediment discharge of most rivers. Topographic control (which in fact serves as a surrogate for tectonic character of the basin), however, is strongly dependent on the erodibility of the substrate — younger sedimentary rocks being more erodible than older crystalline rocks. Climate (particularly precipitation) and human impact also play important roles, often explaining deviations from the load predicted on the basis of topography and basin area alone.

Suspended sediment yield of mountainous rivers increases five- to ninefold for every order of magnitude decrease in basin area, the result of decreased storage capacity, steeper gradients, and greater susceptibility to episodic events, such as floods and landslides. As a result, a large percentage of the suspended sediment discharged to the ocean comes from small mountainous rivers, which generally discharge onto active margins. In contrast, a substantial part of the fluvial sediment load of large rivers, which mostly discharge to passive margins, may be deposited on subsiding lowlands or deltas; as a result, the actual sediment discharge to the ocean may be considerably less for some large rivers than cited in the literature. We therefore have probably overestimated the sediment discharge of large rivers to the ocean while underestimating the discharge of smaller rivers. This situation, however, probably changes with long-term sea-level regression and lowstands.

Dissolved sediment yield of mountainous rivers with moderate to high runoffs also increases with decreasing basin area, but at a lesser rate than for suspended solids. Dissolved yield is demonstrably lower for river basins with low precipitation, indicative of less chemical weathering. While dissolved sediment yield increases with decreasing basin area, its increase is less than that for suspended sediment. The resulting high TSS/TDS ratio in smaller mountainous rivers presumably reflects the decreased effect of chemical weathering owing to the shorter period of sediment storage between erosion and discharge to the ocean.

Evolution of the river basin also can cause changes in fluvial sediment discharge. For instance, as a river progrades and (eventually) merges with other rivers, the suspended sediment yield should decrease and dissolved sediment should increase, a situation not unlike that expected for rivers draining denuded mountains.

Acknowledgments

I thank Bill Ruddiman for his patience as I crept through the analysis of these river data. Over the years Michel Meybeck and Bob Meade have supplied valuable insights that have aided me in gaining a better understanding of rivers and river processes. Earlier versions of the manuscript were read critically by Bill Ruddiman and Betty and Bob Berner; I thank them for their help. Parts of this research were funded by the Office of Naval Research (N00014-94-1-0179) and the National Science Foundation (OCE-9222405).

REFERENCES

1. Fournier, F. (1960). *Climat et Erosion*. Presses Universitaires de France, Paris, 201 p.
2. Holeman, J. N. (1968). *Water Resource Res.* **4**, p. 737.
3. Milliman, J. D., and Meade, R. H. (1983). *J. Geol.* **91**, p. 1.
4. Milliman, J. D., and Syvitski, J. P. M. (1992). *J. Geol.* **100**, p. 525.
5. Milliman, J. D., Rutkowski, C., and Meybeck, M. (1995). River discharge to the sea, A global river index (GLORI). IGBP-LOICZ Report, 125 pp.
6. Meybeck, M., and Ragu, A. (1996). River discharges to the oceans: an assessment of suspended solids, major ions and nutrients. GEMS/EAP Report. UNEP/WHO, 245 pp.

7. Ludwig, W., and Probst, J.-L. (1996). In: *Erosion and Sediment Yields: Global and Regional Perspectives* (D. E. Walling and B. W. Webb, eds.), IAHS Publ. **236**, p. 21.

8. Meade, R. H., Dunne, T., Richey, J. E., Santos, U. deM, and Salati, E. (1985). *Science* **228**, p. 488.

9. Milliman, J. D., Quaraishee, G. S., and Beg, M. A. A. (1984). In: *Marine Geology and Oceanography of the Arabian Sea and Coastal Pakistan* (B. U. Haq and J. D. Milliman, eds.), p. 65. Van Nostrand-Reinhold, New York.

10. Milliman, J. D., and Ren, M.-E. (1994). In: *Climate Change: Impact on Coastal Habitation* (D. Eisma, ed.), p. 57. Lewis, Boca Raton, FL.

11. Livingstone, D. A. (1963). Chemical composition of rivers and lakes. Data on Chemistry. U.S. Geol. Survey Prof. Paper 440G, 64 p.

12. Langbein, W. B., and Schumm, S. A. (1958). *Trans. Am. Geophys. Union.* **39**, p. 1076.

13. Summerfield, M. A. (1991). *Global Geomorphology.* Longman, Singapore.

14. Walling, D. E. (1983). *J. Hydrol.* **65**, p. 209.

15. Pinet, P., and Souriau, M., 1988. *Tectonics* **7**, p. 563.

16. Hicks, D. M., Hill, J., and Shankar, U. (1996). In: *Erosion and Sediment Yield: Global and Regional Perspectives* (D. E. Walling and B. W. Webb, eds.), IAHS Publ. **236**, p. 149.

17. Douglas, I. (1996). In: *Erosion and Sediment Yield: Global and Regional Perspectives* (D. E. Walling and B. W. Webb, eds.), IAHS Publ. **236**, p. 463.

18. Walling, D. E., and Webb, B. W. (1996). In: *Erosion and Sediment Yield: Global and Regional Perspectives* (D. E. Walling and B. W. Webb, eds.), IAHS Publ. **236**, p. 3.

19. Milliman, J. D., Qin, Y.-S., Ren, M. E., and Saito, Y. (1987). *J. Geol.* **95**, p. 751.

20. Meade, R. H. (1996). In: *Sea-Level Rise and Coastal Subsidence: Causes, Consequences, and strategies* (J. D. Milliman and B. U. Haq, eds.), p. 63. Kluwer, Dordrecht.

21. Walling, D. E., and Webb, B. W. (1983). IAHS Publ. **141**, p. 3.

22. Drever, J. I. (1988). *The Geochemistry of Natural Waters*, 437 pp. Prentice-Hall, Englewood Cliffs, N.J.

23. Chester, R. (1990). *Marine Geochemistry*, 698 pp. Unwin-Hyman, London.

24. Meybeck, M. (1981). In: *River Inputs to Ocean Systems* (J. M. Martin, J. D. Burton, and D. Eisma, eds.), p. 18. UNEP/UNESCO, Paris.

25. Meybeck, M. (1988). In: *Physical and Chemical Weathering in Geochemical Cycles* (A. Lerman and M. Meybeck, eds.), p. 247. Kluwer, Dordrecht.

26. Meybeck, M. (1994). In: *Material Fluxes on the Surface of the Earth*, p. 61. Natural Academy Press, Washington, D.C.

27. Degens, E. T., Kempe, S., and Richey, J. E. (1991). *Biogeochemistry of Major World Rivers.* SCOPE 42, 347 pp. John Wiley, Chichester.

28. Edmond, J. M., Palmer, M. R., Measures, C. I., Grant, B., and Stallard, R. F. (1995). *Geochim. Cosmochim. Acta* **59**, p. 3301.

29. Raymo, M. E., and Ruddiman, W. F. (1992). *Nature* **359**, p. 117.

30. Fulthorpe, C. *et al.*, (1990). *Am. Assoc. Petrol. Geol. Memoir.*

31. Alam, M. (1996). In: *Sea-Level Rise and Coastal Subsidence: Causes, Consequences, and Strategies* (J. D. Milliman and B. U. Haq, eds.), p. 169. Kluwer Dordrecht.

32. Whiting, B. M., Karner, G. D., and Driscoll, N. W. (1994). *J. Geophys. Res.* **99**, p. 13791.

11

The Effect of Late Cenozoic Glaciation and Tectonic Uplift on Silicate Weathering Rates and the Marine $^{87}Sr/^{86}Sr$ Record

Joel D. Blum

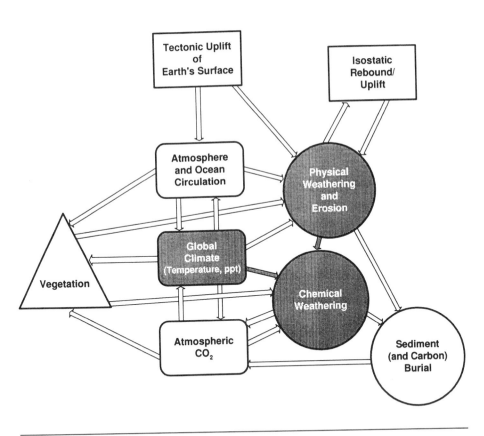

Joel D. Blum • Earth, Ecosystem and Ecological Sciences Program, Earth Sciences Department, Dartmouth College, Hanover, North Hampshire 03755.

Tectonic Uplift and Climate Change, edited by William F. Ruddiman. Plenum Press, New York, 1997.

1. INTRODUCTION

It is widely accepted that the consumption of atmospheric CO_2 by silicate weathering provides a negative climate feedback mechanism, and that this mechanism has played an important role in moderating global climate over geological timescales (see e.g., Berner[1]). Extension of this concept to feedbacks between climate change and tectonic uplift, which are discussed in many chapters of this volume, is dependent on the assumption that glaciation and tectonic uplift lead to increased rates of chemical weathering and CO_2 consumption. Both glaciation and rapid uplift are accompanied by rapid physical denudation, which exposes fresh bedrock minerals to the agents of chemical weathering. In addition, tectonic uplift may create mountain ranges capable of sustaining alpine glaciers and ice sheets, especially during glacial maxima, thus linking uplift and glaciation on geological timescales. While it is generally accepted that glaciation and uplift will lead to accelerated rates of chemical weathering as compared to regions of nonglaciated or stable topography, the magnitude of this effect has proven difficult to estimate. Taylor and Blum[2] investigated the relation between soil age and silicate weathering rates and found a well-defined power law relation for young soils ($\leqslant 138$ thousand years old), suggesting that predictions could be made of the effect of glaciation and tectonic uplift on silicate weathering rates by considering the effect of these processes on average soil ages over large geographic areas.

The marine $^{87}Sr/^{86}Sr$ record is often used as a direct proxy for global silicate weathering rates over geological timescales and has been used by Raymo and Ruddiman,[3] Edmond,[4] and Richter et al.[5] to provide tests of theories relating glaciation and tectonic uplift to silicate weathering rates. This, however, is dependent on the assumption that changes in the riverine flux of silicate weathering-derived base cations to the oceans are directly reflected in marine $^{87}Sr/^{86}Sr$. Blum and Erel[6,7] investigated the Sr isotope systematics of silicate weathering and demonstrated that the Sr isotopic composition released by weathering of a single silicate rock type changes dramatically with the age of the soil. Thus, the marine $^{87}Sr/^{86}Sr$ ratio cannot be considered a direct proxy for silicate weathering rates without considering the ages of soils in river basins, and the effect of this factor on the $^{87}Sr/^{86}Sr$ released by weathering.

Previous studies of the influence of glaciation on weathering rates have focused either on field measurements of cation fluxes from modern alpine glaciers,[8] or estimates of global riverine bicarbonate fluxes utilizing empirical relations between estimated runoff and weathering.[9] While both approaches lead to the general conclusion that weathering rates are accelerated in glacial environments, there are no modern analogs for the Laurentian and Fennoscandian ice sheets, and thus, it is difficult to extrapolate these results to the problem of how late Cenozoic glacial–interglacial cycles have influenced global weathering rates.

Other investigators have tackled the question of the relation between tectonic uplift and silicate weathering rates primarily by studying riverine chemical fluxes of large river basins with contrasting uplift rates.[10–12] However, there is such a complex matrix of factors that affect weathering rates in large basins (e.g., human activities, precipitation, temperature, vegetation, relief, silicate rock mineralogy, uplift rate, and glacial history) that it is extremely difficult to confidently isolate the influence of the various factors affecting weathering rates. In addition, the extreme weatherability of carbonate and evaporite rocks compared to silicates results in a dominance of dissolved cation chemistry by trace amounts of carbonate and evaporite that occur in almost all large basins. While a great deal of insight has been gained from large basin studies, particularly with respect to global geochemical fluxes, it has proven difficult to isolate the effects of specific processes on rates of silicate weathering.

The study of base cation depletion from soil profiles of known age is a methodology that has allowed quantification of long-term weathering rates. Previous studies have investigated soils developed on glacial moraines deposited by continental ice sheets, which provide weathering rates at a single time interval,[13,14] or soils developed on fluvial or marine terraces, which can be problematic because sediments deposited by water are commonly density-sorted and do not represent pristine material.[15-17] Present-day weathering rates have also been measured from catchment solute fluxes,[8,18] but do not provide information on changes in rates with time and can be dramatically influenced by depletion of the soil cation exchange pool by atmospheric pollution.[19-21]

Previous studies of the $^{87}Sr/^{86}Sr$ ratio of Sr released by chemical weathering show it to be a sensitive indicator of the relative rates of weathering of individual silicate minerals within the soil parent material.[22-24] Understanding the Sr isotope systematics of weathering is necessary to make predictions of how tectonic uplift and climate may affect the riverine Sr flux and $^{87}Sr/^{86}Sr$ ratio delivered to the oceans. In this chapter I review the recent studies of Taylor and Blum,[2] and Blum and Erel[6,7] in which silicate weathering rates and Sr isotope release by weathering were investigated for a granitic soil chronosequence developed on six Quaternary glacial moraines with ages varying from 0.4 to ≥297 thousand years. The results of these studies are discussed within the context of their bearing on studies of global glaciation, tectonic uplift, and climate change.

1.1. Description of the Wind River Mountains Soil Chronosequence

Tayor and Blum,[2] and Blum and Erel[6,7] measured base cation and Sr isotope release from six soil profiles developed on alpine glacial moraines with ages varying from 0.4 to ≥297 thousand years, which form a soil chronosequence (as defined by Jenny[25]) in the Fremont Creek drainage basin in the northern Wind River Mountains, Wyoming. The area was chosen because the bedrock and glacial deposits consist entirely of Archean granitoids and gneisses typical of the continental shield rocks that were extensively glaciated by ice sheets during the late Cenozoic glaciations, and because the area has glacial deposits and soils that have already been thoroughly described, mapped, and dated by Richmond,[26] Mahaney and Halvorson,[27] Sorenson,[28] Hall and Horn,[29] and Gosse et al.[30,31] In addition, owing to the remote location of the field site, it has been relatively unaffected by anthropogenic inputs of atmospheric dust and acidity, in contrast to localities in the northeastern United States and northern Europe that have been the main focus of previous research.

Soil profiles from the Gannett Peak, Audubon, and Titcomb Lake moraines were sampled in the Titcomb Basin at elevations of 3440 to 3230 m. These moraines were mapped by Mahaney and Halvorson,[27] who correlated the Gannett Peak till with the Little Ice Age for which the global historic record indicates maximum glacial advances between 0.3 and 0.5 thousand years ago,[32] and the Audubon till with the Audubon glacial advance, which is dated by ^{14}C at 1.5–2.5 thousand years ago.[27] The Titcomb Lake till was dated by [10]Be exposure age at 11.7 $\pm$ 0.6 thousand years ago.[31] Soil profiles from the Pinedale and Bull Lake moraines were sampled at the south end of Fremont Lake at elevations of 2315 and 2285 m. These moraines were mapped by Richmond[26] and dated by [10]Be at 21.7 $\pm$ 0.7 and 138 $\pm$ 4 thousand years ago, respectively.[30,31] Soils of pre-Bull Lake age were sampled at elevations of 2555 m in Surveyor Park to the east of Fremont Lake, and at 2275 m at the south end of Fremont Lake. The pre-Bull Lake soil parent material in the Surveyor Park area was mapped as the Sacagawea Ridge till and at the south end of Fremont Lake was mapped as Sacagawea Ridge outwash by Richmond,[26] but their origin and age are quite uncertain and constrained only to be $\geqslant$297 thousand years old by [10]Be.[30] Samples from soil profiles developed on each of these six moraines will be referred to in this paper (in order of increasing age) as the Gannett Peak, Audubon, Titcomb Lake, Pinedale, Bull Lake, and Sacagawea Ridge soils. Average annual temperature and precipitation are 1.8°C and 230 mm at Fremont Lake, and estimated to be -3.5°C and 500–700 mm at Titcomb Basin.[27,33] The three high-elevation profiles are above the timberline and vegetation is dominated by perennial sedges and grasses as well as lichens. The three low-elevation profiles are below the lower limit of trees, and vegetation is dominated by sagebrush (*Artemisia tridentata*).

The soil parent materials (i.e., tills) for each of the profiles is a mixture of Archean granite, granodiorite, and diorite gneiss, which outcrops throughout the drainage basin.[34] The terminal moraines are a homogenized mixture of boulders, pebbles, and sand and silt-sized grains, on which soils up to 1 m in depth have developed. The primary minerals observed in optical and scanning electron microscope analysis of the deepest (unweathered) sample from each soil profile are (in order of decreasing abundance): plagioclase (An_{7-35}, i.e., oligoclase), quartz, K-feldspar, biotite, hornblende, and magnetite. Minor amounts ($<1\%$) of pyroxene, garnet, apatite, sphene, muscovite, and chlorite were also observed in some samples. Secondary clay minerals identified by X-ray diffraction studies in the <2-μm size fraction include kaolinite, vermiculite, hydrobiotite, illite, and smectite.[27,35] Thin sections were searched exhaustively using both optical and scanning electron microscopes for inclusions of calcite in the granitic gneisses, and it was confirmed that calcite is absent (even at trace levels) in these parent rocks, in contrast to the hydrothermally altered granitic gneiss soil parent material studied at the Loch Vale watershed, Colorado, by Mast *et al.*[18]

Soils were sampled adjacent to soil profiles previously described in detail and, therefore, complete soil descriptions and soil properties will not be repeated here. The Gannett Peak soil profile was collected adjacent to soil pit TB-15 of Mahaney[36] and is a weakly developed Entisol in an area of discontinuous soil cover on a moraine below an active glacier. The Audubon soil profile was collected adjacent to TB-6 of Mahaney[36] and is also a weakly developed Entisol, but is in an area of more continuous soil cover. The Titcomb Lake soil profile was collected adjacent to soil pit TB-2 of Mahaney[36] and TB-13 of Mahaney and Sanmugadas,[33] and was classified by these authors as a cryochrept. The Pinedale soil profile was collected adjacent to soil pit FL-7 of Hall and Shroba[37] and was classified by Sorenson[28] as a typic cryoboroll. The Bull Lake soil profile was collected adjacent to soil pit FL-3 of Hall and Shroba,[37] was classified by Sorensen[28] as a typic cryoboroll, and found to contain up to 1% pedogenic $CaCO_3$. The Sacagawea Ridge soils are highly weathered alfic cryochrept and arguidic cryoboroll[28] on the till and outwash sites, respectively, and contain up to 1% pedogenic $CaCO_3$.

1.2. Sample Collection and Analysis

Soil pits were dug to depths of 60 to 120 cm on flat summit sites of each of the six moraines in order to minimize disturbances owing to creep, erosion, and sediment accumulation. Excavation was ceased when visibly unweathered till was reached that contained clasts without observable weathering rinds. Within each soil pit, samples of about 2 kg were taken over roughly 10-cm intervals and stored in acid-washed low-density polyethylene (LDPE) bags. Soil samples were dried at 50°C, sieved through a 2-mm mesh and weighed by size fraction. Representative aliquots of each <2-mm sample were taken for determination of bulk soil major oxide concentrations, a second aliquot was taken for determination of bulk soil Rb and Sr concentration and $^{87}Sr/^{86}Sr$ ratios, and a third aliquot was taken for determination of the ammonium acetate exchangeable soil fraction Rb and Sr concentration and $^{87}Sr/^{86}Sr$ ratios.

Streamwater and snow samples were collected in acid-washed LDPE bottles. Water samples were collected from the stream draining the Twins Glacier cirque upstream of Titcomb Lakes at elevations of 3505 and 3320 m during July of 1993. The catchment area for these stream samples is covered predominantly by till of Audubon age, although some Gannett Peak age till is also present. Snow was collected from snowfields at an elevation of 3400 m in the Twins Glacier cirque upstream of Titcomb Lakes during July of 1993, and from Union Pass 35 km to the northeast (along the crest of the Wind River Mountains) during March of 1993. Snow was sampled by removing the top 10 cm to minimize dry deposition since the time of snowfall, and then snow was collected over a 1-m depth profile to get an average value for the winter snowpack. Samples of the prevalent granitic-gneiss bedrock were also collected near Titcomb Lake for mineral and whole rock analysis. Polished thin-sections of the soils and fused glass beads produced by melting the soils were examined by optical and scanning electron microscopes and analyzed by energy dispersive X-ray analysis following the methods outlined in Taylor and Blum[2] to determine parent material mineralogy and bulk soil major element composition. Methods of analysis for Rb and Sr concentration and ^{87}Sr/^{86}Sr ratios are given in Blum and Erel.[6,7]

2. RESULTS

2.1. Relation between Weathering Rates and Soil Age

The average composition of the deepest (least weathered) samples in weight percent is: SiO_2 (69.2), Al_2O_3 (15.9), CaO (3.20), Na_2O (3.52), K_2O (2.55), MgO (1.15), FeO_{total} (3.48), and TiO_2 (0.55). In soil profiles developed on the moraines, Ca, Na, K, and Mg show a general decrease in concentration toward the surface, whereas the Ti concentration increases toward the surface, consistent with the assumption that Ti is immobile in these soils.[2] The overall amount of base cation depletion at the top of the profiles increases with increasing soil age and the total amount of base cations lost from each profile also increases with soil age.[2]

Long-term chemical weathering rates (R_{LT}) were calculated for each soil profile by comparing soil analyses with the deepest (unweathered) sample for each profile. Ti was used as the immobile index element (e.g., April et al.,[13] Merritts et al.,[17] Bain et al.[15]). First, the depletion factor ($x_{i,j}$) was calculated for the <2 mm fraction of each soil sample based on the equation

$$x_{i,j} = (C_{i,j}/C_{i,p})/(Ti_j/Ti_p) \tag{1}$$

where $C_{i,j}$ is the concentration of element i in soil slice j, $C_{i,p}$ is the concentration of element i in the parent material (p), and Ti_j and Ti_p are the concentrations of Ti in the weathered soil and parent materials, respectively. Second, the chemical inventory ($S_{i,j}$) was calculated for each element in the <2 mm fraction of each soil slice that would have existed in a 1-m^2 column prior to any weathering on the basis of the equation

$$S_{i,j} = (C_{i,p})(\rho)(F_j)(T_j) \tag{2}$$

where ρ is the parent material density (the average measured ρ of 2.0 g cm^{-3} was used), F_j is the weight fraction of <2 mm material, and T_j is the thickness of the soil slice. Third, values for $S_{i,j}$ were summed over each soil pit to get a total initial inventory for each element (I_i). Fourth, the present-day inventory of each element from each soil slice ($L_{i,j}$) was calculated on the basis of the equation

$$L_{i,j} = (x_{i,j})(S_{i,j}) \tag{3}$$

and $L_{i,j}$ was summed over each soil profile to get the total amount of each element present in each soil profile (P_i). Fifth, the amount weathered (W_i) was calculated

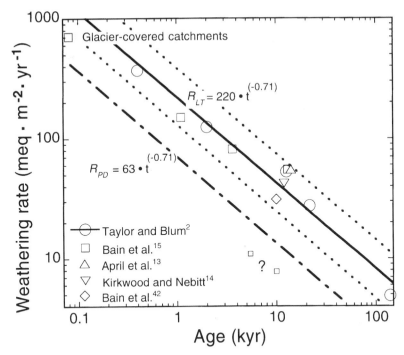

FIGURE 1.   Log–log plot of base cation weathering rates vs. soil age. Data points are long-term weathering rates (R_{LT}) from the references listed on the figure. The dark solid line is a power-law curve fit to the data from Taylor and Blum[2] ($R = 0.99$). Light dotted lines bracket curve fit to data and represent estimated maximum uncertainties of a factor of two in weathering rates. Question mark indicates two points from Bain et al. [15] that are suspect on the basis of high Zr contents of parent materials. Dashed line represents present-day weathering rates (R_{PD}) calculated from solid curve (see text). The shaded area represents the range of fluxes in glacier-covered catchments compiled by Sharp et al.[8] and are plotted at an arbitrary age.

for each element in each profile on the basis of the equation

$$W_i = I_i - P_i \qquad (4)$$

and W_i was divided by the age of the soil profile to yield R_{LT} for each base cation. The sum of base cationic charge in meq m^{-2} year^{-1} is reported as the long-term weathering rate (R_{LT}) and is plotted in Fig. 1.

To calculate long-term weathering rates from soil profile cation denudation, one must choose sampling sites in which there has been a minimum amount of eolian deposition of material with a composition significantly different from that of the soil parent material. Sampling sites were carefully chosen on flat summits of broad terminal moraines on the southwestern side of the Wind River Mountains to minimize eolian deposition. Dahms[38] has suggested that there is evidence for eolian deposition of heavy minerals from the Absoroka volcanics in the soils of the Wind River Mountains. However, data from Dahms[38,39] can be used to demonstrate that this contribution is minor, probably less than 5% of the total soil minerals. In addition, $^{87}Sr/^{86}Sr$ measurements of bulk digests of both B and C horizons of each soil profile indicate that there is no systematic lowering of the $^{87}Sr/^{86}Sr$ ratio toward the surface as would be required by the addition of as little as 5% of the low $^{87}Sr/^{86}Sr$ Absoroka volcanics.[7] Another necessary criterion for the choice of soils is to select sites where there has not been erosion of soil surfaces. The choice of flat sampling sites, combined with the dry climate, well-drained nature of the soils, and systematic behavior of the soil chemistry and grain-size distribution with increasing depth, suggests that erosion has not been significant.

Calculations of long-term weathering rates are additionally dependent on the assumptions that the deepest sample from each profile and each of the >2-mm soil fractions are not significantly weathered, and that Ti is immobile in granitic soils. The unweathered nature of the deepest samples and >2 mm-size fraction was verified by optical and scanning electron microscope observations of soil polished thin sections. The only significant weathering that has taken place is the oxidation of Fe^{2+} from biotite and development of Fe-oxide staining in and around biotite crystals. The immobility of Ti in the soils has been generally documented in the literature (see, e.g., Garden,[40] and April et al.[13]) and is consistent with the steadily decreasing Ti content of the soils with depth in the profiles.[2]

The most important additional variables other than soil age that may influence weathering rates are temperature, precipitation, and vegetation. Each of these factors is relatively constant among the three profiles in Titcomb Basin and among the three profiles near Fremont Lake. However, the Fremont Lake profiles are in an area of lower precipitation and higher temperature than the Titcomb Basin profiles. To assess the importance of this climatic difference, the difference in predicted weathering rates between the two sites was calculated on the basis of the relation between modern-day granitic catchment weathering rates and climate proposed by White and Blum.[41] This relation suggests that the decreased precipitation and increased temperature would approximately offset one another, and thus suggests that there would be no significant climatically induced difference in weathering rates between the two sites.

The weathering rates for all of the base cations decrease with increasing soil age (by factors of 30–100) providing constraints on the relation between R_{LT} and soil ages over the age interval of 0.4 to 138 thousand years. It is estimated that uncertainties in weathering rates (owing to analytical error and the possibilities of soil heterogeneity, eolian deposition, or surface erosion) are less than a factor of two. Values for R_{LT} and soil age (Figs. 1–3) vary from $370\,meq\,m^{-2}\,year^{-1}$ for the 0.4-thousand-year-old profile to $4.9\,meq\,m^{-2}\,year^{-1}$ for the 138-thousand-year-old profile. The R_{LT} value for the Sacagawea soil is $5.3\,meq\,m^{-2}\,year^{-1}$, which is a maximum estimate because the adopted age of 297 thousand years is a minimum age. Nevertheless, the age of the Sacagawea till is unlikely to be greater than about 500 thousand years and thus the weathering rate is bracketed

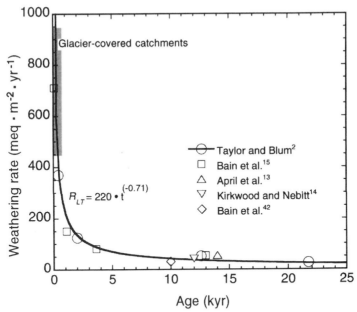

FIGURE 2.   Linear plot of the same data and power law function as in Fig. 1, but with the age axis truncated at 25 thousand years.

between approximately 3 and 5 meq m^{-2} year^{-1}. The rapid decreases in weathering rates with increasing soil age appear to level off to a steady-state weathering rate of 1–5 meq m^{-2} year^{-1} for soil ages greater than about 138 thousand years. Directly comparable cation depletion studies of soils developed on granitic glacial moraines in the Adirondack Mountains (northeastern United States) yielded R_{LT} values of 45 and 62 meq m^{-2} year^{-1} for 14-thousand-year-old soils,[13] studies in Ontario (Canada) yielded a R_{LT} value of 42 meq m^{-2} year^{-1} for 12-thousand-year-old soils,[14] and studies in Scotland yielded a mean R_{LT} value of 31 meq m^{-2} year^{-1} for 10-thousand-year-old soils[42]; all of these plot close to the trend established by Taylor and Blum[2] (Figs. 1 and 2). A cation depletion study of soils developed on six fluvial terraces in Scotland ranging in age from 0.08 to 13 thousand years[15] yielded four values for R_{LT} close to the trend established by the Wind River Mountains data. However, two of the terraces that have anomalously high parent material Zr contents yielded much lower R_{LT} values (Fig. 1), perhaps owing to heterogeneity of zircon in the fluvial sediments.

The data reviewed here suggest that there is a power law relation between R_{LT} and soil age over the 0.4- to 138-thousand-year timescale described by the equation

$$R_{LT} = 220 \cdot t^{(-0.71)} \tag{5}$$

where R_{LT} is in meq m^{-2} year^{-1}, and t is the age in thousands of years. This

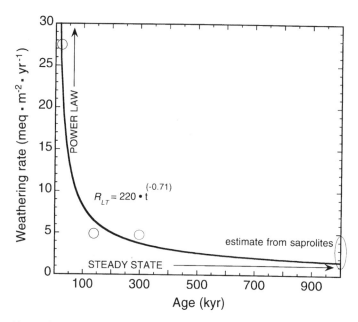

FIGURE 3.   Linear plot of the same data and power-law function as in Figs. 1 and 2, but with the weathering rate axis truncated at 30 meq m^{-2} year^{-1}. Also plotted is the weathering rate from the Sacagawea Ridge moraine, which is absent from Fig. 1 and was not used in the determination of the power-law function because of uncertainties in the moraine age. The "estimate from saprolites" field is derived by "correcting" the saprolite formation rates from Pavich[57] to the climatic conditions of the Wind River Mountains using the empirical relation determined by White and Blum.[43]

relation however, is specific to the precipitation–temperature regimes of the studied soils, which are estimated to vary by less than 50% and 5°C, respectively, among the five sites compared in Fig. 1. These differences in climate translate into a difference of less than a factor of two in predicted cation fluxes on the basis of the model of White and Blum.[43] The data from the literature plotted in Fig. 1 are also consistent with this correlation within an uncertainty in the R_{LT} of less than a factor of two (Fig. 1).

To proceed to a discussion of how present-day (or stream flux-derived) weathering rates (R_{PD}) compare with the empirically derived equation for long-term rates (R_{LT}), we must transform the y-axis of Fig. 1 from R_{LT} to R_{PD}. Given that

$$R_{LT} = k \cdot t^{-n} \tag{6}$$

and

$$R_{LT} = (1/t) \int_0^t R_{PD}(t')dt' \tag{7}$$

by differentiation one can derive the relationship:

$$R_{PD}/R_{LT} = 1 - n \qquad (8)$$

Thus, for the data from this study we get:

$$R_{PD} = 63 \cdot t^{(-0.71)} \qquad (9)$$

(Fig. 1). One implication of Eq. (8) is that present-day silicate weathering rates (R_{PD}) will always be approximately one-third of the long-term average weathering rates (R_{LT}) regardless of how long it has been since the most recent deglaciation. The exponential decrease in R_{LT} with time plotted in Fig. 1 will be mimicked by an exponential decrease in R_{PD}, except that R_{PD} will be lower by a factor of three. We have also plotted estimates of values for the R_{PD} in partially glacier-covered catchments compiled by Sharp et al.,[8] which range from 450 to 950 meq m^{-2} year^{-1}. The "soil age" for the glaciated catchments is clearly very "young" and is plotted at ≤0.1 thousand years for comparison, but this age is not meant to be taken literally, because the weathering surfaces in these catchments are either glacier-covered or mostly barren rock.

2.2. Relation between $^{87}Sr/^{86}Sr$ Released by Weathering and Soil Age

Two types of $^{87}Sr/^{86}Sr$ ratio measurements were completed on soil samples: One B-horizon and one C-horizon sample from each soil pit were totally digested and analyzed as bulk soils; two to five samples that include A, B, and C horizons from each soil pit were exchanged with ammonium acetate buffered at a pH of 7 and the exchange fraction was analyzed (see Blum and Erel[7] for detailed methods and complete data). The bulk soil analyses were used by Blum and Erel[7] to show that the $^{87}Sr/^{86}Sr$ ratios of the soil parent materials for the six soil profiles within the Fremont Creek drainage basin were essentially the same, and the eolian input to the B horizons was sufficiently small as to have no significant effect on the soil $^{87}Sr/^{86}Sr$ ratio. A small decrease in the $^{87}Sr/^{86}Sr$ ratios from the three youngest soils to the three oldest soils was attributed to vermiculitization of biotite from the older soil profiles. Although all other minerals appear essentially unweathered, biotite crystals observed in optical and scanning electron microscope investigations of soil thin-sections of even the deepest C-horizon samples display evidence of alteration with Fe-oxide staining, lower birefringence and pleochroism, and open cleavage planes containing Fe–Ti oxide precipitates. It is largely the weathering of biotite that leads to the "oxidized" appearance of C-horizon samples as deep as 1–2 m in Bull Lake soil profiles.[37]

In comparison to the soil digests, the $^{87}Sr/^{86}Sr$ ratios of the soil cation exchange pool vary to a much greater extent among soil profiles, ranging from 0.71137 to 0.79467 (Figs. 4 and 5). The $^{87}Sr/^{86}Sr$ ratios display only a narrow range within each soil pit, but an enormous range among soil pits.[7] Exchange fraction samples of the B horizons (the zone of most active weathering) are

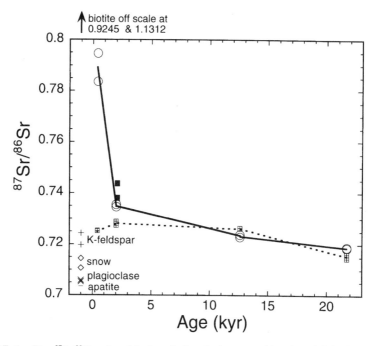

FIGURE 4. The $^{87}Sr/^{86}Sr$ ratio of bulk soil digests (square with $+$) and B-horizon soil cation exchange complex (open circles) for samples from the four youngest soil profiles plotted vs. the age in thousand of years of the moraine on which the soil formed. Solid squares are analyses of streamwater draining predominantly Audubon till. Bedrock K-feldspar ($+$), plagioclase ($\times$), apatite ($-$), biotite (off-scale), and snow values are from Blum and Erel[7] and Naylor et al.[86], and are plotted at an arbitrary negative "age" for comparison. The solid and dashed lines connect the B-horizon soil samples.

considered the best estimate of the $^{87}Sr/^{86}Sr$ presently being released by weathering. Samples of streamwater draining the cirque above the Audubon till (which is predominantly an Audubon-age weathering surface and is hereafter referred to as Audubon streamwater) have $^{87}Sr/^{86}Sr$ ratios that are only slightly higher than the values of the cation exchange pool of the B-horizon Audubon soil (Figs. 4 and 5), presumably owing to some input from the higher $^{87}Sr/^{86}Sr$ Gannett Peak soils, which are also present in a small percentage ($<10\%$) of the drainage basin. Samples of the 1992 snowpack at Union Pass and the 1993 snowpack at Titcomb Basin have Sr concentrations 20 to 50 times lower than the Audubon streamwater and $^{87}Sr/^{86}Sr$ of 0.70164 and 0.71442, considerably lower than most of the soil values.[7] The Sr concentrations of the snow samples are so low compared to streamwater that melt water could only account for some 5% or less of the Sr dissolved in the streamwater samples draining the Audubon-age weathering surface. This and other arguments led Blum and Erel[7] to conclude that atmospheric contributions of Sr to the soil cation exchange pool are quite small in this setting.

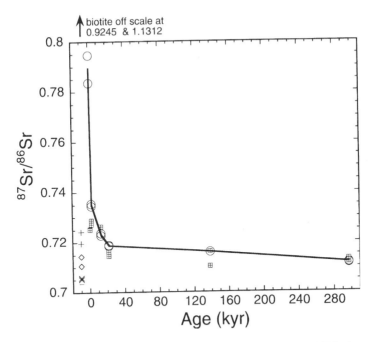

FIGURE 5. The $^{87}Sr/^{86}Sr$ ratio of bulk soil digests (square with $+$) and B-horizon soil cation exchange complex (open circles) for samples from all six soil profiles plotted vs. age in thousands of years of the moraine on which the soil formed. Bedrock K-feldspar ($+$), plagioclase ($\times$), apatite ($-$), biotite (off scale), and snow values are from Blum and Erel[7] and Naylor *et al.*[86], and are plotted at an arbitrary negative age for comparison. The solid line connects the B-horizon cation exchange samples.

It is clear from Fig. 4 that the $^{87}Sr/^{86}Sr$ ratio of the soil exchange complex decreases dramatically with increasing soil age, whereas the bulk soil $^{87}Sr/^{86}Sr$ ratios show little change. For all B-horizon samples the soil cation exchange pool $^{87}Sr/^{86}Sr$ is probably controlled almost entirely by chemical weathering of the dominant Sr-bearing parent minerals, which include plagioclase, K-feldspar, and biotite. The $^{87}Sr/^{86}Sr$ ratios of minerals from bedrock in the Fremont Creek drainage are plotted in Figs. 4 and 5. Because the $^{87}Sr/^{86}Sr$ ratio of the Gannett Peak and Audubon soils is higher than either plagioclase or K-feldspar, biotite weathering must be a major contributor of exchangeable Sr to these soils. Similarly, because the $^{87}Sr/^{86}Sr$ ratios of the Titcomb Lake, Pinedale, Bull Lake, and Sacagawea Ridge soils are lower than either biotite or K-feldspar, plagioclase must be a major contributor of exchangeable Sr to these soils. Thus the only mineral pair that can produce the soil-exchangeable $^{87}Sr/^{86}Sr$ values in each of the soil profiles is plagioclase and biotite. If the plagioclase–K-feldspar mineral pair were responsible for the Titcomb Lake, Pinedale, Bull Lake, and Sacagawea Ridge soil values with no contribution from biotite, K-feldspar weathering would have to be 25, 6, and 4 times faster than plagioclase, an unlikely scenario considering that K-feldspar is known to weather at a much slower rate than plagioclase in laboratory[46] and soil[41] studies, and K-feldspar is also commonly observed to weather from bedrock outcrops at a much slower rate than plagioclase.

Blum and Erel[7] discuss various explanations for the observed $^{87}Sr/^{86}Sr$ in the soils and conclude that the weathering behavior of biotite has a dominant influence on the Rb–Sr isotope systematics of both the soil digests and soil cation exchange pool in the soil chronosequence, thus demonstrating that the weathering rate of biotite is not constant through time in young soils. Clay mineralogical studies have shown that in cool alpine as well as temperate soils biotite weathers first to hydrobiotite (mixed-layer biotite–vermiculite) and then to vermiculite.[44,45] The vermiculitization reaction is perhaps the most rapid weathering reaction that takes place when fresh granitic materials are exposed to the weathering environment, as evidenced by Fe-oxide staining on otherwise fresh outcrops and Fe-oxide staining and open cleavage observed in thin sections of soil biotite in otherwise unweathered parent materials. This is also consistent with mineralogical studies of these soils, which indicate a decrease in the abundance of biotite and increase in the abundance of vermiculite with increasing soil age.[27] Hydrobiotite (mixed-layer biotite–vermiculite) has also been identified by X-ray diffraction,[35] but only in the two youngest soil profiles, which have the highest exchangeable $^{87}Sr/^{86}Sr$ ratios of the soil exchange pool. The presence of the intermediary mineral hydrobiotite is consistent with the high $^{87}Sr/^{86}Sr$ ratios and also suggests that the biotite-vermiculite reaction is currently occurring in these soils.

3. DISCUSSION

The results of several soil chronosequence studies carried out in the Wind River Mountains have been described above. Weathering rates were found to decrease with soil age according to the power law equation: $R_{LT} = 220 \cdot t^{(-0.71)}$, where R_{LT} is in $meq\,m^{-2}\,year^{-1}$, and t is in (thousands of years)^2. In addition, a negative correlation was observed between the $^{87}Sr/^{86}Sr$ ratio of exchangeable Sr in soils and the soil age, indicating that the $^{87}Sr/^{86}Sr$ ratio of Sr released in the early stages of weathering is significantly higher than in later stages owing to the rapid alteration of biotite to vermiculite and coincident release of highly radiogenic Sr.[6,7] The remainder of this chapter will involve a discussion of how these results can be applied to a variety of issues central to the question of how late Cenozoic glaciation and tectonic uplift may have affected rates of silicate weathering and the consequent rate of consumption of atmospheric CO_2.

3.1. Silicate versus Carbonate Weathering

The chronosequence studies summarized above, and the discussions of global weathering rates below, are restricted solely to the silicate weathering fraction of regional chemical denudation and dissolved river fluxes. Calcium carbonate weathers approximately 10 times more rapidly than crystalline silicate mineral,[47,48] and therefore even a small amount of carbonate in a river basin can have a dominant influence on the dissolved-element chemistry of a river. In a detailed study of major element and Sr isotopic compositions of rivers draining the Canadian Shield, Wadleigh et al.[49] estimated that 20% of the weathering flux of Sr was derived from silicates. Based on global river flux data, Berner and Berner[50] arrived at the result that 20% of the weathering-derived Ca dissolved in world-average river water is derived from silicate weathering, as compared to 80% from carbonate and sulfate weathering. Bickle[51] estimated an upper limit for the silicate weathering derived portion of the riverine Sr flux of about 40%. Although silicate weathering accounts for only a small proportion of the cation flux to the oceans, it is this portion that is of most interest in biogeochemistry because it is silicate weathering, and not carbonate weathering, that consumes atmospheric CO_2 and provides a negative feedback mechanism to the global climate system.[50] It is also only the silicate portion of the weathering flux that significantly influences the marine $^{87}Sr/^{86}Sr$ ratio, because Sr derived from marine carbonate and sulfate is generally close to the marine $^{87}Sr/^{86}Sr$ ratio, whereas silicate $^{87}Sr/^{86}Sr$ ratios generally deviate dramatically from marine values.

3.2. Power-Law Relation between Soil Age and Weathering Rates

The power law relation observed between soil age and silicate weathering rates[2] for the Wind River Mountains chronosequence has widespread implications for biogeochemistry because there is reason to believe that it is generally applicable to global silicate weathering processes. The relation probably results from several factors that change with soil age, including depletion of very fine-grained particles and exhaustion of rapidly weathered minor mineral phases. Harden[16] observed a similar power law relation between soil age and increases in the clay content of soils developed on granitic alluvial deposits spanning the 10- to 30-thousand-year age range. Hodder et al.[52] demonstrated that clay formation in soils developed on rhyolitic tephras could be described in its early stages in terms of a power law relation. A power law relation commonly referred to as the "parabolic rate law" has also been used to describe the proportionality of feldspar weathering with the square root of time in the initial stages of laboratory dissolution experiments.[53,54] While this was originally interpreted as indicating that the rate-controlling step in mineral dissolution was diffusion of solutes through a thickening, altered surface layer, it was later shown by Holdren and Berner[55] that a large portion of this nonlinear behavior arises from the grinding procedure used for the preparation of minerals. The parabolic behavior was partially attributed to the exhaustion of fine particles produced by grinding of the mineral specimens. When fine particles were carefully removed, a more linear rate law was observed. This may be directly analogous to what is observed for the glacial soil chronosequence, where the grinding and plucking action of glaciers produces a bimodal distribution of mineral grain sizes.[56]

Based on the chronosequence data presented above, it can be argued that the power law weathering behavior gives way to a linear rate law in the soils after a weathering duration of approximately 138 thousand years (Fig. 3). We note that in old saprolitic mantles in the Appalachian Piedmont, soil formation progresses at a fairly constant rate, presumably having reached a steady state behavior.[57-59] Pavich[57,58] estimated a minimum granitoid weathering rate in the Appalachian Piedmont of 4 mm per million years^{-1} based on solute fluxes and a maximum weathering rate of 20 mm per million years^{-1} based on ^{10}Be inventories, and pointed out that this agreed well with independent geophysical estimates of long-term tectonic uplift rates in this region. These rates correspond to a base cation denudation rate of 5 to 25 meq m^{-2} year^{-1}. Comparing measured granitoid saprolite formation rates directly with the Wind River Mountains soil chronosequence may be problematic because these investigated saprolites occur in warmer and wetter climates than the Wind River Mountains and there is a strong link between granitoid weathering rates and both mean annual temperature and precipitation.

Because the effects of glaciation on weathering discussed in this investigation are largely restricted to northern latitudes and high elevations, where precipitation and temperature are roughly comparable with those in the Wind River Mountains, I have taken the approach of attempting to "correct" the weathering rates determined for the Appalachian Piedmont to correspond to a climatic region similar to the Wind River Mountains or Canadian Shield. Based on the relation between weathering rates and both precipitation and temperature suggested for granitoid catchments by White and Blum,[43] it appears that climatic factors would be expected to result in weathering rates in the Appalachian Piedmont that are about 5 times greater than the Wind River Mountains, giving a "corrected" estimate of steady-state rates for the Wind River Mountains of 1 to 5 meq m^{-2} year^{-1}. This estimate of steady-state rates for saprolites is plotted at a soil age of 1000 thousand years in Fig. 3 along with the power-law relation established by Taylor and Blum[2] for chronosequence samples ranging in age from 0.4 to 138 thousand years. Also included in Fig. 3 is the weathering rate for the approximately 300-thousand-year-old soil from the chronosequence, which was not used by Taylor and Blum[2] to establish the power law because of higher uncertainty in the age of this soil profile, but which resulted in a maximum estimate of the weathering rate for a 300-thousand-year-old soil. Although the comparison of rates from the chronosequence and saprolites is subject to numerous assumptions, there are few other gauges of long-term granitoid weathering rates on passive topographic surfaces. Figure 3 demonstrates that while the power law relation observed for weathering of the chronosequence may significantly accelerate weathering for young soils, beyond about 138 thousand years in age the rates are expected to drop to much lower values of approximately 1 to 5 meq m^{-2} year^{-1} when normalized to the climatic conditions of the Wind River Mountains. Thus, prior to late Cenozoic glaciation, silicate weathering on the tectonically passive Canadian Shield is estimated to have been in the range of 1 to 5 meq m^{-2} year^{-1}.

3.3. Application of Chronosequence Studies to Global Weathering

The research described thus far in this contribution has forged a link between the soil age (or residence time) and both the chemical denudation rate and $^{87}Sr/^{86}Sr$ ratio released by silicate weathering. The results of this work are most relevant to very young (i.e., immature) soils in temperate climates, which generally occur in two settings: (1) late Cenozoic glacial deposits related to both alpine and continental glaciation and (2) rapidly uplifting mountainous areas undergoing high rates of mechanical denudation, in many cases predominantly by mass wasting. In both of these situations, mechanical erosion processes periodically regenerate fresh mineral surfaces and expose them to the agents of chemical weathering.

The results presented here are specific to the soil parent material and climate of the Wind River Mountains, but a strong case can be made that the results are globally representative of glaciated silicate shield terraines and can be extrapolated to a discussion of global weathering. The physical erosion processes occurring in both alpine and continental glacial environments tend to produce tills with a bimodal distribution of grain sizes centered around the silt and pebble size ranges.[56] Thus, weathering of alpine moraines should be similar to ground moraines of continental ice sheets. The mineralogy and bulk chemical composition of the soil parent materials from the Wind River Mountains are characteristic of the average chemical composition of large areas of the continental shields. In Table 1 the major element composition of the Wind River Mountains glacial moraines are listed along with estimates of the average chemical compositions of

TABLE 1. Major Element Composition of Samples from the C-Horizon of Each Soil Profile and Estimates for Large Areas of Continental Crust (from Taylor and McLennan[60] and references therein)

	SiO_2	Al_2O_3	CaO	Na_2O	K_2O	MgO	FeO (total)	TiO_2	P_2O_5
Moraines									
Gannett P.	68.5	15.1	3.51	3.41	2.36	1.53	4.36	0.55	0.63
Audubon	67.3	16.3	3.45	4.29	2.56	1.40	3.58	0.63	0.51
Titcomb L.	68.1	17.2	3.12	3.96	2.97	0.95	2.70	0.48	0.47
Pinedale	69.9	15.2	2.95	4.16	2.73	0.75	3.14	0.68	0.44
Bull L.	72.3	14.7	2.94	2.53	2.08	1.36	3.19	0.52	0.45
Sacagawea R.	69.2	16.1	3.58	2.81	2.60	0.93	3.88	0.44	0.44
Crustal areas									
Baltic and Ukrainian Shield	66.0	15.3	3.7	3.2	3.5	2.4	4.8	0.6	0.2
Canadian Shield	64.9	14.6	4.1	3.5	3.1	2.2	4.0	0.5	0.2
Canadian Shield	65.3	15.9	3.4	2.9	3.9	2.2	4.4	0.5	0.2
Northwest Scottish Highlands	70.4	14.9	3.1	4.5	2.0	1.6	2.9	0.4	0.1

the Baltic and Ukrainian shields, the Canadian Shield, and the northwest Scottish Highlands,[60] and close agreement is found for all major elements. Similarly, the biotite granodiorite modal mineralogy of the Wind River Mountains glacial moraines (approximately 40% plagioclase, 20% K-feldspar, 25% quartz, 10% biotite, and 5% amphibole, pyroxene, chlorite, and oxides) is very close to estimates of the average mineralogical composition of the upper crust, which ranges from 31 to 39% plagioclase, 9 to 13% K-feldspar, 20 to 24% quartz, 8 to 11% biotite, 4 to 8% muscovite, and 0 to 2% amphibole, pyroxene, chlorite, and oxides.[60] The only major crustal mineral lacking from the bedrock of the Wind River Mountains is muscovite, which is extremely resistant to chemical weathering and is thus not an important factor in this discussion. In addition to the compositional and mineralogical similarities, the approximately 2.5-billion-year age of the Wind River Mountains bedrock, which corresponds with a major period of crustal growth globally,[60] is close to the average age of the Canadian Shield.

Both continental and alpine glaciers strip land surfaces of soils and weathered bedrock during advance, and following retreat they leave fresh bedrock surfaces and tills comprised of finely ground minerals. During and following glacial retreats, this fresh bedrock material is exposed to the agents of chemical weathering. As discussed by Kump and Alley,[61] it is not immediately obvious what net effect a full glacial period has on overall chemical weathering rates. During glacial maxima some weathering takes place beneath soft-bed ice sheets, but this may be limited by the supply of CO_2 to the base of the ice sheet. At the glacial margins, large volumes of finely ground silt are exposed on glacial outwash plains and retransported and deposited over large areas as loess. Following glacial retreat, weathering rates should be at their peak as large areas of freshly deposited till are exposed and large volumes of meltwater flux through the glacial outwash areas. Thus, ice-covered areas should experience decreased weathering rates, whereas ice margins and deglaciated areas should experience accelerated weathering rates. Overall, one would expect decreased silicate weathering during glacial maxima and increased silicate weathering during interglacial periods. I will attempt to place some constraints on the magnitude of changes in silicate weathering on the glacial–interglacial timescale, as well as on changes that may have taken place during the onset of late Cenozoic glaciation when periodic glacial–interglacial cycling began in the Northern Hemisphere.

The marine $\delta^{18}O$ record is closely related to global ice volume and can be used to place constraints on the duration of continental glaciations. The periodicity of glacial–interglacial cycles is approximately 100 thousand years, and for the most recent glacial–interglacial cycle (for which the most detailed climate records are available), it is generally accepted that the 100-thousand-year interval between the last two glacial terminations was approximately one-half glacial and one-half interglacial periods.[62] Prior to about 0.9 million years ago the marine $\delta^{18}O$ record suggests that 41-thousand-year orbital cycles may have dominated over 100-thousand-year cycles.[63] For simplicity the entire 2.5-million-year late Cenozoic period of glacial–interglacial cycling will be treated here as alternating periods of 50-thousand-year full glacial periods followed by 50-thousand-year interglacial periods. This is admittedly a major oversimplification of the complex history of ice-sheet advance and retreat, but is in accord with the rough estimates of weathering rates presented here. If 20-thousand-year half-cycles were adopted for the period from 0.9 to 2.5 million years ago, this would have the effect of enhancing the expected increases in CO_2 consumption and marine $^{87}Sr/^{86}Sr$ that we discuss below.

To calculate the effects of glaciation on silicate weathering rates, cation fluxes from land areas deglaciated during interglacial periods must first be estimated. For this I utilize the power law function from Taylor and Blum[2] and adopt a value equivalent to a long-term weathering rate for a soil surface 50 thousand years old, yielding a value of $13\,meq\,m^{-2}\,year^{-1}$. This is only about 40% of the $31\,meq\,m^{-2}\,year^{-1}$ weathering rate estimated for the 15-thousand-year period since the last deglaciation of the Canadian Shield. One must also estimate weathering fluxes from areas when they are covered by continental ice sheets. Whether there is any appreciable weathering flux at all from beneath ice sheets depends largely on whether they have frozen or unfrozen beds. A recent study by Clark et al.[64] based on glacial geology and ice mechanics suggests that the North American ice sheet was unfrozen and moved on a soft bed of till. Weathering beneath ice sheets is probably limited by the low availability of CO_2 and is thus much slower than subareal weathering of freshly ground glacial debris. There are no published direct measurements of weathering rates beneath continental ice sheets, but Kump and Alley[61] discussed the various factors that might control these rates and came to the conclusion that "continental ice sheets reduce chemical weathering significantly — probably by one or more orders of magnitude — compared to preglacial or postglacial rates in the glaciated regions." I have chosen as a best guess of the maximum rate for subglacial weathering a value of approximately $1\,meq\,m^{-2}\,year^{-1}$, which is one order of magnitude less than the average rate estimated for the 50-thousand-year period prior to glacial advance.

Averaging the 50-thousand-year glacial and 50-thousand-year interglacial intervals yields a silicate weathering rate of $7\,meq\,m^{-2}\,year^{-1}$ for the entire period of late Cenozoic glaciation. If it is assumed that no weathering occurred in the subglacial environment (i.e., a frozen bed), one would arrive at a slightly lower weathering rate of $6.5\,meq\,m^{-2}\,year^{-1}$. Prior to the onset of late Cenozoic glaciations 2.5 million years ago, the soils in areas that have experienced late Cenozoic glaciation (in particular the Canadian Shield) probably developed thick saprolitic soils with ages of the magnitude of millions of years, such as are presently found in areas of North America unaffected by late Cenozoic glaciation.[57,59] If the power law relation for weathering rates is extrapolated to 1 million years, an estimated weathering rate of about $2\,meq\,m^{-2}\,year^{-1}$ is calculated, which is consistent with weathering rates for saprolites when corrected for differences in climate as discussed above (Fig. 3). Thus, if one compares the estimate of pre-late Cenozoic glaciation weathering rates with the late Cenozoic glacial period estimate averaged over glacial–interglacial cycles, it is calculated that the onset of late Cenozoic glaciation is likely to have produced an approximately threefold increase in the silicate weathering rate in glaciated areas.

The weathering rates discussed here pertain only to the silicate weathering flux from glaciated areas. To transform this to a change in global silicate weathering flux one must consider that only some 20% of the continental landmass was glaciated by the combination of continental ice sheets and expanded mountain glaciers during the last glacial maximum.[65,66] Up to roughly an additional 10% of the continental landmass may have been covered by fluvial and eolian glacial deposits,[67] but we disregard this in our calculations to be conservative in our estimates of the effects of glaciation on weathering. Based on the study of Wadleigh et al.,[49] one can calculate that the riverine cation flux per unit area from the Canadian Shield is close to the average global value of Palmer and Edmond.[68] Therefore, since some 20% of the continental crust was glaciated, one can assume that about 20% of the flux is subject to the effects of glaciation. Thus, a threefold increase in the average silicate weathering rates in glaciated areas during the late Cenozoic transforms into approximately a 40% increase in global silicate weathering rates (and consequent CO_2 consumption by silicate weathering). This "correction" can also be applied to the glacial–interglacial timescale and leads to the conclusion that global silicate weathering would have increased approximately fivefold during the first 1 thousand years after deglaciation compared to the full-glacial value if deglaciation were an instantaneous process. However, deglaciation of most of the continental landmass occurred over an interval of about 5 thousand years.[66] When deglaciation is integrated over a 5-thousand-year time period, one arrives at an estimate that global silicate weathering rates were approximately two-fold higher during the 5 thousand years of deglaciation compared to the previous full-glacial period. Global silicate weathering rates are also estimated to have been approximately 40% higher during the 5 thousand years of deglaciation compared to the 10-thousand-year period that followed the 5 thousand years of deglaciation. This may have provided a negative feedback mechanism whereby orbitally driven deglaciation enhanced the rate of CO_2 consumption by weathering and increased marine productivity owing to increased riverine nutrient fluxes during interglacial periods. This weathering-induced increase in CO_2 consumption may then have reinforced the orbitally driven cooling leading to the subsequent glacial period.[69]

The recognition that weathering rates do not remain constant with soil age has several implications for estimates of global geochemical fluxes, and affects estimates of base cation residence times in the ocean. One immediate conclusion is that riverine base cation fluxes from regions deglaciated during the Holocene are only representative of the present instant in time, which happens to be about 15 thousand years after deglaciation. The chronosequence data of Taylor and Blum[2] suggest that silicate weathering fluxes were much higher immediately following deglaciation and will continue to decrease with time following a power law function until the next glacial advance. Calculations of marine residence times for base cations are based on present-day (i.e., 15 thousand years after deglaciation) dissolved riverine fluxes and are, therefore, not representative of long-term averages. In the cases of Ca, Mg, and Na, riverine fluxes are strongly buffered by carbonate and evaporite weathering. Because carbonate and evaporite weathering are so much more rapid than silicate weathering and are probably not limited by reactive surface area, it is unlikely that glaciation would significantly affect rates of carbonate or evaporite weathering to the degree observed for silicate weathering. Only some 20% of the global Ca flux, about 40% of the global Mg flux, and about 34% of the global Na flux comes from silicate weathering[50] and, therefore, their residence times are not expected to be greatly affected by glaciation. However, about 95% of K in rivers is derived from silicate weathering, and thus the K flux should be quite sensitive to glaciation and should track the variations in silicate weathering fluxes owing to glaciation described above. One immediate implication of this is that the 10-million-year marine residence time of K based on present-day river fluxes is probably greater than the actual residence time averaged over glacial–interglacial cycles. Based on predicted silicate weathering fluxes averaged over glacial–interglacial cycles, the pre-late Cenozoic glaciation residence time of K is estimated to be about 30% greater and the residence time during the late Cenozoic glacial period about 25% greater than the accepted values based solely on present-day fluxes.

3.4. Implications of Chronosequence Studies for the Marine $^{87}Sr/^{86}Sr$ Record

In order to make estimates of the effects that glaciation may have had on silicate weathering rates, several loosely constrained approximations have been necessary. Ideally one would hope to test, or at least place some constraints on, these conclusions using the geologic record. The best proxy that is available for global cation denudation rates is the record of changes in the marine $^{87}Sr/^{86}Sr$ ratio. Marine $^{87}Sr/^{86}Sr$ ratios are controlled by the balance between fluxes of dissolved Sr derived from: (1) continental weathering, (2) midocean ridge hydrothermal circulation, and (3) diagenesis of ocean floor sediments. Diagenesis does not greatly affect marine $^{87}Sr/^{86}Sr$ because the $^{87}Sr/^{86}Sr$ ratio of this Sr is close to the marine value at any particular time and the Sr fluxes are quite small in comparison to the total. Hydrothermal circulation is important but generally assumed to have changed little during the past 30 million years as evidenced by nearly constant rates of seafloor production.[5] Therefore, the variation in marine $^{87}Sr/^{86}Sr$ is commonly used as a proxy for changes in continental weathering. However, these changes can reside either in the total continental flux of Sr or in the $^{87}Sr/^{86}Sr$ of the Sr released by continental weathering. The average riverine $^{87}Sr/^{86}Sr$ can be altered by changes in the proportion of carbonate, mafic volcanic, and granitoid rocks that contribute solutes to the oceans, or by the age of soils in granitoid terranes.[6]

The discussion here will be limited to the past 3.5 million years of the marine Sr isotope record to evaluate whether the changes in silicate weathering cation flux and $^{87}Sr/^{86}Sr$ ratio predicted to have occurred at the onset of the late Cenozoic glacial period can be accommodated within the marine $^{87}Sr/^{86}Sr$ record. Richter and DePaolo,[70] Capo and DePaolo,[71] and Hodell et al.[72] were the first to investigate the detailed marine $^{87}Sr/^{86}Sr$ record in this time interval. For this study the more recent and detailed data set of Farrell et al.[73] is used for

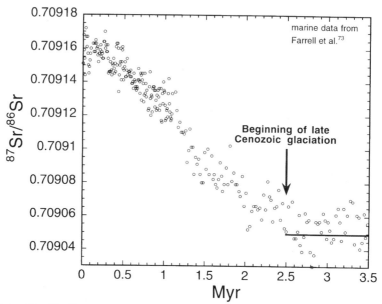

FIGURE 6. The $^{87}Sr/^{86}Sr$ ratio of seawater during the past 3.5 million years derived from measurements of planktonic foraminifera from ODP Site 758 in the Indian Ocean[73]. Note the zero slope prior to late Cenozoic glaciation between 3.5 and 2.5 million years ago, and the abrupt change in slope at the beginning of late Cenozoic Northern Hemisphere glaciation.

reference. Farrell *et al.*[73] analyzed 455 planktonic foraminfera samples in the time interval from 7 million years ago to the present from ODP site 758. The long residence (or replacement) time of Sr (i.e., ~2.4 million years) compared to the short oceanic mixing time (i.e., ~1 thousand years[74]) requires that the Sr isotopic composition of the world's oceans has remained homogeneous at any given time. Therefore, the data of Farrell *et al.*[73] for the Indian Ocean can be considered representative of the world ocean. In the time interval from 3.5 to 2.5 million years ago the marine $^{87}Sr/^{86}Sr$ remained nearly constant at a value of 0.70905 and then 2.5 million years ago began a steady increase to the present day value of 0.70917 (Fig. 6).[72,73] This transition 2.5 million years ago marks a change in the flux and/or $^{87}Sr/^{86}Sr$ ratio of Sr entering the oceans and has been attributed by Raymo *et al.*[75] and Hodell *et al.*[72] to the onset of Northern Hemisphere glaciation, and/or increased rates of tectonic uplift in the Himalaya Mountains and Tibetan Plateau.

It is possible to test the sensitivity of the marine $^{87}Sr/^{86}Sr$ ratio to the changes in cation flux and $^{87}Sr/^{86}Sr$ ratios that are believed to have been caused by the onset of late Cenozoic glaciation. The first assumption that must be made is that the global ocean was in a steady state with respect to the $^{87}Sr/^{86}Sr$ ratio during the time interval from 3.5 to 2.5 million years ago. This is a reasonable assumption because the marine $^{87}Sr/^{86}Sr$ ratio remained relatively constant

TABLE 2. Parameters Used in Sr Seawater Calculations
(from Palmer and Edmond[68])[a]

	Flux (mole/year)	$^{87}Sr/^{86}Sr$
Riverine	$J_R = 33.3 \times 10^{10}$	$R_R = 0.7119$
Hydrothermal	$J_H = 15.6 \times 10^9$	$R_H = 0.7035$
Diagenetic	$J_D = 3.4 \times 10^9$	$R_D = 0.7085$

[a]Amount of Sr in ocean: $N = 1.25 \times 10^{17}$ moles; Sr residence time: $\tau = 2.39 \times 10^{10}$ years.

during the entire 4.5- to 2.5-million-year time period. Instantaneous shifts in the Sr flux and/or $^{87}Sr/^{86}Sr$ ratio of riverine input can then be assumed, and the response of the marine $^{87}Sr/^{86}Sr$ ratio can be calculated and compared with the marine record. Utilizing the equations describing steady-state and transient-state evolution of marine $^{87}Sr/^{86}Sr$ (see Hodell et al.[76] for a good summary) and utilizing the constants given in Table 2, one can also calculate the instantaneous change in either riverine Sr flux or $^{87}Sr/^{86}Sr$ values that would best reproduce the Farrell et al.[73] seawater $^{87}Sr/^{86}Sr$ curve over the past 2.5 million years. The result of this calculation is that either a 0.00027 increase in the riverine $^{87}Sr/^{86}Sr$ ratio or a 10% increase in the global riverine Sr flux 2.5 million years ago would closely reproduce the observed changes in marine $^{87}Sr/^{86}Sr$ values (Fig. 7).

The next step is to evaluate how the changes in riverine Sr flux and/or $^{87}Sr/^{86}Sr$ ratio required to explain the marine record compare with estimates of how these parameters were affected by the onset of late Cenozoic glaciation. In an earlier section I argued that the onset of glaciation was likely to have produced a 40% increase in global silicate weathering rates. Considering that only 20% of the global riverine Ca (and Sr) flux is derived from silicate weathering, this would translate into an 8% increase in global riverine Sr flux. Thus, if the average riverine $^{87}Sr/^{86}Sr$ ratio remained constant, the calculated increase in Sr flux is close to, but lower than, the 10% increase needed to explain the marine $^{87}Sr/^{86}Sr$ curve (Fig. 7). The 8% estimated increase is probably a maximum estimate, however, because large portions of the Canadian Shield are currently covered with peat or are bare rock surfaces,[77] and therefore weathering rates are probably somewhat inhibited. However, it is also very unlikely that the average riverine $^{87}Sr/^{86}Sr$ ratio would have remained constant. It has been argued above that the riverine $^{87}Sr/^{86}Sr$ ratio increased at the onset of late Cenozoic glaciation owing to the generation of young weathering surfaces by glacial action and the resulting acceleration in the weathering rate of biotite.

If the riverine Sr fluxes are held constant, and riverine $^{87}Sr/^{86}Sr$ is allowed to vary with time following deglaciations as indicated by the chronosequence study presented above, one arrives at the estimates that the $^{87}Sr/^{86}Sr$ ratio of rivers draining glaciated Precambrian silicate rocks containing biotite would have increased by an average of 0.004 at the onset of late Cenozoic glaciation.[6] This

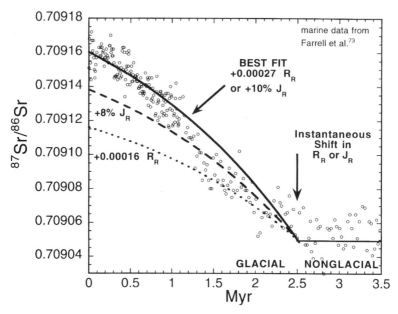

FIGURE 7. The same marine $^{87}Sr/^{86}Sr$ data as in Fig. 6, but with curves indicating the predicted response of marine $^{87}Sr/^{86}Sr$ to abrupt changes in either the riverine flux of $Sr(J_R)$ or the average riverine $^{87}Sr/^{86}Sr$ ratio (R_R) at the beginning of late Cenozoic glaciation. The best fit curve corresponds to either an increase in J_R of 10% or an increase in R_R of 0.00027.

could result in an increase in global riverine $^{87}Sr/^{86}Sr$ of 0.00016 based on the estimate that 20% of the continental area was glaciated and 20% of the riverine Sr flux is from weathering of Precambrian silicate rocks containing biotite. The estimated shift in riverine $^{87}Sr/^{86}Sr$ is only about 60% of the shift of 0.00027 that is required to reproduce the marine $^{87}Sr/^{86}Sr$ record over the past 2.5 million years (Fig. 7). This too is probably a maximum estimate because portions of the Canadian Shield are comprised of younger accreted terranes, which do not contain highly radiogenic biotite and would not, therefore, produce any increase in $^{87}Sr/^{86}Sr$ following deglaciation. Thus, an instantaneous increase in riverine $^{87}Sr/^{86}Sr$ of 0.00016 accompanied by an increase in riverine Sr flux of about 4%, or a somewhat lower increase in riverine $^{87}Sr/^{86}Sr$ and higher increase in Sr flux will match the marine $^{87}Sr/^{86}Sr$ record across the transition marking the onset of late Cenozoic glaciation (Fig. 7). A change in riverine $^{87}Sr/^{86}Sr$ owing to accelerated biotite weathering alone, however, is not seen as being a sufficient explanation for the entire shift in the marine record.

This analysis of the effect of glaciation on the marine $^{87}Sr/^{86}Sr$ record cannot prove one way or the other whether the changes in the marine record are due to glaciation, increased tectonic uplift, changes in global climate patterns, or other factors. The shift in the marine $^{87}Sr/^{86}Sr$ ratio is, however, coincident with the onset of Northern Hemisphere glaciation and does demonstrate that the estimates of the effects of Northern Hemisphere glaciation on silicate weathering (derived from the chronosequence studies), are of the proper magnitude to explain trends in marine $^{87}Sr/^{86}Sr$. One can see from Fig. 6 that the marine $^{87}Sr/^{86}Sr$ ratio appears to still be increasing at the present time, thus implying that the ocean is not currently in a steady state with respect to $^{87}Sr/^{86}Sr$. $^{87}Sr/^{86}Sr$ will likely continue to rise for several million years from the present value of 0.70917 to a steady-state value calculated to be 0.70922. It is important to note that any instantaneous shift in the fluxes and/or $^{87}Sr/^{86}Sr$ ratios into the ocean will incur a transient state that will tend toward a new steady-state value within a period of several residence times. Therefore, the sustained increase in the marine $^{87}Sr/^{86}Sr$ over the past 40 million years cannot be due to an instantaneous shift in either the riverine Sr flux or $^{87}Sr/^{86}Sr$ ratio such as I have proposed for the onset of late Cenozoic glaciation. Instead it must be due to a steadily increasing flux of Sr and/or increase in the $^{87}Sr/^{86}Sr$ ratio, which is probably related to the continual rise and/or dissection of the Himalayan Mountains and Tibetan Plateau.[5,75] Thus, it is my view that glaciation may be responsible for the fine structure of the most recent marine $^{87}Sr/^{86}Sr$ record, but that this is superimposed on long-term tectonically driven trends in marine $^{87}Sr/^{86}Sr$.

3.5. Tectonic Uplift and Silicate Weathering Rates

One of the consequences of tectonic uplift is that mountain ranges are often raised to elevations that can support alpine glaciation. Both alpine and continental glaciation are known to significantly increase physical denudation rates.[78,79] Sharp et al.[8] summarized available data on chemical weathering rates of silicate catchments containing active alpine glaciers and found that they are elevated by at least an order of magnitude compared to nonglaciated catchments (see Fig. 5). The only soil profile cation depletion weathering rates in the literature that approach the values for glaciated catchments are for very young soils 80 and 400 years old, studied by Bain et al[15] and Taylor and Blum,[2] respectively. Thus alpine glaciation can be considered a direct result of mountain uplift and would be expected to significantly increase silicate chemical weathering rates in many high mountain ranges — most notably the Himalaya and the Andes. However, the magnitude of this increase is difficult to estimate. Only a small proportion of the drainage basins of the large rivers draining the Himalaya and the Andes have been covered by glaciers, but when combined with the extensive areas covered by glacially derived fine-grained alluvium in the large foreland basins, one could argue that a considerably larger (but difficult to quantify) proportion of the uplifted region has been affected by the presence of glaciers. Other indirect outcomes of uplift are the changes in mean annual temperatures that accompany high elevation, and changes in precipitation patterns owing to orographic effects. The net effect of these parameters will vary depending on the mountain range, and there are insufficient data to estimate the global magnitude of these effects.

Increased tectonic uplift will clearly result in increased physical erosion rates,[80] but the relation between physical and chemical erosion rates has not been firmly established. Stallard and Edmond[81] discussed the difference between erosion-limited and weathering-limited regimes and established an important conceptual framework for discussion of the influence of uplift on chemical weathering. Although several workers have discussed these relations in detail qualitatively,[10-12] there is no quantitative basis for predicting the magnitude of the expected acceleration of chemical erosion owing to uplift. The chronosequence weathering studies reviewed in this chapter suggest that this link may be related to changes in the average age of soils in and adjacent to mountainous regions. However, the studies of soils developed on glacial moraines may not be directly applicable to soils developed on young surfaces exposed by mass wasting in tectonically active areas, because they will presumably have a somewhat different reactive surface area and a different distribution of grain sizes than glacial deposits. Nevertheless, one would still expect weathering rates to be accelerated in these areas. It appears likely that the rate of change of weathering rates with soil age will be smaller for young surfaces produced by mass wasting because young surfaces will not generally be composed of finely ground unweathered material as in the case of most glacial deposits.

Based on the Wind River Mountains chronosequence study,[2] it is expected that if soil residence times are shorter than about 138 thousand years, chemical weathering will be accelerated over steady-state rates. The concept of soil age is easy to define for young soils developed on glacial deposits of known age, but in most other environments this concept is not simple and may not even be definable. Considerable work has been done defining and measuring the steady-state formation rates and residence times of old saprolitic mantles in the Appalachian Piedmont using solute fluxes and ^{10}Be inventories,[57,58] but there has been very little study of younger soils in tectonically active areas, which is what would be most relevant to the present discussion. In many rapidly uplifting areas, soil ages are controlled largely by the size and frequency of landslides, which strip soil mantles and expose fresh bedrock and talus. Landslide frequency can be controlled by soil characteristics and local hydrology, and/or by the frequency and size of large earthquakes as documented by Keefer.[82] In some locations landslides can be thought of as producing a catchment-wide average soil age. Larsen[83] used aerial photos to map the size and frequency of landslides over a 39-year period in the Luquillo Mountains of eastern Puerto Rico and determined that 0.64% of the land area is disrupted per century, which gives an average recurrence interval of 16 thousand years for any given soil profile. Brown et al.[84] developed a methodology for utilizing in situ produced cosmogenic ^{10}Be in fluvial sediments to estimate denudation rates and applied this to the Luquillo Mountains with results generally consistent with the landslide frequency estimates of Larsen.[83] Unfortunately, this is the only study of this type that I know of from which average soil age can be estimated. There are no similar studies in actively uplifting areas from which one might estimate the relation between uplift rates and average soil age.

The Wind River Mountains chronosequence study[6,7] demonstrates that the presence of biotite with a Precambrian Rb–Sr mineral age in the soil parent material can lead to a correlation between soil age and the $^{87}Sr/^{86}Sr$ released by silicate weathering. Thus, as in the case of silicate weathering rates, one might assume that tectonic uplift rates should correlate somewhat with riverine $^{87}Sr/^{86}Sr$. However, this is probably not an important effect because rapidly uplifting mountain belts do not generally expose rocks with old mineral ages. In most major collisional orogenic belts such as the Himalaya, Appalachians, and European Alps, metamorphism was nearly coincident with rapid uplift, and thus although rocks with old protolith ages are exposed, they generally have Rb–Sr biotite ages that are only tens of millions of years at the time of exposure owing to rapid uplift. Additionally, in major volcanic-arc related mountain belts such as the Andes, Cascades, and Aleutians, the most important rocks undergoing erosion are young volcanics which are neither old nor biotite-bearing. Thus preferential weathering of biotite in young soils, which is clearly an important mechanism controlling riverine $^{87}Sr/^{86}Sr$ in recently glaciated Precambrian continental shields, is probably not significant in most uplifting mountain belts.

4. AREAS FOR FURTHER RESEARCH

The results of the glacial soil chronosequence studies in the Wind River Mountains can be directly applied to questions related to the effect of global glaciation on silicate weathering rates because the composition, mineralogy, and grain-size distribution of the soil parent material is representative of large regions of glaciated continental crust. Because glacial deposits can be easily dated, the relations established in such studies between weathering rates and the $^{87}Sr/^{86}Sr$ released by weathering can be extrapolated to large areas. However, the application of chronosequence data to quantitative estimates of the relation between tectonic uplift and silicate weathering rates and $^{87}Sr/^{86}Sr$ ratios is a much more difficult problem. The chronosequence studies provide insight into the mechanisms responsible for changes in silicate weathering with soil age. This helps to explain the relation between rapid uplift and accelerated silicate weathering, but it will not be possible to make quantitative estimates until more is known about the average age of weathering surfaces in rapidly uplifting areas and the effect of the grain size distribution of nonglacial sediments on weathering rates. Studies of average soil residence times on steepened slopes in mountainous terrain based on landslide frequencies or in situ cosmogenic nuclide production could shed light on the difficult problem of soil ages in mountainous terrain. Investigations of long-term weathering rates on bedrock surfaces and talus deposits would aid in application of the chronosequence data to rapidly uplifting areas.

In the studies of Taylor and Blum[2] and Blum and Erel,[6,7] the goal was to isolate the effects of soil age on silicate weathering rates and $^{87}Sr/^{86}Sr$ ratios released, by careful selection and study of a single soil chronosequence. The results of these studies provide important constraints on this isolated factor, which links mechanical and chemical erosion rates. However, there is a complex matrix of other factors that influence weathering rates including precipitation and

temperature,[43] changes in vegetation,[85] and distribution of exposed rock types —
all of which can also greatly affect weathering rates. Understanding the complex
interplay of these variables on silicate weathering rates provides a significant
scientific challenge for the future.

Acknowledgments

I wish to thank Yigal Erel and Aaron Taylor for invaluable collaborations
without which this work would not have been possible, and Alex Blum and
William Ruddiman for helpful reviews and comments. Funding was provided by
the NSF through Presidential Faculty Fellowship Grant EAR-9350632.

REFERENCES

1. Berner, R. A. (1995). In: *Chemical Weathering Rates of Silicate Minerals.* Reviews in Mineralogy, Vol. 31 (A. F. White and S. L. Brantly, eds.), pp. 565–583. Mineralogical Society of America, Washington, D.C.
2. Taylor, A., and Blum, J. D. (1995). *Geology* **23**, p. 979.
3. Raymo, M. E., and Ruddiman, W. F. (1992). *Nature* **359**, p. 117.
4. Edmond, J. M. (1992). *Science* **258**, p. 1594.
5. Richter, F. M., Rowley, D. B., and DePaolo, D. J. (1992). *Earth Planet. Sci. Lett.* **109**, p. 11.
6. Blum, J. D., and Erel, Y. (1995). *Nature* **373**, p. 415.
7. Blum, J. D., and Erel, Y. (1997). *Geochim. Cosmochim. Acta* **61**, p. 3193.
8. Sharp, M., Tranter, M., Brown, G. H., and Skidmore, M. (1995). *Geology* **23**, p. 61.
9. Gibbs, M. T., and Kump, L. R. (1994). *Paleoceanography* **9**, p. 529.
10. Stallard, R. F. (1995). *Ann. Rev. Earth Planet. Sci.* **23**, p. 11.
11. Lerman, A. (1994). In: *Material Fluxes on the Surface of the Earth*, Chap. 2 (National Research Council, eds.), pp. 28–45. National Academy Press, Washington, D.C.
12. Meybeck, M. (1994). In: *Material Fluxes on the Surface of the Earth*, Chap. 4 (National Research Council, eds.), pp. 61–73. National Academy Press, Washington, D.C.
13. April, R., Newton, R., and Coles, L. T. (1986). *Geol. Soc. Am. Bull.* **97**, p. 1232.
14. Kirkwood, D. E., and Nesbitt, H. W. (1991). *Geochim. Cosmochim. Acta* **55**, p. 1295.
15. Bain, D. C., Mellor, A., Robertson-Rintoul, M. S. E., and Buckland, S. T. (1993). *Geoderma* **57**, p. 275.
16. Harden, J. W. (1987). *U.S.G.S. Bull.* **1590-A**, p. A1.
17. Merritts, D. J., Chadwick, O. A., Hendricks, D. M., Brimhall, G. H., and Lewis, C. J. (1992). *Geol. Soc. Am. Bull.* **104**, p. 1456.
18. Mast, M. A., Drever, J. I., and Baron, J. (1990). *Water Resources Res.* **26**, p. 2971.
19. Kirchner, J. W., Dillon, P. J., and LaZerte, B. D. (1992). *Nature* **358**, p. 478.
20. Miller, E. K., Blum, J. D., and Friedland, A. J. (1993). *Nature* **362**, p. 441.
21. Bailey, S. W., Hornbeck, J. W., Driscoll, C. T., and Gaudette, H. E. (1996). *Water Resources Res.* **32**, p. 707.
22. Blum, J. D., Erel, Y., and Brown, K. (1994). *Geochim. Cosmochim. Acta.* **58**, p. 5019.
23. Åberg, C., Jacks, G., and Hamilton, P. J. (1989). *J. Hydrol.* **109**, p. 65.
24. Bain, D. C., and Bacon, J. R. (1994). *Catena* **22**, 201.
25. Jenny, H. (1980). *The Soil Resource*, 377 pp. Springer-Verlag, New York.
26. Richmond, G. M. (1973). *U.S.G.S. Quadrangle Map GQ-1138*, Scale 1:24,000.

27. Mahaney, W. C., and Halvorson, D. L. (1986). In: *Rates of Chemical Weathering of Rocks and Minerals* (S. M. Colman and D. P. Dethier, eds.), pp. 247–167. Academic Press, Orlando.
28. Sorenson, C. J. (1986). *U.S.G.S. Map I-1800*, Scale 1:24,000.
29. Hall, R. D., and Horn, L. L. (1993). *Chem. Geol.* **105**, p. 17.
30. Gosse, J., Klein, J., Evenson, E., Lawn, B., and Middleton, R. (1994). *U.S.G.S. Cir.* **1107**, p. 114.
31. Gosse, J., Klein, J., Evenson, E., Lawn, B., and Middleton, R. (1995). *Science* **268**, p. 1329.
32. Imbrie, J., and Imbrie, K. P. (1979). *Ice-Ages–Solving the Mystery*, 224 pp. Harvard University Press, Cambridge, Mass.
33. Mahaney, W. C., and Sanmugadas, K. (1983). *Z. Geomorph.* **273**, p. 265.
34. Worl, R. G., Lee, G. K., Long, C. L., and Ryan, G. S. (1984). *U.S.G.S. Map MF-1636-A*, Scale 1:250,000.
35. Douglas, T. A. (1994). Unpublished Senior Honors Thesis. Dartmouth College, Hanover, New Hampshire.
36. Mahaney, W. C. (1978). In: *Quaternary Soils* (W. C. Mahaney, ed.), pp. 227–247. Geoabstracts Ltd., Norwich, U.K.
37. Hall, R. D., and Shroba, R. R. (1995). *Arct. Alp. Res.* **27**, p. 89.
38. Dahms, D. E. (1993). *Geoderma* **59**, p. 175.
39. Dahms, D. E. (1991). Unpublished PhD Thesis, pp. 340. University of Kansas, Lawrence.
40. Gardner, L. R. (1980). *Chem. Geol.* **30**, p. 151.
41. White, A. F., Blum, A. E., Shultz, M. S., Bullen, T. D., Harden, J. W., and Peterson, M. L. (1996). *Geochim. Cosmochim. Acta* **60**, p. 2533.
42. Bain, D. C., Mellor, A., Wilson, M. J., and Duthie, D. M. L. (1994). *Water, Air, Soil, Pollut.* **73**, p. 11.
43. White, A. F., and Blum, A. E. (1995). *Geochim. Cosmochim. Acta.* **59**, p. 1729.
44. Wilson, M. J. (1970). *Clay Miner.* **8**, p. 291.
45. Gilkes, R. J., and Suddhiprakarn, A. (1979). *Clay Miner.* **27**, p. 349.
46. Lasaga, A. C. (1984). *J. Geophys. Res.* **89**, p. 4009.
47. Meybeck, M. (1987). *Am. J. Sci.* **287**, p. 401.
48. Drever, J. I., and Clow, D. W. (1995). In: *Chemical Weathering Rates of Silicate Minerals*, Reviews in Mineralogy, Vol. 31 (A. F. White and A. L. Brantley, eds.), pp. 463–483. Mineralogical Society of America, Washington, D.C.
49. Wadleigh, M. A., Veizer, J., and Brooks, C. (1985). *Geochim. Cosmochim. Acta.* **49**, p. 1727.
50. Berner, E. K., and Berner, R. A. (1996). *Global Environment: Water, Air, and Geochemical Cycles*, 376 pp. Prentice-Hall, Englewood Cliffs, N.J.
51. Bickle, M. J. (1994). *Nature* **367**, p. 699.
52. Hodder, A. P. W., Green, B. E., and Lowe, D. J. (1990). *Clay Minerals* **25**, p. 313.
53. Wollast, R. (1967). *Geochim. Cosmochim. Acta.* **31**, p. 635.
54. Busenburg, E., and Clemency, C. V. (1976). *Geochim. Cosmochim. Acta* **40**, p. 41.
55. Holdren, G. R. Jr., and Berner, R. A. (1979). *Geochim. Cosmochim. Acta* **43**, p. 1161.
56. Small, R. J. (1987). In: *Glacio-Fluvial Sediment Transfer* (A. M. Gurnell and M. J. Clark, eds.), pp. 111–145. Wiley, Chichester.
57. Pavich, M. J. (1986). In: *Rates of Chemical Weathering of Rocks and Minerals* (S. M. Colman and D. P. Dethier, eds.), pp. 551–590. Academic Press, New York.
58. Pavich, M. J. (1989). *Geomorphology* **2**, p. 181.
59. Thomas, M. F. (1994). In: *Rock Weathering and Landform Evolution* (D. A. Robinson and R. B. G. Williams, eds.), pp. 287–301. John Wiley, New York.
60. Taylor, S. R., and McLennan, S. M. (1985). *The Continental Crust: Its Composition and Evolution*, 312 pp. Blackwell, London.
61. Kump, L. R. and Alley, R. B. (1994). In: *Material Fluxes on the Surface of the Earth*, Chap. 3 (National Research Council, eds.), pp. 46–60. National Academy Press, Washington, D.C.
62, Shackleton, N. J., Imbrie, J., and Hall, M. A. (1983). *Earth Planet. Sci. Lett.* **65**, p. 233.
63. Raymo, M. E. (1994). *Rev. Earth Planet. Sci.* **22**, p. 353.
64. Clark, P. U., Licciardi, J. M., MacAyeal, D. R., and Jenson, J. W. (1996). *Geology* **24**, p. 679.
65. Flint, R. F. (1971). *Glacial and Quaternary Geology*, 892 pp. John Wiley, New York.

66. Hughes, T. J., Denton, G. H., Anderson, B. G., Schilling, D. H., Fastook, J. L., and Lingle, C. S. (1981). In: *The Last Great Ice Sheets*, Chap. 6 (G. H. Denton and J. T. Hughes, eds.), pp. 275–318, John Wiley, New York.

67. Taylor, S. R., McLennan, S. M., and McCulloch, M. T. (1983). *Geochim. Cosmochim. Acta* **47**, p. 1897.

68. Palmer, M. R., and Edmond, J. M. (1998). *Earth Planet. Sci. Lett.* **92**, p. 11.

69. Lovelock, J. E., and Kump, L. R. (1994). *Nature* **369**, p. 732.

70. Richter, F. M., and DePaolo, D. J. (1988). *Earth Planet. Sci. Lett.* **90**, p. 382.

71. Capo, R. C., and DePaolo, D. J. (1990). *Science* **249**, p. 51.

72. Hodell, D. A., Mead, G. A., and Mueller, P. A. (1990). *Chem. Geol.* **80**, p. 291.

73. Farrell, J. W., Clemens, S. C., and Gromet, L. P. (1995). *Geology* **23**, p. 403.

74. Broeker, W. S., and Peng, T-H. (1982). *Tracers in the Sea*, 690 pp. Eldigio Press, Columbia University, Palisades, New York.

75. Raymo, M. E., Ruddiman, W. F., and Froelich, P. N. (1988). *Geology* **16**, p. 649.

76. Hodell, D. A., Mueller, P. A., McKenzie, J. A., and Mead, G. A. (1989). *Earth and Planet. Sci. Lett.* **38**, p. 165.

77. Harden, J. W., Sundquist, E. T., Stallard, R. F., and Mark, R. K. (1992). *Science* **258**, p. 1921.

78. Hay, W. D. (1994). In: *Material Fluxes on the Surface of the Earth*, Chap. 1 (National Research Council, eds.), pp. 15–27. National Academy Press, Washington, D.C.

79. Hallet, B., Hunter, J., and Bogen, J. (1996). *Global and Planetary Change* **12**, p. 213.

80. Ahnert, F. (1970). *Am. J. Sci.* **268**, p. 243.

81. Stallard, R. F., and Edmond, J. M. (1983). *J. Geophys. Res.* **88**, p. 9671.

82. Keefer, D. K. (1984). *Geol. Soc. Am. Bull.* **95**, p. 406.

83. Larsen, M. C. (1991). *Geol. Soc. Am. Abstr. with Prog.* **23**, A256.

84. Brown, E. T., Stallard, R. F., Larsen, M. C., Raisbeck, G. M., and Yiou, F. (1995). *Earth Planet. Sci. Lett.* **129**, p. 193.

85. Berner, R. A. (1992). *Geochim. Cosmochim. Acta* **56**, p. 3225.

86. Naylor, R. S., Steiger, R. H., and Wasserburg, G. J. (1970). *Geochim. Cosmochim. Acta* **34**, p. 1133.

12

Himalayan Weathering and Erosion Fluxes: Climate and Tectonic Controls

Louis A. Derry and Christian France-Lanord

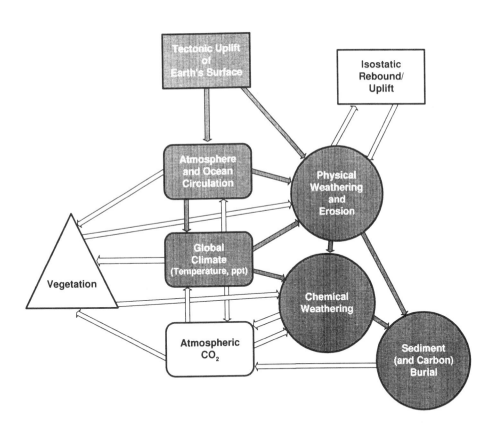

Louis A. Derry • Department of Geological Sciences, Cornell University, Ithaca, New York 14853-1504.   *Christian France-Lanord* • Centre de Recherches Pétrographiques et Géochimiques-CNRS, BP20 54501, Vandoeuvre-les-Nancy, France.

Tectonic Uplift and Climate Change, edited by William F. Ruddiman. Plenum Press, New York, 1997.

1. INTRODUCTION

A long-standing problem in Earth Sciences is understanding the evolution of global biogeochemical cycling through time. Such an understanding requires knowledge of many variables, one of the most important of which is the magnitude and composition of chemical weathering fluxes from the continents to the oceans. It is not possible to sample anything but present-day weathering fluxes, and the geological record does not contain a direct record of dissolved fluxes. Therefore reconstructing past fluxes depends on our ability to relate fluxes to some parameter that is measurable in the geological record. On a global scale, variations of some oceanic tracers such as the isotopic compositions of Sr and Ca are related to erosion and weathering fluxes; however, interpretation of these proxies is rarely straightforward. On a regional scale, records of synorogenic sedimentary accumulation provide information on past weathering and erosion fluxes, and may constrain the interpretation of global proxies.

The Himalaya is a region well suited for the study of the interaction of climate change with erosion and weathering processes. There is strong evidence of significant regional environmental changes superimposed on a very active tectonic history during the Neogene.[1,2] Much of the sediment derived from Neogene erosion of the Himalaya is still preserved, providing an opportunity to examine how clastic sediments record erosion and weathering processes.[3] Both suspended and dissolved fluxes in the major Himalayan rivers are high; therefore, changes in these fluxes may have global significance.[4] Isotopic proxies of past changes in global biogeochemical cycling have received considerable attention in the recent geological and paleoceanographic literature. The Sr and Ca isotopic composition of Himalayan weathering and erosion products are anomalous, and the marine record of these proxies may be sensitive to Himalayan processes.[4-6]

2. SEDIMENTS AS THE RECORD OF EROSION AND WEATHERING

The physical record of erosion and weathering is found in clastic sedimentary deposits. Sediment volume and chronostratigraphy can be used to estimate the mass of material removed from an orogenic belt by physical erosion and transferred to a depositional basin.[7-9] The accuracy of this approach may be limited by uncertainties about the stratigraphic architecture of the basin or by losses to subduction or subsequent uplift and reerosion of the basin sediments. Clastic sediments contain a record of chemical weathering as well. Weathering of silicate rocks is typically incongruent, resulting in the production of solutes and new secondary or residual phases (primarily clays) from the breakdown and partial dissolution of primary minerals . These secondary minerals accumulate in sedimentary basins, and so the mineralogical and chemical compositions of clastic sediments provide important information about chemical weathering of silicate source rocks. Clays are the direct record of silicate weathering. However, clastic sediments do not usually record information about the weathering of carbonate rocks, because they weather congruently (i.e., dissolve without the formation of secondary minerals) and leave no detrital record. Except in instances where detrital carbonate is found in sedimentary basins, little information about this important part of the weathering cycle is preserved.

The relationship between mechanical erosion and chemical weathering is complex. A number of studies have related sediment flux to drainage basin size, mean elevation, climate, river chemistry, and other factors.[10-13] Dissolved fluxes in large river systems typically contain a large component derived from the congruent dissolution of carbonate and evaporite sediments, and atmospherically derived salts, which can significantly mask the contribution of silicate weathering to the dissolved load.[13-17] Yet it is the smaller fraction of dissolved constituents released by silicate weathering that is potentially capable of driving global climate change.[18]

In this paper we present results from an ongoing investigation into the geochemistry of sediments in the Bengal Fan. Our data provide constraints on the provenance and supply rate of Bengal Fan sediments, on their weathering history, and on the isotopic composition of Sr and C in Himalayan erosion and weathering fluxes over the last 20 million years. We compare these data from the major repository of Himalayan erosion products with other constraints on Himalayan tectonic and climate processes during the Neogene.

3. THE BENGAL FAN

The Bengal Fan (Fig. 1) has been a repository for sediment derived from the India–Asia collision zone since the late Eocene.[19,20] The volume of postcollision sediments in the Fan is ca. $12.5 \times 10^6 \, km^3$.[3] Einsele *et al.*[9] have recently estimated a volume of $7.0 \times 10^6 \, km^3$ for sediments in the Bengal Fan less than 20 million years old. They estimate that an additional $3.1 \times 10^6 \, km^3$, of which $2.0 \times 10^6 \, km^3$ is 20 million years old or less, lies in the Bengal foredeep and delta of Bangladesh. An unknown quantity of Bengal Fan sediment has been subducted beneath the Sunda Arc.

DSDP Leg 22 (Site 218) and ODP Leg 116 (Sites 717–719) recovered sediments of Holocene through early Miocene age from the lower Bengal Fan.[21,22] Silty turbidites dominate sedimentation at Sites 218, 717, and 718 from the early (ca. 20 million years ago) to the late (ca. 7 million years ago) Miocene. Accumulation rates at Sites 717 and 718 are quite high from the mid-Miocene to 7.4 million years ago. From 7.4 million years ago to ca. 0.9 million years ago sedimentation shifts to slower accumulation of fine grained mud turbidites (Fig. 2). Near 0.9 million years ago sedimentation returns to silt turbidites deposited at high rates. Grain size also decreases between 7.4 and 0.9 million years ago. The clay mineral composition of the sediments changes with the two major breaks in sedimentological characteristics.[23] Clays in sediments older than 7.4 million years

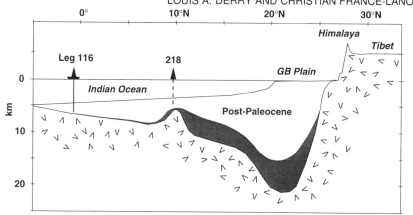

FIGURE 1. Longitudinal cross section of southern Tibet, the Himalaya, and the Bengal Fan, modified from Curray.[3] Locations of ODP Leg 116 and DSDP Site 218 (projected onto line of section) are indicated.

or younger than 0.9 million years are dominantly illite and chlorite (IC assemblage). From 7.4 to 0.9 million years ago clays are mostly smectite with lesser kaolinite (SK assemblage). The mineralogical composition of the coarser fractions is mostly quartz, feldspar, and micas, with minor amounts of a variety of heavy minerals from metamorphic and igneous sources.[24] In contrast to the clay mineralogy, the abundance pattern of the major detrital minerals has changed little over the last 10 million years.[25]

Several processes could have contributed to the observed changes in physical sedimentation and clay mineralogy. Sediment supply rates from the erosion of the Himalaya may have varied.[26] Fluctuating sea level could have varied the rate of transfer of sediment from the delta region to the deep sea.[27] Shifts in depocenters between distal and proximal basins or internal lobe switching could also cause sedimentation rate variations in the fan. While these internal sediment redistribution processes have contributed to accumulation rate variability, similar changes in sedimentation rates in the Bengal Fan and adjacent basins argue that the primary cause of sedimentation rate variations has been changes in sediment supply. Although not well-dated, DSDP Site 218 (1100 km distant) shows similar patterns of sedimentation and clay mineralogy to the Leg 116 sites.[21,23] Sediment accumulation rates do not increase in the Siwalik foreland basin sequence during the late Miocene, ruling out increased sedimentation in the foreland as a cause of decreased sedimentation in the fan.[28] Subsurface data from the Indian portion of the Bengal Delta show a decrease in sediment accumulation in the late Miocene,[29] making it unlikely that the delta grew at the expense of the fan. Distal Indian ocean DSDP and ODP sites receiving only indirect sediment flux from the Himalaya show a late Miocene decrease in clastic sediment accumulation rates, implying that there was a general decrease in sediment supply to the Indian

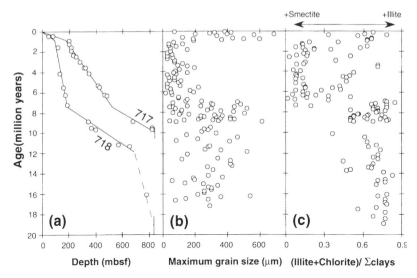

FIGURE 2. Sedimentological data from ODP Leg 116 Holes 717C and 718C.[22,23,82] Ages recalculated using timescale of Cande and Kent.[83]: (a) Low sedimentation rates, (b) fine-grain size, and (c) smectite–kaolinite (SK) clays characterize the interval from 7.4 to 0.9 million years ago. Coarse silts with illite–chlorite clays dominate at other times. The biostratigraphic constraints on sediment accumulation are uncertain for the early Miocene. The apparent increase in sedimentation rate about 12 million years ago could be an artifact of poor biostratigraphic control prior to that time.

Ocean.[30] Data from DSDP Site 222 in the Indus Fan also show a late Miocene–early Pliocene decrease in sedimentation rates,[21] implying that the sediment flux from all major Himalayan rivers decreased around the same time.

We note that a period of increased Himalayan sediment delivery to the Indian Ocean from 4 to 2 million years ago has been inferred from stacking of terrigenous sediment mass accumulation rate (MAR) records from various Indian Ocean ODP and DSDP holes.[30] This conclusion is based primarily on sites on the Owen and 90 E ridges, well above either the Bengal or Indus fans, which have MAR at least an order of magnitude smaller than the fan sites. Because terrigenous inputs to these sites are low, terrigenous MAR analysis of these sites is highly sensitive to changes in real or estimated carbonate content. Physical oceanographic changes may also be important in controlling the delivery of terrigenous sediment to the ridge-top sites. Thus it is not clear how well terrigenous MAR records from these sites record regional patterns in sediment delivery to the oceans. The only site from turbidite sediments listed in Rea[30] as recording a significant increase in terrigenous MAR from 2–4 million years ago is DSDP 222 from the Indus Fan. In reanalyzing the primary data for this site, we have not been able to reproduce a significant increase in MAR around 4 million years ago. The data from cores that actually penetrate Bengal Fan turbidites show only a minor interval of coarse Pliocene sediment and no biostratigraphically resolved sedimentation rate peak. An interval of increased silt sedimentation of probable Pliocene age is most clearly seen at DSDP Site 218, but unfortunately, the age constraints on this site are poor, and it is not possible to make a meaningful estimation of sediment accumulation rates or MAR from this site. In general, the quality of biostratigraphic control in the turbidite sediments is much less than in the ridge carbonate sequences. This presents a significant problem for combining the turbidite and carbonate-rich records, because both calculating MAR and stacking records require accurate and comparable chronologies to avoid aliasing or false peaks.

We conclude that there is as yet no firm basis for inferring a large increase in Indian Ocean sediment delivery in the mid-Pliocene. While there appears to be a brief Pliocene interval of increased sediment flux from the Ganges–Brahmaputra (GB) river system, this change appears to have been minor relative to the late Miocene and Pleistocene variations. Patterns of physical sedimentation from both the Bengal Fan and adjacent basins receiving Himalayan sediment strongly imply that the late Miocene transition in Bengal Fan sedimentation to lower accumulation rates and finer grained material primarily reflects a regional decrease in Himalayan sediment supply, and is not an artifact of changing depocenters or sea-level fluctuations.

4. Nd AND Sr ISOTOPIC CONSTRAINTS ON SEDIMENT PROVENANCE

Three tectonostratigraphic units in the Himalaya are potential sources of the huge volume of sediment in the Bengal Fan. The structurally highest are sediments of the Tibetan Sedimentary Series (TSS), primarily Paleozoic to Mesozoic sediments of the Tethys platform now exposed in southern Tibet. Along the crest of the Himalayan chain the TSS are separated from the underlying High Himalayan Crystalline (HHC) series by the low-angle South Tibetan Detachment System (STDS).[31,32] The HHC consists of high-grade gneisses mostly derived from quartzo-pelitic sediments. High Himalaya leucogranites, derived from the HHC, are intrusive in the TSS and the HHC.[33] Structurally below the HHC, and separated by the Main Central Thrust (MCT), is the Lesser Himalayan sequence of mostly low-grade metasediments of Precambrian age.[34] The Lesser Himalaya (LH) include quartzites, pelitic rocks, and minor dolomitic carbonates and black shales. Other potential source rocks are too small to contribute significantly to the Bengal Fan volume, or are essentially reworked material from the above three sources (i.e., the Siwalik foreland sediments).

We have used Nd and Sr isotopic data to constrain the provenance of Bengal Fan sediment over the last 20 million years.[35,36] The Nd and Sr isotopic characteristics of each of the three major sources are distinctive (Fig. 3). Nd and Sr data from the Bengal Fan sediments clearly fall in the range expected from HHC sources, regardless of mineral type or grain size fraction measured. The

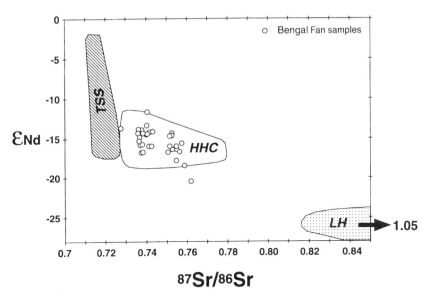

FIGURE 3. Nd and Sr isotope systematics of Bengal Fan clays ($<2\,\mu$m).[35,37] Fields for Tibetan Sedimentary Series (TSS), High Himalayan Crystalline series (HHC), and Lesser Himalaya (LH) from France-Lanord et al.[35] and Galy et al.[36] Bengal Fan clays of all ages (and all other analyzed phases, including feldspars, micas, and whole rocks) plot within a field defined by HHC. Nd isotopic values are equivalent in SK and IC clays, implying that the differing clay assemblages are derived from the same source. The HHC has been the dominant sediment source to the fan since the early Miocene.

dominant source of all detrital material in the Fan in the last 20 million years has been the HHC, with lesser contributions from the LH and TSS. Mixing of LH and TSS sources could mimic an HHC source, but all sample types at all stratigraphic levels, including coarse mica and feldspar separates, whole rocks, and clay fractions, show the same range of isotopic values. If mixing between LH- and TSS-derived material is responsible for the observed pattern, mixing ratios between the different components would have to remain similar through time despite the large changes in tectonic activity, climate, and sedimentary processes known to have occurred during the last 20 million years. We find this alternative hypothesis untenable.

Nd and Sr isotopic mass balance implies that, during most intervals, the fraction of Bengal Fan detrital material derived from the HHC has exceeded 80%.[35] A few samples with low ε_{Nd} values (-19 to -22) suggest increased contributions of LH material. A few very fine-grained samples ($<0.1\,\mu$m clay fractions) have high ε_{Nd} values, near -12, and may contain a component of smectite clays derived from the alteration of volcanics, either in the Paleogene Arc in southern Tibet or in the Deccan Plateau volcanics of India.[37] However the volume of any such component is trivial in comparison to the fan as a whole. Nearly all other clay mineral Nd analyses are consistent with an HHC source, regardless of whether the clays are composed of less than $2\,\mu$m illite-chlorite, less than $2\,\mu$m smectite-kaolinite, or large vermiculite grains.

5. SEDIMENTARY CONSTRAINTS ON EARLY MIOCENE HIMALAYAN ELEVATIONS

A first-order conclusion from the Nd and Sr data from Bengal Fan sediments is that the HHC has dominated the provenance of Himalayan detrital sediment production for at least 20 million years. Possible changes in tectonic expression of the collision process,[38] changes in regional climate,[2] and changes in sedimentation[39] do not appear to have resulted in significant changes in sediment provenance. The few samples that suggest a slightly larger contribution from LH sources are mostly found between 7 and 1 million years ago. A consequence of the observed dominance of HHC material in the flux to the fan is that high-grade metamorphic rocks similar to those currently cropping out in the Himalaya must have been exposed at the surface as early as 20 million years ago and have been undergoing rapid uplift for at least 30 million years.[36] Available data suggest that the Bengal Fan prograded south rapidly in the early Miocene.[27] By this time the HHC was already its primary sediment source. We infer that significant elevation and relief must have existed in the Himalaya as early as 20 million years ago, placing a qualitative but important constraint on the extent of continental thickening in the early Miocene.[35,36] The presence of K-feldspar grains with Ar–Ar cooling ages indistinguishable from their Bengal Fan sedimentation ages at all stratigraphic levels on Leg 116 also implies rapid erosion of at least some region in the Himalaya for the last 20 million years, and therefore the existence of significant elevation and relief.[40] Low-angle normal faults of the STDS were active in the early Miocene,[32,41] which appears to record extensional collapse of an elevated, gravitationally unstable portion of the crust.[42] Our inference of high elevations in the Himalayan range is consistent with an early Miocene extensional event at high elevations related to over-thickening of the crust, while compressional tectonics were active at lower elevations.[43] Although the sedimentary data do not constrain elevations quantitatively, we are not currently aware of any data that are inconsistent with early Miocene elevations along the southern margin of the Tibetan Plateau that were as high as at present.

6. $\delta^{18}O$ AND δD CONSTRAINTS ON WEATHERING PROCESSES

A notable feature of the mineralogical and geochemical record from the Bengal Fan is the observation that clay-sized fractions from both the SK and IC intervals yield essentially the same Nd isotopic compositions. Thus, the change from an illite-chlorite dominated assemblage to a smectite–kaolinite dominated assemblage does not result from changing provenance, but rather reflects different weathering histories of the same material. This conclusion is in contrast to those reached by workers who studied clay mineralogy alone, and who inferred major changes in sediment provenance to explain the shifting clay assemblages.[44,45] The isotopic composition of O and H in the Bengal Fan micas and clays provides further constraints on the origin of the two clay assemblages.[23,35] Samples with the IC assemblage clays have $\delta^{18}O$ values around 12, which imply that these clays are mostly derived from high-temperature micas by simple mechanical grinding (Fig. 4). The IC clays have $\delta^{18}O$ values slightly higher than high-temperature micas from Himalayan outcrops. They appear to be mostly composed of mechanically abraded high-temperature micas, with a minor component consisting of low-temperature smectite and kaolinite.

The isotopic composition of meteoric water in the Himalayan drainage system varies widely because of large contrasts in elevation, temperature, and water sources.[46] Streamwaters generally become increasingly depleted in $\delta^{18}O$ and δD with increasing elevation and distance to the north, with very low values in southern Tibet.[47] Waters at low elevations in the Indo–Gangetic Plain have

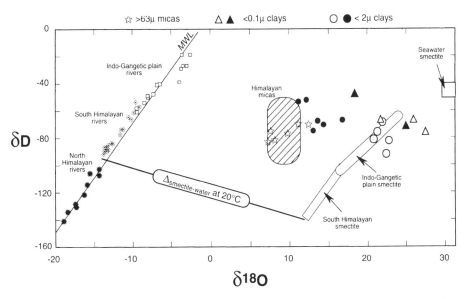

FIGURE 4. δD–$\delta^{18}O$ systematics of Himalayan river waters, Himalayan rocks, and Bengal Fan materials.[37] Data points along the meteoric water line (MWL) are samples from the Ganges or its tributaries (Ramesh and Sarin[46] and unpublished data, CRPG-Nancy). Hatched field is the range of micas from Himalayan outcrops.[84] Symbols: * = Bengal Fan micas > 63 μm, ●, ○ = <2 μm IC and SK clays; △, ▲ = <0.1 μm IC and SK clays. Fields for smectite equilibrium with meteoric water were calculated using published fractionation factors.[85,86] IC clays are mixtures of unaltered micas with high-temperature signals with some low-temperature kaolinite ± smectite. SK clays were equilibrated at low elevations in the GB floodplain.

much higher $\delta^{18}O$ and δD values, which generally increase downstream (Fig. 4). The isotopic compositions of pedogenic Bengal clays can be used to infer the zone of weathering and neoformation. In contrast to the IC clays, SK assemblage samples are characterized by higher $\delta^{18}O$ values (20–25‰) reflecting their pedogenic origin. SK clays are near calculated fields for smectite in equilibrium with water found at low elevations in the Indo–Gangetic plain. They are too rich in D and ^{18}O to have been produced by interaction with waters at higher elevations, and too light to be in equilibrium with seawater or pore water.

Shifts in the oxygen isotopic composition of the Bengal Fan clays are well-correlated with losses of K_2O, implying that chemical weathering occurred under the same conditions as the stable isotopic exchange with meteoric waters (Fig. 5). Thus the locus of chemical weathering in the Ganges–Brahmaputra (GB) drainage system has been the low-elevation Indo–Gangetic Plain.[37] The Nd and Sr isotopic data show that the IC and SK clays have the same provenance. The O and H isotopic data show that the difference between the IC and SK clay assemblages appears to be the extent of weathering experienced by the sediments in this environment. The SK samples have undergone more intense weathering at low elevations, resulting in loss of potassium (and other elements) but enrichment in ^{18}O and D during the formation of pedogenic clays.

The correlated shifts in sediment delivery, grain size, and clay mineralogy and chemistry in the Bengal Fan cores from around 7.4 million years ago appear to mark a prominent change in the weathering environment in the GB basin. Chemical weathering intensity of sediment in the GB system increased at the same

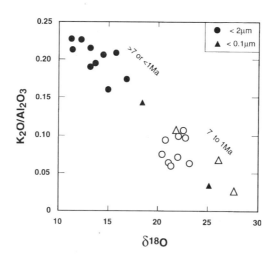

FIGURE 5. Correlation of K_2O/Al_2O_3 with $\delta^{18}O$ for Bengal Fan clays. Open symbols are SK clays, filled symbols are IC clays as determined by XRD. The strong negative correlation implies that potassium losses occurred in the same environment as ^{18}O enrichment, i.e., that weathering occurred primarily in the GB floodplain.

time that sediment transport rates and transport energy were lower overall. Floodplain weathering was apparently most intense during the interval of reduced physical transport of material, suggesting that weathering intensity is related to the residence time of sediment in the foreland basin weathering zone. Intervals of rapid, high energy sediment transport are also intervals of low weathering intensity, during which large volumes of sediment are produced and transported to the sea, but undergo little chemical processing.

7. $\delta^{13}C$ CONSTRAINTS ON THE WEATHERING PALEOENVIRONMENT

Quade et al.[1] have shown that plants using the C_4 photosynthetic pathway underwent a significant ecological expansion in the Himalayan foreland between 7 and 8 million years ago. The $\delta^{13}C$ of carbonate and organic carbon (OC) from Siwalik paleosols both show large increases that are correlated with changes in herbivore fossil assemblages from browsers to grazers. These changes have been interpreted to show rapid expansion of savanna grasses in the sub-Himalayan

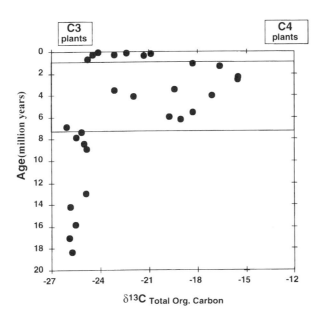

FIGURE 6. $\delta^{13}C$ from organic carbon in Bengal Fan sediments.[6] Organic carbon in the fan is dominated by terrigenous sources. All samples more than 7 million years old have values typical of C_3 plants. Values for samples less than 7 million years old represent varying mixtures of C_3 and C_4 plant sources. These data are consistent with data from paleosols in the Himalayan foreland basin.[1,48]

region during the late Miocene.[48] OC from the Bengal Fan is dominated by terrigenous sources, and its isotopic composition shows a similar sharp increase about 7 million years ago (Fig. 6). $\delta^{13}C$ values increase from ca. $-26\%_0$ (typical of C_3 plants) to as high as $-15\%_0$ (a strong C_4 plant signature) supporting the hypothesis of a regionally extensive floral shift.[6] Furthermore, high $\delta^{13}C$ values indicative of a strong grassland (C_4) component are found in fine-grained samples rich in smectite–kaolinite clays (Fig. 7). The $\delta^{13}C$ values of Bengal Fan organic matter are also positively correlated with percent total organic carbon (TOC). The most chemically weathered samples have the highest TOC values and the largest fraction of C_4-derived carbon, whereas less weathered samples have more C_3-derived carbon. We interpret the $\delta^{13}C$ patterns in sediments younger than 7.4 million years ago to reflect mixing between carbon derived from C_3-dominated upland forests and carbon derived from C_4-dominated lowland savannas.

Rapid transport brings both C_3 carbon and unweathered sediments through the floodplain directly to the Bengal Fan. However, sediments that reside on the floodplain in the pedogenic environment experience carbon turnover and pick up C_4 signatures characteristic of the savanna grasses believed to have occupied this region since the late Miocene. The close correlation of increased chemical weathering intensity and C_4 carbon signatures implies that stabilization and weathering in floodplain soils has been a key control on weathering processes in the GB system. The expansion of savanna grasslands is in part coincident with

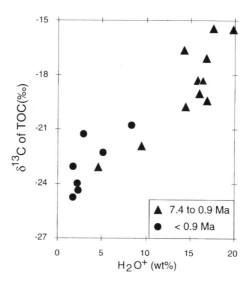

FIGURE 7. $\delta^{13}C$ from organic carbon vs. clay-bound water in Bengal Fan sediments.[6] Increasing water contents result from increasing amounts of hydrated smectite and reflect more heavily weathered samples. Increased weathering is correlated with increasing C_4 plant carbon content, implying that pedogenesis and carbon turnover took place together in lowland soils with C_4 plant cover (savanna grasses).

regional environmental changes that led to increased weathering, but soil stabilization by C_4 grasses may have contributed to increased floodplain residence time and pedogenesis. Thus the carbon isotope geochemistry supports the hypothesis that weathering variations in the Bengal fan sediments largely reflect differing residence times in the floodplain weathering environment.

8. WEATHERING ENVIRONMENTS AND $^{87}Sr/^{86}Sr$ OF HIMALAYAN RIVERS

During the late Cenozoic the marine $^{87}Sr/^{86}Sr$ ratio rose dramatically.[49-51] Rivers draining the Himalaya have both radiogenic $^{87}Sr/^{86}Sr$ ratios and high Sr concentrations when compared to other major river systems[52]; thus increased Sr flux from the Himalaya could explain the change in the marine $^{87}Sr/^{86}Sr$ record.[53,54] However, a quantitative assessment of how much the Himalayan Sr flux has changed requires information on how the isotopic ratio of Sr in Himalayan rivers has changed with time. The marine Sr isotopic mass balance is underconstrained, and cannot be used to uniquely infer past patterns in both the $^{87}Sr/^{86}Sr$ and the Sr flux from paleoceanographic measurements. Richter et al.[54] attempted to avoid this ambiguity by assuming a "reasonable" history for $^{87}Sr/^{86}Sr$ in Himalayan rivers. While "reasonable," their assumptions were untested. Information on past riverine isotopic ratios is only available (if at all) from the sedimentary record.

The late Miocene shift in weathering processes in the GB drainage system resulted in an increase in silicate weathering intensity, followed by a decrease in weathering intensity during the Pleistocene. The changes are recorded in the Bengal Fan clays, which are the residual products of incongruent silicate weathering. Fluxes of dissolved species derived from silicate weathering should also have changed. It is not possible to reconstruct dissolved fluxes directly from the geological record, but the Sr isotopic mass balance can be used to estimate changes in dissolved fluxes from the GB system during the Neogene. Such a calculation requires an estimate of the isotopic composition of the dissolved Sr flux in the GB system as a function of time.

Sr occupies interlayer sites in clay minerals, and is exchanged during pedogenesis or neoformation.[55] The stable isotopic data from pedogenic clays in the Bengal Fan demonstrates that they were produced in the Gangetic Plain, thus their interlayer Sr should record $^{87}Sr/^{86}Sr$ ratios from paleo-floodplain waters. We treated smectite-rich $<2\,\mu m$ and $<0.1\,\mu m$ clay separates and hand-picked vermiculite separates with ion exchange solutions and acetic acid to remove adsorbed and carbonate Sr, and then digested the remaining clays to isolate the interlayer Sr.[37] The pedogenic interlayer Sr data show significant variation in time, which reflects changes in the $^{87}Sr/^{86}Sr$ of GB floodplain water (Fig. 8). The shifts in $(^{87}Sr/^{86}Sr)_{GB}$ are correlated with other changes in sedimentation and weathering history recorded by Bengal Fan sediments. $(^{87}Sr/^{86}Sr)_{GB}$ data are near

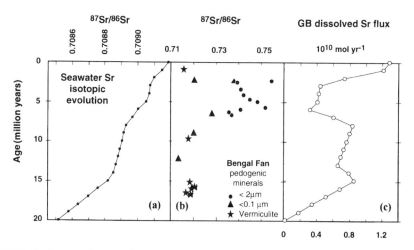

FIGURE 8. Sr isotopic evolution of Ganges–Brahmaputra (GB) river system[37]: (a) Neogene evolution of seawater ^{87}Sr/^{86}Sr[57,87]. (b) Evolution of ^{87}Sr/^{86}Sr from interlayer cations of pedogenic clays in the Bengal Fan. These values were set in the GB floodplain during pedogenesis, and are used to estimate the evolution of river water ^{87}Sr/^{86}Sr. (c) The data from (a) and (b) are used to drive a Sr isotopic mass balance model, which calculates the evolution of the GB Sr flux.[37,54] Sr fluxes fell substantially in the late Miocene, when weathering intensity increased in the GB system. Fluxes returned to high values in the Plio-Pleistocene. Other base cation fluxes are generally correlated with Sr in river systems, and it is probable that their flux patterns were similar.

0.72 prior to 7.4 million years ago, and rise to more than 0.74 in the early Pliocene. The values fall to near 0.72 (the modern value) in the late Pleistocene. Data from pedogenic and freshwater biogenic carbonates in the Siwalik sequence show a very similar pattern of variation,[56] supporting our interpretation of the pedogenic clay Sr as a record of river water variations.

The large temporal variations in the ^{87}Sr/^{86}Sr of the GB dissolved flux have significant implications for the Neogene evolution of the marine Sr isotopic mass balance. With these data on the evolution of the ^{87}Sr/^{86}Sr of the GB, we can model the evolution of the GB Sr flux necessary to satisfy the observed seawater record during the past 20 million years, making the assumption that all other sources of Sr to the ocean remained constant in both flux and isotopic ratio during this interval.[37] Our calculated GB Sr flux 20 million years ago is near zero (Fig. 8), which is probably an artifact resulting from the assumption that all other Sr fluxes (non-GB) have not varied from their modern values. Because of the high Mio–Pliocene $(^{87}$Sr/^{86}Sr$)_{GB}$ values, we find that a *decrease* in the GB Sr flux of more than 50% is required to satisfy the marine Sr mass balance, despite a monotonic *increase* in the seawater ^{87}Sr/^{86}Sr record during this interval, including an almost steplike increase around 6 million years ago.[57] Although the assumption that all other Sr sources to the oceans remained unchanged is unlikely to be strictly true, only a significant (and well-timed) decrease in the average ^{87}Sr/^{86}Sr of the flux of Sr from all other sources could have offset the change in

$(^{87}Sr/^{86}Sr)_{GB}$ and could eliminate the need for a reduced GB Sr flux. It is equally possible that other sources of Sr became, on average, more radiogenic in the Mio–Pliocene, which would require even larger decreases in the GB Sr flux. The common inference that rapid increases in seawater $^{87}Sr/^{86}Sr$ ratios are driven by increased weathering fluxes is clearly contradicted in this instance, whereas it appears to be consistent with the observations over the last 2–3 million years. The relationship between the global river Sr flux and the seawater Sr record, even over the restricted interval of the Neogene, is obviously more complicated than has been assumed by many workers in the past.

The constraints on the GB Sr flux also have implications for the evolution of other elemental fluxes from the GB. The concentration of Sr is positively correlated with those of other cations in most rivers. These correlations are typically highest for Ca and Mg, and lowest for Na and Si, with K intermediate.[58–61] Thus diminished Pliocene Sr fluxes from the GB also imply diminished fluxes of other base cations, and an overall reduced alkalinity flux. It appears unjustified to assume that Neogene increases in marine $^{87}Sr/^{86}Sr$ necessarily imply increased alkalinity fluxes to the oceans.

The large variations in the $^{87}Sr/^{86}Sr$ of the GB dissolved flux could reflect a shift in the silicate source of dissolved Sr, or a shift between the carbonate/silicate weathering ratio for Sr, or changes in the relative weathering rates of different minerals in the silicate source. Before addressing the causes of the Neogene variations in river Sr fluxes, it is worth considering the modern Sr mass balance in the GB system, which is unusual because it is both radiogenic and at high concentration when compared to other major river systems.[5] Various sources for the radiogenic Sr have been proposed. Krishnaswami et al.[4] argued that weathering of the very radiogenic LH was responsible. Palmer and Edmond[52] argued that the dissolution of radiogenic marbles from the LH was the primary source. Edmond et al.[61] later proposed that weathering of recently metamorphosed Ca–Na feldspars from the HHC was important. Examination of the published data for Sr and major ions from the Himalayan rivers shows that it is not possible to resolve the mass balance adequately to discriminate among these hypotheses. Permissible choices of end members for a mixing calculation can show that less than 10% or more than 40% of the modern dissolved Sr flux is derived from silicate weathering, with the balance derived from carbonates and evaporites. For example, possible carbonate end members include Tethyan sediments ($^{87}Sr/^{86}Sr \approx 0.708$ if fresh and ≈ 0.725 if altered) and LH dolostones ($^{87}Sr/^{86}Sr \approx 0.8$). Possible silicate end members include HHC feldspars ($^{87}Sr/^{86}Sr \approx 0.735$) and LH metasediments ($^{87}Sr/^{86}Sr \geqslant 0.9$). Various choices and weighting of end members are compatible with the available data, yet the implications of these scenarios for weathering flux interpretations based on Sr isotopes differ greatly.

Given the uncertainty associated with the modern Sr isotopic mass balance for the GB system, we must approach the causes of past variations in $(^{87}Sr/^{86}Sr)_{GB}$ cautiously. A few Bengal Fan samples of Pliocene age show low ε_{Nd} values, suggesting that LH contributed a slightly larger fraction of the erosion flux starting about 7 million years ago. The LH also has extremely radiogenic

$^{87}Sr/^{86}Sr$ values, and enhanced weathering of the LH could increase the GB Sr isotopic ratio. However, as discussed above, ε_{Nd} values from most samples are not different before and after 7 million years ago. If enhanced weathering of the LH contributed to the highly radiogenic Pliocene GB Sr flux, this weathering was not accompanied by a significant increase in detrital sediment derived from the LH.

Clay mineral compositions in the Bengal Fan record a marked change in weathering intensity, which is associated with the shift toward high $(^{87}Sr/^{86}Sr)_{GB}$ values. HHC rocks contain abundant modal biotite ($> 30\%$) and less muscovite. These minerals have high $^{87}Rb/^{86}Sr$ values, and Sr isotopic ratios near 0.8.[36,62] Weathering of micas, especially biotites, can release highly radiogenic Sr to solution.[63] Sr abundances in HHC and Bengal Fan biotites are 5–20 ppm, while muscovites are 30–200 ppm, and feldspars are 40–150 ppm.[36] HHC rocks are very micaceous, and micas are a quantitatively significant reservoir of radiogenic Sr, even given the relatively low Sr contents of biotites. We propose that intensified weathering of micas contributed significantly to the high Pliocene $(^{87}Sr/^{86}Sr)_{GB}$ values. The increase in weathering intensity and the increase in $(^{87}Sr/^{86}Sr)_{GB}$ values are causally related. While weathering intensity was low, feldspar weathering dominated the silicate component of Sr derived from the HHC ($^{87}Sr/^{86}Sr \approx 0.74$). When weathering intensity was high, smectite clays were produced from primary radiogenic micas, in the process releasing very radiogenic Sr.

Finally, the silicate to carbonate weathering ratio in the GB system could have changed. Carbonate weathering is congruent, and leaves no physical evidence in the stratigraphic record, so it is very difficult to evaluate this possibility quantitatively. However, the size of the shift in $(^{87}Sr/^{86}Sr)_{GB}$ values was large enough that we suspect that the silicate/carbonate weathering ratio increased in the Mio–Pliocene, although we cannot estimate the magnitude of this effect directly.

As noted above, the Sr isotopic mass balance does not constrain the modern silicate/carbonate weathering ratio in the GB system very closely. The same problem is evident for the base cations Ca^{2+}, Mg^{2+}, and Na^+, which are predominantly derived from carbonate (Ca^{2+}, Mg^{2+}), evaporite (Ca^{2+}, Mg^{2+}, Na^+), or cyclic salt (Na^+) sources. The dissolved transport index (DTI) is the fraction of an element transported by a river in the dissolved form.[64] For K^+ and Si, nearly all or all of the river load is derived from silicate weathering. Water chemistry data from the Ganges mainstem[65] are available for K^+ and Si, but we are not aware of any published analyses of suspended sediment chemistry in the GB. We can estimate K^+ and Si values for the GB suspended load using the average composition of the HHC, because the HHC dominates the sediment flux. K^+ and Si values in recent deltaic sediments from Bangladesh are very similar to average HHC values (ca. 3.0% K_2O and 70% SiO_2), indicating that this approximation is certainly adequate for our present purpose (France-Lanord and Derry, unpublished data). Combining these data with water and sediment flux estimates,[11] we estimate the DTI for $K^+ = 9.3\%$, and for Si $= 0.33\%$. The estimated global average DTI values are 14% for K^+, and 4.4% for Si.[64] The DTI for Si in the Ganges is among the lowest of any major river system. This very low value indicates that present-day silicate weathering intensity is very low, and suggests that a very high proportion of the base cation flux is derived from nonsilicate sources. The moderate DTI for K^+ is consistent with the high mica content of the predominant HHC source, and with the IC clay mineral assemblage. Chlorite and vermiculite, derived from incongruent weathering of biotite, are K^+-poor. The conversion of biotite to vermiculite may be a major source of K^+ to the GB dissolved flux.

The low chemical weathering intensity of the GB system has implications for the global budget of CO_2 consumed by weathering reactions. Only alkalinity derived from silicate weathering of Ca, Mg, and to a minor extent Na will significantly affect the long term CO_2 budget. Because the base cation flux from the modern GB system is strongly dominated by carbonate and evaporite weathering, it has been difficult to make an accurate estimate of the magnitude of the silicate-derived alkalinity flux from the Himalaya. The dissolved silica flux provides a measure of silicate weathering processes. Diatom growth can significantly modify dissolved Si fluxes in some rivers, but the GB is highly turbid with low autochthonous production rates. The discharge-weighted Si fluxes from the Ganges and Brahmaputra rivers are very similar to those of other large river systems (Fig. 9). This relationship implies that silicate weathering in the GB drainage is not anomalous with respect to the global average, but is in fact quite

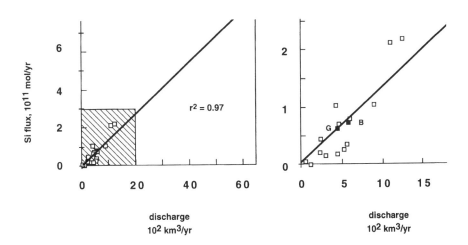

FIGURE 9.   (a) Water discharge vs. Si flux for the 17 largest rivers in the world, which account for 31% of global water discharge.[66] The r_2 value for linear correlation is 0.975. (b) Inset showing position of Ganges (G) and Brahmaputra (B) on global correlation line (solid symbols). Si flux correlates well with river discharge across a wide range of weathering environments and suspended-load/dissolved-load ratios. The discharge-weighted Si flux from the Ganges–Brahmaputra river system is similar to other large drainage systems.

typical for a river system of that size. The GB accounts for ca. 2.7% of global river discharge, and ca. 2.0% of global river Si flux (data from Berner and Berner[66]). Independent estimates of modern and long-term CO_2 consumption based on watershed studies in Nepal and on the major element chemistry of Himalayan sediments also suggest that the Himalaya accounts for approximately 2% of global CO_2 consumption by silicate weathering,[67,68] Silicate weathering in the Himalaya appears to make a small contribution to the global CO_2 balance today, and this has probably been true throughout the late Cenozoic.

9. A MODEL FOR NEOGENE WEATHERING PROCESSES IN THE HIMALAYA

We can now summarize the inferences about the last 20 million years of weathering history derived from the Neogene Himalayan sedimentary record. Chemical weathering intensity in the GB system, currently and during most of the past 20 million years, is low. Weathering intensity and sediment delivery to the Bengal Fan appear to be inversely correlated. Available data show that the GB system is currently strongly weathering-limited (following the terminology of Stallard[69]). The sedimentological and geochemical observations imply that the GB weathering environment was similar to the present during most of the Miocene, with rapid sediment delivery and low weathering intensity. Similar conditions have also prevailed since at least the mid-Pleistocene. However, from the late Miocene to Pleistocene, the weathering environment was characterized by lower sediment delivery rates, lower transport energies, and intensified chemical weathering. The locus of this chemical weathering was at low elevations, in the Indo–Gangetic floodplain. Sediments derived from the High Himalaya were stored in the floodplain long enough to undergo pedogenesis. Evidence for this pedogenesis comes from the abundance of pedogenic clays, and the abundance of organic carbon derived from lowland C_4 plants, probably savanna grasses. Thus the sediments experienced carbon turnover in the soil environment, while undergoing chemical weathering, including K loss, Sr exchange, and O and H exchange. A consequence of this enhanced weathering intensity was the release of very radiogenic Sr to solution from the weathering of micas. This process contributed to very high $^{87}Sr/^{86}Sr$ values in the GB dissolved flux during the early Pliocene. Shorter-lived events continued to supply relatively unweathered sediment associated with C_3 plant carbon to the Bengal Fan. Associated with the increase in weathering intensity and $(^{87}Sr/^{86}Sr)_{GB}$ ratios, and decrease in sediment supply, was a decrease in dissolved cation flux from the GB system.

Both climate and tectonic processes may have been active in driving the dramatic changes in weathering environments and fluxes in the GB system over the last 20 million years. Prell et al.[70] used the appearance of marine plankton species associated with modern summer upwelling to infer the onset of strong seasonal upwelling in the Arabian Sea beginning 7–9 million years ago. They proposed that this upwelling marked a major intensification of the Indian monsoon system. Quade et al.[1,48] argued that the expansion of C_4 plants in the Himalayan foreland and a change in soil carbonate $\delta^{18}O$ values between 9 and 6 million years ago was caused by a shift toward a stronger monsoon climate. The proposed intensification of the monsoon is roughly coincident with the changes in weathering style observed in the Bengal Fan record. While the effect of the proposed monsoonal intensification on weathering processes is not certain, a shift toward a monsoon climate is a seemingly attractive hypothesis for explaining the increase in weathering intensity and floral changes in the Himalayan foreland. However, the increase in precipitation and seasonality expected from a monsoonal climate is not consistent with the decrease in sediment supply, the decrease in grain size, the decrease in base cation flux, and the increase in sediment residence time in the floodplain inferred from the observations. If anything, such a climate shift would predict enhanced erosion and transport of sediment, and yet there is no evidence to suggest such enhancement occurred. Furthermore, erosion rates in the GB system were high for at least 10 million years before the late Miocene events. River transport was rapid and energetic enough to transport large volumes of unweathered, coarse sediment to the distal Fan throughout the interval from 8 to 20 million years. The sedimentological data imply that the strong seasonality necessary to drive episodic or periodic flooding was characteristic of the climate in the Himalaya even earlier than 8 million years ago. Recently, Fluteau et al.[71] have argued that significant monsoonal circulation in south Asia began much earlier, and was influenced by the closure of the eastern Tethys Ocean. The presence of an early Miocene monsoonal climate would be consistent with the sedimentological observations. The Pleistocene return to an environment of low weathering intensity and high-flux, does not coincide with any known major change in the monsoon cycle. Thus, the extent of monsoon influence over Himalayan weathering and erosion appears to be variable in time.

A potential driver for the late Miocene monsoon intensification is the uplift of a large region of Tibet.[72] According to one hypothesis, Tibet reached a maximum elevation during the late Miocene, which triggered extensional tectonism within the southern Tibetan Plateau, and monsoon activity.[38,73] However, the sedimentological data do not provide support for this hypothesis, because a prediction of the model is that erosion rates and sediment supply should be highest beginning in the late Miocene, as a result of high elevations, high relief and high precipitation fluxes. Rather, the available data imply the existence of a high Himalayan range since the early Miocene (although the sedimentological data do not directly constrain the elevation history of Tibet, because Tibet contributes little sediment to the GB system). Furthermore, the decrease in sediment supply beginning in the late Miocene could be construed to reflect a decrease in mean elevation of the Himalayan source region. Given the evidence for increased monsoon activity around this same time, it is hard to see how clastic sediment flux and dissolved cation fluxes could both decrease in the presence of constant or increasing elevation and relief. If mechanical and chemical erosion decrease, but tectonic uplift rates remain constant, mean elevation must increase, unless some coincident tectonic extension is sufficiently large. Any increase in mean elevation of the Himalaya should, under most circumstances, result in increased erosion. The absence of evidence for such an increase in Mio–Pliocene erosion rates suggests that both tectonic uplift and mean elevations were reduced during this interval.

Sediment fluxes increased rapidly in the Pleistocene, suggesting that both a renewal of tectonic uplift and the onset of major Northern Hemisphere glaciation could have played a role. Rapid sea-level variations on the order of 100 m could also have reworked sediment from the delta to the deep sea, although this cannot be a major cause of sedimentation rate changes over an area the size of the Bengal Fan. Glacial erosion rates can be, at least locally, very high.[74] Expansion of glaciation in the Himalaya during the Pleistocene could have contributed to the high sediment fluxes and low weathering intensity observed in the last million years. Low chemical weathering intensity also appears characteristic of glacial conditions. A recent comparison of silica fluxes from glaciated and nonglaciated drainages shows that glaciated drainages have high overall fluxes but significantly lower discharge-weighted base cation and silica fluxes than nonglaciated drainages.[75] For silica, the difference is approximately a factor of five, implying that silicate weathering under glacial conditions is inefficient. These patterns are consistent with the temporal evolution of the GB sediment flux in the late Pliocene and Pleistocene, and the inverse relationship we infer between sediment flux and weathering intensity. Both expanded glaciation in the Himalaya and decreased sediment residence time in the floodplain probably contributed to the return to low chemical weathering intensity in the GB system during the Pleistocene. Climate cooling and glaciation in the Pleistocene are associated with rapid mechanical erosion, high sediment flux, high base cation flux, but low weathering intensity in the GB system.

We note that a number of earlier workers in the Himalaya proposed that the great present-day elevations of the range are a geologically young feature.[76–78] The evidence they cited has been shown not to be diagnostic.[79] However, new evidence which *requires* continuously high Himalayan elevations since the late Miocene has not become available. We find that the simplest interpretation of the sedimentological data is consistent with reduced Himalayan elevations and erosion rates beginning in the late Miocene, and renewed uplift and erosion beginning in the Plio–Pleistocene.[26,80,81] However, available data do not directly constrain the elevation history of the Himalaya, and additional evidence is required to better address this very important issue. We wish to reopen the discussion of earlier hypotheses that called for renewed Plio–Pleistocene uplift of the Himalaya because: (a) we find such a hypothesis consistent with current sedimentological observations, and (b) we are not aware of any data that rule out such a hypothesis, despite the fact that some recent workers have regarded such views as outdated. We would welcome any new observation or experiment that could rule in — or rule out — such a hypothesis.

10. CONCLUSION

Crystalline rocks of the High Himalaya have been the primary source of detritus in the Ganges–Brahmaputra system at least since 20 million years ago, but fluxes of suspended and dissolved materials have varied widely. Changes in clay mineralogy record a late Miocene increase in chemical weathering intensity, which is coupled with a regional decrease in sediment supply from the Himalaya. Stable isotopic data for O, H, and C show that the locus of intensified weathering was the Indo–Gangetic floodplain, where Himalayan sediments resided in a pedogenic environment. We suggest that the weathering differences recorded in sediments from the Bengal Fan primarily reflect changes in the transport dynamics of Himalayan sediment, and that the late Miocene increase in weathering intensity reflects increased residence time in floodplain soils and the development of pedogenic mineral assemblages. The Pleistocene return to lower weathering intensity and higher sediment fluxes marks a decrease in floodplain residence time, and an expansion of glacial weathering environments. By using the record of interlayer cation $^{87}Sr/^{86}Sr$ from pedogenic minerals as a proxy for paleo-river Sr isotope ratios, we can use the Sr isotopic mass balance to estimate the GB Sr flux as a function of time. We find that $^{87}Sr/^{86}Sr$ and Sr flux are inversely correlated, as is true in most modern river systems. Dissolved $^{87}Sr/^{86}Sr$ values peaked in the Pliocene, but Sr fluxes were reduced by ca. 50%. Both the Himalayan Sr flux and alkalinity flux decreased during an interval of increasing marine $^{87}Sr/^{86}Sr$ values. The late Miocene shift in weathering approximately coincides with the onset of strong monsoonal circulation in the region inferred from paleoceanographic and paleobotanical indicators. Although chemical weathering intensity increased during the late Miocene, weathering fluxes and sediment fluxes decreased. The modern silica budget of the GB system is consistent with low chemical weathering intensity, and suggests that silicate weathering in the Himalaya plays a limited role in contributing to the global CO_2 balance.

Thus we find that at least three commonly held assumptions about Himalayan weathering are contradicted: (1) Himalayan weathering fluxes of Sr (and alkalinity) can decrease significantly and still cause an increase in the seawater $^{87}Sr/^{86}Sr$ ratio, demonstrating the strong control river isotopic variability has over the oceanic Sr isotope mass balance. (2) The inferred late Miocene onset of the Asian monsoon system was followed by a decrease in erosional sediment flux and in chemical weathering flux. A possible resolution of the apparent paradox posed by these observations is that tectonic uplift rates, and elevations, decreased during the Mio-Pliocene, and were renewed only in the last 1–3 million years. (3) While suspended sediment flux from Himalayan rivers is highly anomalous compared to other large river systems, silicate weathering fluxes are not, because of low chemical weathering intensity. The interplay of tectonic, climatatic, and ecological changes in the Himalayan region has exerted strong control over Himalayan chemical weathering and erosion processes, and the export of sediment and solutes to the oceans. Much work remains to be done to better understand these interactions and their history.

REFERENCES

1. Quade, J., Cerling, T. E., and Bowman, J. R. (1989). *Nature* **342**, p. 163.
2. Prell, W. L., and Kutzbach, J. E. (1992). *Nature* **360**, p. 647.
3. Curray, J. R. (1994). *Earth Planet. Sci. Lett.* **125**, p. 371.
4. Krishnaswami, S., Trivedi, J. R., Sarin, M.M., Ramesh, R., and Sharma, K. K.(1992). *Earth Planet. Sci. Lett.* **109**, p. 243.
5. Palmer, M. R., and Edmond, J. M. (1992). *Geochim. Cosmochim. Acta* **56**, p. 2099.
6. France-Lanord, C., and Derry, L. A. (1994). *Geochim. Cosmochim. Acta* **58**, p. 4809.
7. Wetzel, A. (1993). *AAPG Bull.* **77**, p. 1679.
8. Beck, R. A., *et al.* (1995). *Nature* **373**, p. 55.
9. Einsele, G., Ratschbacher, L., and Wetzel, A. (1996). *J. Geol.* **104**, p. 163.
10. Pinet, P., and Souriau, M. (1988). *Tectonophysics* **7**, p. 563.
11. Milliman, J. D., and Syvitski, P. M. (1992). *J. Geol.* **100**, p. 525.
12. Summerfield, M., and Hulton, N. J. (1994). *J. Geophys. Res.* **99**, p. 13871.
13. Gaillardet, J., Dupré, B., and Allègre, C. (1995). *Geochim. Cosmochim. Acta* **59**, p. 3469.
14. Gibbs, R. J. (1967). *Geol. Soc. Amer. Bull.* **78**, p. 1203.
15. Stallard, R. F., and Edmond, J. M. (1981). *J. Geophys. Res.* **86**, p. 9841.
16. Stallard, R. F., and Edmond, J. M. (1983). *J. Geophys. Res.* **88**, p. 9671.
17. Meybeck, M. (1987). *Am. J. Sci.* **287**, p. 401.
18. Raymo, M. E., and Ruddiman, W. F. (1992). *Nature* **359,** p. 117.
19. Gansser, A. (1966). *Ecolog. Geol. Helvet.* **59**, p. 831.
20. Curray, J. R., and Moore, D. G. (1971). *Geol. Soc. Am. Bull.* **82**, p. 563.
21. Shipboard Scientific Party (1974). In: *Init. Repts. DSDP*, Vol. 22 (C.C. von der Borch and J.C. Sclater, eds.), pp. 325–367. U.S. Government Printing Office, Washington, D.C.
22. Shipboard Scientific Party (1989). In: *Proc. ODP Init. Rep.*, Vol. 116 (J. R. Cochran and D. A. V. Stow, eds.), Ocean Drilling Program, College Station, TX.
23. Bouquillon, A., France-Lanord, C., Michard, A., and Tiercelin, J.-J. (1990). In: *Proc. ODP, Sci. Res.*, Vol. 116 (J. R. Cochran and D. A. V. Stow, eds.), pp. 43–58. Ocean Drilling Program, College Station, TX.
24. Ingersoll, R. V., and Suczek, C. A. (1979). *J. Sed. Petrol.* **49**, p. 1217.
25. Yokoyama, K., Amano, K., Taira, A., and Saito, Y. (1990). In: *Proc. ODP, Sci. Res.*, Vol. 116 (J. R. Cochran and D. A. V. Stow, eds.), pp. 59–73. Ocean Drilling Program, College Station, TX.
26. Amano, K., and Taira, A. (1992). *Geology* **20**, p. 391.
27. Cochran, J. R. (1990). In: *Proc. ODP, Sci. Res.*, Vol. 116 (eds. J. R. Cochran and D. A. V. Stow, eds.), pp. 397–414. Ocean Drilling Program, College Station, TX.
28. Burbank, D. W., Derry, L. A., and France-Lanord, C. (1993). *Nature* **364**, p. 48.
29. Rao, A. T., and Raman, C. V. (1986). *Indian J. Marine Sci.* **15**, p. 20.
30. Rea, D. K. (1992). In: *Synthesis of Results from Scientific Drilling in the Indian Ocean*, Vol. 70 (R. A. Duncan, D, K. Rea, R. B. Kidd, U. von Rad, and J. K. Weissel, eds.), pp. 387–402. American Geophysical Union, Washington, D.C.
31. Burg, J. P., and Chen, G. M. (1984). *Nature* **311**, p. 219.
32. Pêcher, A. (1991). *Tectonics* **10**, p. 587.

33. Le Fort, P., Cuney, M., Deniel, C., France-Lanord, C., Sheppard, S. M. F., Upreti, B. N., and Vidal, P. (1987). *Tectonophysics* **134**, p. 39.

34. Colchen, M., Le Fort, P., and Pcher, A. (1980). Carte géologique Annapurna-Manaslu-Ganesh, Himalaya du Népal. Echelle 1:200.000. Centre National de la Recherche Scientifique, Paris.

35. France-Lanord, C., Derry, L., and Michard, A. (1993). In: *Himalayan Tectonics*, Vol. 74. (P. J. Treloar and M. Searle, eds.), pp. 603–621. Geological Society London, London.

36. Galy, A., France-Lanord, C., and Derry, L. A. (1996). *Tectonophysics* (in press).

37. Derry, L. A., and France-Lanord, C. (1996). *Earth Planet. Sci. Lett.* **142**, p. 59.

38. Molnar, P., England, P., and Martinod, J. (1993). Rev. Geophys. **31**, p. 357.

39. Stow, D. A. V., *et al.* (1990). In: *Proc. ODP, Sci. Res.*, Vol. 116. (J. R. Cochran and D. A. V. Stow, eds.), pp. 377–396. Ocean Drilling Program, College Station, TX.

40. Copeland, P., and Harrison, T. M. (1990). *Geology* **18**, p. 354.

41. Hodges, K. V., Parrish, R. R., Housh, T. B., Lux, D. R., Burchfield, B. C., Royden, L. H., and Chen, Z. (1992). *Science* **258**, p. 1466.

42. Burchfiel, B. C., Chen, Z., Hodges, K. V., Liu, Y., Royden, L. H., Deng, C., and Xu, J. (1992). *Spec. Paper Geol. Soc. Am.* **269**, p. 41.

43. Hubbard, M. S., and Harrison, T. M. (1989). *Tectonics* **8**, p. 865.

44. Brass, G. W., and Raman, C. V. (1990). In: *Proc. ODP, Sci. Res.* (J. R. Cochran and D. A. V. Stow, eds.), pp. 35–41. Ocean Drilling Program, College Station, TX.

45. Aoki, S., Kohyama, N., and Ishizuka, T. (1991). *Mar. Geol.* **99**, p. 175.

46. Ramesh, R., and Sarin, M. M. (1992). *J. Hydrol.* **139**, p. 49.

47. France-Lanord, C., Derry, L. A., Le Fort, P., and Gajurel, A. P. (1995). In: Proc. of the Goldschmidt Conference, p. 46. Geochemical Society, Pennsylvania State University.

48. Quade, J., Cater, J. M. L., Ojha, T. P., Adam, J., and Harrison, T. M. (1995). *Geol. Soc. Am. Bull.* **107**, p. 1381.

49. DePaolo, D. J. (1986). *Geology* **14**, p. 101.

50. Hess, J., Bender, M., and Schilling, J. G. (1986). *Science* **231**, p. 979.

51. Koepnick, R. B., Burke, W. H., Denison, W. E., Hetherington, E. A., Nelson, H. F., Otto, J. B., and Waite, L.E. (1985). *Chem. Geol.* **58**, p. 55.

52. Palmer, M. R., and Edmond, J. M. (1989). *Earth Planet. Sci. Lett.* **92**, p. 11.

53. Hodell, D. A., Mueller, P. A., McKenzie, J. A., and Mead, G. A. (1989). *Earth Planet. Sci. Lett.* **92**, p. 165.

54. Richter, F. M., Rowley, D. B., and DePaolo, D. J. (1992). *Earth Planet. Sci. Lett.* **109**, p. 11.

55. McBride, M. B. (1994) *Environmental Chemistry of Soils*, Oxford University Press, New York.

56. Quade, J. (1993). *Geol. Soc. Am. Abstr. Progs.* **25**, p. 175.

57. Farrell, J. W., Clemens, S. C., and Gromet, L. P. (1995). *Geology* **23**, p. 403.

58. Meybeck, M. (1979). *Rev. Géol. Dyn. Géogr. Phys* **21**, p. 220.

59. *Goldstein, S. J., and Jacobsen, S. B.* (1987). *Chem. Geol.* **66**, p. 245.

60. Andersson, P. S., Wasserburg, G. J., Ingri, J., and Stordal, M. C. (1994). *Earth Planet. Sci. Lett.* **124**, p. 195.

61. Edmond, J. M., Palmer, M. R., Measures, C. I., Grant, B., and Stallard, R. F. (1995). *Geochim. Cosmochim. Acta* **59**, p. 3301.

62. Deniel, C. (1985). Apport des isotopes du Sr, Nd et du Pb a la connaissance de l'age et de l'origine des leucogranites himalayenes. Thesis. Université de Clermont-Ferrand, France.

63. Blum, J. D., and Erel, Y. (1995). *Nature* **373**, p. 415.

64. Martin, J. M., and Meybeck, M. (1979). *Mar. Chem.* **7**, p. 173.

65. Sarin, M. M., Krishnaswami, S., Dilli, K., Omayajulu, B. l. K., and Moore, W. S. (1989). *Geochim. Cosmochim. Acta* **53**, p. 997.

66. Berner, E. K., and Berner, R. A. (1996). *Global Environment: Water, Air, and Geochemical Cycles*, Prentice-Hall, Upper Saddle River, NJ.

67. France-Lanord, C., and Derry, L. A. (1996). *EOS, Trans. Am. Geophys. Union* **77**, p. 796.

68. Derry, L. A., and France-Lanord, C. (1996). *EOS, Trans. Amer. Geophys. Union* **77**, p. 796.
69. Stallard, R. F. (1985). In: *The Chemistry of Weathering* (J. I. Drever, ed.), p. 293.   Reidel, Hingham, MA.
70. Prell, W. L., Murray, D. W., and Clemens, S. C. (1992). In: *Synthesis of Results from Scientific Drilling in the Indian Ocean*, Vol. 70 (R. A. Duncan, D. K. Rea, R. B. Kidd, U. von Rad, and J. K. Weissel eds.), pp. 447–469. American Geophysical Union, Washington, D.C.
71. Fluteau, F., Besse, J., Ramstein, G., and Valet, J. P. (1996). *EOS, Trans. Amer. Geophys. Union* **77**, p. 300.
72. Ruddiman, W. F., and Kutzbach, J. E. (1989). *J. Geophys. Res.* **94**, p. 18409.
73. Harrison, T. M., Copeland, P., Kidd, W. S. F., and Yin, A. (1992). *Science* **255**, p. 1663.
74. Hallet, B., Hunter, L., and Bogen, J. (1996). *Glob. Planet. Change* 12, p. 213.
75. Anderson, S. P., Drever, J. I., and Humphrey, N. I. (1996). In: *Proc. 4th International Conference on the Geochemistry of the Earth's Surface*. Ilkley, England.
76. Gansser, A. (1983). In: Mountain Building Processes (K.J. Hsü, ed.), p. 221.   Academic Press, London.
77. Xu, R. (1984). In: *The Evolution of the East Asian Environment; Paleobotany, Paleozoology, and Paleoanthropology* (R. O. Whyte, ed.), pp. 426–432. University of Hong Kong Press, Hong Kong.
78. Powell, C. M. (1986). *Earth Planet. Sci. Lett.* **81**, p. 79.
79. England, P., and Molnar, P. (1990). *Geology* **18**, p. 1173.
80. Debrabant, P., Fagel, N., Chamley, H., Bout, V., and Caulet, J. P. (1993). *Paleogeogr. Palaeo Clim. Palaeoecol.* **103**, p. 117.
81. Fagel, N., Debrabant, P., and André, L. (1994). *Mar. Geol.* **122**, p. 151.
82. Gartner, S. (1990). In: *Proc. ODP, Sci. Res.*, Vol. 116 (J. R. Cochran and D. A. V. Stow, eds.), pp. 165–187. Ocean Drilling Program, College Station, TX.
83. Cande, S. C., and Kent, D. (1992). *J. Geophys. Res.* **97**, p. 13917.
84. France-Lanord, C., Sheppard, S. M. F., and Le Fort, P. (1988). *Geochim. Cosmochim. Acta* **52**, p. 513.
85. Yeh, H. W., and Savin, S. M. (1977). *Geol. Soc. Amer. Bull.* **88**, p. 1321.
86. Sheppard, S. M. F., and Gilg, A. H. (1996). *Clay Minerals* **31**, p. 1.
87. Oslick, J. S., Miller, K. G., Feigenson, M. D., and Wright, J. D. (1994). *Paleoceanography* **9**, p. 427.

13

Late Cenozoic Vegetation Change, Atmospheric CO$_2$, and Tectonics

Thure E. Cerling

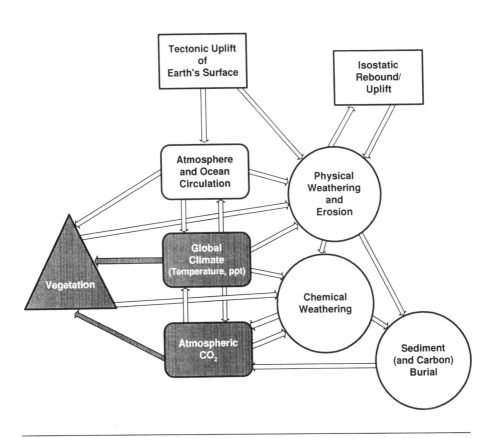

Thure E. Cerling • Department of Geology and Geophysics, University of Utah, Salt Lake City, Utah 84112.

Tectonic Uplift and Climate Change, edited by William F. Ruddiman. Plenum Press, New York, 1997.

1. INTRODUCTION

Great changes in the geochemistry of the oceans, in climate, in vegetation, and in tectonics have taken place in the Cenozoic. The $^{87}Sr/^{86}Sr$ and $^{187}Os/^{186}Os$ ratios in the ocean have undergone substantial alternations, which have been attributed to modifications in weathering related to the uplift of the Himalayas.[1-11] Global climate has undergone cooling through most of the Tertiary as documented by $\delta^{18}O$ records in marine foraminifera, which have increased by about 4‰ from the Eocene to the present.[12] The increase in ^{18}O is due to both temperature and the ice-volume effect.

The collision of the Indian and Asian plates beginning in the Eocene produced the Himalayas and eventually the Tibetan Plateau, which is the world's largest continental topographic anomaly. As such it disrupts the normal zonal flow, and a low-pressure center is fixed over it in the summer and a high-pressure center in the winter. The plateau enhances the natural conditions that would set up a monsoon in the region,[13] i.e., the rising of air over the warm landmass, which draws in moist air from the Indian Ocean to the south in summer, and the reversal of this pattern in the winter. The summer monsoon is enhanced by the release of latent heat to the atmosphere during precipitation. The tectonics and monsoon have been linked, and it is suggested that the development of the Asian monsoon was related to a threshold effect.[14] Global circulation models (GCMs) suggest that the monsoon has long-distance effects because of the disruption of the normal east-to-west flow pattern.[15,16] A marked strengthening of the monsoon is noted by the increase in the foraminifera *Globigerina bulloides*, an indicator of upwelling, between 8.6 and 7.4 million years ago in the western Arabian Sea.[17] Harrison *et al.*[18] suggest that Tibet reached its maximum elevation about 8 million years ago.

Tectonics, climate change, geochemistry of the oceans, and evolution of the atmosphere are linked because tectonics provides material for weathering. The chemistry of the ocean is related to the relative input sources from the continents or the midocean ridges and subduction zone volcanoes and the removal processes on continents. CO_2 is consumed during weathering of the terrestrial silicates and only part of it is returned to the atmosphere during subsequent formation of carbonates in the oceans. For example, it has been proposed that the uplift of the Himalayas has had important effects on the ocean $^{87}Sr/^{86}Sr$ record[8] and on the evolution of CO_2 in the atmosphere.[19,20]

There have been major ecologic changes in the Cenozoic. The evolution of grasses and the development of grasslands and steppes are particularly prominent. Cerling *et al.*[21] noted that major changes in global ecology occurred at similar times over widespread regions, suggesting an atmospheric control on ecology.

In this paper I will review the evidence for global ecologic change in the late Cenozoic. Major changes in paleoecology are indicated in the sedimentary sequence of the Siwaliks in Pakistan[21-23] and in the nearby marine environment.[24,25] Similar changes occurred at similar times in Africa,[26] North America,[27] and South America.[28] Taken together, these indicate a major change in global terrestrial ecology, with the present ecology in the late Neogene being unlike any previous period.

2. CHARACTERISTICS OF TERRESTRIAL PLANTS AND ECOSYSTEMS

There are three principal photosynthetic pathways used by terrestrial plants, referred to as the C$_3$ pathway, the C$_4$ pathway, and CAM. The plants using these pathways have distinctive ecology and isotope systematics, which are due to their different methods for photosynthesis. They lead to characteristic δ^{13}C values for organic matter, for soil carbonate *in situ* in soils, and for δ^{13}C preserved in fossils as collagen or as carbonate apatite in biological tissues such as bones and teeth.

2.1. C$_3$ Pathway

C$_3$ plants make up most of the trees and shrubs, so that all forests from the tropics to the boreal regions are dominated by C$_3$ plants. In addition, plants using the C$_3$ pathway are physiologically adapted to cooler climates and low water stress so that they are the dominant ecosystem in high latitudes and at high altitudes. The δ^{13}C of C$_3$ plants is quite variable; under normal conditions C$_3$ plants range from about -25 to $-27\%_{oo}$. However under conditions of high water stress and high light irradiance, they close their stomata, which results in an enrichment of a few $\%_{oo}$, so that C$_3$ plants in semidesert to desert environments can have δ^{13}C values as high as $-22\%_{oo}$.[29,30] On the other hand, in forest environments where exchange with the open troposphere is poor, and where light levels are very low, very depleted δ^{13}C values for plants have been observed; Medina *et al.*[31] and van der Merwe and Medina[32] report δ^{13}C values of $-37\%_{oo}$ in the Amazon rain forest.

In C$_3$ plants photosynthesis is catalyzed by Rubisco (ribulose-1,5-biphosphate carboxylase/oxygenase) so that CO$_2$ and RuBP produce two 3-carbon (hence the name "C$_3$") PGA molecules (phosphoglycerate). PGA is then reduced by ATP and NADPH to sugar, which is converted to starch and regenerates Rubisco. The net process releases O$_2$ to the atmosphere. However, Rubisco can also catalyze the oxygenation of RuBP,[33] the net reaction consuming O$_2$ and producing CO$_2$, which reduces net photosynthesis. The relative reaction rates of carboxylation versus oxygenation is dependent on the ratio of CO$_2$ to O$_2$ in the atmosphere (presently ca. 1:6000; the pre-1850 value was ca. 1:7500). The ratios internal to the leaf are greater because CO$_2$ partial pressures are lower and O$_2$ levels are higher than the atmospheric level because of CO$_2$ uptake and O$_2$ release. The lower limit for C$_3$ photosynthesis is reached at CO$_2$ concentrations (assuming modern atmospheric O$_2$ levels) between about 150 to 50 ppm CO$_2$(CO$_2$:O$_2$ between 1:15,000 to 1:40,000). Because stomata serve as conduits for CO_2 uptake and water loss, decreasing atmospheric CO_2 levels (especially at high temperatures) cause a decrease in photosynthetic efficiency in C_3 plants.

2.2. C_4 Pathway

The C_4 pathway is present in many plant families (the Monocotyledons Cyperaceae, Gramineae, and Liliaceae; and the Dicotyledons Aizoaceae, Amaranthaceae, Asteraceae, Boraginaceae, Capparadaceae, Caryophyllaceae, Chenopodiaceae, Euphorbiaceae, Molluginaceae, Nyctaginaceae, Polygonaceae, Portulacaceae, and Zygophyllaceae), and these families make up an important part of the biomass (as sedges, grasses, saltbush, and euphorbia) in temperate to tropical grasslands, savannas, and semideserts. Although many families have genera that use the C_4 pathway, the family with the greatest C_4 biomass globally is the Gramineae, the grasses. It is estimated that about half of the more than 10,000 grass species use the C_4 pathway.[34] The $\delta^{13}C$ of C_4 plants ranges from about -11 to $-14\%_0$.

C_4 plants have a different anatomy from C_3 plants. Kranz anatomy, unique to C_4 plants, consists of a mesophyll cell in contact with a bundle sheath cell. Photosynthesis involves initial CO_2 fixation in the mesophyll cell to form a 4-carbon acid (hence the name "C_4"). The acids are transported to the bundle sheath cell, where CO_2 is released and then fixed by Rubisco in the same way as in C_3 photosynthesis. However, the internal concentration of CO_2 in the bundle sheath cell is many times higher (>2000 ppmV) than the atmospheric CO_2 concentration,[35] and the C_4 pathway is commonly referred to as a "CO_2-pump." In this way, by spatially separating the sites of CO_2 uptake and CO_2 reduction, C_4 plants enrich internal CO_2 concentrations before Rubisco catalyzation and therefore are adapted to lower atmospheric CO_2 levels than C_3 plants. A benefit associated with the enrichment in CO_2 is an increased water-use efficiency compared to C_3 plants at high temperatures.

2.3. CAM Plants

CAM plants are also present in many families (e.g., the Monocotyledons Agavaceae, Bromeliaceae, Liliaceae, and Orchidaceae; and Dicotyledons Aizoaceae, Asclepiadaceae, Asteraceae, Bataceae, Cactaceae, Caryophyllaceae, Chenopodeceae, Crassulaceae, Euphorbiaceae, and Portulacaceae) and tend to make up a minor component of the biomass (as cacti, agave, and other desert succulents) in savannas, semideserts, and deserts. The $\delta^{13}C$ values for CAM plants range from about -12 to $-25\%_0$, depending on what fraction of C_4 versus C_3 photosynthesis takes place.

CAM plants have a different strategy for CO_2 enrichment. Rather than a spatial separation of CO_2 fixation and photosynthesis, they have a temporal separation so that CO_2 is fixed at night when humidity is higher and temperatures are lower, and is photosynthesized during the day when light levels are high. However, the net photosynthetic rate is much lower than either C_3 or C_4 plants.

2.4. Soils: $\delta^{13}C$ and $\delta^{18}O$

Cerling and others[36-38] have discussed the theoretical basis for the $\delta^{13}C$ in soils. Diffusion and equilibrium isotopic fractionation result in a 14 to 17‰ enrichment in ^{13}C in soil carbonate compared to soil organic matter. Therefore $\delta^{13}C$ values in C_3 ecosystems for organic matter and soil carbonate are about -24 to -27‰ and -10 to -12‰, respectively; a C_4 ecosystem has $\delta^{13}C$ values for organic matter and soil carbonate of about -11 to -13‰ and 0 to $+2$‰, respectively. The $\delta^{18}O$ of soil carbonate is determined by the $\delta^{18}O$ of soil water, which is largely controlled by the isotopic composition of infiltrated meteoric water, sometimes modified by evaporation. Because the range in $\delta^{18}O$ of meteoric water is much greater than the range caused by temperature-dependent isotope fractionation, the isotopic composition of meteoric water controls the $\delta^{18}O$ of soil carbonate. Therefore, in general, more positive $\delta^{18}O$ values for soil carbonate indicate warmer climates.[37,38]

2.5. Diet

The isotopic enrichment from diet in herbivores to tooth enamel is about 14‰ based on the estimated diets of wild animals[39] and captive animals with a known diet.[27] Therefore the $\delta^{13}C$ of a C_3-dominated diet ranges from about -8 to -15‰ (to -22‰ in closed canopies),[40] and the $\delta^{13}C$ of a C_4-dominated diet from about 0 to $+3$‰. Wang and Cerling[41] showed that tooth enamel is less susceptible to diagenesis than dentine or bone.

3. GLOBAL OBSERVATIONS

3.1. Indian Subcontinent

The Siwalik Sedimentary Sequence is one of the most complete sedimentary terrestrial sequences in the world. Its rich vertebrate fossil remains have made it the target of paleontological expeditions for over 80 years.[42] It has an aggregate thickness of several kilometers and has been well-dated by magnetostratigraphy.[43,44] Detailed isotopic and stratigraphic studies have been carried out in Pakistan and, more recently, in Nepal. Studies of paleosol carbonate,[22,23] mammalian tooth enamel,[26,45,46] and fossil ratite eggshell[46] indicate a major expansion of C_4 biomass beginning between 8 and 7 million years ago (Fig. 1). The signal is first seen in mammalian tooth enamel, as is to be expected because of the selective nature of the grazing by certain mammals. The oldest equids in the Siwaliks have a C_3-dominated diet but equids switched to a C_4-dominated diet by 5 to 6 million years ago.[26,47] By 7 million years ago there is a clear indication of C_4 biomass in the overall ecosystem as determined using paleosol carbonates, although a greater than 90% C_4 biomass is not reached until about 6 million years ago.[22,23] Similar isotope shifts are found in sediments 1000 km further east, in central Nepal.[48] The regional nature of this change from C_3 to C_4 biomass is shown in the $\delta^{13}C$ record of the Bengal Fan,[24,25] which undergoes a

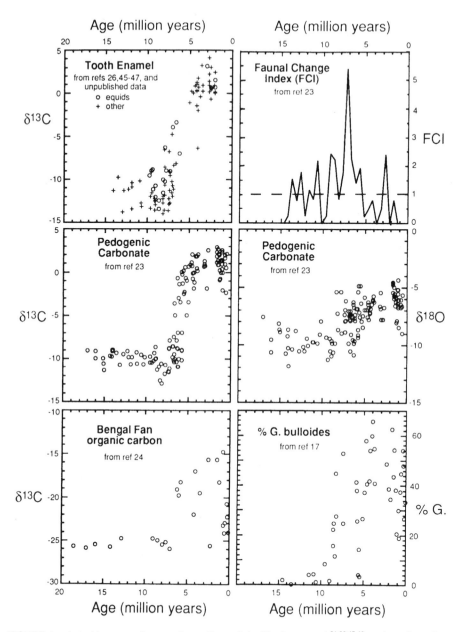

FIGURE 1.   Late Neogene changes in southern Asia. Tooth enamel,[21,26,45,46] pedogenic carbon-ates,[22,23] and faunal change[23] data are terrestrial evidence for global change between 8 and 6 million years ago from the Siwaliks in Pakistan; the Bengal Fan[24] shows changes in detrital $\delta^{13}C$ and *G. bulloides* data[17] from the Arabian Sea (an indicator of monsoon-related upwelling) document global change in the marine environment between 8 and 6 million years ago.

significant shift starting about 7 million years ago. Figure 1 shows the various isotope records of ecologic and climate change in the Himalayan foreland. Besides the records mentioned above, it shows that the soil carbonates underwent a significant shift in $\delta^{18}O$ about 8 million years ago, approximately the same time as a marked increase in *G. bulloides* is noted in the Arabian Sea.[17] *G. bulloides* is considered to be an important indicator of upwelling related to the present Asian monsoon system.

There are other striking changes at this time in the fossil record of the Siwaliks. Main periods of faunal change were noted by Barry *et al.*[49] at about 13, 9, and 7 million years ago. Unfortunately, a distinct change in the character of sediments is found in the Siwaliks starting about 7.5 million years ago, and fossils are not as abundant after 7 million years as compared to earlier times. In part because of this, and because of the interest in *Sivapithecus* (once thought to be a possible ancestor of man but now recognized as an ancestor of orangutans), most expeditions have concentrated on the older fossil record. To remove this collecting bias, Quade and Cerling[23] devised a faunal change index that is simply a record of "change"/"stasis" for time intervals using published faunal lists. They found that the greatest period of faunal change was about 7 million years ago (Fig. 1). *Sivapithecus* becomes extinct in the Siwalik record, and forest-dwelling mammals are replaced by a fauna adapted to more open habitats.[49]

The original interpretation of the shift to C_4 biomass, accompanied by a 3 to 4‰ enrichment in $\delta^{18}O$ of soil carbonate, was that the Asian monsoon system originated or intensified starting in the late Miocene about 8 million years ago.[22] The finding of an increase in *G. bulloides* seems to confirm this interpretation. However, as is seen below, this is part of a much larger change in which C_4 biomass increased globally at this time.

The Siwalik sediments in the Himalayan foreland provide an unparalleled record of faunal and ecologic change in the late Neogene. The isotopic record provides evidence independent of the fossil record of major ecologic and climate change. Further studies of the detailed record of faunal change linked to isotope studies will be very important.

3.2. East Africa

Isotopic studies of paleosols and diets from East Africa also document ecologic change in the late Neogene (Fig. 2).[26,47,50-54] Data from tooth enamel show that mammals did not have a significant fraction of C_4 biomass in their diet until 8 million years ago,[26,47,54] and most data from fossil soils[50-52] do not show a significant fraction of C_4 biomass until late Pliocene or Pleistocene times. Only the data from Kingston *et al.*[53] have been interpreted as indicating abundant C_4 biomass prior to 8 million years ago; fossils from the same deposits do not show any indication of having a C_4 component in their diet,[25,55] so that study is problematical.

The oldest equids in East Africa appear about 10.5 million years ago, the time of the well-documented "Hipparion datum," and as those in Pakistan and Europe,

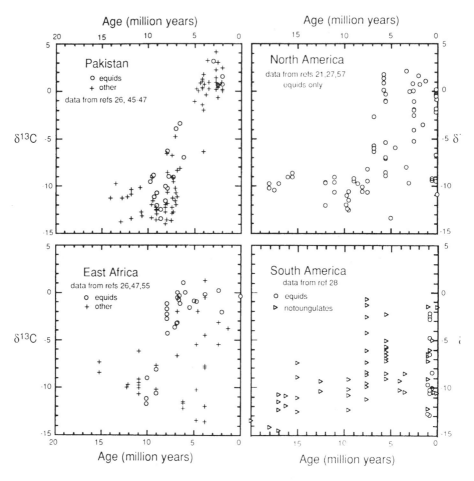

FIGURE 2. $\delta^{13}C$ values of fossil tooth enamel from southern Asia (Pakistan),[26,45-47] Africa,[26,47,54] North America,[21,27] and South America.[28] In each region, all $\delta^{13}C$ values older than 8 million years are -8 to $-14\%_0$, but reach values as high as $0\%_0$ by about 6 million years ago.

they have a C_3-dominated diet; by about 8 million years ago they have a C_4-dominated diet, as do gomphotheres (Cerling *et al.*, unpublished data).

The detailed paleosol records from the Turkana Basin[52] and the Olduvai Gorge[51] regions are in marked contrast to that of the Siwaliks. The East African paleosol record shows a gradual change toward increasing C_4 biomass,[21] with pure C_4 ecosystems indicated in the Olduvai region only in the last 1 million years, and reaching only about 75% C_4 biomass in the Turkana region. In both these areas, no paleosols older than 1 million years are as C_4-rich as the modern soils. In contrast, the transition to pure C_4 ecosystems in the Siwaliks of southeast Asia was reached by 5 million years ago.

3.3. South America

The South American record is not as well known, but MacFadden *et al.*[28] recently published $\delta^{13}C$ values for mammals in Argentina. They find a transition to a C$_4$-based diet taking place about 7.9 million years ago (Fig. 2). Equids are not found in South America until about 2 million years ago, after the Great American Interchange, so that notoungulates were used in that study.

3.4. North America

Cerling *et al.*[56] measured $\delta^{13}C$ values in more than 300 equids from widespread parts of North America to study the transition to C$_4$-based diets. All samples older than 7 million years have a C$_3$-dominated diet (Fig. 2), even from a 12- to 13-million-year-old locality that has well-documented C$_4$ plants.[57] This implies that C$_4$ plants were present in North America by 12 million years ago although they were a minor part of the flora. Cerling *et al.*[56] find a significant C$_4$ fraction in the diet by 6.8 million years ago at the Coffee Ranch locality, and a pure C$_4$ diet in equids from Mexico between 5.5 and 6.0 million years ago. The modern distribution of C$_4$ biomass in North America, including the south to north gradient from C$_4$ to C$_3$ biomass in the Great Plains and the gradient across the Sonoran Desert (C$_4$) to the Mojave Desert (C$_3$), is recognizable by about 4 million years ago.

3.5. Western Europe

Modern Europe has very few C$_4$ plants, and very few C$_4$ plants are found in the Mediterranean region. In spite of the hot summers, little rain falls during the Mediterranean summers so C$_4$ plants are rare. Similarly, Europe is far enough north and cool enough in the summer that C$_4$ plants are not found. The $\delta^{13}C$ of mammalian tooth enamel from Spain and France does not indicate any C$_4$ component at any time in the last 20 million years.[47] Equids from the time of the Hipparion datum (10.5 million years ago) to the present have $\delta^{13}C$ values from -9 to $-14\%_0$. Preliminary data from Mediterranean North Africa (Cerling, unpublished data) also indicate little, if any, C$_4$ biomass in the diets of equids.

3.6. Global Summary

Stable isotope studies from paleosols and mammalian tooth enamel show that C$_4$ plants were not major components of ecosystems anywhere in the world until between 8 and 6 million years ago when they expanded their range. By 6 million years ago C$_4$ plants (or ecosystems) were abundant enough that they made up more than 90% of the diets of some equids from southern Asia, East Africa, North America, and South America. For each of these regions, all $\delta^{13}C$ values for tooth enamel older than 8 million years are between -8 and $-14\%_0$, and by 6 million years ago $\delta^{13}C$ values of $0\%_0$ are reached. In northern North

America ($>37°N$) the signal is not as strong as further south[56]; in Europe no C_4 component is detected in the late Neogene.[47]

The modern distribution of C_4 plants has a strong latitude gradient[58-60] with C_4 grasses, the most common of C_4 plants, being dominant below about 30° latitude. The transition to C_3 grasses takes place between 30 and 45° latitude, depending on seasonality of rainfall and temperature, with regions of summer rainfall having more C_4 biomass. Above about 45° latitude very few C_4 grasses are found. This global ecology, the "C_4-world," was emplaced between 8 and 6 million years ago. The previous "C_3-world," where even tropical and subtropical grasses used the C_3 photosynthetic pathway, was fundamentally different as far as the physiology of plants was concerned.

4. GLOBAL CHANGE IN THE NEOGENE: CHANGES IN ATMOSPHERIC CO₂?

It is clear from the above discussion and from Fig. 2 that the change in C_4 ecosystems was a global event. Cerling et al.[21,47] have suggested that atmospheric CO_2 levels gradually declining through a threshold important to C_3 plants could have given C_4 plants a competitive edge in tropical through warm-temperate climates. The near synchronous nature of the C_4 expansion favors such an atmospheric link. Figure 3 shows that the higher levels of CO_2 in the Mesozoic must have crossed through the threshold for photorespiration in C_3 plants, estimated to be between about 500 and 600 ppm. This scenario is compatible with the model of Raymo et al.[19] and Raymo and Ruddiman[20] linking decreasing atmospheric CO_2 with increased weathering during Cenozoic mountain building events.

It is important to note that the timing of the global C_4 expansion is very close to the indications of increased upwelling in the Arabian Sea, and is also close to the time of a significant shift in the $\delta^{18}O$ of soil carbonates in the Siwalik sediments (Fig. 1). The $\delta^{18}O$ of pedogenic carbonate in East Africa also changes in the late Neogene,[50] but the section is a composite one and is not as definitive as the record from the Siwaliks. This means that the conditions causing the increase in C_4 biomass may have also caused regional or possibly global climate change. More continuous sedimentary sequences need to be studied to determine the extent and magnitude of changes in the meteoric water cycle. Changes in atmospheric CO_2 levels are likely to be associated with changes in the meteoric water cycle.

5. RELATION TO THE LATE NEOGENE 6 CARBON ISOTOPE SHIFT IN THE OCEAN

A significant change in the $\delta^{13}C$ of the ocean occurs in the late Neogene[61] with the oceans being more positive in dissolved inorganic carbon by about 0.5‰

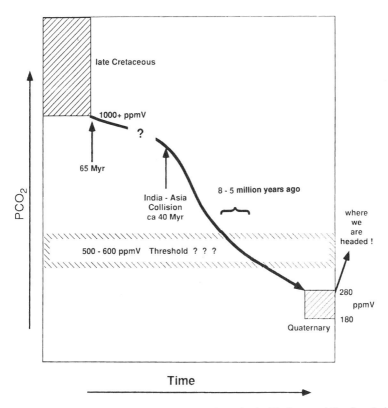

FIGURE 3.   Model for the evolution of the atmosphere in the Tertiary and the threshold for C_3 photorespiration. The high levels reached in the Mesozoic (>1000 ppmV) declined in the Tertiary to the modern levels of ca. 180 ppmV (glacial) to 280 ppmV (interglacial).

in the late Miocene compared to the more negative Pliocene values. While I do not suggest that a global terrestrial ecosystem change is totally responsible for this shift, it could contribute to it. Figure 4 shows the total terrestrial carbon storage, net primary production, and the residence time of carbon in different terrestrial ecosystems. It is apparent that C_3-dominated ecosystems tend to have higher net carbon storage, higher net productivity, and a longer residence time of carbon than do C_4 ecosystems. This has important implications concerning the storage of terrestrial carbon owing to changes in ecosystems and to the isotopic composition of carbon distributed between the labile pools in the atmosphere (700 Gt), the biosphere (2200 Gt), and the ocean (35,500 Gt) (Gt = 10^{15} g).

In the following discussion I consider possible changes in the $\delta^{13}C$ of the ocean in response to changes in the biomasses and net primary productivities of ecosystems shown in Fig. 4. I assume an average fractionation of 5.5‰ between the atmosphere and average dissolved inorganic carbon in the ocean (DIC), and a discrimination of 20.0‰ for C_3 plants. Figure 5 shows that changing the C_4 biomass (-13‰) back to C_3 biomass (-26‰) makes the oceans more positive

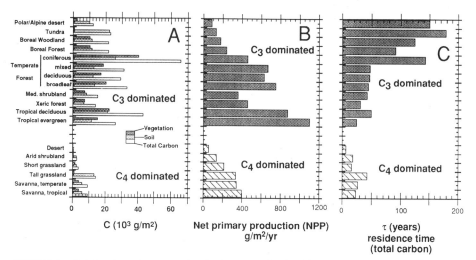

C (10^3 g/m²)

Net primary production (NPP)
g/m²/yr

τ (years)
residence time
(total carbon)

FIGURE 4.   Total carbon storage and net primary production (data from Adams *et al.*,[62] Schlesinger,[63] and Melillo *et al.*[64]), and calculated residence time of carbon in different terrestrial ecosystems.

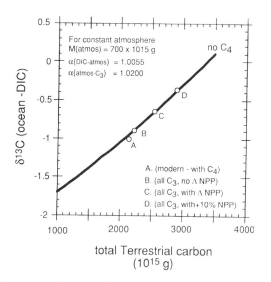

total Terrestrial carbon
(10^{15} g)

FIGURE 5.   Calculated carbon isotope effect in the oceans using different scenarios for terrestrial ecosystem change. It is assumed that the labile carbon pool is constant and consists of the atmosphere (700 Gt), the biosphere (2200 Gt), and the ocean (35,500 Gt). The atmosphere is held constant for different scenarios. The different scenarios illustrate that the carbon isotopic composition of DIC in the ocean can be affected by reasonable changes in terrestrial ecosystems: (A) Modern $\delta^{13}C$ of oceans and modern terrestrial biomass. (B) $\delta^{13}C$ of oceans assuming C_4 biomass is converted to C_3 biomass (xeric forests or shrublands) with accompanying increase in carbon storage owing to greater total carbon in C_3 ecosystems (from Fig. 4a). (C) $\delta^{13}C$ of oceans assuming scenario B, but with an increase in net primary productivity (NPP) from the C_4 ecosystems to the respective C_3 ecosystems (changes in NPP from Fig. 4b). This scenario results in a significant increase in the total terrestrial biomass. (D) As in C, except that all ecosystems have an increase of 10% NPP (values taken from Fig. 4b).

by only about 0.1‰, far less than the observed 0.5‰ shift. However, Fig. 4 shows that C$_3$ ecosystems have a higher productivity and higher carbon storage than C$_4$ ecosystems. Changing C$_4$ ecosystems back to C$_3$ ecosystems, and using a conservative choice of replacement by xeric forests or shrublands, causes increased stored carbon resulting in an additional 0.25‰ shift in the observed direction. Furthermore, it has been repeatedly noted that increased CO$_2$ concentrations in the atmosphere serve to fertilize C$_3$ plants, causing an increase in their productivity. Assuming an increase of 10% in the net primary productivity of the terrestrial biosphere causes a net increase in carbon storage (assuming a constant residence for each ecosystem) and an additional 0.25‰ shift in the δ^{13}C of DIC (Fig. 5). This exercise shows that changes in the labile carbon isotope pools of terrestrial ecosystems can have an important effect on the isotopic composition of the atmosphere and ocean.

6. SUMMARY

In this paper I showed that a major change in terrestrial biomass took place in the late Neogene, between about 8 and 6 million years ago, with C$_4$ biomass becoming significant in the hotter parts of the Earth from the tropics to the warmer temperate regions. This change is recorded in southern Asia, East Africa, North America, and South America, all in a geologically very short period of time. Grasses are the most abundant C$_4$ plants on Earth today. Prior to 8 million years ago, virtually all plants used the C$_3$ photosynthetic pathway. The "C$_3$-world" is thought to be favored by high CO$_2$:O$_2$ atmospheric ratios, on the order of 1:4000. In the present low-CO$_2$ world (the "C$_4$-world") CO$_2$:O$_2$ ratios are on the order of 1:6000 for the pre-1850 atmosphere. Evidence from the Arabian Sea shows an increase in monsoon-related upwelling, and stable oxygen isotope studies of pedogenic carbonate suggest fundamental changes in the meteoric cycle in southern Asia, and possibly East Africa, at this time.

These observations are compatible with CO$_2$ drawdown by weathering related to tectonics, but do not provide unequivocal evidence for a link between tectonics and global ecological change. The consequence of replacing C$_3$ biomass by C$_4$ biomass could include changes in the isotope composition of DIC in the ocean.

REFERENCES

1. DePaolo, D. J., and Ingram, B. L. (1985). *Science* **227**, p. 938.
2. Hess, J., Bender, M. L., and Schilling, J. (1986). *Science* **231**, p. 979.
3. Hodell, D. A., Mueller, P. A., and Garrido, J. R. (1991). *Geology*, **19**, p. 24.
4. Pegram, W. J., Krishnaswami, S., Ravizza, G. E., and Turekian, K. K. (1992). *Earth Planet. Sci. Lett.* **113**, p. 569.
5. Ravizza, G. (1993). *Earth Planet. Sci. Lett.* **118**, p. 335.

6. Peucker-Ehrenbrink, B., Ravizza, G., and Hofmann, A. W. (1995). *Earth Planet. Sci. Lett.* **130**, p. 155.

7. Palmer, M. R., and Edmond, J. M. (1989). *Earth Planet. Sci. Lett.* **92**, p. 11.

8. Edmond, J. M. (1992). *Science* **258**, p. 1594.

9. Krishnaswami, S., Trivedi, J. R., Sarin, M. M., Ramesh, R., and Sharma, K. K. (1992). *Earth Planet. Sci. Lett.* **109**, p. 243.

10. Hodell, D. A., Mead, G. A., and Mueller, P. A. (1990). *Chem. Geol. (Isotope Geosci. Sect.)* **80**, p. 291.

11. Richter, F. M., and Turekian, K. K. (1993). *Earth Planet. Sci. Lett.* **119**, p. 121.

12. Miller, K. G., Fairbanks, R. G., and Mountain, G. S. (1987). *Paleoceanography* **2**, p. 1.

13. Young, J. A. (1987). In: *Monsoons* (J. S. Fein and P. L. Stephens, eds.), p. 211–243. John Wiley, New York.

14. Prell, W. L., and Kutzbach, J. E. (1992). *Nature* **360**, p. 647.

15. Kutzbach, J. E., Guetter, P. J., Ruddiman, W. F., and Prell, W. L. (1989). *J. Geophys. Res.* **94**, p. 18393.

16. Kutzbach, J. E., Prell, W. L., and Ruddiman, W. F. (1993). *J. Geol.* **101**, p. 177.

17. Kroon, D., Steens, T., and Troelstra, S. R. (1991). *Proc. Ocean Drilling Program, Scientific Results* **117**, p. 257.

18. Harrison, T. M., Copeland, P., Kidd, W. S. F., and Yin, A. (1991). *Science* **255**, p. 1663.

19. Raymo, M. E., Ruddiman, W. F., and Froelich, P. N. (1988). *Geology* **16**, p. 649.

20. Raymo, M. E., and Ruddiman, W. F. (1992). *Nature* **359**, p. 117.

21. Cerling, T. E., Wang, Y., and Quade, J. (1993). *Nature* **361**, p. 344.

22. Quade, J., Cerling, T. E., and Bowman, J. R. (1989). *Nature* **342**, p. 163.

23. Quade, J., and Cerling, T. E. (1995). *Palaeogeogr. Palaeoclim. Palaeoecol.* **115**, p. 91.

24. France-Lanord, C., and Derry, L. A. (1994). *Geochim. Cosmochim. Acta* **58**, p. 4809.

25. Derry, L. A., and France-Lanord, C. (1996). *Earth Planet Sci. Lett.* **142**, p. 59.

26. Morgan, M. E., Kingston, J. D., and Marino, B. D. (1994). *Nature (London)* **367**, p. 162.

27. Wang, Y., Cerling, T. E., and MacFadden, B. J. (1994). *Palaeogeogr. Palaeoclim. Palaeoecol.* **107**, p. 269.

28. MacFadden, B. J., Cerling, T. E., and Prado, J. (1996). *Palaios* **11**, p. 319.

29. Ehleringer, J. R., Field, C. B., Lin, Z. F., and Kuo, C. Y. (1986). *Oecologia* **70**, p. 520.

30. Ehleringer, J. R., and Cooper, T. A. (1988). *Oecologia* **76**, p. 562.

31. Medina, E., Montes, G., Cuevas, E., and Rokzandic, Z. (1986). *J. Tropical Ecol.* **2**, p. 207.

32. van der Merwe, N. J., and Medina, E. (1989). *Geochim. Cosmochim. Acta* **53**, p. 1091.

33. Osmond, C. B., Winter, K., and Ziegler, H. (1982). In: *Encyclopedia of Plant Physiology* (O. L. Lange, P. S. Nobel, C. B. Osmond, and H. Ziegler, eds.), pp. 479–547. Springer-Verlag, Berlin.

34. Hattersley, P. W., and Watson, L. (1992). In: *Grass Evolution and Domestication* (G. P. Chapman, ed.), pp. 38–116. Cambridge University Press, Cambridge.

35. Furbank, R. T., and Hatch, M. D. (1987). *Plant Physiol.* **85**, p. 958.

36. Cerling, T. E. (1984). *Earth Planet. Sci. Lett.* **71**, p. 229.

37. Cerling, T. E., and Quade, J. (1993). *Am. Geophys. Union Monograph* **78**, p. 217.

38. Cerling, T. E., and Wang, Y. (1996). In: *Mass Spectrometry of Soils* (T. W. Boutton and S.-I. Yamasaki, eds.), pp. 113–131. Marcel Dekker, New York.

39. Lee-Thorp, J., and van der Merwe, N. J. (1987). *South Afr. J. Sci.* **83**, p. 712.

40. Cerling, T. E., unpublished data.

41. Wang, Y., and Cerling, T. E. (1994). *Palaeogeogr. Palaeoclim. Palaeoecol.* **107**, p. 281.

42. Pilgrim, G. E. (1913). *Records Geol. Surv. India* **43**, p. 264.

43. Tauxe, L., and Opdyke, N. D. (1982). *Palaeogeogr. Palaeoclim. Palaeoecol.* **37**, p. 43.

44. Johnson, N. D., Opdyke, N. D., Johnson, G. D., Lindsay, E. H., and Tahirkheli, R. A. K. (1982). *Palaeogeogr. Palaeoclim. Palaeoecol.* **37**, p. 17.

45. Quade, J., Cerling, T. E., Morgan, M. M., Pilbeam, D. R., Barry, J., Chivas, A. R., Lee-Thorp, J. A., and van der Merwe, N. J. (1992). *Chem. Geol. (Isotope Geosci. Sect.)* **94**, p. 183.

46. Stern, L. A., Johnson, G. D., and Chamberlain, C. P. (1994). *Geology* **22**, p. 419.

47. Cerling, T. E., Harris, J. M., MacFadden, B. J., Ehleringer, J. R., Leakey, M. G., Quade, J., Eisenmann, V. *Nature* (in press).
48. Quade, J., Cater, J. M. L., Ojha, T. P., Adam, J., and Harrison, T. M. (1995). *Geol. Soc. Am. Bull.* **107**, p. 1381.
49. Barry, J. C., Johnson, N. M., Raza, S. M., and Jacobs, J. L. (1985). *Geology* **13**, p. 637.
50. Cerling, T. E. (1992). *Global Planet. Change* **5**, p. 241.
51. Cerling, T. E., and Hay, R. L. (1986). *Quat. Res.* **25**, p. 63.
52. Cerling, T. E., Bowman, J. R., and O'Neil, J. R. (1988). *Palaeogeogr. Palaeoclim. Palaeoecol.* **63**, p. 335.
53. Kingston, J. D., Marino, B. D., and Hill, A. (1994). *Science* **264**, p. 955.
54. Leakey, M. G., Feibel, C. S., Bernor, R. L., Harris, J. M., Cerling, T. E., Stewart, K. M., Stoors, G. W., Walker, A., Werdelin, L., and Winkler, A. J. (1996). *J. Vert. Paleo.* **16**, p. 556.
55. Hill. A. (1995). In: *Paleoclimate and Evolution with Emphasis on Human Origins* (E. S. Vrba, G. H. Denton, T. C. Partridge, and L. H. Burckle, eds.), pp. 178–193. Yale University, New Haven.
56. Cerling, T. E., Harris, J. M., and MacFadden, B. J. In: *Stable Isotopes and the Integration of Biological, Ecological, and Geochemical Processes* (H. Griffiths, D. Robinson, P. Van Gardingen, eds.). Bios Scientific Publishers, Oxford (in press).
57. Tidwell, W. D., and Nambudiri, E. M. V. (1989). *Rev. Palaeobot. Palynology* **60**, p. 165.
58. Teeri, J. A., and Stowe, L. G. (1976). *Oecologia* **23**, p. 1.
59. Hattersley, P. W. (1983). *Oecologia* **57**, p. 113.
60. Hattersley, P. W., and Watson, L. (1992). In: *Grass Evolution and Domestication* (G. P. Chapman, ed.), pp. 38–116. Cambridge University Press, Cambridge.
61. Hodell, D. A., and Kennett, J. P. (1986). *Paleoceanography* **1**, p. 285.
62. Adams, J. M., Faure, H., Faure-Denard, L., McGlade, J. M., and Woodward, F. I. (1990). *Nature* **348**, p. 711.
63. Schlesinger, W. H. (1991). *Biogeochemistry: An Analysis of Global Change*, 443 pp. Academic Press, New York.
64. Melillo, J. M., McGuire, A. D., Kicklighter, D. W., Moore, B., III, Vorosmarty, C. J., and Schloss, A. L. (1993). *Nature* **363**, p. 234.

14

Chemical Weathering Yields from Basement and Orogenic Terrains in Hot and Cold Climates

John M. Edmond and Youngsook Huh

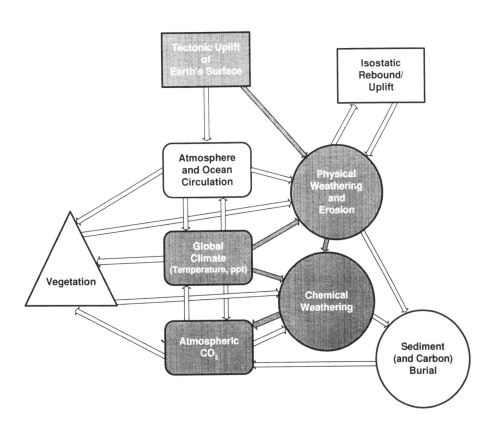

John M. Edmond and Youngsook Huh ● Department of Earth, Atmospheric, and Planetary Sciences, Massachusetts Institute of Technology, Cambridge, Massachusetts 02139.

Tectonic Uplift and Climate Change, edited by William F. Ruddiman. Plenum Press, New York, 1997.

1. INTRODUCTION

The ^{18}O record in benthonic foraminifera over the last 70 million years can only be interpreted in terms of a more-or-less continual climatic deterioration over the period of record.[1,2] Abyssal water temperatures record winter conditions in the oceanic areas where climatic extremes generate the densest waters. The isotopic temperatures in the abyssal waters in the early Eocene, 50 million years ago, were about 15°C higher than at present. It is unlikely that a Mediterranean-type "warm salty bottom water" could have dominated the global circulation in the face of freezing temperatures as ice-brine formation should have prevailed as the major source of convection, then much as it does today. Hence the mean surface temperature must have been much higher than at present, and the latitudinal climatic gradient much lower; this is in accord with the fossil record. The only feasible way to achieve this climatic optimum appears to be through an enhanced greenhouse effect involving greatly elevated atmospheric pCO_2, despite the fact that general circulation model (GCM) runs following this approach have been less than successful in simulating the conditions inferred from the geologic observations.[3-6] The problem then is to explain the mechanisms responsible for large oscillations (order tenfold) in pCO_2 on geologic timescales.

The inorganic sources of CO_2 are from the mantle, via oceanic hydrothermal activity and arc volcanism, and from the continents by metamorphic regeneration from buried carbonates via the "Urey reaction":

$$CaCO_3 + SiO_2 \Rightarrow CaSiO_3 + CO_2 \tag{1}$$

Over roughly the last 100 million years there has been some, probably highly variable, contribution from the recycling of planktonic carbonates in subduction zones also caused by this reaction.[7,8] Oxidative weathering of organic-rich black shales exposed in orogenesis represents an additional source, also sporadic. The predominant sinks are the weathering of aluminosilicate rocks (the fixation of CO_2 as HCO_3^-, bicarbonate) and subsequent formation of limestones and, in addition, the burial of organic carbon:

$$CO_2 + H_2O \leftrightarrow \underset{\text{Carbonic acid}}{H_2CO_3} \leftrightarrow H^+ + \underset{\text{Bicarbonate}}{HCO_3^-} \leftrightarrow 2.H^+ + \underset{\text{Carbonate}}{CO_3^{2-}} \tag{2}$$

$$\underset{\text{Albite}}{7NaAlSi_3O_8} + 6.CO_2 + 26.H_2O \Rightarrow 6.Na^+ + \underset{\text{Bicarbonate}}{6HCO_3^-}$$

$$+ 10.\underset{\text{Silica}}{Si(OH)_4} + 3.\underset{\text{Montmorillonite clay}}{Na_{0.33}Al_{2.33}Si_{3.67}O_{10}(OH)_2} \tag{3}$$

$$Ca^{2+} + 2.HCO_3^- \Leftrightarrow \underset{\text{Limestone}}{CaCO_3} + CO_2 + H_2O \tag{4}$$

$$CO_2 + H_2O \Leftrightarrow \underset{\text{Carbohydrate}}{C(H_2O)} + O_2 \tag{5}$$

The strengths of these sources and sinks have different time constants of variability. Here only the inorganic aspects of the problem will be considered.

The predominant view of the controls on greenhouse climate originates from the "thermostat" model of Walker *et al.*[9] Chemical weathering, hence CO_2 consumption, is assumed to be limited by the availability of water from precipitation and therefore by mean atmospheric temperature (controlled by the CO_2 greenhouse effect) since the saturation vapor pressure of H_2O at tropical temperatures increases by $\sim 6\%/1°C$. High temperatures enhance the weathering sink for CO_2 and low ones diminish it; the climate system drifts around the habitable optimum. This hypothesis has led to a great deal of theoretical activity and model simulations of pCO_2 levels over the last few hundred million years,[10] but surprisingly there has been little observational effort to test the core assumptions. It has also been challenged by the "Raymo hypothesis," which says, in effect, that episodes of orogenic activity can override the "thermostat" and drive climate deterioration through increased CO_2 consumption by weathering caused by increased exposure, independent of the climatic regime.[11] This deterioration would persist until the cessation of the tectonic episode, i.e., tens of millions of years, consistent with the geologic record of the duration of past glacial epochs.[12,13] However, the factors controlling chemical weathering, climatic or otherwise, are not well understood. Here an attempt will be made to summarize those existing data sets that might support some generalizations concerning the large-scale mechanisms involved in inorganic consumption of atmospheric CO_2 by aluminosilicate weathering. Using the range of environmental and tectonic conditions on the existing Earth, one can investigate the effects on $ØCO_2$, the aluminosilicate weathering yield, of climate, i.e., the mean annual temperature and runoff, as a function of tectonic regime and thereby constrain the possible influences of both the Walker and the Raymo hypotheses.

2. SOME STRATEGIC CONSIDERATIONS

The 20 largest rivers contribute only about 30% of the global discharge[14]. Hence, attempts to derive an exact contemporary geochemical budget based on sampling of *all* rivers are impossible *a priori*. One can only hope to generalize from environmentally representative drainages as to the possible changes in the weathering regime and flux that can result from climatological and tectonic events. In effect one can look for relative responses but not absolutes. The present is a poor time to attempt to study natural weathering processes across a representative range of environments since so many of them have been strongly perturbed by human activity, often on continental scales. By definition, pristine environments are remote, making detailed study logistically difficult. Cartography, hydrology, meteorology, and geology are sciences driven to a very significant degree by social demand and so, again by definition, the physical observational infrastructure is usually weak in remote locations.

Access to suitable field areas can be difficult owing to logistical, political, and other factors; in addition these considerations vary over time in any particular region. Because of large-scale atmospheric contamination, remoteness alone may not be a sufficient criterion. Clearly the widespread occurrence of acid precipitation greatly complicates fundamental studies; in addition other contaminants, e.g., sulfate, are imported into the basins and can be difficult to separate from the primary weathering yield. Small basins are particularly susceptible to this effect and also have unstable flow conditions that respond entirely to short-term weather events. This greatly increases the sampling effort required to form a representative picture. Here, therefore, the discussion will be restricted to large, reasonably pristine systems. While acknowledging the drawbacks, often only one-time sampling that is usually biased toward the falling river stage, denial of continued access for follow-up studies owing to changing circumstances, etc., we know that large rivers and their major tributaries integrate the locally generated hydrologic signals and thus have stable and characteristic hydrographs; contain a range of geological, climatological and vegetational environments; and themselves contribute significantly to the global geochemical flux. Influences from outside these basins, given their scale and location, are usually minor. Such studies are the only potential basis for the generalization of the responses of the global weathering environment and CO_2 fixation rates to externally forced tectonic and climatic change.

Hydrologic information from the regions studied (essential for the calculation of weathering fluxes) can be nonexistent or difficult to obtain. In some (rare) cases data are included in the original publications on the river measurements. In others they are contained in the UNESCO numerical compilations for "selected" rivers of the world.[15] As a last resort the discharge can be estimated from the regional maps of areal runoff[16] because these are based on the interpolation and contouring of the available hydrologic measurements (themselves usually unpublished).

Calculation of the parameter of interest from these data sets, the basin-scale consumption rate of CO_2 by aluminosilicate weathering ($\text{Ø}CO_2$ in moles km year^{-1}) is subject to numerous difficulties. In the simplest case, weathering of a catchment composed entirely of igneous and metamorphic rocks, the relationship is obvious from Eqs. (4) and (6), ($\text{Ø}CO_2 = \text{Ø}TZ^+$, where TZ^+ is the total cationic concentration in equivalents (corrected for sea salt aerosols) in the riverine samples. This formulation comes from the requirement for electroneutrality; one proton from carbonic acid is exchanged for every positive cationic charge dissolved and the corresponding bicarbonate anion is "fixed" in solution (see the reaction scheme above). Unfortunately, the direct measurement of this fixed CO_2, the bicarbonate alkalinity derived from acid titration, is often ambiguous owing to the presence of unanalyzed organic acids. These can be so prevalent, e.g., in "black" tropical rivers, that the measured alkalinity is *negative* despite significant concentrations of cations.[17] However, in basins contaminated by salt from mining, pulping, etc., the alkalinity has to be used, $\text{Ø}CO_2 = \text{Ø}ALK$, although this is a minimum estimate for the reasons given above. In a basin of mixed lithology containing limestones and dolomites, as identified by the $^{87}Sr/^{86}Sr$ ratio, which these rapidly soluble rocks buffer at between 0.7075 and 0.7090 depending on age,[18] recourse must be made to $\text{Ø}Si$, the flux of dissolved silica. Garrels[19] investigated the HCO_3^-:Si ratios in groundwaters from a variety of igneous and metamorphic rock types and found a range from about 0.1 to 1.0 with most of the values between 0.3 and 0.5. This ratio is essentially ALK:Si, the inverse of Si:TZ^+. As a conservative estimate in such situations, $\varnothing CO_2 = 2.\varnothing Si$. This is also the estimated ratio for aluminosilicate weathering in general, based on the global fluvial data set.[20] However, this formulation is also ambiguous. In severely weathered terrain, where the soluble cations have been completely released, much of the silica comes from the degradation of kaolinite to gibbsite:

$$6.\text{Montmorillonite} + 2.CO_2 + 25.H_2O \Rightarrow 2.Na^+ + 2.HCO_3^-$$
$$+ 8.Si(OH)_4 + 6.\underset{\text{Kaolinite}}{Al_2Si_2O_5(OH)_4} \qquad (6)$$

$$\text{Kaolinite} + 10.H_2O \Rightarrow 2.Al(OH)_3 + 2.Si(OH)_4 \qquad (7)$$
$$\underset{\text{Gibbsite}}{}$$

The extreme reaction [Eq. (7)] does not involve protons and therefore does not contribute to $\varnothing CO_2$; Si:TZ^+ is then high, (>3) (Edmond *et al.*[17]). In frost-dominated regimes where the weathering is very superficial, since the disaggregated rock detritus is rapidly transferred to the mechanical load, a cation exchange reaction appears to occur involving only the outermost few unit cells of the crystal lattice, with retention of silica. In these cases the ratio Si:TZ^+ is low, (<1) and the relation $\varnothing CO_2 = \varnothing TZ^+$ is used. Lakes and artificial impoundments along a river channel act as traps for silica owing to large-scale fixation by diatoms and subsequent sedimentation; they are especially prevalent in formerly glaciated areas, where overdeepened valleys contain lakes and often provide attractive dam sites for hydropower. In this case $\varnothing Si$ is not reflective of the actual weathering regime. Then recourse, again, has to be made to the formulation, $\varnothing CO_2 = \varnothing ALK$, because carbonate precipitation is usually minor in this situation.

Supporting diagnostic evidence for the relative importance of aluminosilicate weathering in a particular basin comes from the Sr isotope data. Limestones and unmetamorphosed basalts weather at similar rates, much faster than massive igneous and metamorphic rocks and shales. Shales lack mechanical integrity and are rapidly transferred to the physical load with little weathering once the relatively labile matrix cements, Ca–Mg–Fe carbonates, sulfides, and organic carbon, have been removed. For example, most of the suspended load in the Mississippi derives from the Badlands, fissile shale terrains of the Dakotas, and that of the Orinoco from similar lithologies in the Guaviare and Meta.[21,22] The Sm/Nd model ages of fluvial suspended material and eolian dust fall around 2 billion years, regardless of immediate provenance indicating the persistence of fine-grained detritus in a relatively unaltered state through numerous cycles of erosion and redeposition.[23] The $^{87}Sr/^{86}Sr$ values in limestones reflect the evolution of the seawater value over the last few hundred million years, in the range 0.707–0.709.[18]

Carbonate rocks are rapidly soluble and contain much higher Sr concentrations than most other rocks with the exception of marine evaporites; limestones have concentrations of several thousand ppm, ($1 \, ppm = \sim 12 \, \mu moles/kg$) and their diagenetic product, dolomites, several hundred ppm. Basalts carry the nonradiogenic mantle value, 0.703–0.705, with tens to a few hundred ppm Sr. Most massive igneous rocks have similar concentrations; the average crustal isotope ratio is estimated at 0.730 but individual basement rock types span a wide range from less than 0.705 in mafic terrains up to about 1.0 in acidic rocks.[24] In addition, Sr and Rb are strongly fractionated between minerals during the crystallization of igneous rocks, Sr into the labile Na–Ca-feldspars and the radioactive parent Rb into the much more refractory K-feldspars and micas. Thus it is not possible for isotopic values significantly different from the limestone range to persist in the presence of carbonate (or evaporite) weathering; the latter, involving rapidly soluble rocks with a characteristic, relatively narrow range in the isotope ratio and high concentrations, acts to dominate the isotopic systematics wherever it occurs. Normally, weathering cannot release the radiogenic Sr stored in igneous and metamorphic rocks in high yield; the nonradiogenic Sr in the Na–Ca-feldspars is released preferentially.[25] The radiogenic Sr produced in the K-feldspars and, especially, the micas is transported in the mechanical load after weathering breakdown of the labile, nonradiogenic Na–Ca matrix minerals at the outcrop and during sediment transport.

With information on the geologic, hydrologic, and chemical systematics (geologic maps, discharge measurements, overall composition, TZ^+, Si, alkalinity, Sr, and $^{87}Sr/^{86}Sr$) for river systems from representative environments, it is possible to derive estimates of the driving greenhouse parameter $\emptyset CO_2$ with enough accuracy to identify the mechanisms that control its value.

3. THE EXISTING DATA SET

Geochemical models of the cycle of atmospheric CO_2 have used, for the weathering component, $\emptyset CO_2$, global compilations of fluvial chemical data such as those of Livingstone[26] and Maybeck[27] or assemblages of routine water-quality data, e.g., the U.S.G.S. reports for the United States.[14] These data sets are extremely heterogeneous in quality (many are of variable age and completeness and lack information on discharge) and in choice of river basins with respect to size, geology, environment, degree of perturbation, and seasonality of sampling (dry season data are overrepresented in the global compilations). The data also consist largely of point samples taken at the mouths of rivers or major tributaries draining a wide range of environments, making it difficult to deconvolve the chemical signals in terms of mechanisms. The results of these efforts have been a number of broad, and conflicting, correlations but little understanding (see Summerfield and Hulton[28] for a critique).

The systematic study of large rivers on the basinal scale began with the pioneering work of Gibbs[29] on the Amazon. On the basis of conductivity measurements, he deduced that the dominant weathering yield was from the

Andes. He also produced a time series of chemical data from the lower river, which demonstrated that, despite the perfect sinusoidal form of the hydrograph, there were important asymmetries in the compositions over the seasons.[30] The first comprehensive *chemical* data set from a large basin was that of Reeder *et al.*[31] for the Mackenzie. This was followed by a similar study of the Amazon,[32] of the Himalayan tributaries of the Ganges–Brahmaputra,[33] of the northern part of the Congo,[34,35] of the Frazer,[36] of the Orinoco,[17,37] and of the lower St. Lawrence.[38] Recent work has extended this global reconnaissance to the basement, platforms, and collision zones of eastern Siberia (Lena, Yana, Indigirka, Kolyma; Y. Huh, thesis in preparation). The data sets are of variable completeness. The Mackenzie data include the major ions and silica but no Sr isotopes. The data from the Congo often lack alkalinity or silica values. The Amazon, Orinoco, Frazer, St. Lawrence, and Siberian data are complete in this regard. The Himalayan data set often lacks silica. The geologic and hydrologic coverage is uneven in all the data sets. Despite these deficiencies, there is now reasonably comprehensive information on the continental arc of the western Americas as a function of latitude, 15°S–70°N, on the Indo–Tibet continental collision zone, probably the largest such tectonic event since the Pan-African some 550 million years ago, and on tropical and subarctic basement and other terrains. This makes it possible to develop tentative generalizations about the response of chemical weathering yields and atmospheric CO_2 drawdown to climatic and tectonic changes.

3.1. Basement Terrains

The accessible basement terrains of the Northern Hemisphere, the Canadian and Baltic Shields, have been pervasively impacted by Plio–Pleistocene glaciations. Thus they are covered with lithologically heterogeneous debris transported by ice streams often from outside the existing drainage basins. Their weathering yields are therefore influenced by an additional variable rather than simply the contemporary climate.[39] The Aldan Shield and Trans-Baikal Highlands in southeastern Siberia in the drainage of the Lena are basement terrains situated in a region remote from the predominantly oceanic sources of water vapor. They experience a severe continental climate with seasonal temperature variations in excess of 75°C. The regime is semiarid with most of the precipitation occurring in a few monsoonal events in midsummer. Thus, even at the highest elevations there are few permanent snow fields. At the glacial maxima ice streams appear to have been restricted to high valleys; there was no large-scale formation of ice sheets.[40] The landscapes are therefore in close to a climatically unperturbed state and the effects of temperature and runoff on weathering yields can be compared directly with the tropical regimes. In the latter the most comprehensive data sets are for the Guayana Shield and, to a much lesser extent, the Congo Shield. Data for the Brazilian Shield are less detailed largely because the main channel flows in a failed Cretaceous rift flanked by exposures of marine limestones and evaporites.[32] Access to the interior shield terrains is generally difficult. All three areas are in the humid tropics with stable annual temperatures of about

28°C, precipitation rates in excess of 1500 mm, and runoff values of around 1000 mm.

The Guayana Shield is ideal for a comparative study as, similar to the Siberian analogues, it is completely devoid of chemical and biogenic sediments of marine or continental origin. The fluvial geochemistry of the Guayana Shield and its relationship to lithology and topography has been discussed in detail elsewhere.[17] Briefly, the Shield occupies the right bank of the Orinoco drainage. It is composed of a wide range of rock types from greenstone belts to highly alkalic rapakivi granites. These basement rocks are generally of Trans-Amazonian age, some 2.1 billion years old. The only platform cover is the Rorima quartzite ($\sim$1.7 billion years old) with fluvio-lacustrine and deltaic facies. This cover is most extensive in the east, where it attains a thickness of about 2 km; isolated outliers occur as mesas all the way to the western limits of the Shield exposures. The topography slopes from near sea level in the west to elevations in excess of 1,500 m around the southern and eastern rim of the basin. Precipitation is strongly seasonal producing runoff rates of more than 1,000 mm. At low and intermediate elevations the basement is mantled with ferruginous laterite with thicknesses of tens of meters.

The rivers are generally "black" with dissolved organic material, the pH values are in the range 4–6 and the compositions are sufficiently dilute (40–600 μeq) that for some streams H^+ is the major cation. The chemical systematics show that the soluble cations are completely mobilized from the basement rock. This results in the accumulation of a refractory residual cover of gibbsite, kaolinite, quartz, and ferric oxyhydroxides — laterite. Over time this cover develops into an effective seal separating the fresh rock from the surface environment. The weathering yields are controlled by the flux of water through the subsurface weathering "front" and are extremely low. The pervasive nature of this slow reaction is demonstrated by the observation of "fluvial isochrons"; in basins of homogeneous lithology, the $Rb–Sr–^{87}Sr/^{86}Sr$ systematics of the river data agree with reported whole-rock isochrons. The Sr isotopic data range from a high of 0.9217 in streams draining the Parguaza rapakivi granite to 0.7139 in streams from the western, lowland shield. In addition the cation data, as displayed on a ternary diagram, follow the evolutionary trend displayed by average igneous rock types. Well over half the molar dissolved load is silica (18–308 μM) derived from the degradation of kaolinite to gibbsite [Eq. (7)]. If the CO_2 fixation rate is taken as the flux of cations in equivalents, $\varnothing CO_2 = \varnothing TZ^+$, the range of values (Table 1) is between 205×10^3 moles km^{-2} year^{-1} for the upland shield (Ventuari) and 16×10^3 moles km^{-2} year^{-1} for the lowland areas (Atabapo).

The rivers draining the Brazilian Shield in the Amazonian lowlands were sampled in the flood season above their confluences with the main channe.[32] The rivers are similar to those of the Guayana Shield, "black," acid (pH 5–7), extremely dilute (TZ$^+$ 54–260 μeq), and with Si making up much of the dissolved load (71–192 μM). The Sr is quite radiogenic, 0.7161–0.7378. The calculated values for $\varnothing CO_2$ (as $\varnothing TZ^+$) are between 88 and 268×10^3 moles km^{-2} year^{-1} at the high end of the range for the Orinoco drainages (Table 1).

The data for the Congo are restricted to the lower part of the main stem and the basin of the Oubangi, the major right bank tributary, and as such are much less extensive than for the Orinoco.[33,34] Runoff is intermediate, (~ 600 mm), but with no strong seasonality. The basement terrain is similar to that of the Guayana Shield, intrusive and metamorphic rocks mantled by thick laterite. However there are significant outcrops of limestones, about 10% of the catchment area studied, and these exert a strong influence on the dissolved load. Only a few of the tributaries have compositions indicative solely of basement weathering. These are very dilute, "black" streams with relatively radiogenic $^{87}Sr/^{86}Sr$ (> 0.730) and high silica. The rivers are very potassic with K:Na ratios between 0.5 and 1; weathering is thus severe since the K-phases are much more refractory than the Na-feldspars. The ratio $Si/[Na + K]$, an index of weathering severity, is also very high, 2.5–6.1; the latter is by far the highest observed and indicates a large preponderance of kaolinite decomposition relative to contributions from cation-containing phases. The chemistry of the lower Congo is quite similar to that of the Orinoco above its delta,[37] indicating that, overall, the importance of chemical and biogenic sediments relative to basement rocks is similar in the two basins with the lithologies being that of marine platform cover and Andean continental arc, respectively. The data sets reported by Négrel et al[34] and Dupré et al[35] (their Tables 1 and 1 and 4, respectively) have been combined to calculate flux estimates ($\varnothing CO_2 = \varnothing TZ^+$). The $\varnothing CO_2$ values (Table 1) range from 15 to 73×10^3 moles km^{-2} $year^{-1}$; the data for the Zaire main stem and Kasai are not used as the importance of the carbonate contribution in these very large basins is not known. The values are similar to those from the lower lying areas of the Guayana Shield, consistent with the topography of their basins.

The cratonic areas in the Southern Hemisphere, southern Africa, India, and western Australia, now semiarid to arid, have geomorphologies similar to that developed on the humid Guayana Shield.[41,42] Thus a transition to full-scale pluvial conditions in these regions would not result in markedly increased CO_2 fixation rates. There would be a significant increase in the silica flux owing to the breakdown of relict kaolinite to gibbsite [Eq. (7)]; this reaction does not involve proton consumption or CO_2 fixation. It is probable that similar stable cratons in the Northern Hemisphere resembled their southern counterparts in preglacial times before their cover was removed by ice action of various types.[41] It can be concluded that the weathering yield from stable basement terrains is not sensitive to climate change save in extreme conditions; subtropical aridity as in the Southern Hemisphere, seasonal freezing, and the associated ice-induced disruption of the protective mantle and recent full-scale glacial erosion as in the Northern Hemisphere.

The weathering of basement rocks at midlatitude can only be examined using the data from the relatively small left bank tributaries of the lower St. Lawrence in southeastern Canada.[38] These drain the Canadian Shield in a temperate, maritime climatic environment with an annual precipitation of about 1200 mm and runoff of some 550 mm. The six left bank tributaries for which data are reported all have lakes or impoundments on their courses. Thus, because of

TABLE 1

	Area (10^3 km^2)	Discharge (10^9 m^3 year^{-1})	Na <	K	Mg	Ca	Cl	SO$_4$
						(10^9 moles year^{-1})		
Guayana Shield								
Upper Orinoco	60	130	2.7	2.0	0.72	1.3	—	—
Ventuari	40	94	2.9	1.7	0.80	1.0	—	—
Atabapo	10	13	0.07	0.03	0.008	0.020	—	—
Guainia	26	72	0.38	0.29	0.094	0.15	—	—
Parguaza	5.4	13	0.20	0.21	0.014	0.059	—	—
Cuchivero	18	10	0.65	0.25	0.17	0.23	—	—
Caura	50	85	2.7	1.4	0.77	1.2	—	—
Aro	14	9.4	0.65	0.19	0.26	0.31	—	—
Caroni	93	130	1.8	1.0	0.85	1.2	—	—
Amazonian Shield								
Trombetas	81	220	7.9	4.0	2.0	2.9	3.5	1.1
Tapajos	476	440	14	8.8	4.0	9.7	4.4	1.3
Xingu	501	330	25	8.9	7.9	10.6	6.3	0.33
Negro	669	1000	17.5	10	6.5	9.2	7.7	2.2
Congo Shield								
Obangui	475	110	6.9	2.9	4.5	7.2	2.4	1.4
Lobaye	31	11	0.25	0.21	0.42	0.53	0.10	0.10
Zaire	1600	534	72	29	52	36	25	13
Likouala	60	5.0	0.065	0.15	0.20	0.20	0.080	0.040
Sangha	250	53	1.8	2.0	1.6	3.3	0.95	0.53
Alima	50	19	0.65	0.70	0.63	0.72	0.34	0.23
Kasai	900	345	18	11	13	15	9.0	4.5
Canadian Shield								
Ottawa	90	37.7	4.7	0.79	3.1	8.4	4.8	3.1
St. Maurice	42	23.6	2.3	0.31	0.52	1.3	1.4	1.0
Aldan/ Trans-Baikal								
Aldan at Tammot	46	9.2	0.58	0.058	1.4	2.8	0.13	0.42
Gt. Seligri	2.3	0.46	0.13	0.001	0.021	0.055	0.004	0.006
Sumnayn	2.5	0.50	0.016	0.001	0.019	0.010	0.002	0.004
Jelinda	0.63	0.13	0.006	0.001	0.071	0.10	0.002	0.001
Illymach	3.4	0.68	0.025	0.003	0.10	0.20	0.003	0.034
Timpton	42	8.4	0.40	0.036	0.29	0.88	0.077	0.19
Vitim	13	5.2	0.25	0.057	0.29	0.72	0.063	0.20
Olekma	12	3.6	0.25	0.036	0.16	0.35	0.086	0.13

removal by diatoms, the silica data (56–76 μM) may not be representative of the actual weathering yield. Owing to their proximity to the North Atlantic, their chloride levels are high (216–487 μM); in addition to this sea salt input there is a probable road salt component as the rivers were sampled along the developed corridor between Montréal and Quebéc City.[38] Overall the rivers are quite dilute (alkalinities between 57 and 315 μeq); the Ca:Mg ratios range from 2.5 to 4. The Sr concentrations are low (0.5–0.2 μM) and the isotopic compositions relatively nonradiogenic (0.71068–0.71104); the authors calculate a shield end-member ratio

alk	Si	$^{87}Sr/^{86}Sr$	Dissolved Flux (10^9 moles year^{-1})	Dissolved flux (10^6 moles km^{-2} year^{-1})	Net CO$_2$ flux 10^3 (moles km^{-2} year^{-1})
					$\varnothing CO_2 = \varnothing TZ^+$
—	13	0.7415	20	0.33	146
—	12	0.7415	19	0.47	205
—	0.39	0.7461	0.52	0.050	16
—	3.6	0.7543	4.5	0.17	45
—	1.4	0.8535	1.8	0.34	103
—	1.8	0.7397	3.1	0.17	94
—	12	0.7350	18	0.36	161
—	2.1	0.7285	3.5	0.25	141
—	10	0.7322	15	0.16	74
					$\varnothing CO_2 = \varnothing TZ^+$
14	26	0.7161	61	0.75	268
48	65	0.7322	155	0.33	107
58	63	0.7292	180	0.36	142
9	67	0.7378	129	0.19	88
					$\varnothing CO_2 = \varnothing TZ^+$
26	25	0.7197	76	0.16	64
—	2.3	0.7154	—	—	70
200	101	0.7179	528	0.32	—
—	0.76	0.7162	—	—	15
6.9	11	0.7163	28	0.11	51
—	3.2	0.7163	—	—	73
53	—	0.7204	—	—	—
					$\varnothing CO_2 = \varnothing ALK$
10	2.9	0.7110	37.8	0.42	156
5.5	1.7	—	14.0	0.33	129
					$\varnothing CO_2 = \varnothing TZ^+$
6.6	1.1	0.7123	13.1	0.28	194
0.13	0.042	0.7202	0.39	0.17	121
0.11	0.049	0.7174	0.21	0.08	29
0.31	0.017	0.7134	0.51	0.81	555
0.51	0.069	0.7089	0.94	0.28	184
2.5	0.92	0.7120	5.29	0.13	64
1.6	0.42	0.7128	3.60	0.28	179
0.87	0.36	0.7134	2.24	0.19	102

of 0.713, very low for basement rocks. It can be concluded that weathering is quite superficial with very little of the radiogenic Sr contained in the refractory K-silicates being released. The degree to which the yields are influenced by a cover of glacial debris is not clear. Discharge/area data are available for two of the tributaries measured, the Ottawa and the St. Maurice.[15] In this case $\varnothing CO_2$ cannot be estimated from the cation flux because of the uncertainties associated with the origin of the chloride. The alkalinity flux, $\varnothing ALK$, is used in the calculations (Table 1). The results are 156 and 129 $\times$ 10^3 moles km^{-2} year^{-1},

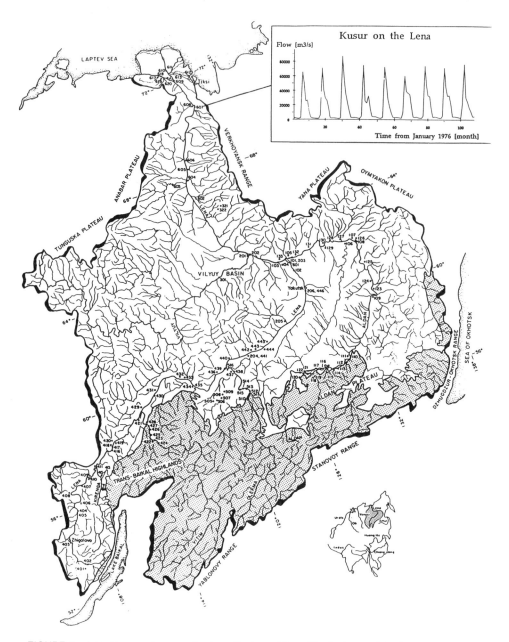

FIGURE 1.   Location map for samples from the Lena River. The hydrograph in the upper right is
from Kusar (Station 607) above the canyon that leads to the delta. The stippled area indicates
continuous outcrop of the igneous and metamorphic rocks of the Trans-Baikal Highlands and the
Aldan Shield (Huh, thesis in preparation).

respectively, higher than those from the Congo and in the midrange observed in the Orinoco.

The Trans-Baikal Highlands and Aldan Shield constitute a continuous outcrop of Proterozoic to early Archean basement rocks that forms the southern rim of the drainage of the upper Lena in southeastern Siberia (Fig. 1; Zonenshain et al.[43]). The area is rich in gold and other nonferrous mineral resources that have been exploited with increasing intensity since the beginning of this century. Thus, unusually for such a remote and inhospitable area, there have been sustained geological investigations. In addition the rivers are major transportation arteries so that the hydrologic and meteorological data are extensive. With the exception of scattered marble and banded iron formation in the strongly metamorphosed fold belts, the terrain is devoid of chemical sediment of marine or continental origin. This is in striking contrast to the Lena Platform to the north, which is one of the largest and oldest stable sedimentary basins in the world and has existed essentially undisturbed, except for oscillations in sea level, since at least the latest Precambrian.[43] The regional hydrographs are dominated by a meltwater spike in April–May and are complicated by a highly variable rainwater pulse in July–August. Headwater streams of the Aldan that drain the shield terrain have been sampled, as have the major tributaries from the Highlands, the Vitim, and Olekma. Logistics dictate midsummer collection times. The rivers are relatively dilute with alkalinity values between 200 and 700 μeq. The Sr is variably radiogenic with values ranging from 0.7089 in the Rb-poor, granulite facies of the ultramafic shield terrains to 0.7237 in the highlands. The Si:TZ$^+$ ratios are low, 0.05–0.37, with silica values between 73 and 127 μM (most values around 100 μM). The large data set from these terrains (>75 samples) is still being worked on (Y. Huh, thesis in preparation); representative values for $\emptyset CO_2$ (taken as $\emptyset CO_2 = \emptyset TZ^+$; Table 1) are in the range $29–555 \times 10^3$ moles km^{-2} year^{-1} with most values between 100 and 200, indistinguishable from those on the other shield terrains.

From this survey of all the available systematic data for basement terrains, admittedly sparse and often incomplete, it is difficult to identify any influence of climate (temperature/runoff) on the fixation rate of CO_2 by aluminosilicate weathering. If exposure were equal in all terrains presumably there would be such a relationship, but this is not the case. In the tropics weathering is self-limiting. The outcropping rocks are mantled by an impermeable residual of kaolinite and gibbsite cemented by ferric oxyhydroxides, laterite, and, in extreme cases, amorphous silica, silcrete. On a timescale of a few million years these systems seal themselves shut. Solifluction, lateral creep of this mantle, serves to level the preexisting surface topography (although considerable relief may be maintained in the bedrock in the subsurface) thus lowering the hydraulic gradient available for physical erosion.[41,42] In the presence of freezing temperatures, e.g., at contemporary middle and high latitudes, this process is short-circuited by ice action, gelifluction and frost heave, which disrupts any accumulating mantle and allows its rapid physical removal. In addition, frost cracking at all scales continually generates fresh rock surfaces available for weathering reactions. The result is that the controlling variable on CO_2 fixation rates is not climate *per se*, but the *physical* mechanisms that determine exposure. One cannot look to the climatically variable weathering of stable cratons for a contribution to the "thermostat."[9]

3.2. The Continental Arc of the Western Americas, 15 °S–70 °N

Good coverage exists for the eastern slope of the tropical Andes in the drainage of the Amazon and the Orinoco[32,37] and for the eastern and western slopes of the Rockies in the drainages of the Mackenzie, Frazer, and Yukon rivers extending to subarctic and arctic Canada.[31,36,37] The only significant environmental perturbations at the time these data sets were collected were well-defined point sources of NaCl contamination associated with pulp mills on two of the Frazer tributaries, a hydropower dam in the Yukon headwaters at Whitehorse and many impoundments on the Frazer tributaries. This very large data set from the continental arc of the western Americas (> 450 samples) displays an enormous range of chemistries reflective of the great diversity of lithologies exposed in the orogen.

The Andean drainage is dominated by a handful of major tributaries to the Amazon–Orinoco system: the Madeira, Ucayali, Huallaga, and Marañón (which combine to form the Solimões or upper Amazon, above the Negro confluence), the Napo and Japura in the Amazon drainage and the Guaviare, Meta, Arauca and Apure, the large left bank tributaries of the Orinoco. The fluvial geochemistry of these two systems has been discussed in detail elsewhere.[32,37,44] The eastern slope of the Andes receives in excess of 2000 mm of rainfall annually (in some areas > 4000 mm) with resulting runoff rates in excess of 1000 mm.[16] There are permanent snow fields at the highest elevations (> 4500 m) and a few small valley glaciers. These expanded significantly during the Ice Age maxima but did not occupy large areas of the drainages.[45] The chemical signatures (Fig. 2) range from those of the Madeira headwaters, which are dominated by the sulfuric acid weathering of black shales,

$$FeS_2 + 3.75.O_2 + 3.5.H_2O \Rightarrow Fe(OH)_3 + 2.SO_4^{2-} + 4.H^+ \tag{8}$$

to the Huallaga, where piercement structures of halite result in extremely high dissolved loads. Incidentally it is this very unusual source that generates the high conductivities originally noted by Gibbs.[29] The other Andean tributaries reflect the weathering of marine limestones and continental red beds and, in the Iça, andesites. Overall the contribution of aluminosilicate weathering is quite variable as reflected in the $^{87}Sr/^{86}Sr$ ratios, which range from 0.7185 for the Madeira, to 0.7086 for the Solimões at Iquitos below the confluence of its major Andean tributaries, to a low of 0.7075 for the Iça reflecting the andesitic component. The values for the Orinoco tributaries are very similar to those of the Madeira. Extracting $\emptyset CO_2$ from this data set is difficult because in general the fluxes are dominated by nonsilicate weathering and since, in an exposure-limited regime,

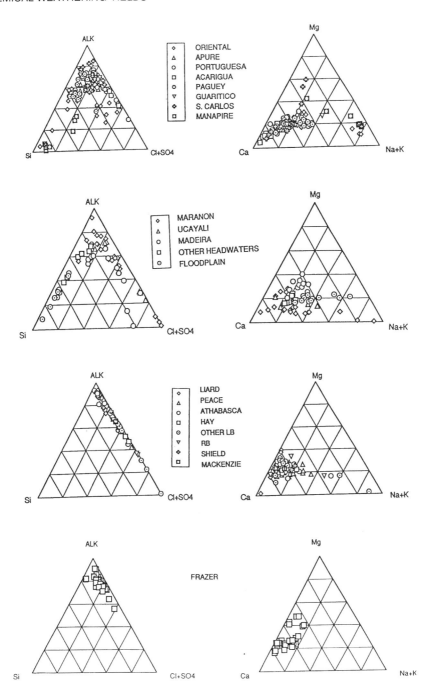

FIGURE 2. Ternary plots for the streams draining the Andes (Orinoco, Amazon) and the Rockies (Mackenzie, Frazer). The Yukon is not shown since the data form just one very tight grouping.

TABLE 2. Flux Estimates

	Area (10³ km²)	Discharge (10⁹ m³ year⁻¹)	Na <	K	Mg	Ca (10⁹ moles year⁻¹)	Cl	SO₄	alk
Orinoco Andes									
Guaviare	74	160	8.3	2.2	3.4	10.9	2.4	1.8	28.8
Meta	94	140	6.6	2.2	6.2	18.2	4.2	9.1	29.4
Arauca	16	32	1.8	0.80	1.7	3.2	0.16	1.6	8.0
Apure	167	94	13.0	3.8	9.4	42.0	0.95	9.5	90.5
Amazonian Andes									
Solimões	717	1,200	312	33	118	624	204	102	1,380
Ica	132	280	7.6	3.9	4.3	14	2.1	0.76	24
Japura	236	500	38	11	21	88	22	9.5	212
Madeira	997	770	50	24	49	82	10	28	234
Mackenzie									
Peace	280	53.4	12.8	1.23	18.9	58.5	4.06	19.8	120
Athabasca	145	22.0	11.8	0.44	6.8	18.7	7.46	5.5	43.3
Liard	250	47.1	4.3	0.61	15.9	43.0	0.80	12.5	91.9
Mackenzie	1,700	300	91.2	6.0	96.0	269	64.2	98.1	521
Yukoh at Eagle	294	157	12.5	4.7	39.2	102	1.6	47.3	236
Frazer									
McGregor	5.49	6.8	0.103	0.046	1.33	3.93	0.061	0.360	9.72
Nechako	32.2	9.1	0.846	0.155	1.23	2.88	0.109	0.355	8.07
Quesnel	12.9	6.8	0.258	0.068	0.625	3.05	0.041	0.449	6.70
Chilcoten	19.3	3.4	0.316	0.061	0.394	0.870	0.027	0.224	2.47
Bridge	5.6	3.4	0.384	0.044	1.64	1.57	0.092	0.571	5.61
Thompson	57.2	23.9	1.65	0.478	1.86	7.07	0.454	1.65	16.1

mineral weathering is only partial, there are potentially large fractionations between the major ions relative to silica. As discussed above, the CO_2 drawdown caused by aluminosilicate weathering can be crudely estimated under these conditions as $\emptyset CO_2 = 2.\emptyset Si$. The values for the Andean headwaters (Table 2) range from $14^3 \times 10^3$ moles km^{-2} year^{-1} for the Apure, 219 for the Madeira, 500 for the tributaries of the Solimões to as high as 1000 for the Arauca. These area-based calculations are underestimates because they include the variably sized floodplains in the area measurement. However, they overlap the high end of the rates computed for the shield terrains and range to substantially greater values.

The Mackenzie Basin lies between approximately 52° and 70°N in subarctic and arctic Canada. Regional elevations in the left bank Rocky Mountain drainages can be in excess of 2000 m. There is a large latitudinal gradient in runoff from about 1000 mm in the more temperate southern headwaters to less than 200 mm in the arid north. The geomorphology is heavily influenced by glacial action. The chemical data are remarkable in being completely dominated by the weathering of limestones and evaporites[31] (Fig. 2); in this respect they are rather similar to the tributaries of the Solimões. However the Si values are very much

TABLE 2. (*Continued*)

Si	$^{87}Sr/^{86}Sr$ >	Dissolved flux Total 10^9 moles year^{-1}	Dissolved flux Areal 10^6 moles km^2/year^{-1}	Arc (km)	Dissolved flux per km of arc (10^6 moles km^{-1} year^{-1})	Net CO_2 flux $\emptyset CO_2 = 2.\emptyset Si$ 10^3 moles km^{-2}/year^{-1}
24	0.7171	82	1.1	150	547	649
14	0.7164	90	0.96	450	200	298
8.0	0.7195	25	1.6	100	250	1,000
12	0.7138	180	1.1	600	300	143
192	0.7086	2,965	4.1	1,600	1,853	536
32	0.7075	89	0.67	100	890	485
65	0.7088	467	2.0	300	1,557	551
109	0.7185	586	0.59	1,100	532	219
2.78	—	238	0.85	720	330	19
1.65	—	96	0.66	100	960	23
3.77	—	173	0.69	600	288	30
17.4	0.7110	1,163	0.68	2,000	580	20
23.6	0.7137	466	1.6	800	583	161
						$\emptyset CO_2 = \emptyset ALK$
0.258	0.7191	15.8	2.9	—	—	1,750
0.710	0.7050	14.4	0.45	—	—	250
0.537	—	11.7	0.91	—	—	523
0.340	0.7043	4.70	0.24	—	—	128
0.401	0.7048	10.3	1.8	—	—	1,008
1.77	0.7122	31.0	0.54	—	—	282

lower; the few high values do not exceed $130\,\mu M$ and values under $50\,\mu M$ are common. The single reported Sr isotope value, from the mouth of the river, (0.7110) reflects this. While it was not possible to sample the right bank drainages from the Canadian Shield,[31] it is clear that, as in the tropics, their relative contribution is very small. $\emptyset CO_2$, computed as for the Andean streams (Table 2), defines the low extreme in the available data sets (19–30 $\times$ 10^3 moles km^{-2} year^{-1}).

The Frazer drains the western slope of the Rockies between about 48° and 56°N across the drainage divide from the southern Mackenzie.[36] The runoff rates again have a strong longitudinal variation with an average value of about 1000 mm. The headwaters are in clastic and carbonate sediments but the rest of the basin is underlain by metamorphic and granitic rocks and Tertiary volcanics. The basin is much more heavily populated than that of the Mackenzie, although strict pollution regulations are in force to protect the enormous resources of salmon. However, most of the tributaries contain artificial impoundments or lakes, many of them large. Hence the Si data are certainly low owing to removal by diatoms. The river has a relatively high concentration of dissolved species with alkalinities between about 500 and 2500 μeq. In contrast to the Mackenzie, the Si values are around 100 μM (range 32–140 μM) with values as high as 407 μM recorded for one tributary, the Blackwater. The low values are characteristic of the summer flood stage; winter values, taken just before freeze-up, are between 25 and 100% higher. The waters are relatively Mg-rich; the Ca–Mg data define two relationships, one with Ca:Mg $\sim$ 3 and the other with a value of unity. The latter streams drain the mafic terrains. The Sr values are intermediate, ($\sim$ 1 μM); the ^{87}Sr/^{86}Sr values range from as high as 0.7513 in the headwaters to 0.7043 in the volcanic belt. Most of the values cluster around 0.715. This precludes significant contributions to the dissolved load from limestones and evaporites, consistent with the geology. $\emptyset$CO$_2$ (calculated as $\emptyset$CO$_2$ = $\emptyset$ALK; Table 2) ranges from 1750 $\times$ 10^3 moles km^{-2} year^{-1} on the small McGregor headwater tributary (0.7191) to 1008 for the Bridge (0.7048) to 128 for the Chilcoten (0.7043). These values are much higher than for the Mackenzie at equivalent latitudes and similar to values observed in the Andes.

The upper Yukon drains the western slope of the Rockies between latitudes 58° and 65°N. A one-year time series (67 samples) is available from Eagle, Alaska, just west of the U.S.–Canada border.[37] The region has a subarctic climate with a runoff of about 200 mm. The data are remarkable in that the Si and K concentrations exactly track the discharge spike at spring ice breakup while all the other measured parameters show a perfect inverse, dilution-controlled relationship. The Si values are relatively high, about 110 μM during under-ice flow rising to 190 μM at peak discharge. The ^{87}Sr/^{86}Sr ratio is 0.7137, significantly more radiogenic than the basin averages for the Mackenzie and Frazer. In the former case limestones are responsible for the lower ratios, in the latter the mafic rocks. $\emptyset$CO$_2$ (calculated as $\emptyset$CO$_2$ = 2.$\emptyset$Si), is 161 $\times$ 10^3 moles km^{-2} year^{-1}, much higher than for the Mackenzie and greater than that of the Apure in the Orinoco Basin.

In order to normalize for the variable inclusion of floodplains in the estimates of areal consumption it is perhaps more useful to express $\emptyset$CO$_2$ in these curvilinear mountain belts in terms of moles per year per length of arc drained. In the large tributaries discussed above, the location of the main stem that they feed is, in general, at the distal end of the alluvial fans produced by the erosion of the ranges. Hence, secondary weathering processes in the fans themselves,[46] which are strictly related to the orogeny, are included in this representation of $\emptyset$CO$_2$. The range in values is about a factor of 100 from 640 $\times$ 10^3 moles per year per km arc length in the Içá drainage of the Amazon to a value of 7 for the Peace. Within each arc segment the values vary substantially, from 200 to 640 for the Amazonian Andes, from 40 to 320 for the Eastern Andes and from 7 to 33 for the Mackenzie. The Frazer has the highest northern value (459 $\times$ 10^3 moles per km of arc length), comparable to the tropical fluxes. Thus the regional variability in $\emptyset$CO$_2$ is large no matter how it is calculated.

Several features stand out in the results presented above. The variability in $\emptyset$CO$_2$, either areal or in terms of arc length, is as great within arc segments as between them, with the striking exception of the Mackenzie. This is presumably a reflection of lithologic and topographic variations between basins. There is no discernible relationship of $\text{\O}CO_2$ to climate despite the enormous range in conditions, from the tropics to the polar desert. There is no evidence for the often-claimed correlation of Si concentration with temperature (as opposed to lithology): witness the Yukon (see Berner and Berner[47] for discussion of the Si:T hypotheses).

3.3. The Himalayan Collision Zone

The Himalayan orogeny constitutes the largest continental collision event since the Pan-African in the latest Precambrian. Vast areas have been affected by underthrusting of the Archean basement of the Indian plate and its metamorphism and subsequent tectonic and erosional exhumation.[48] Most of the existing geochemical information on the rivers draining this orogen comes from the work of Krishnaswami, Sarin, and their colleagues at Ahmedabad.[33,49] The Himalayas are remarkable, not only for their scale and that of the Tibetan Plateau associated with them, but also for the rapid tectonic exposure of recently metamorphosed and partially melted continental lithosphere.[50] In these rocks, which form the crystalline core of the range, the Sr isotopic systematics have been completely reset. Normally, mantle-derived rocks retain the "initial" Sr isotopic systematics characteristic of their origin in major mineral phases such as the Na–Ca-feldspars that exclude the radioactive parent Rb from their lattice sites. Values usually range around 0.703. The radiogenic Sr is contained in the K-minerals that accept Rb. These are refractory to weathering and thus the radiogenic Sr is preferentially partitioned into the mechanical load, principally as micas. In the Himalaya the Na–Ca-feldspars have accepted the metamorphically remobilised radiogenic Sr and constitute a labile source resulting in dissolved Sr with both very high isotopic ratios and high *flux*, something impossible in conventional circumstances[25] (Fig. 3). The ratios in the tributaries for which discharge data are available range from 0.7149 to 0.7400; the highest observed ratios, (>0.8) are at the high end of those in the tropical shield terrains but with Sr concentrations over an order of magnitude greater. Thus, the relationship between the Sr concentrations and the isotopic compositions defines a steep relationship that is unique among the rivers of the world[44,51] (Fig. 3). The collision has, without doubt, been responsible for the rapid rise in the seawater $^{87}Sr/^{86}Sr$ ratio over the last 20 million years.

Calculating the CO_2 drawdown as $\text{\O}CO_2 = 2.\text{\O}Si$ (Table 2) results in fluxes between 104 and 322×10^3 moles km^{-2} $year^{-1}$, in the midrange of the Andean values. Expressed in terms of arc length the fixation rates are between 50 and 304×10^6 moles km^{-1} $year^{-1}$ comparable to the Eastern Andes in the Orinoco drainage but lower than observed in the Amazonian segment. Prior to the Indo–Tibetan collision, the northern boundary of the Neo-Tethys was a continental active margin located over the old subducting seafloor and now represented by the late Cretaceous Gangdese magmatic arc in southern Tibet.[52] Topographically it appears to have been a Sumatran-type low-altitude feature

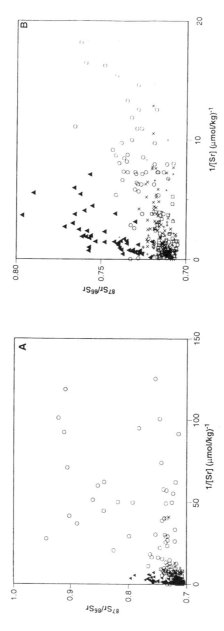

FIGURE 3. Plots of $^{87}Sr/^{86}Sr$ vs. $1/Sr$ (in μM) using all available published data ($\sim$425 points; upper) and in expanded scale (lower) to show the relation for streams where the Sr concentrations are significant.[25] Note the complete separation of the Himalayan tributaries from the main data field. Shields — open circles, North America (non-shield terrains) — open squares, Himalayan — filled triangles, Andean Arc — X, other samples — + .

rather than an elevated Andean-type plateau.[53,54] From the calculations given above, the effect on $\emptyset CO_2$ from the Himalayan uplift alone would have been minor in the Andean case; one mountain belt replaced another. If the collision occurred at an Indonesian-type arc, which seems much more likely, the effect would have been far greater. In either case the change in the $^{87}Sr/^{86}Sr$ composition of the global fluvial Sr flux to the oceans was very large, for the reasons discussed earlier. Comparison of data from the other active mountain belts (Fig. 3; Table 2) thus demonstrates that the $^{87}Sr/^{86}Sr$ record in marine limestones cannot be interpreted unambiguously as an index of global weathering intensity.[10] It can be used in such a way on stable cratons but the flux, where radiogenic, is very low. In active orogenies, where physical weathering and transport predominate, any radiogenic Sr must be present in a very labile form if it is to be released during the relatively short exposure time of the parent rock. On the present Earth, only the Himalayan metamorphics and associated leucogranites meet this criterion.

4. CONCLUSIONS

The reasonably comprehensive data sets discussed here lend little support to the "thermostat" hypothesis.[9] Lithology is the dominant variable determining the chemical yields from orogenic belts. The physical processes that generate exposure of fresh rock control the weathering fluxes from stable cratonic areas. The data do support the "Raymo hypothesis"[11] in that tectonically active mountain belts are the loci for accelerated drawdown of atmospheric CO_2; hence their initiation and evolution have a direct influence on global climate. The parallel development of the continental arc in the western Americas and the closure of the Neo-Tethys and subsequent collisions constitute an era of intensive mountain building probably sufficient to account for the observed climatic deterioration over the past 50 million years. In the absence of orogenic activity, however, it is not clear what controls the *upper* limit on atmospheric pCO_2, i.e., what one might call the "reverse Raymo hypothesis," the Eocene climatic optimum. Apparently the mantle processes ultimately responsible for orogenesis are active enough so that continental arcs are developed and collisions occur with sufficient frequency to preclude a runaway greenhouse effect. Over the last few tens of millions of years both, essentially random, processes have been very active. The resulting CO_2 drawdown is now close to complete, such that the Earth has entered into what will likely be a prolonged glacial epoch with the atmospheric pCO_2 held at a kinetic minimum by the resulting accelerated exposure of aluminosilicate rocks.[17]

Acknowledgments

Our work on the tropical rivers of South America and on the subarctic and arctic rivers in Alaska and Eastern Siberia has been supported for over 20 years by the Ocean and Earth Sciences Divisions of the NSF. We thank our colleagues in the U.S., France, Canada and, especially, India and Russia for long-term collaboration and exchange of data and samples.

REFERENCES

1. Savin, S. M. (1977). *Ann. Rev. Earth Planet. Sci.* **3**, p. 319.
2. Zachos, J. C., Stott, L. D., and Lohmann, K. G. (1944). *Paleoceanography* **9**, p. 353.
3. Crowley, T. J., and North, G. R. (1991). *Paleoclimatology*, Oxford University Press, New York.
4. Crowley, T. J. (1993). *Tectonophysics* **122**, p. 277.
5. Greenwood, D. R., and Wing, S. L. (1995). *Geology* **23**, p. 1044.
6. Rayner, R. J. (1995). *Palaeogeog. Palaeoclim. Palaeoecol.* **119**, p. 385.
7. Edmond, J. M., and Huh, Y. (1997). *Rev. Geophys.* **35**, in press.
8. Rea, D. K., and Rull, L. J. (1996). *Earth Planet. Sci. Lett.* **140**, p. 1.
9. Walker, J. C. G., Hays, P. B., and Kastings, J. F. (1981). *J. Geophys. Res.* **86**, p. 9776.
10. Berner, R. A. (1994). *Am. J. Sci.* **294**, p. 56.
11. Raymo, M. E., and Ruddiman, W. F. (1992). *Nature* **359**, p. 117.
12. Eyles, N. (1993). *Earth-Sci. Rev.* **35**, p. 1.
13. Frakes, L. A., Francis, J. E., and Sytus, J. I. (1992). *Climate Modes of the Phanerozoic*, Cambridge University Press, New York.
14. Holland, H. D. (1978). *The Chemistry of the Atmosphere and Oceans*, Wiley-Interscience, New York.
15. UNESCO (1971). *Discharge of Selected Rivers of the World*, UNESCO Press, New York.
16. UNESCO (1977). *Atlas of World Water Balance*, UNESCO Press, New York.
17. Edmond, J. M., Palmer, M. R., Measures, C. I., Grant, B., and Stallard, R. F. (1995). *Geochim. Cosmochim. Acta* **59**, p. 3301.
18. Burke, W. H., Denison, R. E., Hetherington, E. A., Koepnik, R. B., Nelson, H. F., and Otto, J. B. (1982). *Geology* **10**, p. 516.
19. Garrels, R. M. (1967). In: *Researches in Gechemistry*, Vol. 2 (P. H. Abelson, ed.), pp. 405–420. John Wiley, New York.
20. Berner, R. A., Lasaga, A. C., and Garrels, R. M. (1983). *Am. J. Sci.* **283**, p. 641.
21. Meade, R. H., Yuzyk, T. R., and Day, T. J. (1990). In: *The Geology of North America*, Vol 0-1, pp. 255–280. Geological Society of America, Denver.
22. Meade, R. H. (1994). *Quat. Intern.* **21**, p. 29.
23. Goldstein, S. L., O'Nions, R. H., and Hamilton, P. J. (1984). *Earth Planet Sci. Lett.* **70**, p. 221.
24. Faure, G. (1986). *Principles of Isotope Geology*, John Wiley, New York.
25. Edmond, J. M. (1992). *Science* **258**, p. 1594.
26. Livingston, D. A. (1963). U.S.G.S. Prof. Paper 400G, p. 1.
27. Meybeck, M. (1979). *Rev. Geol. Dyn. Geog. Phys.* **21**, p. 215.
28. Summerfield, M. A., and Hulton, N. J. (1994). *J. Geophys. Res.* **99**, p. 13871.
29. Gibbs, R. J. (1967). *Geol. Soc. Am. Bull.* **78**, p. 1203.
30. Gibbs, R. J. (1972). *Geochim. Cosmochim. Acta* **36**, p. 1061.
31. Reeder, S. W., Hitchon, B., and Levinson, A. A. (1972). *Geochim. Cosmochim. Acta* **36**, p. 825.
32. Stallard, R. F., and Edmond, J. M. (1983). *J. Geophys. Res.* **88**, p. 9671.
33. Sarin, M. M., Krishnaswami, S., Dilli, K., Soyamajula, B. L. K., and Moore, W. S. (1989). *Geochim. Cosmochim. Acta* **53**, p. 997.
34. Négrel, P., Allègre, C. J., Dupré, B., and Lewin, E. (1993). *Earth Planet. Sci. Lett.* **120**, p. 59.
35. Dupré, B., Gaillardet, J., Rousseau, D., and Allègre, C. J. (1996). *Geochim. Cosmochim. Acta* **60**, 1301.
36. Cameron, E. M., Hall, G. E. M., Veizer, J., and Krouse, H. R. (1995). *Chem. Geol.* **122**, p. 149.

37. Edmond, J. M., Palmer, M. R., Measures, C. I., Brown, E. T., and Huh, Y. (1996). *Geochim. Cosmochim. Acta* **60**, p. 2949.

38. Yang, C., Telmer, K., and Veizer, J. (1996). *Geochim. Cosmochim. Acta* **60**, p. 851.

39. Newton, R. M., Weintraub, J., and April, R. (1987). *Biogeochemistry* **3**, p. 21.

40. Velochko, A. A. (1992). *Late Quaternary Environments of the Soviet Union*, University of Minnesota Press.

41. Twidale, C. R. (1982). *Granite Landforms*, Elsevier, New York.

42. Twidale, C. R. (1990). *J. Geol.* **98**, p. 343.

43. Zonenshain, L. P., Kuzmin, M. I., and Natapov, L. M. (1990). *Geology of the USSR: A Plate Tectonic Synthesis*, American Geophysical Union, Geodynamics Series, Vol. 21.

44. Palmer, M. R., and Edmond, J. M. (1992). *Geochim. Cosmochim. Acta* **56**, p. 2099.

45. Clapperton, C. (1993). *Quaternary Geology and Geomorphology of South America*, Elsevier, New York.

46. Johnsson, M. J., Stallard, R. F., and Lundberg, N. (1991). *Geol. Soc. Am. Bull.* **103**, p. 1622.

47. Berner, E. K., and Berner, R. A. (1987). *The Global Water Cycle*, Prentice-Hall, Engelwood Cliffs, N.J.

48. Molnar, P., and Chen, W. (1978). *Nature* **273**, p. 218.

49. Krishnaswami, S., Trivedi, R., Sarin, M. M., Ramesh, R., and Sharma, K. K. (1992). *Earth Planet. Sci. Lett.* **109**, p. 243.

50. Le Fort, P., Cuney, M., Deniel, C. France-Lanord, C., Sheppard, S. M. F., Upretti, B. N., and Vidal, P. (1987). *Tectonophysics* **134**, p. 39.

51. Palmer, M. R., and Edmond, J. M. (1989). *Earth Planet. Sci. Lett.* **92**, p. 11.

52. Einsele, G., Ratschbacher, L., and Wetzel, A. (1996). *J. Geol.* **104**, p. 163.

53. Dürr, S. B. (1996). *Geol. Soc. Am. Bull.* **108**, p. 669.

54. Liu, G., and Einsele, G. (1994). *Geol. Rundsch.* **83**, p. 32.

15

Silicate Weathering and Climate

Robert A. Berner and Elizabeth K. Berner

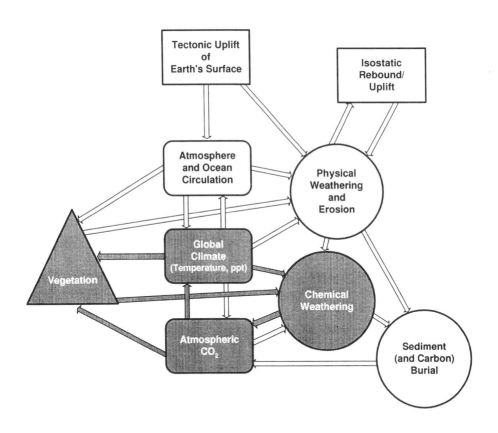

Robert A. Berner and Elizabeth K. Berner ● Department of Geology and Geophysics, Yale University, New Haven, Connecticut 06520-8109.

Tectonic Uplift and Climate Change, edited by William F. Ruddiman. Plenum Press, New York, 1997.

1. INTRODUCTION

There is no doubt that the temperature at the surface of the Earth has not varied excessively since the origin of life. Certainly the stability limits of liquid water have not been exceeded. If CO_2 is an important greenhouse gas, as is commonly accepted, this means that its level in the atmosphere has not varied enough to cause excessively low temperatures (resulting, e.g., in the complete freezing of the oceans) or excessively high ones (resulting, e.g., in the sterilization of life). There must be processes regulating CO_2 that thermostat the earth and that have prevented it from ending up like Mars and Venus.

On a million-year timescale geologic processes dominate the addition of CO_2 to the atmosphere as well as its removal. (For a summary of the long-term or geochemical C cycle, consult Holland[1] or Berner[2]). CO_2 is added by global degassing from a variety of sources including midocean ridge volcanism, midplate volcanism (plumes), subduction zone volcanism, thermal metamorphism, and the diagenetic breakdown of organic matter and $CaCO_3$ in sediments during deep burial. It is also added by the continental oxidative weathering of old organic matter in shales (kerogen). CO_2 is removed from the atmosphere via the weathering of Ca and Mg silicate minerals on the continents with the Ca^{2+}, Mg^{2+}, and HCO_{3-} formed via weathering carried in solution by rivers to the oceans, where the ions are ultimately buried as Ca and Mg carbonate minerals. Generalized overall reactions are[3]:

$$CaSiO_3 + CO_2 \rightarrow CaCO_3 + SiO_2$$

$$MgSiO_3 + CO_2 \rightarrow MgCO_3 + SiO_2$$

CO_2 is also removed via the burial of organic matter in sediments and (possibly) via the low-temperature alteration of submarine basalt (for the latter, however, see below).

On a million-year timescale the fluxes of CO_2 via degassing and silicate weathering, along with organic matter weathering and burial, are very large compared to the mass of C at the surface of the earth. A recent estimate[4] is that the global degassing flux is approximately 8×10^{18} moles per million years. The mass of C in the atmosphere is about 0.06×10^{18} moles. Thus, if degassing were to greatly exceed uptake via weathering, let us say by a factor of 2 (and the organic subcycle were held constant), then it would take only about 7500 years for atmospheric CO_2 to double. In only 10 million years the atmospheric CO_2 level by this scenario would rise to about 1000 times that at present (assuming that $CaCO_3$ solubility equilibrium in the ocean were maintained with bottom sediments and that increased atmospheric CO_2 did not lead to increased consumption via silicate weathering on the continents). This CO_2 level is obviously excessive and it means that rates of CO_2 release via global degassing plus organic matter weathering, on the one hand, and CO_2 uptake via Ca–Mg-silicate weathering and organic matter burial, on the other, must be close to one another.[1]

The assumption of appreciable input and output imbalance over millions of years is untenable.

The requirement that degassing and weathering must be closely balanced means that there must be some stabilizing negative-feedback mechanism for maintaining atmospheric CO_2 within reasonable bounds. If mountain uplift were to accelerate the silicate weathering rate, then for constant degassing and organic burial/weathering, a counterbalancing factor would have to come into play to maintain C mass balance. The same is true if the rate of global degassing changed. The process normally called upon to provide the necessary negative feedback is the response of silicate weathering to changes in climate.[2,4,5–15] Simply stated the idea is that higher global temperatures and greater rainfall on the continents, caused, for example, by higher CO_2 owing to greater degassing, should bring about enhanced CO_2 removal via faster silicate weathering. Conversely, greater silicate weathering by uplifted mountains would lower the atmospheric CO_2 level, which would in turn lower CO_2 uptake by weathering, thereby counterbalancing the effect of the mountains.

Climate–silicate weathering feedback provides a ready mechanism to explain the constancy of Earth's surface temperature over geologic time in the presence of a warming sun. Standard models for the evolution of the sun indicate a linear increase in radiation from a value about 30% less than today in the Archean to that at present.[13,16] In order to counter intense cooling in the past that would have led to complete freezing of the oceans, it has been suggested that formerly the level of atmospheric CO_2 was much higher and that this contributed sufficient greenhouse warming to counteract the diminished solar input. Then over geologic time as solar radiation increased, the rate of silicate weathering correspondingly increased, which lowered CO_2 and thereby maintained a balance between solar input and the greenhouse effect in order to prevent excessive warming.[5,13]

Over geologic time rainfall may be the most important component in any weathering–climate feedback effect. Certainly rainfall has a major influence on vegetation, which itself is an important factor in rock weathering.[17,18] On a global basis over geological time rainfall should also correlate with atmospheric CO_2 and global mean temperature. A warmer Earth, owing to an enhanced CO_2-greenhouse, should be a wetter Earth. This is because at higher temperatures the vapor pressure of seawater is higher, resulting in more evaporation, greater transport of water vapor over the continents, and greater rainfall and river runoff on the continents. An example of the effect of increases in atmospheric CO_2 is shown by the GCM calculations of Manabe and Stauffer,[19] who calculate important increases in precipitation at high latitudes owing to global warming. For periods of intense global warming, as occurred in the Cretaceous, this should have led to greater rainfall, more vegetation, and greater silicate weathering at high latitudes.

Volk[10] has suggested another negative feedback mechanism involving the weathering of silicates on the continents—the fertilization of plant growth by atmospheric CO_2. From experimental studies (see summary by Bazzaz[20]) it has been found that with sufficient water, nutrients, and light, plants will grow faster at higher CO_2 levels. If true, faster plant growth entails more rapid weathering of minerals to provide the necessary nutrients for growth, and this can entail faster uptake of CO_2 and conversion to dissolved HCO_3^-. The quantitative significance of this process is not well known, but it should be considered as another possible weathering–feedback mechanism for stabilizing the environment. It has been incorporated in C cycle modeling.[2,4]

A rather different stabilizing feedback mechanism has been suggested — the uptake of CO_2 during the conversion of Ca-silicate minerals to $CaCO_3$ during the low-temperature reaction of submarine basalts with seawater.[21,22] The problem with this process as a stabilizing feedback is that the rate of reaction of seawater-dissolved CO_2 with basalt is very insensitive to the level of atmospheric CO_2.[15] Even if this submarine seawater–basalt "weathering" is quantitatively important, (which has not been conclusively demonstrated on a global basis[15]) it provides a much weaker feedback than continental weathering.

It has been stated (Edmond et al.[23,24] and Chapter 14) that there is essentially no field evidence of a climate–weathering feedback mechanism, as discussed above, on the present Earth and that the environment is not stabilized at high CO_2 levels except when sufficient mountain building brings about a drop in CO_2 owing to enhanced erosion and weathering. According to this hypothesis, in the absence of extensive mountain building atmospheric CO_2 drifts upward until tectonics ultimately brings about a decrease. It is the purpose of this paper to refute these assertions based not only on the necessity for having a close weathering feedback as stated above but also on a different approach to the study of present-day silicate weathering.

2. SILICATE WEATHERING RATE ON THE PRESENT EARTH

2.1. Studies of River Water Chemistry

One way to study the rate of rock weathering is to examine the chemical composition of river water (e.g., Holland[1]). For Ca–Mg silicates the riverine flux of dissolved HCO_3^- is a direct measure of the uptake of atmospheric CO_2 owing to weathering. However, it is important to verify that the rivers are not also affected by the weathering of the carbonate minerals calcite and dolomite. *Weathering of carbonate minerals does not affect atmospheric CO_2 on a multimillion-year timescale because CO_2 consumed in carbonate weathering is returned shortly thereafter upon carbonate precipitation in the oceans* (e.g. Berner[2]). It is well established[25–27] that even traces of $CaCO_3$ in a drainage basin can greatly affect the chemical composition of river water. This is because carbonate minerals dissolve so much faster than silicate minerals.[27]

Based on investigations into river chemistry some studies have concluded that there is little evidence for enhanced weathering as a function of climate (Edmond and Huh,[24] Summerfield and Hulton[28] and Chapter 14). However, in these studies the possibility of input of dissolved species from carbonates was not excluded. Summerfield and Hulton[28] were not concerned with the presence or absence of carbonates and state that "a factor not considered in the correlation and regression analysis presented here is that of the lithologic control of denudation rates" (p. 13880). Actually, all the rivers considered by Summerfield and Hulton are impacted to varying degrees by carbonate weathering as can be seen from the data summarized by Meybeck[27] and Berner and Berner.[29]

Study of rivers from orogenic belts is important in that silicate weathering is enhanced by the removal of weathering products by erosion on steep topography.[23,30] For several large rivers cited by Edmond and Huh (Chapter 14) the authors state that "lithology is the dominant control determining the chemical yields from orogenic belts" and it obscures any effects owing to climate. They demonstrate this via analysis of the following major river systems: the Mackenzie, Fraser, and upper Yukon rivers and the Andean headwaters of the Amazon and Orinoco. Since all these rivers are impacted to varying degrees by carbonate weathering, if one wishes to study the effect of climate in areas of high relief on the weathering of silicates alone, one cannot use these rivers and must turn elsewhere.

Evidence for a presumed lack of a climate signal is also presented by Edmond and Huh[24] and Chapter 14 for relatively stable shield areas. This includes the tropical Guayana, Congo, and Brazilian shields, and the cold eastern Siberian Aldan Shield and Trans-Baikal Highlands. The surfaces of the tropical shields have had most of the cationic silicates removed by intense weathering over a long period of time so that groundwaters intercept only relatively cation-free weathering products. The result is that CO_2 uptake by silicate weathering is very low. This is an excellent example of the effect of relief on weathering. CO_2 uptake via silicate weathering on a low-relief land surface is limited because of the inability to erosively remove the protective cover of highly weathered, Ca- and Mg-free, clay-rich residuum.[23,30]

The Aldan Shield river water data show a similar flux of CO_2 to those for the much warmer tropical shields. However, the river chemical data of Edmond and Huh[24] and in Chapter 14 are not what would be expected for silicate weathering. The ratio of dissolved cations to silica is very high, cations are dominated by Ca and Mg, and there is little evidence for any clay weathering product. This is the usual evidence for carbonate weathering. Also, the $^{87}Sr/^{86}Sr$ ratio is mostly in the range 0.712–0.713 indicating a major contribution from carbonates in an otherwise old radiogenic silicate terrain. In agreement with the idea that carbonate weathering is a major contributor to water chemistry, the study of Gordeev and Sidorov[31] shows an abundance of outcropping Paleozoic and Precambrian carbonate rocks in the same general area, especially near to where Edmond and Huh took many river water samples. Furthermore, Rosen *et al.*[32] state that on the Aldan Plateau itself many different crystalline rock types, including greenstones and metasediments, are made up partly of carbonates.

Edmond[24,33] ascribes the absence of aluminous clay weathering products in the Aldan drainage to an unusual type of weathering whereby only the surfaces of the primary silicate minerals are involved in the exchange of cations for H^+.

This exchange phenomenon has been studied extensively in the laboratory[34] and, in the case of feldspar, it involves one or two of the outermost unit cells of the feldspar structure, in other words roughly the outermost nanometer. Further inward attack of protons results in the formation of residual Al-enriched weathering product, which is presumably not forming under Aldan-type weathering.

A simple calculation shows that Edmond's hypothetical mechanism for silicate weathering in the Aldan region is untenable. Ion exchange reaction of the outermost nanometer of a typical 0.2-mm-diameter sand grain (there is an abundance of sand-sized material in the Aldan drainage) results in a ratio of mass remaining to mass reacted of about 30,000. This value (for a spherical grain) is obtained as follows:

$$\text{Mass of grain} = \tfrac{4}{3}\pi r^3 \rho \qquad \text{Mass reacted} = 4\pi r^2 \Delta r \rho$$

where r is the radius, Δr the outermost thickness reacted, and ρ the density.

$$\frac{\text{Mass remaining}}{\text{mass reacted}} = \frac{\tfrac{4}{3}\pi r^3 \rho - 4\pi r^2 \Delta r \rho}{4\pi r^2 \Delta r \rho}$$

Since only a small mass is removed, this expression can be simplified to

$$\frac{\text{Mass remaining}}{\text{Mass reacted}} = \frac{r}{3\Delta r}$$

Substituting $r = 10^{-2}$ cm and $\Delta r = 10^{-7}$ cm we obtain a ratio of 30,000.

Now, if only the outermost nanometer is reacted, for continued weathering more and more grains must be attacked by H^+ ions. The dissolved yield flux of cations (mainly Ca and Mg) in the Aldan rivers is approximately 200×10^3 moles km^{-2} year^{-1} (Table 3 of Edmond and Huh[24]) equivalent to about 4×10^4 kg km^{-2} year^{-1} of weathered rock (based on the idealized formula $CaMgSi_2O_6$). For a ratio of mass remaining to mass reacted of 30,000, this means that about 1.2×10^9 kg per km^2 per year must represent remaining material that has been attacked to provide the surficial cations. For an approximate density of 3 this converts to 4×10^{-4} km^{3}km^{-2} year^{-1} or a thickness of about 40 cm. Every year 40 cm of rock are needed to supply the required cationic flux. For 1000 years 400 m are needed. After 1 million years 400 km are required. This is far too great a thickness to supply the cationic flux if reaction involves only grain surfaces and no clay residue is formed!

When studying silicate weathering the best way to avoid the problems attending carbonate weathering is to study small rivers whose drainage basins do not contain any mapped carbonate. This has been done in a study by White and Blum[35] in which they analyzed results from 68 small watersheds underlain only by granites and high-grade gneisses. In contrast to the conclusions of Edmond and Huh,[24] White and Blum found a positive correlation of weathering rate with

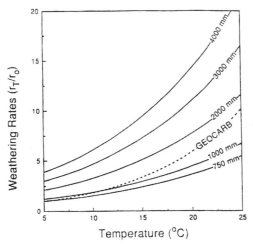

FIGURE 1. Generalized plot of the results of White and Blum[35] for the rate of weathering of granites and gneisses as a function of mean annual temperature and rainfall (shown as contours). Results are derived from data on the fluxes of dissolved sodium and silica from small mono-lithologic drainage basins. Superimposed is the weathering feedback expression for GEOCARB 2[4] (after White and Blum[35] and Berner[18]).

mean annual temperature and a strong positive correlation with rainfall. Their results are shown in Fig. 1. (Because of the importance of vegetation in storing and releasing Ca, Mg, and K in small watersheds, White and Blum were unable to use fluxes of these elements to deduce weathering rates and were thereby restricted to using only dissolved silica and sodium.)

In a study of weathering of the same siliceous rock type as a function of elevation in a small region of the southern Swiss Alps, Drever and Zobrist[36] found that the rate of weathering as revealed by cationic (and HCO_3^-) fluxes decreased strongly and continuously with elevation. The chemical cationic denudation flux at 2400 m was 25 times lower than that at 300 m. The higher-elevation locations were cooler and less heavily vegetated than the 300-m location that was covered with deciduous forest. Here the effect of climate, in terms of a combination of temperature and vegetation, has a strong positive influence on the rate of silicate weathering.

It has been stated that riverine chemical data from Iceland and Hawaii refute the importance of climate on weathering.[24] Riverine fluxes for (carbonate-free) basaltic rocks on the islands of Hawaii and Iceland reported by Bluth and Kump[37] are similar even though Iceland has a much lower mean annual temperature than does Hawaii. However, there are several problems with this interpretation. Bluth and Kump emphasize that the deciphering of a climatic signal between the two sites is masked by other factors such as differences in topography and clay weathering product cover. More importantly Gislason *et al.*,[38] in an extensive study of Iceland water chemistry, show that Iceland in many places is underlain by hyaloclastite, a rock type dominated by shattered, finely divided volcanic glass. This noncrystalline, high-surface area material weathers much faster than ordinary basalt. Further, Gislason *et al.* state that lower loss rates Ca and Mg to solution adjacent to Icelandic glaciers, compared to those for nonglaciated and vegetated portions of the island, provide evidence for a positive correlation between climate and weathering, in agreement with our conclusions here.

2.2. Studies of Clay Weathering Products

The above discussion indicates that dissolved components of river water can often be misleading as a measure of silicate weathering. A better guide is the presence of clay weathering product. Clay minerals cannot form from carbonates and their accumulation over time in soils or in depositional regions records the long-term weathering of silicates (for a good example consult the paper by Derry and France-Lanord[39]). Standard works on silicate weathering[40–42] indicate that clay formation by weathering is most favored under tropical regimes and, secondarily, under temperate climates. The major controls on how much clay is formed are rainfall followed by temperature. Little clay is formed at high latitudes, which are both cold and dry.

The classic book by Jenny[40] is most instructive as to the quantitative role of climate in silicate weathering. Jenny, from the study of thousands of soils in the United States, has constructed plots of soil clay content (defined as the fraction with grainsize $<2 \mu m$) as a function of rainfall and temperature for constant

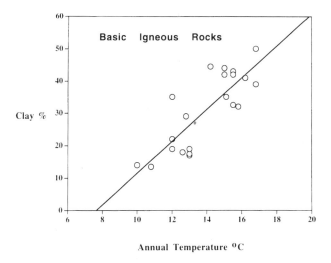

FIGURE 2.   Plot of clay content of weathered basic igneous rocks along a north–south trend ranging from central New Jersey to Georgia, as a function of mean annual temperature. Each dot represents the clay content ($<0.2 \mu m$) in the top meter of a different soil (after Jenny[40]). (Normalization for small differences in rainfall/evaporation makes little difference in the trend).

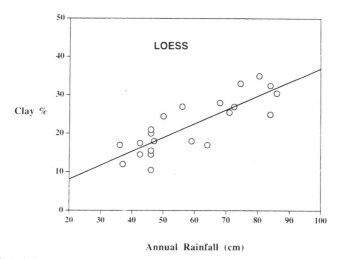

Annual Rainfall (cm)

FIGURE 3.   Plot of clay content of weathered glacial loess in Kansas and Missouri, along the 11 °C isotherm, as a function of mean annual rainfall. Each dot represents clay content in the top 25 cm of a different soil (after Jenny[40]).

lithology. Some of his results are shown in Figs 2 and 3 for basic igneous rocks of the eastern United States and the loess belt of Kansas and Missouri, respectively. Jenny also constructed a simplified three-dimensional plot of clay content for granite–gneiss weathering versus both temperature and the ratio of precipitation to evaporation. The plot shows an increase of clay content with temperature that is greater with higher precipitation ratio. The diagram is uncannily similar to the three-dimensional plot of rate of weathering as a function of temperature and rainfall deduced by White and Blum[35] from streamwater chemistry alone (see above discussion).

A study of the clay mineralogy of the sediments of the Atlantic Ocean by Biscaye[43] indicates a correlation between fine-fraction mineralogy and climate. The minerals kaolinite and gibbsite are good indicators of silicate weathering, whereas the minerals chlorite and illite represent, to a large extent, physically eroded detritus from preexisting sedimentary and low-grade metamorphic rocks.[39] If there is little chemical weathering, the detrital minerals should dominate over the weathering products in the fine fraction of sediments. Biscaye showed that the ratios of kaolinite/chlorite and gibbsite/illite decrease considerably with latitude as one moves both north and south in the Atlantic. Again, this is what one would expect if weathering rates decrease with decreasing temperature.

2.3. Experimental Studies

Experimental studies of silicate mineral dissolution rates (for a recent summary consult the book by White and Brantley[44]) indicate a strong positive correlation of rate with temperature. Such data have been used in the GEOCARB II model of Berner[4] and the calculations of others.[45,46] The weathering feedback expression of the GEOCARB II model, which is based partly on laboratory dissolution results, agrees rather well with the field-deduced results of White and Blum[35] (see Fig. 1). In attempting to model Earth processes, fundamental physicochemical considerations, derived from laboratory and theoretical studies,[46] should not be neglected.

3. SILICATE WEATHERING OVER GEOLOGIC TIME: CLIMATE, TECTONICS, AND EVOLUTION

Although we have so far emphasized the importance of climate as a control on silicate weathering, there are other important factors involved in weathering that have major effects on atmospheric CO_2 over geologic time. First of all, there is tectonics. We have no quarrel with the assertion of Raymo and Ruddiman[47] that mountain uplift is an important control on rock weathering and on atmospheric CO_2. We are convinced from the arguments of Stallard[31] and Edmond et al.[23] that to have appreciable silicate weathering it is necessary to have sufficient relief so as to enable erosive removal of any protective clay overburden. The importance of tectonics has been acknowledged by one of us by the incorporation of a special term for topographic relief in GEOCARB modeling.[4] However, *given sufficient relief*, we then insist that climate also plays an important role in weathering. This includes the effects of both rainfall and temperature and also how together they control vegetation. (In fact, uplift itself can have a major effect on climate.[47])

There are other important factors besides climate and tectonics. The role of the evolving sun has already been discussed in the introduction. The rise of vascular land plants in upland areas during the Devonian should have had a major effect at that time on rock weathering and, therefore, on atmospheric CO_2.[17,18] The development of a rooting system with a large surface area for reaction with soil minerals must have brought about an acceleration of silicate decomposition. Vascular plants (plus associated symbiotic microflora) accelerate weathering by: (1) secreting organic acids and chelates around their roots in order to obtain nutrient elements from silicate minerals; (2) producing organic litter which decomposes to additional organic acids and carbonic acid; (3) bringing about, on a regional scale, more water recycling through the soil owing to greater rainfall resulting from transpiration; and (4) anchoring clay-rich soil against erosion and thereby allowing retention of water and continued weathering of primary minerals between rainfall events. Quantitative estimates of the effect of trees on silicate weathering rates have been made in a number of controlled field studies[18] and results indicate accelerations ranging from factors of 2 to 10. Added to this are the unpublished preliminary results of K. Moulton and R. A. Berner on adjacent small vegetated and unvegetated areas of western Iceland (in a forest preserve) where the stream flux of HCO_3^- is enhanced severalfold by both birch trees and evergreens.

It has been maintained that plants do not accelerate weathering because they merely recycle nutrients from their own litter in nutrient-poor soils.[24] This is the case for low-lying tropical rain forests, but on sloping topography erosion episodically removes the weathered clayey mantle during storms and especially after trees have been decimated by windstorms, drought, disease, and fires. In this case the plants require nutrients for further growth and they obtain the nutrients from primary silicates.

Acceleration of weathering by plants in the Devonian does not mean that the rate of uptake of CO_2 by silicate weathering was necessarily higher at that time than previously. This is because of the necessity for atmospheric C mass balance. If degassing remained constant, then the accelerating effect of plants must have been balanced by another factor. The most reasonable balancing factor is a concurrent drop in atmospheric CO_2 with resulting lower global rainfall and mean surface temperature countering the plant-accelerating effect. This drop in CO_2, predicted by GEOCARB modeling[2,4] is believed to have contributed to the subsequent Permo–Carboniferous glaciation. (Additional concurrent CO_2-lowering processes were the increased burial of microbially resistant plant-derived organic remains in sediments and possibly collisionally induced mountain uplift). This predicted major drop in atmospheric CO_2 has been corroborated by independent estimates of paleo-CO_2 based on the study of paleosols. This is shown in Fig. 4.

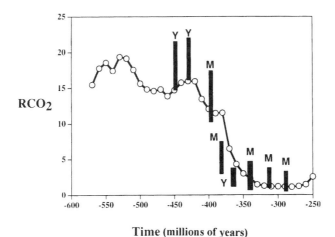

FIGURE 4. Plot of RCO_2 vs. time for the Paleozoic era based on GEOCARB modeling[4] and independent studies of paleosols. RCO_2 is the ratio of the mass of atmospheric CO_2 at a past time to that at present. Y refers to the paper by Yapp and Poths[48] and M to that by Mora et al.[49]

4. CONCLUSIONS

Based on the above discussion we support the modeling approach of a number of researchers[2,4,5-15]: that is, that the effect of atmospheric CO_2 on global weathering rate has served over geologic time as a negative feedback that has stabilized the global environment. Studies of river chemistry that purport to show no effect of climate on silicate weathering rate are compromised by variable contributions from the weathering of highly reactive carbonate minerals. Weathering studies confined to the determination of clay weathering products clearly illustrate a positive effect of both rainfall and temperature on silicate weathering rate.

The necessity for having a negative feedback on weathering does not imply that higher global weathering rates cannot accompany drops in atmospheric CO_2. If both global CO_2 degassing and CO_2 consumption by mountain uplift increased, but the uplift weathering flux increased faster, then C mass balance could be maintained by a drop in atmospheric CO_2 with its decelerating effect on weathering rate balancing the uplift effect. In this way a higher weathering flux (equal to the higher degassing flux) could be accompanied by a lowering of atmospheric CO_2, a situation that has been suggested has existed for the past 40 million years.[47]

Any model for CO_2 over geologic time should take into account *all* factors affecting CO_2. Although the effect of mountain uplift on weathering has been important, so also have the evolution of the sun, biological evolution as it affects weathering by land plants, changes in the weathering and/or burial of organic matter, changes in the rate and mode of global degassing, changes in the lithology of the rocks exposed to weathering. and of course changes in climate. We feel that focusing only on mountain uplift, with neglect of all these other factors, is unnecessarily narrow. Earth system science requires that all factors be evaluated and first-order attempts at quantification at least be attempted.

Acknowledgments

The authors have benefited from discussions with K. Caldeira, A.C. Lasaga, and L. R. Kump and a review by W. F. Ruddiman. Research supported by NSF Grant EAR 9417325 and DOE Grant DE-FGO2-95ER14522.

REFERENCES

1. Holland, H. D. (1978). *The Chemistry of the Atmosphere and Oceans.* John Wiley, New York.
2. Berner, R. A. (1991). *Am. J. Sci.* **291**, p. 339.
3. Urey, H. C. (1952). *The Planets: Their Origin and Development*, Yale University Press, New Haven.
4. Berner, R. A. (1994). *Am. J. Sci.* **294**, p. 56.
5. Walker, J. C. G., Hays, P. B., and Kasting, J. F. (1981). *J. Geophys. Res.* **86**, p. 9776.

6. Berner, R. A., Lasaga, A. C., and Garrels, R. M. (1983). *Am. J. Sci.* **283**, p. 641.

7. Kasting, J. F. (1984). *Am. J. Sci.* **284**, p. 1175.

8. Lasaga, A. C., Berner, R. A., and Garrels, R. M. (1985). In: *The Carbon Cycle and Atmospheric CO₂: Archean to Present* (E. Sundquist and W. S. Broecker, eds.), pp. 397–411. American Geophysical Union Geophysics Monograph 32.

9. Marshall, H. G., Walker, J. C. G., and Kuhn, W. R. (1988). *J. Geophys. Res.* **93**, p. 791.

10. Volk, T. (1987). *Am. J. Sci.* **287**, p. 763.

11. Volk, T. (1989). *Geology* **17**, p. 107.

12. Kump, L. R. (1989). *Am. J. Sci.* **289**, p. 390.

13. Caldeira, K., and Kasting, J. F. (1992). *Nature* **360**, p. 721.

14. Godderis, Y., and Francois, L. M. (1995). *Chem. Geol.* **129**, p. 169.

15. Caldeira, K. (1995). *Am. J. Sci.* **295**, p. 1077.

16. Kasting, J. F., and Ackerman, T. P. (1986). *Science* **234**, p. 1383.

17. Berner, R. A. (1992). *Geochim. Cosmochim. Acta* **56**, p. 3225.

18. Berner, R. A. (1995). In: *Chemical Weathering Rates of Silicate Minerals* (A. F. White and S. L. Brantley, eds.), pp. 565–583. Mineralogical Society of American Reviews of Mineralogy 31.

19. Manabe, S., and Stauffer, R. J. (1993). *Nature* **364**, p. 215.

20. Bazzaz, F. A. (1980). *Ann. Rev. Evol. Systematics* **21**, p. 167.

21. Staudigel, H., Hart, S. R., Schmincke, H. U., and Smith, B. M. (1989). *Geochim. Cosmochim. Acta* **53**, p. 3091.

22. Francois, L. M., and Walker, J. C. G. (1992). *Am. J. Sci.* **292**, p. 81.

23. Edmond, J. M., Palmer, M. R., Measures, C. I., Grant, B., and Stallard, R. F. (1995). *Geochim. Cosmochim. Acta* **59**, p. 3301.

24. Edmond, J. M., and Huh, Y. (1997). *Rev. Geophys.* (under review).

25. Drever, J. I., and Hurcomb, D. R. (1986). *Geology* **14**, p. 221.

26. Meybeck, M. (1979). *Rev. Geol. Dyn. et Geol. Phys.* **21**, p. 215.

27. Meybeck, M. (1987). *Am. J. Sci.* **287**, p. 401.

28. Summerfield, M. A., and Hulton, N. J. (1994). *J. Geophys. Res.* **99**, p. 13871.

29. Berner, E. K., and Berner, R. A. (1996). *Global Environment: Water, Air and Geochemical Cycles*, Prentice-Hall, Upper Saddle River, NJ.

30. Stallard, R. F. (1992). In: *Global Biogeochemical Cycles* (S. S. Butcher, R. J. Charlson, G. H. Orians, and G. V. Wolfe, eds.), pp. 93–121. Academic Press, New York.

31. Gordeev, V. V., and Sidorov, I. S. (1993). *Mar. Chem.* **43**, p. 33.

32. Rosen, O. M., Condie, K. C., Natapov, L. M., and Nozhkin, A. D. (1994). In: *Archean Crustal Evolution* (K. C. Condie, ed.), pp. 411–460. Elsevier, New York.

33. Edmond, J. M. (1993). *Geol. Soc. Am. Ann. Meet. Abst.* p. 414.

34. Blum, A. E., and Lasaga, A. C. (1991). *Geochim. Cosmochim. Acta* **55**, p. 2193.

35. White, A. F., and Blum, A. E. (1995). *Geochim. Cosmochim. Acta* **59**, p. 1729.

36. Drever, J. I., and Zobrist, J. (1992). *Geochim. Cosmochim. Acta* **56**, p. 3209.

37. Bluth, G. J. S., and Kump, L. R. (1994). *Geochim. Cosmochim. Acta* **58**, p. 2341.

38. Gislason, S. R., Arnorsson, S., and Armannsson, H. (1996). *Am. J. Sci.* **296**, p. 837.

39. Derry, L. A., and France-Lanord, C. (1996). *Earth. Planet. Sci. Lett.* **142**, p. 59.

40. Jenny, H. (1941). *Factors of Soil Formation*, McGraw-Hill, New York.

41. Loughnan, F. C. (1969). *Chemical Weathering of the Silicate Minerals.* Elsevier, New York.

42. Birkeland, P. W. (1984). *Soils and Geomorphology.* Oxford University Press, New York.

43. Biscaye, P. (1965). *Geol. Soc. Am. Bull.* **76**, p. 803.

44. White, A. F., and Brantley, S. L. (1995). *Chemical Weathering Rates of Silicate Minerals.* Mineralogical Society of American Reviews of Mineralogy. 31.

45. Brady, P. V. (1991). *J. Geophys. Res.* **96**, p. 18101.

46. Lasaga, A. C., Soler, J. M., Ganor, J., Burch, T. E., and Nagy, K. (1994). *Geochim. Cosmochim. Acta* **58**, p. 2361.

47. Raymo, M. E., and Ruddiman, W. F. (1992). *Nature* **359**, p. 117.

48. Yapp, C. J., and Poths, H. (1996). *Earth Planet. Sci. Lett.* **137**, p. 71.

49. Mora, C. I., Driese, S. G., and Colarusso, L. A. (1996). *Science* **271**, p. 1105.

16

Carbon Cycle Models: How Strong Are the Constraints?

Maureen E. Raymo

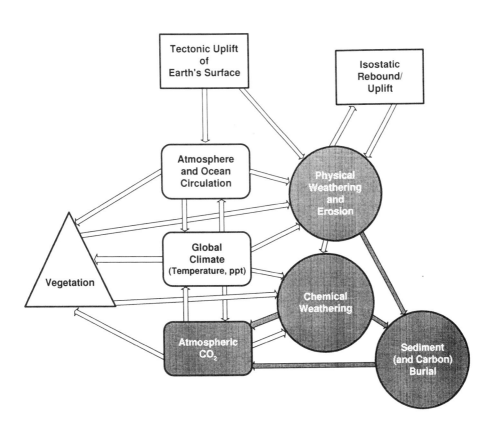

Maureen E. Raymo • Department of Earth, Atmospheric, and Planetary Sciences, Massachusetts Institute of Technology, Cambridge, Massachusetts 02139.

Tectonic Uplift and Climate Change, edited by William F. Ruddiman. Plenum Press, New York, 1997.

1. INTRODUCTION

In 1982, Walker et al.[1] published a landmark paper suggesting that the evolution of the Earth's climate over 4.5 billion years was controlled by a negative feedback loop involving surface temperature, CO_2, and chemical weathering that moderated the strength of the Earth's "greenhouse effect" over time. As solar luminosity increased over time, surface temperatures on the Earth rose, accelerating the rate of surface chemical weathering reactions (which consumed atmospheric CO_2) resulting in a weaker greenhouse — i.e., a negative feedback system for the Earth's surface temperature, which prevented a runaway greenhouse effect in the face of increasing solar output.

This idea, proposed as a solution to the "faint young sun" paradox, was further developed by Berner et al.,[2] who proposed that climate evolution on multimillion year timescales was controlled by a similar negative feedback loop that balanced the supply of CO_2 to the atmosphere from the Earth's interior and the amount of atmospheric CO_2 consumed by chemical weathering reactions. Because the atmospheric CO_2 reservoir is so small relative to the fluxes in and out, a balance between these fluxes must somehow be maintained within the Earth's C cycle or the atmospheric CO_2 reservoir would quickly build up to unreasonably high values or be depleted to near zero values (runaway greenhouse as on Venus or an icehouse planet as on Mars). Geologic evidence suggests that neither of these scenarios has ever been played out on our planet.

In the more detailed mode of Berner et al.[2] (hereafter BLAG), high rates of seafloor spreading (mantle degassing) in the Cretaceous resulted in a buildup of CO_2 in the atmosphere, a strengthening of the Earth's greenhouse effect, and an increase in surface temperatures. As climate warmed, a proposed increase in global chemical weathering driven by the elevated temperatures and a more vigorous hydrologic cycle acted as a negative feedback preventing the buildup of CO_2 in the atmosphere. As spreading rates fell through the Cenozoic, the CO_2 input dropped, the atmospheric CO_2 level therefore dropped, and surface temperatures dropped. In this scenario, the drop in temperature results in a drop in weathering rates, both of which continue until the input of CO_2 from degassing is balanced by the output of CO_2 from chemical weathering. This paper was significant not only because an entire community of scientists started thinking about long-term climate change in terms of the tectonic C cycle and the greenhouse effect, but because the BLAG model made testable predictions. In particular, if the concept of a global temperature–weathering feedback was correct, then global chemical weathering rates should have dropped through the Cenozoic in concert with falling mantle degassing rates and surface temperature.

It was the apparent mismatch of this prediction when compared to geologic evidence for weathering rates that led Raymo et al.[3] to propose that the late Cenozoic uplift and weathering of the Tibetan–Himalayan region led to the drawdown of atmospheric CO_2 and hence to global cooling. This view, however, just traded one set of problems for another; namely, how do you support an enhanced removal rate of C from the atmosphere over the last 40 million years when the C input from the mantle is presumed to have remained relatively constant? Where does the CO_2 that supports this hypothesized increase in chemical weathering come from once the relatively small atmospheric reservoir is depleted? Lastly, is there a negative feedback to atmospheric CO_2 levels, other than the hypothesized temperature–weathering link, which prevents rapid depletion of CO_2? Three obvious possibilities come to mind: either the evidence for increased chemical weathering has been misinterpreted[4] (see also Chapter 18); the assumption of constant mantle CO_2 input over the Neogene is incorrect[5,6]; or another significant flux of C into the atmosphere, which is sensitive to the ocean–atmosphere CO_2 levels (or climate), must exist.

A number of such "alternative" negative feedbacks have been proposed including compensating imbalances in the organic C subcycle[7,8] and changes in basalt weathering.[9] However, these proposals are largely theoretical in nature rather than being based on convincing geologic proof of their existence. The point made here is that many researchers use C cycle mass balance models to bolster and illustrate their views, but few new geochemical proxy data have been added to the debate over the last 8 years [important exceptions include Os isotope data (Chapter 17) and phosphorus accumulation estimates].[10] Additionally, geochemical models are sometimes presented as "confirmation" that the system must behave in a certain way to be consistent with mass balance constraints. In this paper, the case of the organic C subcycle is examined in a series of sensitivity tests; the results illustrate how very different interpretations can be supported by the same model with slightly different assumptions for the input parameters. It is concluded that models cannot adequately constrain our understanding of the global C cycle but are useful for guiding further empirical studies.

2. BACKGROUND

The organic C subcycle, which transfers C between oxidized and reduced reservoirs, is summarized by the generalized reaction:

$$CH_2O + O_2 \leftrightarrow CO_2 + H_2O \tag{1}$$

The formation and burial of organic matter (OM) is balanced over geologic time by weathering and oxidation of OM exposed at the Earth's surface and by metamorphism of OM buried at depth. The flux of C buried in OM, approximately 20–25% of the total C burial flux, is the same order of magnitude as, and indeed is nearly equal to, the flux of C consumed by silicate weathering ($\sim 20\%$ of total C burial[2]). Thus, opposing changes in the organic C cycle have the potential to compensate for imbalances in the rate of silicate weathering (e.g., act as a negative feedback). Indeed, Raymo and Ruddiman[7] proposed that the increased consumption of CO_2 via silicate weathering suggested by the Neogene Sr isotope record could have been offset by a flux of CO_2 from the increased weathering of OM relative to its burial.

In support of this idea, Raymo and Ruddiman cited earlier work by Lasaga et al.[11] and Shackleton,[12] which showed that the mean isotopic composition of carbonate buried has dropped over the last 20 million years. Because the formation of OM is accompanied by about a 24–28‰ isotope fractionation relative to the inorganic C reservoir, any change in the relative amount of OM buried versus weathered (all else being equal) would result in a change in the $\delta^{13}C$ of the inorganic carbonate reservoir. Hence, Lasaga et al.[11] and Shackleton[12] inferred that the late Cenozoic $\delta^{13}C$ decrease reflected decreased burial of OM relative to weathering (causing the observed buildup of ^{12}C in the ocean–atmosphere system). Such an imbalance would obviously be driving Eq. (1) to the right and, hence, be adding CO_2 to the atmosphere (where it could be used for silicate weathering).

This interpretation has been challenged by a number of investigators, who suggest that the decrease in mean ocean $\delta^{13}C$ over the last 20 million years reflects a decrease in the photosynthetic fractionation of OM formed in the ocean, not a decrease in the fraction of OM buried. The most complete treatment of this view (see also Chapter 18) is presented by Derry and France-Lanord[13] (hereafter DFL96), who argue based on isotopic mass balance constraints that a net addition of C to the sedimentary organic C reservoir occurred over the Neogene. If true, this would further compound the problem of the "missing" CO_2 needed for enhanced silicate weathering. The DFL96 study and the studies of Raymo[8] and Shackleton,[12] which are further developed here, provide the focal point for the present discussion primarily because they present contrasting views. Another recent study by Compton and Mallinson[14] presents an interpretation of the organic C cycle similar to that of Raymo[8] and Shackleton.[12] In particular, we need to ask how these similar mass balance models can lead to such different conclusions. How do the assumptions made by these studies differ and are they supported by geological evidence?

3. MASS BALANCE CONSTRAINTS

An excellent discussion of the relevant mass balance equations for the sedimentary C reservoir can be found in the DFL96 study. As described in that and other modeling studies, the change in the size of the sedimentary organic C reservoir with time can be written as

$$dM_{org}/dt = J_{bur}X_{org_{bur}} - J_{er}X_{org_{er}} \qquad (2)$$

where J_{bur} is the burial flux of total C (in carbonate and OM) and J_{er} is the erosion flux of same. $X_{org_{bur}}$ describes the mass fraction of total C buried (or weathered in the case of $X_{org_{er}}$) as OM. For the purpose of evaluating the transfer of C between the organic and inorganic reservoir, the further assumption is made that the erosional flux of total C equals the burial flux (i.e., $J_{bur} = J_{er}$), reducing

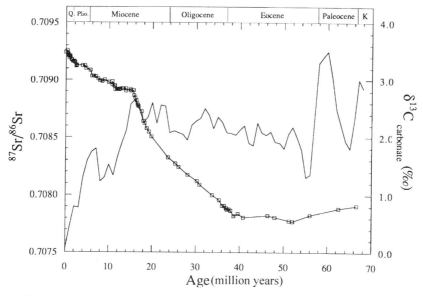

FIGURE 1. Sr isotopic composition of ocean (square symbols) over the Cenozoic from which the estimate of weathering input J_{er} was derived by Richter et al.[16] The solid line with no symbols is the assumed time history of the $\delta^{13}C$ of buried carbonate based on data collected and published by Shackleton.[12]

Eq. (2) to

$$dM_{org}/dt = J_{er}(X_{orgbur} - X_{orger}) \tag{3}$$

Hence, the growth of, or decline in, the size of the organic C reservoir (dM_{org}/dt) and by extension the flux of O_2 and CO_2 from the sedimentary C reservoir depends on the erosion rate of crustal C and the difference between the fraction of organic C in rocks weathered versus buried.

Because organic and inorganic sedimentary C have very different mean $\delta^{13}C$ values (by approximately 24–28‰), we can in theory use the $\delta^{13}C$ value of sedimentary inorganic carbon to constrain both X_{orgbur} and X_{orger} with time (see, for instance, Broecker[15]). As discussed more completely in the DFL96 study:

$$X_{orgbur} = (\delta_{carb} - \delta_{riv})/\Delta B \tag{4}$$

and

$$X_{orger} = (\delta_{ave} - \delta_{riv})/\Delta E \tag{5}$$

where δ_{carb} is the isotopic composition of carbonate removed from the ocean at time t and is given by the data of Shackleton[12] in most studies (Fig. 1). Here, the

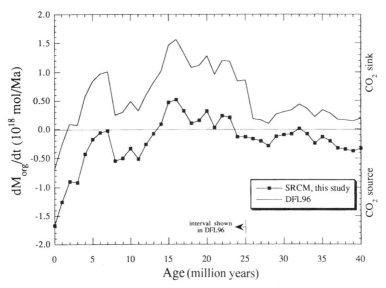

FIGURE 2. Two possible scenarios for the changes in the size of the sedimentary organic C reservoir over the Cenozoic. They approximate (nearly exactly) the dM_{org}/dt history proposed by Derry and France-Lanord[13] (DFL96) and that derived or implied by the studies of Shackleton,[12] Raymo,[8] and Compton and Mallinson[14] (SRCM, this study). When the growth of the sedimentary organic C reservoir is positive, the organic C cycle acts as a sink for atmospheric CO_2 (and visa-versa for negative net growth rates).

data have been interpolated at 1-million-year increments; δ_{riv} is the assumed isotopic composition of all C inputs to the ocean; and ΔB is the mean isotopic difference between carbonate and OM deposited at time t (controlled primarily by the magnitude of the photosynthetic fractionation factor). In the equation for $X_{org_{er}}$, δ_{ave} describes the mean isotopic composition of carbonate rocks weathered through time, while ΔE is the mean isotopic difference between carbonate and OM being weathered.

Using these five parameters and an assumption about the burial/erosion rate, one can evaluate dM_{org}/dt and determine whether the organic C sedimentary reservoir has increased or decreased in size over the Neogene. Two such calculations are shown in Fig. 2 with the assumed values upon which each calculation was based shown in Table 1. The curve labeled "DFL96" closely approximates a calculation by Derry and France-Lanord,[13] differing only in that a slightly different time history for J_{er} was assumed. The curve labeled "SRCM" closely reflects the time history of dM_{org}/dt implied by or calculated in the studies of Shackleton,[12] Raymo,[8] and Compton and Mallinson.[14] The climatic implications of the two curves shown in Fig. 2 are very different. In the DFL96 curve the organic C sedimentary reservoir is acting as a sink for atmospheric CO_2 throughout the last 40 million years (with the exception of the last 2 million years). By contrast, the second curve (SRCM) suggests that the organic C

TABLE 1. Parameters Used for Estimates Shown in Fig. 2

	DFL96	SRCM, this study
$J_{er}{}^a$	Variable	Variable
δ_{riv}	$-5.0\%_{oo}$	$-3.8\%_{oo}$
δ_{ave}	$1.8\%_{oo}$	$2.0\%_{oo}$
ΔB^b	28.0 to 24.5 to 23.5$\%_{oo}$	28.0 to 23.5$\%_{oo}$
ΔE	28$\%_{oo}$	25$\%_{oo}$

[a]Similar weathering rate histories (after Richter et al. [16]) are used for J_{er} in both cases shown above.
[b]The DFL96 value for ΔB decreases in two steps over 40 million years (at 25 and at 7 Ma) while SRCM, this study, assumes a continuous decrease over same 40 million year interval.

subcycle acted as a source of CO_2 to the atmosphere for a large part of the Neogene. In the following section, the sensitivity of the dM_{org}/dt calculation to each of the five parameters in the model is examined.

4. SENSITIVITY TESTS

In this section, the value of each parameter tabulated in Table 1 is systematically changed and dM_{org}/dt recalculated. For each parameter investigated, all parameters other than the one of interest are held at the values given for SRCM in Table 1.

4.1. Rate of Sedimentary Cycling (J_{er})

As mentioned earlier, the magnitude and direction of weathering rate changes over the Neogene are at the root of the controversy over BLAG-type models. Have chemical weathering and erosion rates increased, decreased, or remained constant over the last 40 million years? More to the point, how does this uncertainty affect estimates of dM_{org}/dt inferred from $\delta^{13}C$ records? The answer is: not by much. As pointed out by DFL96, changes in weathering rates act to amplify (or dampen) existing imbalances in the transfer of C between the oxidized and reduced reservoirs.

In Fig. 2, both scenarios assume the same time history for J_{er} (shown in Fig. 3; note that this time history for J_{er} is slightly different from that used but not tabulated in DFL96). Hence, different assumptions about this parameter cannot be the cause of the difference between the two scenarios shown in Fig. 2. The weathering rate used in Fig. 2 was derived by scaling J_{er} to the change in river Sr flux calculated by Richter et al.[16] (Fig. 3) such that the present weathering flux value is approximately equal to the modern C burial flux estimate of Wilkenson and Algeo.[17] In Fig. 4, dM_{org}/dt is also calculated with two additional weathering rate assumptions: specifically, a constant weathering rate of 12.5×10^{18} moles/

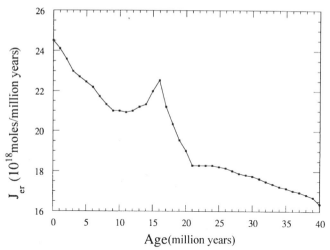

FIGURE 3. The assumed weathering rate J_{er}, which was scaled to the estimated river flux of Sr as determined by Richter et al.[16]

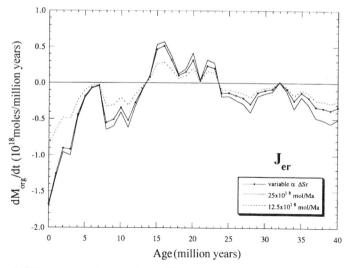

FIGURE 4. Influence of variations in J_{er} on dM_{org}/dt. A low constant weathering rate and a high constant weathering rate are shown for comparison to the Sr-derived rate shown in Fig. 3.

million years and a constant rate of 25×10^{18} moles/million years (all other parameters are as assumed by SRCM, this study, in Table 1). Clearly, at higher weathering rates any imbalance in the partitioning of C between the oxidized and reduced reservoirs is amplified. The crossover points between net positive and net negative growth of the sedimentary organic C reservoir do not change from the variable weathering rate assumption.

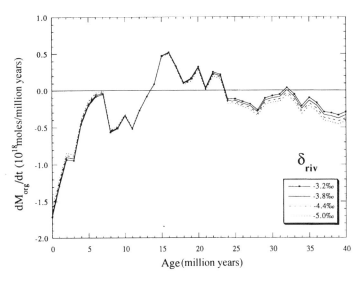

FIGURE 5. Influence of variations in δ_{riv} on dM_{org}/dt.

4.2. River Isotopic Value (δ_{riv})

The calculation of dM_{org}/dt shown in Fig. 2 is also relatively insensitive to the assumed value of the $\delta^{13}C$ input to the ocean (δ_{riv}). In Fig. 5, dM_{org}/dt is shown for four assumed values of δ_{riv}. The $-3.8\%_0$ value is after Garrels and Lerman,[18] while the $-5.0\%_0$ value (assumed by DFL96) is after Craig.[19] Two other values, $-3.2\%_0$ and $-4.4\%_0$, were chosen arbitrarily (although the $-4.4\%_0$ value is that used by Garrels and Perry[20]). The assumptions for the value of this parameter cannot explain the differences between the two scenarios shown in Fig. 2.

4.3. Mean Isotopic Composition of Sedimentary Carbonate Reservoir (δ_{ave})

To calculate the fraction of C eroded as OM [X_{orger}, Eq. (2)] one needs an estimate for both the average $\delta^{13}C$ of the sedimentary carbonate rocks undergoing weathering (δ_{ave}) as well as the relative isotopic offset of the OM being weathered (ΔE). The value of δ_{ave} is estimated from the observed variations in the $\delta^{13}C$ of marine carbonates over 600 million years (Fig. 6), often adjusted for the long residence of carbonate rocks and/or the age–mass distribution of rocks (e.g., DFL96). DFL96 assumed $1.8\%_0$ for this parameter. Compton and Mallinson[14] use a slightly higher estimate (closer to $2.1\%_0$), while this study assumes $2.0\%_0$ (Table 1). Any of these values would be a reasonable assumption given our imperfect knowledge of the geologic record (although assuming a single unchanging value is probably an oversimplification as it is unlikely that δ_{ave} has been constant over 40 million years).

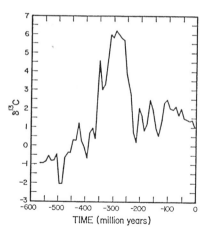

FIGURE 6.   The $\delta^{13}C$ of carbonate deposited over the Phanerozoic (reprinted from Lasaga[29]).

The value chosen for δ_{ave} has a relatively significant effect on dM_{org}/dt (Fig. 7). All else being constant, when 1.6‰ is used for δ_{ave}, the sedimentary organic C reservoir is growing or constant for all but 12 of the last 40 million years. By contrast, when 2.2‰ is used, the organic C reservoir decreases in size for all but 6 million years. The choice of 1.8‰ (DFL96) and 2.0‰ (SRCM) produces intermediate results accounting for some, but not all, of the difference between the two curves shown in Fig. 2.

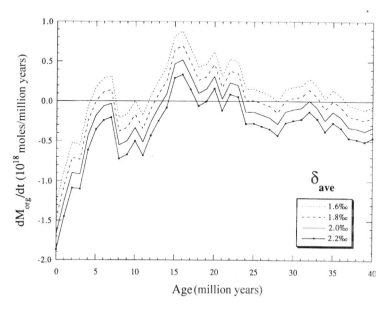

FIGURE 7.   Influence of variations in δ_{ave} on dM_{org}/dt.

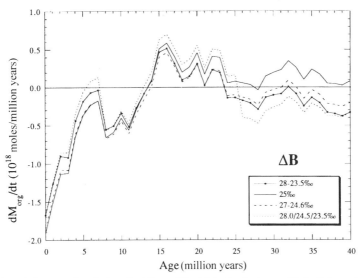

FIGURE 8. Influence of variations in ΔB on dM_{org}/dt. In the last case (dotted) the value for ΔB is stepped down at 25 million years and at 7 million years (after DFL96). In the other two variable ΔB cases (dashed line and solid line with squares), the change indicated occurs linearly over the interval shown.

4.4. Mean Isotope Fractionation between Carbonate and OM Being Buried (ΔB)

The choice of ΔB, the isotopic fractionation between carbonate and OM being buried at time t, has less of an effect on the calculation of dM_{org}/dt than the choice of δ_{ave}. Figure 8 illustrates the influence of different assumptions for ΔB; these assumptions range from constant at 25‰ (e.g., Shackleton[12]) to three more likely scenarios, a gradual decrease from 28 to 23.5‰, a gradual decrease from 27 to 24.6‰, and a stepped decrease from 28 to 23.5‰ (as in DFL96). The latter three scenarios appear to be supported by the data of Popp et al.[21] which suggest that a pronounced decrease in ΔB has occurred since the late Eocene. The conclusion of DFL96 that dM_{org}/dt depends "significantly on the timing of the change in ΔB" is not supported by the sensitivity tests shown in Fig. 8. In fact, the major cause of the difference between the two scenarios shown in Fig. 2 is the value assumed for a related variable, ΔE, as discussed below.

4.5. Mean Isotope Fractionation between Carbonate and OM Being Eroded (ΔE)

As shown in Fig. 9, different assumptions for the isotope fractionation between organic and inorganic reservoirs undergoing weathering (25‰ vs. 29‰) are the major reason for the difference between the two scenarios shown in Fig.

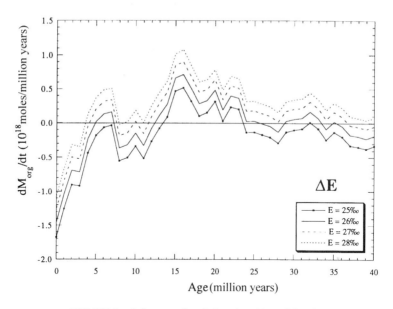

FIGURE 9. Influence of variations in ΔE on dM_{org}/dt.

2. Shackleton,[12] Raymo,[8] and Compton and Mallinson[14] all assumed a value of 25‰, while DFL96 assumed a value of 28‰. To determine this value exactly, one would need to know the age of all C-rich sediment being eroded at any specific time as well as its isotopic content. It is also unlikely that ΔE would be constant with time; for instance, the current Tethyan–Himalayan orogeny is probably associated with the uplift and weathering of late Mesozoic and early Cenozoic formations, while the early Cenozoic may have been dominated by the weathering of early Mesozoic formations or terrain uplifted during the widespread orogenies of the Paleozoic era.

An early attempt to constrain the isotopic composition of buried OM was made by Craig,[19] who determined that modern land plants showed a mean $\delta^{13}C$ value of -25‰ (Pleistocene fossil wood was -24‰). Further, coal values from the Pennsylvanian to Pleistocene showed a mean $\delta^{13}C$ of -24‰ while slate and shale samples from Precambrian to Recent showed a mean $\delta^{13}C$ value of -28‰. Craig also estimated that the mean carbonate $\delta^{13}C$ over the Phanerozoic was 0‰. It was from these results (in particular the estimate for modern land plants) that the typically assumed value of -25‰ for ΔE was derived. A follow-up study by Degens[22] showed slightly modified results (based on a much larger data base); shales and sandstones had a mean $\delta^{13}C$ value of -26‰ while coals had a mean isotope value of -25‰. He also found the mean carbonate $\delta^{13}C$ to approximate 0‰, implying a ΔE value between 25 and 26‰.

Since these early studies, additional field and laboratory work, in particular by Lindh,[23] has suggested that mean carbonate values were actually higher over much of the Phanerozoic, averaging about 1.5‰ over the last 200 million years but ranging between -2 and $+6$‰ over the last 600 million years (Fig. 6). By the late 1980s geochemical models of the C cycle were using these higher values for δ_{ave} but still assumed 25‰ for the organic–inorganic fractionation (e.g., they did not adopt a corresponding increase in the value of ΔE).

More recent analytical studies of ancient OM[21,24] suggest that ΔE could be much larger (e.g., closer to 28‰ than 25‰). Lewan[24] found two distinct types of OM preserved over the Phanerozoic; a more abundant class with typical $\delta^{13}C$ values of -28 to -29‰ and a less abundant type with heavier values around -22‰. Very light $\delta^{13}C$ values were also found by Popp *et al.*[21] for late Mesozoic kerogens. Both studies found the heavy $\delta^{13}C$ OM concentrated in sediments younger than 25 million years of age. If one assumed that the bulk of OM weathered today was buried at -28‰, then this would imply typical ΔE values as high as 30‰. However, the disagreement between average Phanerozoic OM $\delta^{13}C$ values found by Degens[22] and Lewan[24] remains as an unresolved issue. We still may not have a representative survey of the $\delta^{13}C$ of OM contained in the Earth's crust, and subsequently weathered, over geologic time.

5. DISCUSSION

From the above sensitivity tests, one can see that the first-order structure of all model-generated curves follows the basic form of the Shackleton[12] $\delta^{13}C$ of carbonate record. The assumption one makes for the magnitude of the river input (e.g., Richter's Sr curve[16]) imprints a longer-term modulation to model estimates. The sign of dM_{org}/dt, which reflects whether the organic C reservoir is acting as a net source or sink of CO_2 to the atmosphere, depends most heavily on the values assumed for ΔE and δ_{ave}. Both of these variables are determined from accumulated $\delta^{13}C$ measurements of Phanerozoic sediments, although the most recent systematic study[24] was done over 10 years ago. Given the apparent discrepancies among different studies (e.g., Lewan[24] versus Degens[22] and Craig[19]) and the importance of these data to understanding the mass balance of C, it is imperative that these data be reassessed. For instance, if one assumes 28‰ for ΔE and 1.5‰ for δ_{ave}, one would observe an increase in the size of the organic sedimentary reservoir (and therefore net removal of CO_2 from the atmosphere) for all but roughly the last 3 million years. By contrast if one assumed δ_{ave} was 2.5‰ instead (and $\Delta E = 28$‰), then dM_{org}/dt would be negative for most of the last 13 million years (and therefore the organic C reservoir would have acted as a source of atmospheric CO_2 since the mid-Miocene). The implications of each scenario for Neogene climate change are clearly different.

In addition, this study has assumed that the $\delta^{13}C_{carb}$ record shown in Fig. 1 is accurate. In fact, a 0.1‰ change in this $\delta^{13}C$ value would result in a change in dM_{org}/dt of just under 0.1×10^{18} moles/million years. Because Compton and Mallinson[14] used a slightly different $\delta^{13}C_{carb}$ record in their model, they did not observe a late Miocene increase in the burial of OM relative to erosion, as seen in both of the studies shown in Fig. 2. The mean $\delta^{13}C$ of carbonate buried in the Cenozoic is thus another data set that should be evaluated and updated.

An observation on which most agree is that the observed photosynthetic fractionation factor (ΔB) has decreased over the last 60 million years[21] and that this decrease was most likely driven by a decrease in ocean–atmosphere CO_2 levels (see also Chapter 13). If one assumes relatively high values for ΔE, suggested by Lewan's compilation of Phanerozoic data,[24] then the size of the organic sedimentary reservoir would appear to have increased, rather than decreased, over much of the Neogene. Such a shift may have been due to more effective OM burial associated with increased clastic fluxes.[25] However, net burial of organic matter would not only lead to a drop in atmospheric CO_2 but would require an additional CO_2 source to support the excess OM burial. An obvious source is a decrease in the rate of silicate weathering. However, the simplest interpretation of the $\delta^{87}Sr$ data suggests increased silicate weathering.[3,7–9,14,16,26] Further, the Os isotope data (Chapter 17) implies increased chemical weathering rates over the Neogene (because it seems unlikely that just organic-rich deposits would undergo enhanced weathering). Lastly, could the pronounced increase in clastic erosion observed for the late Cenozoic[27,28] have occurred without an accompanying increase in chemical weathering (Chapter 14)? If not, we still have a mass balance problem for C.

In conclusion, I suspect we do not even know the sign, over the last 40 million years, of one of the largest fluxes in the global C cycle. The geochemical mass balance models are best used to make testable predictions and help focus data collection efforts. Because they can only mirror our own incomplete understanding of the Earth's geochemical cycles, it is unclear whether they provide any important "constraints." They cannot provide a constraint if their underlying assumptions are false or even slightly inaccurate. Yet, luckily, some of the most important, and most uncertain, parameters in C cycle models can be measured. Data collection has not kept up with modeling efforts in this field and more data, not more models, are what is needed.

Acknowledgments

Thank you to J. Kauffman and W. Ruddiman for reviews and suggestions which improved this paper. This research was supported by the Petroleum Research Fund of the American Chemical Society.

REFERENCES

1. Walker, J. C. G., Hays, P. B., and Kasting, J. F. (1981). *J. Geophys. Res.* **86**, p. 9976.
2. Berner, R. A., Lasaga, A. C., and Garrels, R. M. (1983). *Am. J. Sci.* **283**, p. 641.
3. Raymo, M. E., Ruddiman, W. F., and Froelich, P. N. (1988). *Geology* **16**, p. 649.
4. Berner, R. A., and Rye, D. M. (1992). *Am. J. Sci.* **292**, p. 136.

5. Caldeira, K. (1992). *Nature* **357**, p. 578.
6. Kerrick, D. M., and Caldeira, K. (1993). *Chem. Geol.* **108**, p. 201.
7. Raymo, M. E., and Ruddiman, W. F. (1992). *Nature* **359**, p. 117.
8. Raymo, M. E. (1994). *Paleoceanography* **9**, p. 399.
9. François, L. M., and Walker, J. C. G. (1992). *Am. J. Sci.* **292**, p. 81.
10. Delaney, M. L., and Filippelli, G. M. (1994). *Paleoceanography* **9**, p. 513.
11. Lasaga, A. C., Berner, R. A., and Garrels, R. M. (1985). In: *The Carbon Cycle and Atmospheric CO$_2$: Natural Variations Archean to Present* (E. T. Sundquist and W. S. Broecker, eds.), p. 397. Geophysics Monograph 32, Merical Geological Union, Washington, D.C.
12. Shackleton, N. J. (1987). In: *Marine Petroleum Source Rocks* (J. Brooks and A. J. Fleet, eds.), pp. 423–434. Special Publication of Geological Society of London 26.
13. Derry, L. A., and France-Lanord, C. (1996). *Paleoceanography* **11**, p. 267.
14. Compton, J. S., and Mallinson, D. J. (1996). *Paleoceanography* **11**, p. 431.
15. Broecker, W. S. (1974). *J. Geophys. Res.* **75**, p. 3553.
16. Richter, F. M., Rowley, D. B., and DePaolo, D. J. (1992). *Earth Planet. Sci. Lett.* **109**, p. 11.
17. Wilkenson, B. H., and Algeo, T. J. (1989). *Am. J. Sci.* **289**, p. 1158.
18. Garrels, R. M., and Lerman, A. (1984). *Am. J. Sci.* **284**, p. 989.
19. Craig, H. (1953). *Geochim. Cosmochim. Acta* **3**, p. 53.
20. Garrels, R. M., and Perry, E. A. (1974). In: *The Sea* (E. D. Goldberg, ed.), pp. 303–316. John Wiley, New York.
21. Popp, B. N., Takigiku, R., Hayes, J. M., Louda, J. W., and Baker, E. W. (1989). *Am. J. Sci.* **289**, p. 436.
22. Degens, E. T. (1969). In: *Organic Geochemistry Methods and Results* (G. Eglinton and M. T. J. Murphy, eds.), pp. 304–329. Springer-Verlag, New York.
23. Lindh, T. B. (1983). Temporal variations in ^{13}C and ^{34}S and global sedimentation during the Phanerozoic, M.S. thesis, University of Miami, p. 98.
24. Lewan, M. D. (1986). *Geochim. Cosmochim. Acta* **50**, p. 1583.
25. Keil, R. (1994). *Nature* **370**, p. 549.
26. Delaney, M. L., and Boyle, E. A. (1988). *Paleoceanography* **3**, p. 137.
27. Ronov, A. B. (1980). In: *The Earth's Sedimentary Shell: Quantitative Patterns of Its Structure, Compositions, and Evolution* (A. A. Yaroshevskii, ed.), p. 80. Nauka, Moscow.
28. Budyko, M. I., Ronov, A. B., and Yanshin, A. L. (1987). *History of the Earth's Atmosphere*, 139 pp. Springer-Verlag, Berlin.
29. Lasaga, A. (1989). *Am. J. Sci.* **289**, p. 411.

17

Os Isotope Record in a Cenozoic Deep-Sea Core: Its Relation to Global Tectonics and Climate

Karl K. Turekian and William J. Pegram

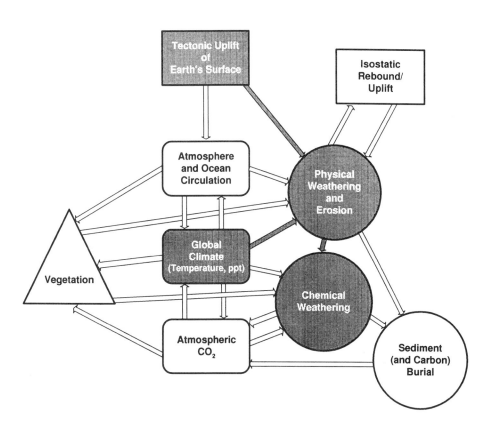

Karl K. Turekian and William J. Pegram • Department of Geology and Geophysics, Yale University, New Haven, Connecticut 06520-8109.

Tectonic Uplift and Climate Change, edited by William F. Ruddiman. Plenum Press, New York, 1997.

1. INTRODUCTION

The oceanic variation of $^{87}Sr/^{86}Sr$ over at least the last 100 million years has been established through the study of calcareous biogenic deposits found principally in deep-sea sediments. The large picture is that in sediments from about 100 million years ago until about 38 million years ago $^{87}Sr/^{86}Sr$ shows only small variations but from that time forward there has been an increase. This variation has been ascribed to the changing relative importance with time of two mixing components: Sr with relatively low $^{87}Sr/^{86}Sr$ derived primarily by the weathering of ridge basalts and added to the oceans at hydrothermal vents and more radiogenic Sr derived from the weathering of continental crust. The approximately 4-million-year residence time of Sr is about a thousand times the mixing time of the oceans, so a marine $^{87}Sr/^{86}Sr$ record anywhere in the deep ocean basins speaks to global changes. The fine details of the sources and types of the actual continental rocks undergoing weathering continue to be the subject of active discussion.

The quest for oceanic variations with time of other isotopic signatures to relate to continental weathering history has included $^{143}Nd/^{144}Nd$ and the Pb isotopic systems. Nd and Pb, however, have residence times in the oceans of only hundreds of years; therefore their marine isotopic records are strongly regional. The only other element that has the potential for recording a global isotopic signature tracking large-scale tectonic events and their associated weathering is Os. The variations in rocks of the amounts of radiogenic ^{187}Os produced from the radioactive decay of ^{187}Re (half-life = 4.6×10^{10} years) relative to non-radiogenic isotopes of Os provides this prospect. This possibility was first suggested by Luck and Turekian[1] in their exploration of the potential variations in oceanic $^{187}Os/^{186}Os$ over time. They carried out an exercise in setting the possible range of marine $^{187}Os/^{186}Os$ over time in order to put limits on the effectiveness of using the Os isotope signature at the Cretaceous–Tertiary boundary as a clue to the source of the Os and Ir enrichment found there, whether terrestrial or meteoritic in origin.[2] Potential variations in the $^{187}Os/^{186}Os$ of seawater are very large, given the enormous range of isotopic compositions of rocks in the Earth's crust; for example, peridotites and basalts recently derived from the mantle have $^{187}Os/^{186}Os$ ratios close to 1 while currently eroding continental crust has an average $^{187}Os/^{186}Os$ of 10–15. Values between these two limits would reflect combinations of Os supplied by these two sources of weathering.

The actual variation of $^{187}Os/^{186}Os$ in the ocean with time could only be determined, however, if a suitable time record could be found. Os, unlike Sr, is not incorporated to any significant extent in calcareous biogenic deposits; therefore the record must be sought in nonbiogenic hydrogenous deposits. The site of primary removal of Os from the ocean is reducing sediments, as shown by Ravizza and Turekian[3] and Ravizza et al.[4] Os was also found to be concentrated in marine ferromanganese oxide nodules by Luck and Turekian[1] who measured their Os isotopic compositions as surrogates for seawater.

The possibility that hydrogenous components of sedimentary deposits could provide a record of the $^{187}Os/^{186}Os$ changes in the oceans led to the search for a suitable core for such a study. Long Lines 44–Giant Piston Core 3 (commonly called "GPC3") raised by C. Hollister from pelagic clay sediments of the North Pacific provided such an opportunity. GPC3 (30° 19.9′N, 157° 49.9′W) was raised from a 5705-m water depth and is 24 m in length. Kyte *et al.*[5] have comprehensively studied the core for chemistry and sedimentology; other studies have been made to assist in dating using ^{10}Be and fish debris micropaleontology.[6–8]

The initial Os isotope study of GPC3 published by Pegram *et al.*[9] has been extended with greater detail by Pegram and Turekian.[10] We interpret the $^{187}Os/^{186}Os$ record of the oceans based on these results and invoke additional work done on ^{3}He and ^{4}He on the same samples supplied by us to Farley.[11] The combination of these sets of data provides an entry into the problem of the relation of mountain-forming processes to changes in climate.

2. METHODS AND RESULTS

The procedures for the analysis of GPC3 for Os concentration and $^{187}Os/$ ^{186}Os by H_2O_2 leaching is described in Pegram *et al.*[9] and in greater detail in Pegram and Turekian.[10] The $^{187}Os/^{186}Os$ data from these two papers obtained by the 6% H_2O_2 leach of sections of GPC3 are plotted against depth in the core in Fig. 1. The assignment of dates along the core is a complex exercise and is discussed by Pegram *et al.*[9] and Pegram and Turekian.[10] Time anchor points are established for the past 5 million years or so by magnetic reversal stratigraphy and at 65 million years ago by the high Ir concentration observed at depth in the core, which is uniquely characteristic of the Cretaceous–Tertiary boundary

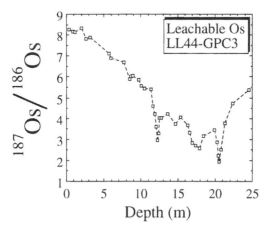

FIGURE 1. $^{187}Os/^{186}Os$ in 6% H_2O_2 leach of core segments in the North Pacific pelagic clay core LL44-GPC3, based on data from Pegram *et al.*[9] and Pegram and Turekian.[10] Analytical uncertainty is less than the symbol size.

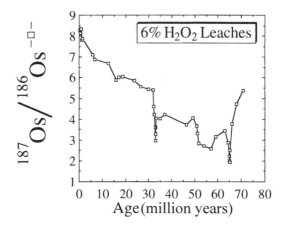

FIGURE 2. $^{187}Os/^{186}Os$-vs.-age curve for GPC3 based on the depth data on 6% H_2O_2 leachable Os and the chronometry discussed in the text.

around the world. Other constraints are set by fish stratigraphy,[7,8] and the time interpretation of boundaries by Berggren et al.[12] Figure 2 is the resulting plot of $^{187}Os/^{186}Os$ against time in GPC3.

Aside from chronometry, the other serious question is whether the H_2O_2 leaching process yields an accurate representation of the ocean value at the time of deposition. The $^{187}Os/^{186}Os$ results of Ravizza[13] and Peucker-Ehrenbrink et al.[14] on metalliferous sediment cores are in general agreement with the results shown in Fig. 2 for the past 20 million years. Between 20 and 50 million years, however, some discrepancies exist, up to a maximum of 0.6–0.8 units

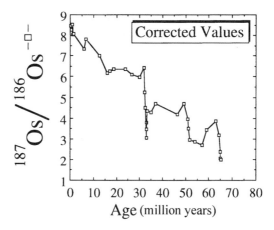

FIGURE 3. Modified $^{187}Os/^{186}Os$-vs.-age curve for GPC3 based on corrections for possible meteoritic contributions. Corrections were made only to the Cretaceous–Tertiary boundary.

of $^{187}Os/^{186}Os$. These differences are due in part to the ambiguity of chronometry in the two cores for these earlier times and in part to differences in leaching methods of the core material by the two laboratories. Although the fine structure may be in dispute, the gross features are similar for the various cores.

Pegram and Turekian[10] made a study of H_2O_2 leaching protocols and concluded that milder H_2O_2 leaching releases Os with a slightly different, generally higher, $^{187}Os/^{186}Os$ relative to Os released by stronger leaches. For sediments with the lowest accumulation rates and therefore with the highest cosmic dust concentration, a mild leach (0.15% H_2O_2) yields higher $^{187}Os/^{186}Os$ values but no more than 10% higher than the leach used in our normal procedure (6% H_2O_2). Levels in the core with high sediment accumulation rates show much less of an effect. The reason for these differences is discussed in detail by Pegram and Turekian.[10] Figure 3 is our best estimate of the $^{187}Os/^{186}Os$ variation over time with the chronometry used in constructing Fig. 2 and the corrections for leaching differences where appropriate.

The features of the $^{187}Os/^{186}Os$ plots of Figs 1, 2, and 3 are the following:

1. Over the past approximately 30 million years there is an almost mono-tonic increase of $^{187}Os/^{186}Os$ from 6 to about 8.6, the latter being effectively the same as values obtained for modern oxic hydrogenous deposits analyzed by Esser and Turekian[15,16] and on modern anoxic deposits analyzed by Ravizza and Turekian.[3] The sense of this change, indicating an increasing relative importance of continental weathering for the Neogene, is similar to that observed for $^{87}Sr/^{86}Sr$ (Fig. 4). For the Os curve, significant increases in slope occur around 17 million years ago (matched by a change in slope for the Sr curve) with the rate of increase over the past 7 million years greater than before that time. We interpret these patterns for Sr and Os isotope ratios as indicating changes in the types of continental rocks undergoing weathering. The 17-million-year change in

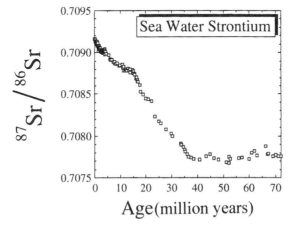

FIGURE 4. The $^{87}Sr/^{86}Sr$-vs.-time curve over the past 70 million years in carbonate deep-sea cores. Data are from a number of sources and are summarized in Pegram *et al.*[9]

$^{87}Sr/^{86}Sr$ and $^{187}Os/^{186}Os$, for example, may be due to a switch from predominantly Antarctic glacial erosion and dissolution of rock flour to intense weathering of the Himalayas. The $^{187}Os/^{186}Os$ change about 7 million years ago we ascribe to the increased weathering of black shales or similar high-Re source rocks in the Himalayas, which affects the Os isotope delivery signature but not the Sr isotope signature of the rivers draining the Himalayas.

2. Kyte et al.[5] located the Cretaceous–Tertiary boundary (65 million years ago) in GPC3 by its profound Ir anomaly. Our detailed study of this region of the core yields a minimum leachate $^{187}Os/^{186}Os$ ratio of about 1.9 at the peak of the Ir anomaly (Fig. 1). We ascribe this low value to the dissolution of debris resulting from the impact of a large meteorite. The meteoritic Os supplied during impact would be about 10^{24} pg Os for a carbonaceous chondritic 10-km-diameter body with an isotopic ratio of about 1. The higher $^{187}Os/^{186}Os$ value observed in the leach, assumed to represent the seawater value of that time, is due in part to the vaporization of some of the more radiogenic target rock Os but also to the relatively radiogenic $^{187}Os/^{186}Os$ ratio of the oceans at the time of the impact. If the Os concentration of seawater at the Cretaceous–Tertiary boundary were 1 pg liter^{-1} with an oceanic volume of 1.35×10^{21} liters, the total amount of Os in the oceans would be 1.35×10^{21} pg. If we assign an $^{187}Os/^{186}Os$ of ocean water of about 3.5 just before the impact, interpolated from leachate $^{187}Os/^{186}Os$ values immediately above and below the Ir anomaly, the amount of meteoritic Os with an $^{187}Os/^{186}Os$ of 1 that would be needed to lower the seawater $^{187}Os/^{186}Os$ to 1.9 is about 2.4×10^{21} pg. This amount of vaporized and dissolved Os represents about 0.25% of the total meteoritic Os. If the Cretaceous ocean had an

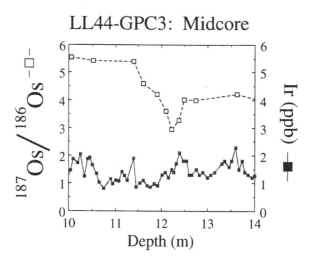

FIGURE 5.   Comparison of bulk Ir concentration data from Kyte et al.[5] and $^{187}Os/^{186}Os$ from Fig. 1 at around a 12-m depth (in the region of the Oligocene–Eocene boundary) in GPC3.

^{187}Os/^{186}Os of 5, then there was 0.5% vaporization. We infer that high bulk Ir and Os concentrations with a low, almost meteoritic, ^{187}Os/^{186}Os in the leach fraction is the hallmark of a meteorite impact in which the vaporized meteorite Os fraction is imprinted on the ocean water at the time of impact. The Os isotopic anomaly at the Cretaceous–Tertiary boundary is superimposed on an overall decline of seawater ^{187}Os/^{186}Os continuing from the late Cretaceous and extending into the Paleocene (Fig. 3). This rapid readjustment of the dissolved Os is compatible with a residence time of about 10^3 to 10^4 years.

3. At middepth in the core corresponding to an age of 32 to 33 million years, there is a dip in the ^{187}Os/^{186}Os that, although not as great a dip as at the Cretaceous–Tertiary boundary, is still distinctive. The Ir data of Kyte et al.,[5] however, show no noticeably high values at the depth in the core where the drop in the ^{187}Os/^{186}Os occurs (Fig. 5). We conclude that the evidence for a meteoritic impact with global consequences is not strong at the horizon, based on the Pt group elements. This observation, in conjunction with other types of evidence, provides the basis for seeking a relationship among the deep-sea ^{187}Os/^{186}Os record over time, global tectonics, and climate change in the Cenozoic.

3. DISCUSSION

The proper interpretation of the ^{187}Os/^{186}Os oceanic variation with time depends on comparisons with other parameters that have been measured in deep-sea sediments. Ir, as we noted above, has proven to be a useful index of a meteoritic impact when confirmed with the ^{187}Os/^{186}Os signature and other sedimentological properties. There are three isotopic measurements that have been extensively utilized in deep-sea sediments, which, along with ^{187}Os/^{186}Os, provide us with useful information on the tectonic and climatic history of the Earth during the Cenozoic. These are ^{87}Sr/^{86}Sr, ^{18}O/^{16}O(δ^{18}O), and the concentration of ^{3}He in the sediments associated with interplanetary dust particles reaching the ocean bottom. We discuss each isotopic parameter in conjunction with the ^{187}Os/^{186}Os record in GPC3. For ^{87}Sr/^{86}Sr and δ^{18}O we depend on the chronology of GPC3 as discussed above to correlate with the vast amounts of ^{87}Sr/^{86}Sr and δ^{18}O data that have been accumulated on a number of cores over the years. For ^{3}He we rely on the measurements made by Farley[11] on the same sequence in GPC3 analyzed for ^{187}Os/^{186}Os so that the comparison is independent of chronology. We do depend on the chronology of GPC3, however, to relate the coupled ^{187}Os/^{186}Os and ^{3}He signatures to large-scale features of Earth history.

3.1 The Relation of ^{87}Sr/^{86}Sr to ^{187}Os/^{186}Os

A simple comparison of the marine ^{87}Sr/^{86}Sr (Fig. 4) and ^{187}Os/^{186}Os (Fig. 3) patterns over time during the Cenozoic reveals that both isotope ratios are driven by the intensity of continental weathering relative to other (in these

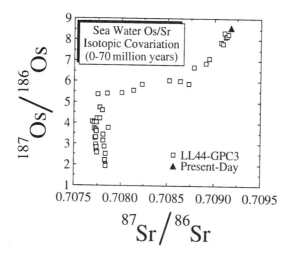

FIGURE 6. ^{187}Os/^{186}Os in 6% H_2O_2 leach converted to time-dependent variation (Fig. 2) vs. the ^{87}Sr/^{86}Sr variation with time (Fig. 4) showing the three regions discussed in the text.

cases, less radiogenic) sources in supplying these elements to the oceans. In Fig. 6 we plot ^{187}Os/^{186}Os versus ^{87}Sr/^{86}Sr to highlight the differences between the two isotopic signatures. However the main relationship between ^{187}Os/^{186}Os and ^{87}Sr/^{86}Sr is insensitive to nuances in chronometry.

There are three major types of relationships between these isotope ratios evidenced by the slopes of the lines connecting the points (Fig. 6). Generally these trends are associated with time segments although this is not explicitly indicated in Fig. 6. The first region lies between 65 and about 35 million years ago. The second is between 35 and 17 million years ago. The third is between 17 million years ago and the present.

The first region, 65 to 35 million years ago, shows virtually no change in ^{87}Sr/^{86}Sr with large variations in ^{187}Os/^{186}Os. The ^{187}Os/^{186}Os ratios range between 1.9 and 5.5 (for the 6% H_2O_2 leach) and include the two major regions of low ^{187}Os/^{186}Os identified in the previous section. Clearly, whatever is affecting the ^{187}Os/^{186}Os of the oceans during that time does not have a parallel expression in ^{87}Sr/^{86}Sr. If low ^{87}Sr/^{86}Sr is due to the dominance of basaltic alteration in submarine hydrothermal regions, the ^{187}Os/^{186}Os variation must be due to other sources or else it would parallel the Sr isotopic pattern. One such independent effect is the meteorite impact at the Cretaceous–Tertiary boundary as already discussed. The other low values we believe are due to variations in the supply of Os with a low ^{187}Os/^{186}Os ratio by the alteration of ultramafic Os-rich rocks found at the oceanic margins.

The second region, between 35 and 17 million years ago, shows increases in both ^{187}Os/^{186}Os and ^{87}Sr/^{86}Sr. We ascribe this pattern to the increased weathering supply of more radiogenic Os and Sr from continental weathering. Initially, Antarctica may have been the predominant source of this continental weathering. The slope of the line of this region is roughly what would be expected if the weathering source increasingly supplies Sr with a $^{87}Sr/^{86}Sr$ of about 0.720 and Os with an $^{187}Os/^{186}Os$ of about 10–15, the average ratios for the continental crust.

The third region is demarked by a sharp increase in $^{187}Os/^{186}Os$ with a more modest increase in $^{87}Sr/^{86}Sr$. The break in slope occurs about 17 million years ago. We ascribe this change in slope to the increased weathering in the Himalayas of Re-rich source rocks. The most obvious rocks fulfilling this condition are black shales, although any rock rich in Re and sufficiently old to provide a high $^{187}Os/^{186}Os$ supply would be acceptable. The intensity of black shale weathering seems to have increased from 7 million years ago to the present. Pegram et al.[17] have shown that streams draining regions with large black shale exposures indeed have high $^{187}Os/^{186}Os$ and that the Ganges River draining the Himalayas is similarly high in $^{187}Os/^{186}Os$.

These patterns must be related to the history of tectonics and climate and we will address this matter after the relationships of $^{187}Os/^{186}Os$ to the other geochemical patterns are reviewed.

3.2. The Relation of ^{3}He Concentration and $^{187}Os/^{186}Os$

Figure 7 shows a plot of the ^{3}He concentration of layers of GPC3 measured by Farley[11] imposed on the plot of $^{187}Os/^{186}Os$ of 6% H_2O_2 leachable Os from Fig. 1. From a depth of about 10 m in the core to the top there is an inverse correlation between ^{3}He concentration and $^{187}Os/^{186}Os$ of the leachable fraction. If the ^{3}He-carrying interplanetary dust particles were accumulating at a constant

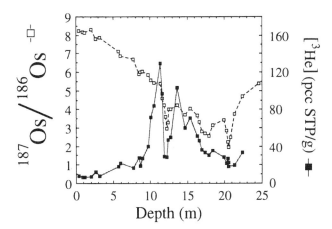

FIGURE 7. The plot of leachable $^{187}Os/^{186}Os$ data from Fig. 1 and the ^{3}He concentration measurements on the same segments by Farley.[11] From 0 to 11.5 m there is an inverse correlation between leachable $^{187}Os/^{186}Os$ and ^{3}He concentration and from 11.5 to 21 m there is a direct correlation between leachable $^{187}Os/^{186}Os$ and ^{3}He concentration.

rate, the decrease in ^{3}He concentration in the sediments as the top of the core is approached would indicate that the rate of sediments supplied to the North Pacific has been increasing with time. The contemporaneous increase in marine ^{187}Os/^{186}Os is compatible with the idea that increased exposure and weathering of the Himalaya and the Tibetan Plateau is responsible for the ^{3}He dilution by sediments. As most of the detrital sediment at the GPC3 site has been derived from windblown material from Asia, the sediment accumulation rates are indicators of the changing geomorphic conditions associated with the beginning of uplift prior to mountain building. The weathering of these continental rocks results in the increase in ^{187}Os/^{186}Os in the seawater. Thus for approximately the past 35 million years the ^{3}He signature and both the ^{87}Sr/^{86}Sr and ^{187}Os/^{186}Os are compatible in their messages. The rise and weathering of mountains increase the availability of sediments to the North Pacific Basin mainly through wind transport off the plateaus to the north of the Himalayas with soluble Os from the weathering of Antarctica and the Himalayas. This coupling implies a complex climatological response. Wind transport to the North Pacific from Asia was increasing even while the major source of continental Sr supply was from Antarctica. The proto-Himalayas provided a sufficient moisture shield to China to enhance aridity and eolian transport with the prevailing westerlies.

A different relationship between ^{3}He concentration and ^{187}Os/^{186}Os of leachable Os occurs from the 11.5-m depth to the Cretaceous–Tertiary boundary (~ 20.5 m). The general pattern appears to be a positive correlation of ^{3}He concentration with ^{187}Os/^{186}Os, whereas further up in the core it shows a negative correlation. The most striking feature is a sharp trough at about 12.2 m of both ^{3}He concentration and ^{187}Os/^{186}Os and we address this feature next.

If we assume that the interplanetary dust particles carrying the ^{3}He were depositing at a constant rate, this trough implies that there was a sharp increase in the accumulation of nonmeteoritic sediment at the time represented by the 12.2-m depth in the core. This depth corresponds to a period some 32 to 33 million years ago based on ichthyolith paleochronology and encompasses the Eocene–Oligocene boundary. At the same depth at which the ^{3}He concentration plummets, the ^{187}Os/^{186}Os ratios drop to a value of about 2.9, lower than the values of 4 to 6 (Fig. 3) on either side of the valley but not as low as the value of 1.9 at the Cretaceous–Tertiary boundary. There are other marked correlations of ^{3}He concentration and ^{187}Os/^{186}Os between 12 and 21 m, implying that there may be a significant common cause for both parameters. The fact that the low ^{187}Os/^{186}Os does not correlate with high Ir concentration as it does at the Cretaceous–Tertiary boundary (see above) implies that the cause of these changes must be attributable to terrestrial phenomena.

What phenomenon could increase the sediment accumulation rate over short periods, as evidenced by the ^{3}He concentration pattern, while at the same time lower the ^{187}Os/^{186}Os ratio of ocean water? We believe that it is continental plate collisions energizing explosive marginal volcanism on the one hand and the exposure of ancient oceanic crust as ophiolite sections at the same time. We impute the sharp increases in sediment accumulation to episodic increases of plate

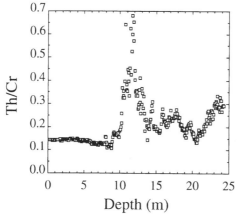

FIGURE 8. A plot of Th/Cr for GPC3 from the data of Kyte *et al.*[5] showing a continental crustal value for sediments from 0–11.5 m and higher values typical of acidic volcanic ash in the lower parts of the core.

margin volcanism of rhyolitic to andesitic compositions based on the chemistry of the sediments of GPC3. For example, Th/Cr is an index of the types of sedimentary components (Fig. 8): Low values, typical of the top 11 m of the core, are characteristic of the erosion of continental crust; higher values found below that depth are typical of acidic volcanic debris. The low [4]He concentrations of these imputed volcanogenic sedimentary layers as measured by Farley[11] are compatible with this source.

The collisions in question involve the closure of the Tethys Sea and subsequent further compression resulting in the string of volcanic mountains and exposed ophiolite sections at the ancient sutures. The source of $^{187}Os/^{186}Os$ decrease we then ascribe to the enhanced weathering of peridotites within exposed ophiolite sections. Pegram *et al.*[17] showed that the $^{187}Os/^{186}Os$ of streams draining ophiolite sections in California and Oregon were lower than the average continental crust (~ 10.5) attaining values as low as 1.4.

We therefore ascribe the changes in both ^{3}He concentration and $^{187}Os/^{186}Os$ throughout the Eocene and Paleocene as evidence of the supply of volcanic debris reaching the North Pacific Basin from the west (although augmented at times from the activity in North America as well) with the weathering of the ophiolites exposed at the suture formed at the site of the former Tethys Sea.

At the present time half of the Os supplied to the oceans comes from continental crust weathering with an $^{187}Os/^{186}Os$ value of around 10–15[16,17] and the other half from Os with an $^{187}Os/^{186}Os$ of 1. This proportion is inferred from the present-day seawater value of about 8.6. If continents did not contribute Os with a radiogenic signature to any great extent prior to the Oligocene as evidenced by the relatively low marine $^{187}Os/^{186}Os$ ratios, then any changes in ophiolite weathering intensity would be sensed directly by the $^{187}Os/^{186}Os$ of the contemporary oceans.

In summary, we believe that the ^{187}Os/^{186}Os and the ^{3}He concentration coupling prior to the Oligocene (imputed to be demarked by the 12.2-m level) in GPC3 shows the effect of plate collision on intense airborne volcanics and ophiolite exposure subject to weathering. The intense volcanism undoubtedly influenced the climatic regime much as Pinatubo's eruption in 1992 did for several years afterward. We assume that the frequency and ubiquity of dominantly rhyolitic explosive events persisted in modifying climate with consequent strong feedbacks that ultimately caused the climate to deteriorate at that time.

3.3. Oxygen Isotopes, ^{187}Os/^{186}Os, and ^{3}He Concentration Correlations

We have related both the ^{187}Os/^{186}Os and the ^{3}He concentration records to tectonic, volcanic, and weathering regimes. Coupling of these parameters to climate is done through the O isotope record of deep-sea deposits. GPC3 has no carbonate component so we translate depth in core in GPC3 to a chronometric record as described above. We then compare the Os isotope and ^{3}He records to the oceanic O isotope data observed from the study of carbonate deep-sea cores. The most comprehensive compilation of δ^{18}O data is that of Miller *et al.*[18] for Tertiary benthic foraminifera from the North Atlantic (Fig. 9). Similar curves have been constructed for the other oceans. The δ^{18}O curve of Fig. 9 implies that

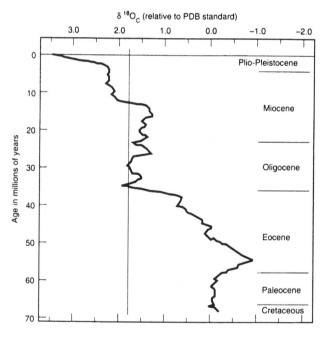

FIGURE 9.   Plot of δ^{18}O variation with time in benthic forams over the Cenozoic.[18] Similar plots have been obtained for cores from the other oceans. The relationship of δ^{18}O to the leachable ^{187}Os/^{186}Os curve of Fig. 3 is striking.

through the Eocene the ocean temperature was decreasing and that some storage of snow in Antarctica in the form of valley and piedmont glaciers was occurring. We contend that volcanism may have played a significant role in this climatic deterioration. Kennett[19] has discussed the role of volcanism in climate change through the Cenozoic. We do not think it is fortuitous that the sharpest increase in $\delta^{18}O$ occurred at the putative Eocene–Oligocene boundary at which massive volcanism and marine deposition lowered the ^3He concentration at the same time that a sharp drop in $^{187}Os/^{186}Os$ occurred because of a dominance of ophiolite exposure and weathering. Changes in ocean circulation and the location of continental masses have also aided in this climate deterioration.

As significant continental glaciation developed on Antarctica starting at the Eocene–Oligocene boundary, the climatic regime deteriorated further. The sedimentary regime in the North Pacific began to be dominated by eolian continental dust. As the elevation of the Himalayas started about 17 million years ago, the supply of continentally imprinted Sr and Os via streams also impacted the oceans. The effects were different for the Sr and Os isotopic signatures because of changes in the types of continental rocks exposed. The most recent epoch of change in weathering material type is seen by the accelerated $^{187}Os/^{186}Os$ increase about 7 million years ago. This Os isotopic change seems to be connected with the time of increased uplift of the Himalayas and the radical change of climate on the Asian and African continents owing to the establishment of the strong monsoonal circulation that has continued with oscillations in intensity to the present day.

To what extent the type and intensity of weathering has influenced the decrease in atmospheric CO_2 is subject to debate, but there is no doubt that increasing the intensity of weathering would have driven the CO_2 abundance in the atmosphere down unless compensated for by supply from other tectonically driven sources. The rise of the Himalayas would also have changed the water vapor distribution on the Earth as large desert areas were formed and the water vapor decreased over them. The decrease in the column of water vapor in large parts of the continents would allow more of the infrared radiation to escape the Earth's lower atmospheric layer and lead to further global cooling. This change in the abundance of atmospheric water vapor as a radiatively important gas may be as important an influence in climate deterioration as the drop in CO_2 as the result of weathering.

4. SUMMARY

We have proposed that the $^{187}Os/^{186}Os$ record in deep-sea sediments, tracking the changing oceanic composition, bears on the history of tectonism and climate during the Cenozoic. We have done this by combining $^{187}Os/^{186}Os$, $^{87}Sr/^{86}Sr$, ^3He concentration, and $\delta^{18}O$ data in deep-sea cores.

We infer that the close coupling of ^3He concentration and leachable $^{187}Os/^{186}Os$ in the Paleogene implies that sedimentation variation in the North Pacific is due to changes in the rate of volcanism and that the drop in the

$^{187}Os/^{186}Os$ ratio is due to intensified ophiolite weathering collaterally. We link both of these to the gradual closing of the Tethys and the consolidation of southern Asia.

We suggest the outrageous hypothesis that massive volcanism, when linked to changing continental and oceanic geography, cooled the Earth sufficiently to begin glaciation on Antarctica, culminating in a continental ice cap by the beginning of the Oligocene. The changing geography of southern Asia resulted in increased supply of windblown continental dust to the North Pacific, overwhelming the volcanic flux typical of the Eocene.

As the Himalayas began to rise and influence the isotopic chemistry of the oceans about 17 million years ago, there was a major change in slope of the $^{87}Sr/^{86}Sr$ and $^{187}Os/^{186}Os$ increase with time. About 7 million years ago a second marked increase in the $^{187}Os/^{186}Os$-versus-time plot occurred. This change in slope for Os corresponds to the increase in the uplift rate of the Himalayas and the establishment of a strong monsoonal climate for Asia and Africa as a consequence. The intense dissection of the Himalayas resulted in the weathering of high-Re rocks such as black shales with high $^{187}Os/^{186}Os$.

The monsoonal circulation affecting the water vapor distribution in the atmosphere may have contributed to the general cooling that started the Northern Hemisphere glacial cycles when coupled to the drop in CO_2 – a natural consequence of increased weathering owing to uplift. Additional effects of the increased weathering of black shales that started 7 million years ago would be the increase of the U concentration in the oceans as observed by Broecker and Peng[20] and a significant new delivery of phosphorus to the oceans. These effects were noted by Pegram et al.[9] If the latter occurred, it would act to lower CO_2 by increasing productivity, as suggested in another context by Broecker.[21]

Acknowledgments

This research was supported in part by the National Science Foundation, Earth Sciences Division. We thank Stan Hart and Greg Ravizza of the Woods Hole Oceanographic Institution for the use of the mass spectrometer on which these measurements were made.

REFERENCES

1. Luck, J. M., and Turekian, K. K. (1983). *Science* **222**, p. 613.
2. Turekian, K. K. (1982). Geological Society of America Special Paper 190, p. 243.
3. Ravizza, G., and Turekian, K. K. (1992). *Geophys. Res. Letters* **19**, p. 1383.
4. Ravizza, G., Turekian, K. K., and Hay, B. J. (1991). *Geochim. Cosmochim. Acta* **55**, p. 3741.
5. Kyte, F. T., Leinen, M. J., Heath, G. R., and Zhou, L. (1993). *Geochim. Cosmochim. Acta* **57**, p. 1719.
6. Mangini, A., Segl, M., Bonani, G., Hofmann, J. J., Morenzoni, E., Nessi, M., Suter, M., Wolfi, W., and Turekian, K. K. (1984). *Nucl. Instr. Meth. Phy. Res.* **B5**, p. 353.

7. Doyle, P. S., and Reidel, W. R. (1979). *Micropaleontology* **25**, p. 337.
8. Gottried, M. D., Doyle, P. S., and Riedel, W. R. (1984). *Micropaleontology* **30**, p. 71.
9. Pegram, W. J., Krishnaswami, S., Ravizza, G. E., and Turekian, K. K. (1992). *Earth Planet. Sci. Lett.* **113**, p. 569.
10. Pegram, W. J., and Turekian, K. K. (1997). *Geochim. Cosmochim. Acta* (in press).
11. Farley, K. A. (1995). *Nature* **376**, p. 153.
12. Berggren, W. A., Kent, D. V, Swisher, C. C. III, and Aubry, M-.P. (1995). Geochronology Time Scales and Global Stratigraphic Correlation, SEPM Sp. Pub. No. 54, p. 129.
13. Ravizza. G. (1993). *Earth Planet. Sci. Letters* **118**, p. 335.
14. Peucker-Ehrenbrink, B., Ravizza, G., and Hoffman, W. W. (1995). *Earth Planet. Sci. Lett.* **130**, p. 155.
15. Esser, B. K., and Turekian, K. K. (1988). *Geochim. Cosmochim. Acta* **52**, p. 1383.
16. Esser, B. K., and Turekian, K. K. (1993). *Geochim. Cosmochim. Acta* **57**, p. 3093.
17. Pegram, W. J., Esser, B. K., Krishnaswami, S., and Turekian, K. K. (1994). *Earth Planet. Sci. Lett.* **128**, p. 591.
18. Miller, K. G., Fairbanks, R. G., and Mountain, G. S. (1987). *Paleoceanography* **2**, p. 1.
19. Kennett, J. (1982). *Marine Geology*, 813 pp. Prentice-Hall, Englewood Cliffs, NJ.
20. Broecker, W. S., and Peng, T. H. (1982). *Tracers in the Sea*, 690 pp. Eldigio Press, Palisades, NY.
21. Broecker, W. S. (1982). *Geochim. Cosmochim. Acta* **46**, p. 1689.

18

Global Chemical Erosion during the Cenozoic: Weatherability Balances the Budgets

Lee R. Kump and Michael A. Arthur

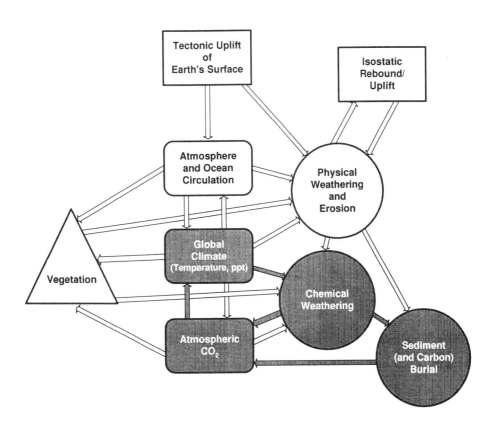

Lee R. Kump and Michael A. Arthur • Department of Geosciences and Earth System Science Center, Pennsylvania State University, University Park, Pennsylvania 16802.

Tectonic Uplift and Climate Change, edited by William F. Ruddiman. Plenum Press, New York, 1997.

1. INTRODUCTION

The question addressed here is whether global chemical weathering and erosion rates have increased over Cenozoic time in response to uplift of the Himalayas.[1,2] Chemical weathering of the continents is a process whereby carbonic acid (derived from the atmosphere or from soil respiration and decomposition) is consumed. Thus it represents a sink for atmospheric CO_2 and provides one link between uplift and climate change. The prevailing hypothesis concerning the cause of the long-term cooling of Cenozoic climates states that as a result of increased rates of silicate weathering, atmospheric CO_2 levels have fallen, the magnitude of the greenhouse effect has been reduced, and thus the globally averaged climate has cooled. Such a scenario is consistent with the secular record of seawater Sr isotopic composition preserved in carbonates[1,3-6] (Fig. 1a), which becomes increasingly radiogenic with time through the Cenozoic. This trend implies a growing influence of continental weathering, relative to seafloor hydrothermal activity, on the Sr isotopic composition of seawater.[7]

Without elaboration, however, the hypothesis is inconsistent with constraints imposed by mass balance.[8,9] As can be easily demonstrated with models of the carbonate–silicate geochemical cycle, rates of silicate weathering must be balanced, on timescales greater than 10^4 years, by rates of volcanic and metamorphic release of CO_2, so that the atmospheric–oceanic CO_2 reservoir does not experience excessively large fluctuations.[10-13] The indication from calculated rates of seafloor production and spreading[14,15] and subduction[16] is that the rate of

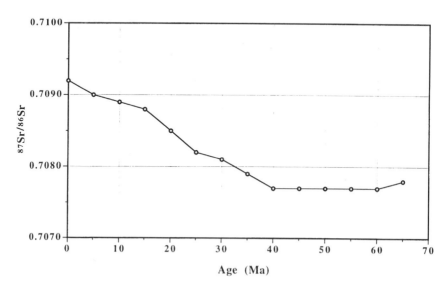

FIGURE 1a. Sr isotopic composition of Cenozoic carbonates. The curve represents a 5-million-year interpolation of various sources of data and indicates the resolution of the numerical calculations presented here.

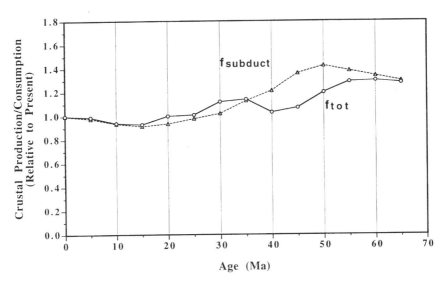

FIGURE 1b. Global ocean crustal production (f_{tot}; data from Larson,[15] modified by Tethys contributions, Scotese *et al.*[40]) and subduction rate ($f_{subduct}$; data from Engebretson *et al.*[16]). Curves are presented as "factors" relative to the present rates.

volcanism (and release of CO_2) has in general decreased, rather than increased, through the Cenozoic (Fig. 1b). Thus, unless it can be demonstrated that the CO_2 flux associated with collision and subduction has increased through the Cenozoic despite a decrease in the rate of subduction,[17-19] and that these compensatory mechanisms have been tightly coupled to chemical erosion rates, we must reject the hypothesis of increased *global* weathering rates as the cause of Cenozoic cooling and of the Sr isotope trend toward more radiogenic values. However, rates of chemical weathering and erosion probably did increase in the Himalayan region following the collision of India and Asia during the Tertiary. Mass balance then requires that chemical erosion rates decreased elsewhere. This teleconnection was likely provided through atmospheric CO_2 and its greenhouse effect on climate, and thus, weathering.

Mass-balance models of the C and Sr cycles can be used to explore alternative hypotheses for the secular trends in seawater Sr isotopic composition. Two possibilities remain: either the flux of nonradiogenic Sr from seafloor hydrothermal alteration has decreased, or the riverine $^{87}Sr/^{86}Sr$ has increased through the Cenozoic.

The evidence for only modest reductions in rates of seafloor spreading, production and subduction, cited above, does not support the first of these hypotheses. Further arguments against it can be made based on the modeling efforts of François and Walker,[12] who were unable to match the Cenozoic Sr isotopic record with a process-based model that included weathering and seafloor hydrothermal processes. They concluded that something was fundamentally wrong with the paradigmatic concept of regulation of atmospheric CO_2 on long timescales by the carbonate–silicate geochemical cycle, and proposed that sea-floor basalt weathering, via a purported dependence on the pH and ΣCO_2 of bottom waters, provides the feedback that regulates atmospheric CO_2. This conclusion was challenged by Caldeira,[20] who pointed out that the required dependence of silicate weathering on pH did not exist in the normal pH range expected for deep marine environments.

If, indeed, increased weathering rates and decreased hydrothermal activity were not responsible for the Cenozoic Sr isotopic trend, we are left with increased riverine $^{87}Sr/^{86}Sr$ as the cause.[21,22] Berner and Rye[23] developed a Sr model that was forced by the C-cycle model results of Berner[24] and allowed for variability in the weathering contribution of Sr from radiogenic and nonradiogenic source rocks. Their model results demonstrated that the Sr isotopic trend could be reproduced without violation of mass-balance constraint, while preserving the carbonate–silicate cycle as the feedback mechanism for atmospheric CO_2

Our modeling is an outgrowth of the approach taken by Berner and Rye.[23] As they did, we allow the riverine Sr isotopic composition to vary, based on the requirements of C and Sr mass and isotope balance. Unlike them, we use a fully coupled model of the C and Sr cycles, similar to the approach taken by Goddéris and François[25] in their study of the Cenozoic Sr trend. Riverine Sr isotopic compositions vary in two ways in the model presented herein: (1) by adjustments in the basalt–granite ratio in the silicate weathering flux calculation based on the riverine Sr isotopic composition required to balance the Sr cycle, and (2) by adjustments in the relative proportions of carbonate–silicate exposure areas, and thus weathering rates. The former has no effect on the C cycle, whereas the latter affects the net global rate of CO_2 consumption (because carbonate weathering does not provide a long-term sink for CO_2) and thus the steady-state atmospheric pCO_2. Goddéris and François[25] did not allow the riverine Sr isotopic composition to vary in this way. Their study instead focused on the sensitivity of the Sr isotopic composition of the oceans to various uplift scenarios for the Himalaya and the rest of the world, and explored the link between mechanical and physical weathering. The model we present is able to reproduce the Cenozoic Sr isotope trend without violating C balance constraints. Because we can predict pelagic and global carbonate deposition rates, we are able to test our model results against observed changes over Cenozoic time in the oceanic carbonate compensation depth (CCD), inferred changes in the global $CaCO_3$ burial rate, and observed changes in the Sr/Ca ratio of marine carbonates.

2. DESCRIPTION OF THE MODEL

The calculations presented here are based on a coupled C–Sr model in its simplest construct. Successive steady states are determined based on mass and isotopic balance at each time-step (duration of 5 million years). In this description we begin with the present-day steady state, needed to establish rate constants and baseline fluxes, and then introduce the form of the flux relationships and the forcing functions for the model, which allow us to perform calculations for the geological past.

2.1. Present-Day Steady State

In constructing the steady-state configurations of the C and Sr cycles, we have not attempted to provide new estimates of modern or preindustrial fluxes. Rather we have used "round numbers" that are generally consistent with values presented elsewhere (e.g., Berner[24]). We risk little by taking this approach, because what we are really interested in are trends through time and not absolute values of the fluxes. The general applicability of the chosen values is further substantiated by their ability to satisfy constraints of both mass and isotopic balance.

The values we have chosen for the C cycle are presented in Fig. 2. Four reservoirs of C are included: (1) organic C in shales, (2) carbonate C in limestones and dolomites, (3) atmospheric CO_2, and (4) oceanic HCO_3^-. The fluxes are chosen such that the input to and output from each reservoir are balanced.

Our version of the present-day steady-state Sr cycle (Fig. 3) has been adapted from Veizer.[26] Values for the isotopic compositions of the overall riverine ($^{87}Sr/^{86}Sr = 0.7101$) and hydrothermal ($^{87}Sr/^{86}Sr = 0.7035$) fluxes and seawater ($^{87}Sr/^{86}Sr = 0.7092$) are used directly. We neglect the diagenetic flux of Sr from

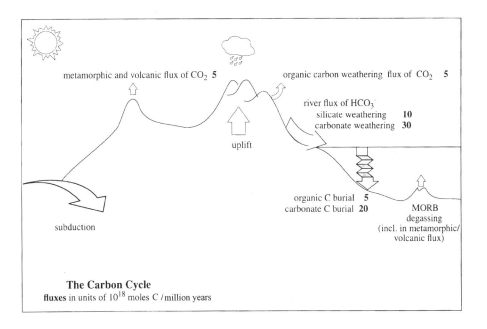

metamorphic and volcanic flux of CO_2 **5**

organic carbon weathering flux of CO_2 **5**

river flux of HCO_3^-
silicate weathering   **10**
carbonate weathering   **30**

uplift

organic C burial   **5**
carbonate C burial   **20**

MORB
degassing
(incl. in metamorphic/
volcanic flux)

subduction

The Carbon Cycle
fluxes in units of 10^{18} moles C / million years

FIGURE 2.   The global, long-term, present-day, steady-state C cycle. Fluxes are shown in bold.

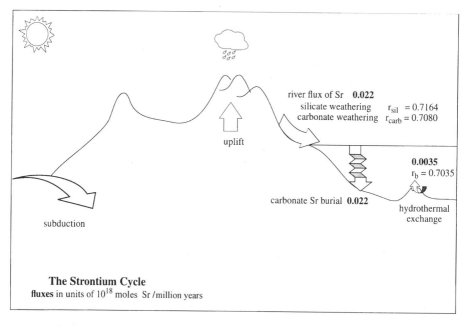

FIGURE 3.   The global, long-term, present-day, steady-state Sr cycle. Fluxes are shown in bold.
$^{87}Sr/^{86}Sr$ ratios are identified as r_i.

sediments; its isotopic value is always close to the oceanic value, giving it little leverage to effect isotopic change.[22] Other flux values are set by data and mass-balance constraints, as discussed below.

The value for the riverine flux given by Veizer (3.1×10^{10} moles/year) requires, at steady state, an average carbonate sedimentation Sr/Ca ratio of 1.9×10^{-3} (assuming a carbonate flux of 20×10^{12} moles/year and a Ca/(Ca + Mg) ratio of 0.8). If his estimate of the groundwater flux of Sr is added, the required ratio becomes even larger (3.3×10^{-3}), much larger than the observed ratio for pelagic carbonates of about 1.4×10^{-3} (Graham *et al.*[27]). The question becomes, then, should we match the steady-state model to the pelagic Sr/Ca ratio, or accept Veizer's fluxes and begin with a mismatch in Sr/Ca ratio, which complicates later comparison with the pelagic record.

There is considerable uncertainty in the groundwater flux, as it is based on only a few observations.[28] The flux may not be nearly so large, but it may in fact be larger, requiring a Sr/Ca ratio for sedimenting carbonates that is larger than the pelagic value. Indeed, the average Sr/Ca ratio of sedimenting carbonate is undoubtedly larger than the pelagic carbonates ratio, because the global average includes platform and reefal carbonates, many of which were originally aragonitic and thus enriched in Sr. Nevertheless, for purposes of later comparison, we match the sedimentation flux to the present-day Sr/Ca ratio in pelagic sediments by specifying a riverine flux of 2.2×10^{10} moles Sr/year. We will return, however, to the significance of the Sr flux associated with shallow-water carbonate deposition.

The riverine flux of Sr is derived from the weathering of carbonate- and silicate-bearing rocks. For the present-day we specify that 75% of the Sr in streams is derived from carbonates and 25% from silicate rocks, after Holland.[29] The carbonate isotopic value is set to $^{87}Sr/^{86}Sr = 0.7080$, and is held at this value for the model calculations performed. The silicate contribution, which must yield an average isotopic composition of 0.7164 for reasons of isotope balance, stems from both radiogenic ($^{87}Sr/^{86}Sr \cong 0.720$) and nonradiogenic ($^{87}Sr/^{86}Sr \cong 0.706$) sources. Changes in the relative proportions of these two sources will drive temporal variations in the riverine isotopic composition in our model; for the present steady state we require that 24% of the total silicate weathering flux be derived from nonradiogenic sources. These proportions and isotopic values are reasonable given our current understanding of weathering rates[30,31] and isotopic compositions of igneous rocks.[32]

The rate of hydrothermal exchange (0.35×10^{10} moles/year) is set by the steady-state constraint. This value is well within the range of values given by Veizer.[26] In keeping with previous workers' observations,[33,34] we assume that the interaction of seawater with oceanic crust at high temperatures modifies the Sr isotopic composition of the fluid, without providing a net source or sink for Sr. The fluid enters with seawater isotopic composition and leaves with that of the basalt ($^{87}Sr/^{86}Sr = 0.7035$).

2.2. Flux Relationships

The predictive ability of geochemical cycling models, i.e., the sensitivity to external forcing factors, resides in the prescribed functional relationships for the fluxes between reservoirs. Following Berner,[11,24,35] we express these dependencies using *correction factors* that are scaled to present-day values. The effect of changes in these factors is assumed to be linear, given the lack of any information to support specific nonlinear relationships. Changes in land area[36] available for weathering (f_a; Fig. 4) are modified for differential exposure of carbonate versus silicate rocks using the paleogeological work of Bluth and Kump,[37,38] expressed as $f_{a_{carb}}$ (Fig. 4).

As others have done before, we presume that changes in seafloor production rate are a proxy for subduction-related CO_2 degassing and midocean ridge hydrothermal exchange rates. Bickle[39] questions this relationship and argues that much of the CO_2 degassing must be related to carbonate metamorphism to explain the relatively nonradiogenic Sr isotopic composition of silicate sediments. If so, our calculations will miss any large atmospheric CO_2 fluctuation caused by metamorphic events decoupled from rates of ocean crust production.

We use the compilation of Larson,[15] amended to account for seafloor production in the Tethys Sea.[40] This is somewhat of a departure from the approach used by Berner in his GEOCARB modeling.[24,35] Because he addresses problems in the Paleozoic, for which direct measurements of past rates of seafloor

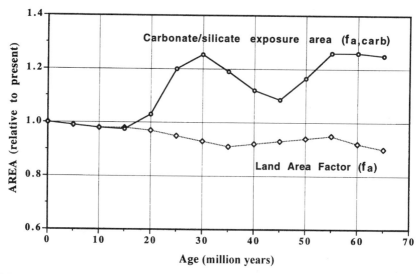

FIGURE 4. Curves showing relative changes in land area (f_a, after Barron et al.[36]) and carbonate exposure area ($f_{a_{carb}}$, after Bluth[37]) for the Cenozoic, expressed as factors relative to the present values.

production cannot be made based on seafloor magnetic anomalies, Berner uses various multiplicative factors of Gaffin's[41] inversion of the eustatic sea-level curve as a proxy for seafloor spreading and production throughout the Phanerozoic. Because Gaffin uses first-order sea-level variation in his inversion, details of shorter-term ($\sim 10^6$ years) crustal production rate changes may be missed.

Another factor that drives the model is the organic C fraction of the total C sedimentation rate, f_{org}, which is calculated using steady-state isotope balance constraints:

$$f_{org} = \frac{\delta_o - \delta_w}{\Delta^{13}C} \tag{1}$$

Here δ_o is the oceanic value of $\delta^{13}C$, as monitored by the average $\delta^{13}C$ of pelagic carbonates (Fig. 5a), δ_w is the riverine value, and $\Delta^{13}C$ (Fig. 5c) is the globally averaged difference in isotopic composition between contemporary carbonate (Fig. 5a) and organic C (Fig. 5b) in sediments. In the calculations we hold the riverine value constant at -4 per mil, a value that gives the canonical 4:1 ratio for carbonate:organic C weathering for the present. [29] We use our own unpublished compilation of isotopic compositions of limestones and kerogens for the late Mesozoic and Cenozoic to generate the age curve for $\Delta^{13}C$. Changes in this value are thought to have arisen from a number of factors, including differing dominant source contributions (e.g., plant or algal type), diagenetic effects, or

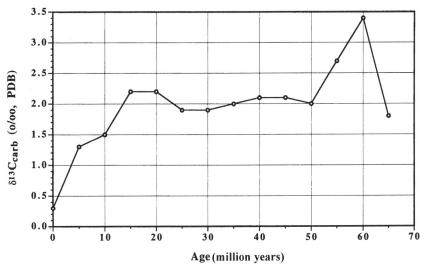

FIGURE 5a.   Globally averaged, whole-rock, C isotopic composition of marine pelagic carbonates ($\delta^{13}C_{carb}$; unpublished compilation) through the Cenozoic at 5-million-year intervals, as used in the model.

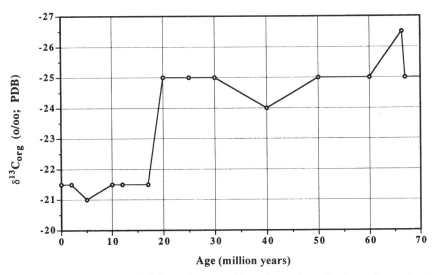

FIGURE 5b.   Globally averaged C isotopic composition of organic matter in marine sediments ($\delta^{13}C_{org}$; unpublished compilation) through the Cenozoic at 5-million-year intervals, as used in the model.

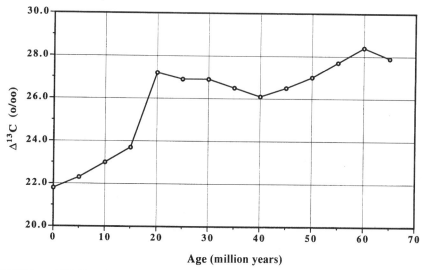

FIGURE 5c. $\Delta^{13}C$ (the difference between carbonate and organic carbon) for the Cenozoic, based on the data in Fig. 5a,b.

physiological effects.[42-46] Current research demonstrates that physiological effects are often most important.[45-47] Specifically, the effect of growth rate and ambient CO_2 concentration variations are invoked to explain variability in $\Delta^{13}C$.

Weathering rates of carbonate and silicate rocks are assumed to depend on climatic factors, especially temperature and net precipitation differences. The effects are captured in the correction factor f_{wr}. As shown below, f_{wr} is determined by the model through inversion of the C mass-balance equations; it is not a forcing factor.

Weathering, metamorphic/volcanic, and hydrothermal exchange fluxes are calculated as the product of the present-day rates shown in Figs. 2 and 3 (F_{ji}^0) and the appropriate correction factors. For carbonate and silicate weathering rates the expressions are of the form

$$F_{w_{carb}} = F_{w_{carb}}^0 \times f_{a_{carb}} \times f_a \times f_{wr}$$

and

$$F_{w_{sil}} = F_{w_{sil}}^0 \times (1 - f_{a_{carb}}) \times f_a \times f_{wr}$$

The weathering of organic C in shales is assumed to be independent of CO_2-related effects on weathering rate[48]

$$F_{w_{org}} = F_{w_{org}}^0 \times f_a$$

Metamorphism of carbonates and release rates of CO_2 are expressed as

$$F_{m_{carb}} = F^0_{m_{carb}} \times f_{tot}$$

The exchange of Sr isotopes at the midocean ridges are both assumed to vary linearly with f_{tot} (e.g., $F_{x_{Sr}} = F^0_{x_{Sr}} \times f_{tot}$).

2.3. Calculating Steady States

Given the flux relationships described above we are able to calculate steady states for the C–Sr cycles for modified conditions of the correction factors, that is, for times in the past when land area, differential rock exposure, and seafloor production rate were different than they are today. The forcing functions are thus $f_a, f_{a_{carb}}$, and f_{tot}.

Knowledge of these factors, combined with steady-state assumptions for the CO_2 content of the atmosphere and HCO_3^- content of the ocean, allows us to calculate f_{wr}. In discussion of these assumptions, we begin with the steady-state expressions for oceanic HCO_3^- and atmospheric pCO_2. The first of these expressions indicates that the rate of carbonate deposition in the world ocean must compensate for 50% of the riverine flux of carbonate and silicate weathering-derived bicarbonate. The other 50% is converted to CO_2, which in essence returns to the atmosphere during carbonate precipitation, based on the following stoichiometry:

$$Ca^{2+} + 2HCO_3^- \rightarrow CaCO_3 + CO_2$$

$$\tfrac{1}{2} f_A f_{wr} (F^0_{w_{carb}} + F^0_{w_{sil}}) = F_{d_{carb}} \qquad (2)$$

(Carbonate and silicate weathering = Carbonate deposition)

(In this representation we simplify the arithmetic by neglecting $F_{a_{carb}}$). The steady state for atmospheric pCO_2 expresses the balance between consumption during weathering and organic C deposition ($F_{d_{org}}$) and production during metamorphism and organic C (in shales) weathering.

$$F_{d_{carb}} + f_{tot} F^0_{m_{carb}} + f_a F^0_{w_{org}} = \tfrac{1}{2} f_a f_{wr} (F^0_{w_{carb}} + 2F^0_{w_{sil}}) + F_{d_{org}} \qquad (3)$$

| Carbonate + | Carbonate | + organic | = Carbonate and | + Organic C |
| deposition | metamorphism | C weathering | Silicate weathering | deposition |

Equations (2) and (3) can be combined to yield an expression for f_{wr}. This expression can be further simplified by recognizing that

$$F_{d_{org}} = \frac{f_{org} F_{d_{carb}}}{1 - f_{org}}$$

The result is

$$f_{wr} = \frac{f_{tot}F^0_{m_{carb}} + f_a F^0_{w_{org}}}{\frac{1}{2}f_a\{F^0_{w_{sil}} + [f_{org}/(1 - f_{org})](F^0_{w_{sil}} + F^0_{w_{carb}})\}} \tag{4}$$

an expression that is dependent only on forcing factors and flux values that have already been established. Expressed in this fashion, f_{wr} is seen to be the factor that ensures that silicate weathering and organic C burial, as CO_2 sinks, balance the CO_2 supply from carbonate metamorphism, volcanism, and organic C weathering.

This factor has been linked theoretically and observationally to atmospheric CO_2, so we can use our calculation of f_{wr} to estimate paleo-CO_2 levels. Berner et al.[11] combined calculated global temperatures and runoffs from global climate model experiments run under conditions of elevated pCO_2 with observations of modern riverine chemistry as functions of runoff and temperature to estimate this relationship. Volk[49] extended this to include "CO_2 fertilization," i.e., the relationships among atmospheric CO_2, terrestrial plant productivity, and soil pCO_2. The strength of the dependence of f_{wr} on atmospheric pCO_2 in the Volk relationship is stronger than that of Berner et al. but weaker than a simple linear relationship.[50] We thus will use the Volk formulation (Fig. 6) to calculate pCO_2 from f_{wr}, both because it explicitly includes a CO_2 fertilization effect, and because it is intermediate in its strength of feedback.

At this point in the description it is appropriate to introduce the Sr cycle

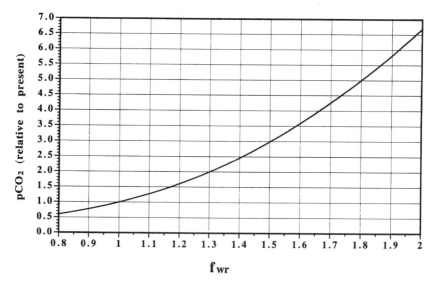

FIGURE 6.   The relationship between the climatic effect on chemical weathering (f_{wr}, normalized to the present-day's effect) and atmospheric pCO_2, based on the relationship developed by Volk.[49] This curve is used to calculate paleo-pCO_2 levels based on calculated values of f_{wr}.

calculations. The Sr isotope balance equation of Kump[22] can be rearranged to solve for the riverine $^{87}Sr/^{86}Sr$ (r_r)

$$r_r = \frac{r_0 - B}{1 + B} \tag{5}$$

where r_0 is the oceanic $^{87}Sr/^{86}Sr$. B is a term that represents the relative effect of hydrothermal isotope exchange on r_0 compared to that of riverine input and is defined as

$$B = \frac{r_b - r_0}{1 + r_b} \frac{f_{tot}F^0_{xSr}}{f_a f_{wr} F^0_{wSr}} \tag{6}$$

where r_b is the isotopic composition of seafloor basalt, and all other parameters have been defined previously. In the model, then, the riverine isotopic composition r_r is set by the requirement for isotopic and mass balance in the Sr system. Its value depends only on specified parameters and f_{wr}. Note that B is a negative number; r_b has been less than r_0 through the Cenozoic. Therefore, if the hydrothermal flux decreases relative to the weathering flux, B becomes less negative, and if r_r remains constant, the oceanic ratio r_o increases. If, however, B is only changing slightly, then an increase in r_o signals an increase in r_r.

3. RESULTS

In this section we present the calculated changes in organic and carbonate C deposition rates, riverine Sr isotopic composition, and atmospheric pCO_2. These changes result from the imposition of temporal variations in volcanic/metamorphic CO_2 outgassing (f_{tot}) and silicate and carbonate exposure area (f_a and $f_{a_{carb}}$), and are consistent with the observed secular trends in oceanic $\delta^{13}C$ and $^{87}Sr/^{86}Sr$.

3.1. Organic and Carbonate C Deposition

Equation (1) and the data shown Fig. 5 are used to calculate changes in f_{org}, the fraction of organic C buried relative to total C burial, through the Cenozoic (Fig. 7). In this figure we compare calculations made with constant $\Delta^{13}C$ (assuming a constant value of 25 per mil) and variable $\Delta^{13}C$ derived from our $\delta^{13}C$ data compilation. The differences are not great but are neverthless significant. In particular, with constant $\Delta^{13}C$ the isotopic trends do require a substantial decrease in f_{org} over the Cenozoic, whereas with variable $\Delta^{13}C$ there is no apparent trend (although values do fluctuate).

The rates of organic and carbonate C deposition (expressed as fluxes rather than as proportions; Fig. 8a) are determined by solving the full set of equations. The rate of carbonate deposition declines somewhat over the Cenozoic in

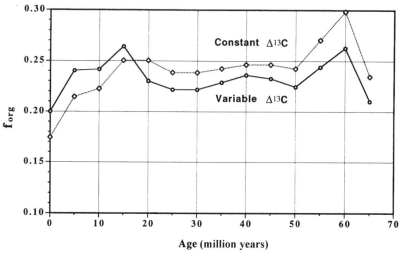

FIGURE 7.   Curves of f_{org} (the fraction of C output as organic C) through the Cenozoic, comparing calculations based on a constant $\Delta^{13}C$ ($-25\%_0$) to calculations based on variable $\Delta^{13}C$ (Fig. 5c).

response to a general decrease in the rates of carbonate and silicate weathering (Fig. 8b). The rate of organic C burial also decrease somewhat through the Cenozoic, mainly as a function of relatively constant f_{org} but decreasing total C fluxes from weathering. The important result, however, is that there is very little change in either of these fluxes through the Cenozoic.

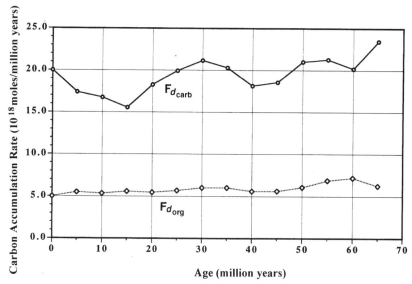

FIGURE 8a.   Model output showing global carbonate ($F_{d_{carb}}$) and organic C ($F_{d_{org}}$) accumulation rates for the Cenozoic, expressed in units of 10^{18} moles C/million years.

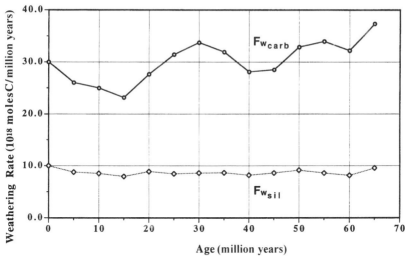

FIGURE 8b. Model output showing global carbonate ($F_{w_{carb}}$) and silicate ($F_{w_{sil}}$) weathering rates for the Cenozoic, expressed in units of 10^{18} moles C/million years.

3.2. Riverine Sr Isotopic Composition

For reasons of mass balance in the C cycle, the model calculates no increase in carbonate and silicate weathering through the Cenozoic. In the model, then, the observed rise in the oceanic value of $^{87}Sr/^{86}Sr$ is accommodated not by increased rates of continental weathering but, instead, by a shift in the isotopic composition of the weathering input (r_r). The required change in this value is from about 0.7088 in the early Cenozoic to 0.7092 today (Fig. 9). According to these calculations, rivers in the early Cenozoic were markedly less radiogenic than at present. We have presumed in the calculation that changes in riverine isotopic composition result from changes in the relative exposure areas of carbonates and silicates ($f_{a_{carb}}$) and from changes in the isotopic composition of weathering silicates (r_{sil}), rather than from a change in the isotopic composition of weathering carbonates, which we fix at $r_{carb} = 0.708$. The change in r_{sil} required to account for the change in r_r is from about 0.711 early in the Cenozoic to 0.716 today.

In the next section we will consider the C–Sr model results in terms of the observed record of oceanic and atmospheric composition and climate in the Cenozoic.

4. IMPLICATIONS FOR CENOZOIC OCEANS AND CLIMATE

4.1. Carbonate Partitioning and the CCD

At first glance the lack of a trend in carbonate accumulation calculated by the model is at odds with inferences based on trends of increasing pelagic

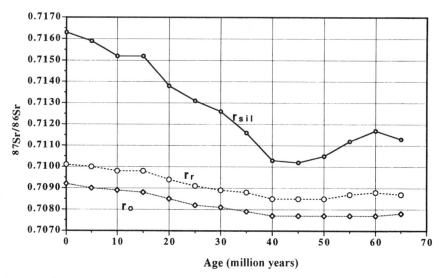

FIGURE 9.   Curves for the Cenozoic Sr isotopic composition of the model global average river Sr flux (r_r), the Sr isotopic composition of the oceans (r_o), and the average Sr isotopic composition of silicate rocks being weathered to produce the riverine value, assuming that the carbonate contributions have a constant Sr isotopic ratio, $r_{carb} = 0.708$.

carbonate accumulation rates[51] and CCD deepening over the Cenozoic (see Delaney and Boyle[52] for a general CCD curve). However, these inferences neglect the substantial fraction of carbonate deposition that takes place in hemipelagic and shelf settings.[53,54] On million-year timescales, the ocean can be considered as being very near saturation with respect to $CaCO_3$; what goes in (in terms of alkalinity) must come out, as the ocean has very little capacity for alkalinity storage. Any tendency toward decreased deposition in shallow-water settings will upset this balance, leading to an increase in oceanic alkalinity (presuming there is no coincidental decrease in riverine delivery). Increased alkalinity will tend to drive the CCD to greater depths. This creates a greater area of seafloor accumulating pelagic carbonates, causing an increase in accumulation rates that offsets the decrease in shallow-water settings and restores the alkalinity balance of the ocean.

We use the paleogeographic estimates of areas of low-latitude, carbonate shelves through time from Opdyke and Wilkinson[53] (Fig. 10a) to partition the total carbonate accumulation (Fig. 8a). To do so we need to specify vertical accumulation rates for the shelf and pelagic settings. For the shelf rate we use a value of 25 m/million years (the average Phanerozoic rate reported by Opdyke and Wilkinson[53]) and a porosity of 10%. For the pelagic rate we used a range of values from 4–12 m/million years and a porosity of 50%, with the "preferred" rate of 8 m/million years corresponding to an average carbonate rain rate of $1.2 \, g/cm^2/$ year (well within the range of values given by Broecker and Peng[55]). The result (Fig. 10b) is a Cenozoic trend of increasing pelagic and decreasing shelf carbonate

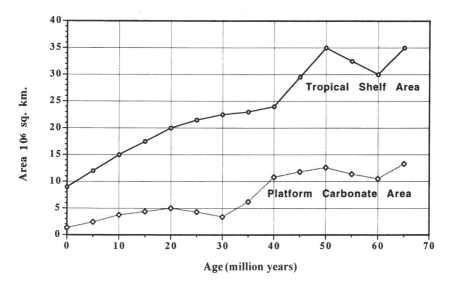

FIGURE 10a. Tropical shelf area and platform carbonate area at 5-million-year intervals for the Cenozoic based on the compilation of Opdyke and Wilkinson.[53] The platform carbonate area is used by us to partition accumulation rates between platform (shelf) and deep-sea (pelagic) settings.

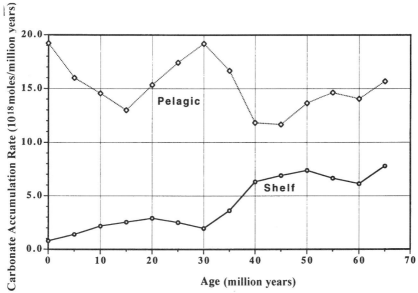

FIGURE 10b. Calculated carbonate accumulation rates on shelves and in deeper pelagic environments, with fluxes expressed in units of 10^{18} moles C/million years.

accumulation, a result broadly consistent with observations of pelagic carbonate accumulation.[51] This shift from shelf to basinal carbonate accumulation through the Cenozoic was a consequence of the reduction in tropical shelf area caused by eustatic sea-level fall, coupled with paleogeographic and paleoenvironmental factors that favored tropical carbonate shelves earlier in the Cenozoic.[53]

These changes in pelagic carbonate accumulation also have consequences for CCD variations. Using the vertical accumulation rate and porosity for pelagic environments noted above, one can calculate the area of seafloor required to accommodate a given pelagic carbonate accumulation rate. Then the hypsometric curve can be used to estimate the depth in the ocean that has a cumulative area at shallower depths that is equal to the area required. (We use today's hypsometric curve in lieu of an estimate of how this has changed in the geological past.) This is the depth we call the CCD although in fact it represents a depth that would fall between the lysocline and the CCD.

Despite the rather restrictive simplifications of constant vertical accumulation rate and seafloor hypsometry, the comparison between observed and modeled CCD variations is quite satisfying (Fig. 11). Not only is the trend of Cenozoic CCD deepening matched, but so too are the second-order patterns of shoalings and deepenings. Figure 11 shows a bracketing of this curve by vertical accumulation rates 50% larger or smaller than the "best fit" value (i.e., the 4, 8, and 12 m/million years values cited above). The 8 m/million years value for average pelagic carbonate accumulation produces the "best fit" to the Cenozoic

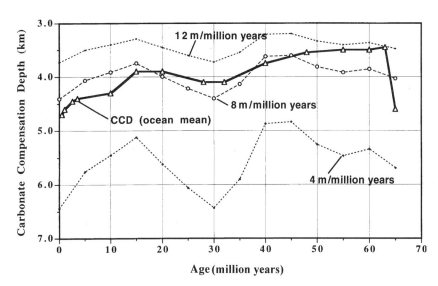

FIGURE 11.   Cenozoic average carbonate compensation depth (CCD) variations compiled from the literature (solid line labeled "CCD ocean mean"; unpublished compilation) compared to the model predicted CCD. The sensitivity to assumed constant rate of pelagic carbonate sedimentation per unit area is demonstrated by the dashed curves, which compare calculations made using 4, 8, and 12 m/million years vertical accumulation rates at 50% average porosity.

CCD curve. The trend does not require changing the average accumulation rate during the Cenozoic, something that was entertained by Opdyke and Wilkinson[53] on the basis of the recalculation of the Davies of Worsley[51] data set. This correspondence between model and data indicates that shelf–basin partitioning is the dominant factor affecting Cenozoic variations in CCD depth. Other factors, such as fluctuations in riverine alkalinity delivery or vertical accumulation rates have been of secondary importance (and small). Certainly a large increase in pelagic carbonate accumulation, required by current "conventional wisdom" regarding the relationship between orogeny and chemical erosion, is inconsistent with the CCD trend after consideration of the effects of shelf–basin partitioning. Note that the effect of increasing the pelagic accumulation rate would be to shoal the CCD, all other factors being constant (Fig. 11).

4.2. Sr Mass Balance and Sr/Ca Ratios

Another test of the model is to compare the required ratio of Sr to Ca deposition in the model to the Sr/Ca ratio of Cenozoic foraminifera (Fig. 12). This test is made possible by the fact that the major sink for both Sr and Ca is carbonate sedimentation (Sr is a minor component of biogenic and abiogenic carbonates). Thus at steady state the Sr/Ca ratio of sedimenting carbonates is set, once corrections are made for the Ca added to seawater as the result of

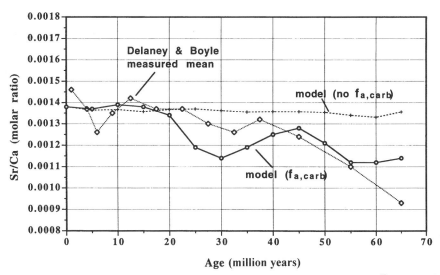

FIGURE 12.   Sr/Ca molar ratio data for Cenozoic foraminifera ("Delaney and Boyle[52] measured mean" shown as thin, solid line) compared to model calculated values for the total carbonate output. The curve labeled "model ($f_{a_{carb}}$)" was generated by the model using ($f_{a_{carb}}$) (the relative exposure area of carbonate rocks compared to silicate rocks), while the curve labeled "model (no $f_{a_{carb}}$)" presumes no change in the relative exposure area of carbonate and silicate rocks through the Cenozoic.

hydrothermal exchange for Mg. If we assume that the Ca/Mg ratio in weathering carbonates and silicates is 0.8 and 0.5, respectively (after Holland[29] and Berner and Berner[56]), then steady state requires a hydrothermal flux of Ca of 1.5×10^{18} moles/million years. For times in the past the hydrothermal flux of Ca (F_{xCa}) is presumed to be the product of this value and f_{tot}. The Sr/Ca ratio in sedimenting carbonates then becomes the ratio of the input flux of Sr from rivers to the total input rate of Ca from rivers and hydrothermal activity:

$$\frac{f_a f_{wr} F^0_{wSr}}{0.5 f_a f_{wr}(0.8 F^0_{w_{carb}} + 0.5 F^0_{w_{sil}}) + 1.5 f_{tot}} = (Sr/Ca)_{carb} \tag{7}$$

The calculation of $(Sr/Ca)_{carb}$ compares favorably with observations of the Sr/Ca ratio in planktonic foraminifera[27,52] (Fig. 12). Both curves show a general increase in the ratio through the Cenozoic, with a dip in the ratio centered on 30 million years. The correspondence between the two curves is largely the consequence of varying the carbonate/silicate exposure area ($f_{a_{carb}}$) in the weathering calculations; if $f_{a_{carb}}$ is fixed at 1.0 (i.e., equal to the current ratio) the predicted Sr/Ca ratio fluctuates between a restricted range of values (from 0.00133 to 0.00138). The implication is that the trend in Sr/Ca ratio through the Cenozoic is a paleogeological signal, the result of a decrease in the rate of carbonate weathering relative to silicate weathering. The match between the curves lends some credence to the general approach taken by Bluth and Kump[38] to estimate the geographical distribution of weathering lithologies in the geological past.

The fit is surprising, given that there was no consideration of the significant difference in Sr/Ca ratio between pelagic carbonates (represented by the observations) and shelf carbonates.[57] If shallow-water carbonates (originally aragonite and thus Sr-rich) preserved their high Sr/Ca ratio there should be a large misfit between our calculated, overall Sr/Ca ratio (which represents the sum of shelf and pelagic carbonate deposition) and the observed ratio in pelagic foraminifera. The correspondence between these curves suggests that, after early diagenesis and upon burial, i.e., on longer timescales, the difference in Sr/Ca between shelf and pelagic carbonates is sufficiently small as to have little differential effect on the Sr/Ca ratio of seawater.

4.3. Atmospheric pCO₂

The C mass-balance equations allow for the calculation of the climate–weathering factor f_{wr} via Eq. (4). Then, using the relationship between f_{wr} and atmospheric pCO₂ shown in Fig. 6, one can calculate the inferred changes in atmospheric pCO₂ resulting from the forcings applied to the model (Fig. 13). The result is that very little change is calculated for the Cenozoic. The apparent inconsistency between this result and the known climate history of the Cenozoic will be discussed below.

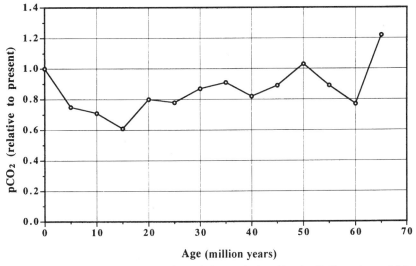

FIGURE 13.   Cenozoic atmospheric pCO_2 variations calculated by the C–Sr cycle model, based on the Volk[49] formulation of sensitivity to f_{wr} (Fig. 6).

4.4. General Implications for the Relationship between Tectonics and Climate

Raymo *et al.*[1] used a variety of observations to support the logical hypothesis that global chemical erosion rates increased progressively through the Cenozoic in response to uplift and erosion in the Himalayas and elsewhere. These included the post-40-million-year nearly continuous rise in oceanic $^{87}Sr/^{86}Sr$, the late Cenozoic increase in apparent carbonate accumulation rates, and the overall deepening of the CCD. They rejected the alternative interpretation of the Sr isotopic trend presented here — that it was the result of an increase in the $^{87}Sr/^{86}Sr$ ratio of rivers — because it could not explain the other observations. We have shown here that the trends in pelagic carbonate accumulation and CCD can indeed be explained by a model that does not require increasing chemical erosion rates through the Cenozoic.

Raymo and Ruddiman[58] recognized the problem of C mass balance in the original hypothesis and presented a scenario of decreasing organic C deposition with increasing carbonate sedimentation to avoid the large C imbalances that would otherwise result from increased chemical erosion. The large decrease in organic C deposition required by their model and inferred by them from the δ_0 record of the Cenozoic (Fig. 5a) is not produced by our model. This is largely the result of the offsetting effect of declining $\Delta^{13}C$ through the Cenozoic. Derry and France-Lanord[59] produced a similar result and are pursuing the consequences of this sort of modeling for the evolution of atmospheric O_2.

The only conclusion consistent with both mass-balance principles and the present state of knowledge concerning the long-term trends in organic C cycling and volcanic and metamorphic CO_2 supply is that *global* chemical erosion rates did not increase significantly through the Cenozoic. This result is counterintuitive but is nevertheless consistent with the observations listed above if one accepts that the locus of carbonate deposition has shifted to the deep sea and that the global riverine $^{87}Sr/^{86}Sr$ has become more radiogenic with time. It is also consistent with the lack of any Cenozoic increase in phosphate accumulation rates. Delaney and Filippelli[60,61] considered this trend a contradiction based on their acceptance of the notion that uplift caused a Cenozoic increase in riverine fluxes of the elements. Of course, if one is arguing for no large change in Cenozoic riverine fluxes, there is no contradiction.

The required trend in riverine Sr isotopic composition can be tested in a number of ways. The trend may reflect an overall reduction through time in the relative exposure area of the abundant marine carbonates and nonradiogenic silicates (e.g., basalts) deposited on the land surface in the Cretaceous. This inference is consistent with the decrease in the proportion of basalt and carbonate exposures through the Cenozoic.[37,38] In addition, it is reasonable to presume that the large rivers of the Himalayan–Tibetan drainage became more radiogenic in character as crystalline rocks became exposed through faulting in the Himalayas.[62,63] Interestingly, the Sr isotopic composition of the world's rivers falls to approximately the value produced by our model if the streams draining the Himalayan region are removed from the average (pers. comm., K. Caldeira and L. Francois). Thus, the overall trend in riverine Sr isotopic composition may represent a combination of a new source of radiogenic Sr from the Himalayas as well as the petering out of the source of nonradiogenic Mesozoic source rocks.

The lack of any substantial CO_2 trend in our model is more of a problem. One might expect that if CO_2 were the important climate factor on geologic timescales,[64,65] significant Cenozoic decreases would be revealed by these calculations, indicating their consistency with the observed Cenozoic cooling. That this trend was not produced is doubly distressing in that the observed decline in $\Delta^{13}C$ (Fig. 5c) implies declining pCO_2, if one accepts the argument that the fundamental control on $\Delta^{13}C$ is the availability of CO_2, and thus atmospheric pCO_2.[66] Thus there is a potential inconsistency between the model results and one of the imposed constraints.

We can remove this inconsistency by calculating a climatic feedback factor that is consistent with the isotopic proxy. Recall that the weathering correction factor f_{wr} was calculated from the C mass-balance equations [(Eq. (4); Fig. 14a)]. Then, in calculating pCO_2, f_{wr} was presumed to be solely a climate factor, reflecting the tendency for enhanced chemical erosion under warmer, wetter climates. In actuality, f_{wr} incorporates all factors other than land area (f_a) and carbonate exposure area ($f_{a_{carb}}$) that might affect chemical erosion rates. In other words, f_{wr} is better considered to be the product of a number of weathering correction factors, including not only climatic factors, but the effects of relief,[67] glaciation,[68,69] and plant coverage[70–72] as well. To isolate the climate effect, we first take our data for $\Delta^{13}C$ (Fig. 5c) and the calibration presented by Freeman and Hayes[66] to calculate a pCO_2 proxy record (Fig. 15). The record differs

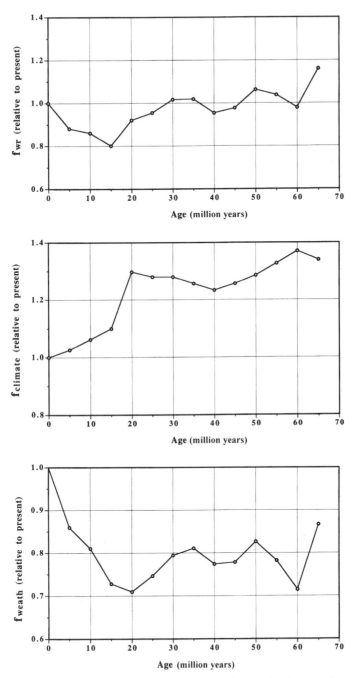

FIGURE 14. Calculated values of weathering correction factors. (a) the overall weathering rate correction factor f_{wr}; (b) the climate correction factor, $f_{climate}$, calculated from the proxy pCO_2 record (Fig. 15); (c) the weatherability correction factor f_{weath}, calculated as the ratio $f_{wr}/f_{climate}$, presuming that these effects are multiplicative.

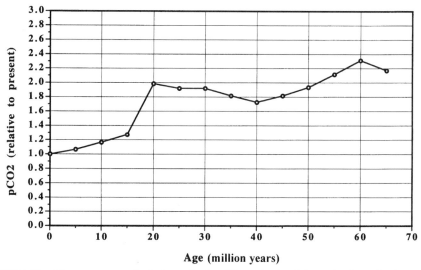

FIGURE 15. The proxy record of atmospheric pCO_2, calculated from the isotopic data presented in Fig. 5c and the calibration curve of Freeman and Hayes[66] (the line 2 calibration from their Table 2).

somewhat from their record because the isotopic data were compiled from different sources. This CO_2 record can then be inverted, using Fig. 6, into a climate–weathering correction factor ($f_{climate}$; Fig. 14b). The residual, which we will call the weatherability factor after François and Walker,[12] f_{weath}, can be determined as follows:

$$f_{weath} = \frac{f_{wr}}{f_{climate}} \tag{8}$$

Figure 14c indicates that this factor shows no significant trend prior to 20 million years ago, and then displays a marked and continual rise to the present.

What does f_{weath} represent? Most simply, it represents the product of the factors that affect chemical erosion listed above other than climate change. That it increased from 20 million years ago to the present is intriguing because that coincides with the reconstructed uplift curve, and thus relief, for the Himalayas.[1] Essentially, the continents became more susceptible to chemical erosion, or more weatherable, from the Miocene to the Recent. This increase in weatherability required compensation, because there was no comparable increase in CO_2 supply to the ocean–atmosphere system. The result was a climatic cooling (reflected in decreased $f_{climate}$).

Uplift of the Himalayas clearly generated high relief. This promoted high rates of chemical erosion *in that region*, and an increase in the "weatherability" of the continents as a whole. At first, the C imbalance created by high Himalayan

CO_2 consumption initiated an atmospheric pCO_2 drawdown. As it did, however, the global climate cooled, and the weathering-limited regions of the world, on average, began to experience reduced rates of chemical erosion. (These are the regions of the world where the factors that control mineral dissolution and weathering, including temperature, water supply, and lithology, rather than the removal rate of soils, determine the chemical erosion rate.[73]) In addition, the biological enhancement of chemical weathering via CO_2 fertilization[49] may have been reduced as atmospheric CO_2 levels fell. This trend continued as long as there was an inbalance in the C cycle. Given that the timescale of adjustment is on the order of 10^5 years, this adjustment was essentially instantaneous.

A conceptual model of the distinction between weatherability and weathering rate is shown in Fig. 16. The atmospheric CO_2 content is represented by a tank

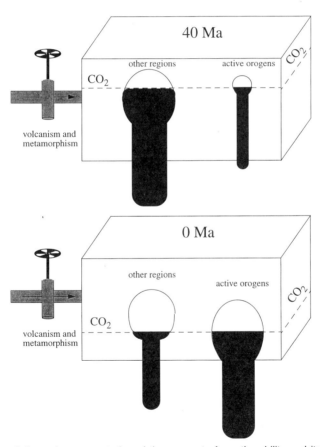

FIGURE 16. Schematic representation of the concept of weatherability and its distinction from weathering rate. Weatherability is reflected in the size of the drain holes, atmospheric pCO_2 is shown as the "water level," the inflow is the volcanic and metamorphic release of CO_2, and the outflow is silicate weathering. As weatherability increases, under constant inflow, the pCO_2 level falls, maintaining a nearly constant silicate weathering rate.

of water. The input to the tank, volcanic and metamorphic CO_2, is held constant. The output, silicate weathering and erosion, depends on the "hydrostatic head," i.e., atmospheric pCO_2, via the greenhouse effect, and on the size of the drain holes (representing the weatherability of the continents). A process that gradually increased the size of the drain holes (i.e., increased continental weatherability) would gradually drain the reservoir, while maintaining a very close balance between input and output. After some time the atmospheric CO_2 level would be lower and the weatherability would be greater, but the weathering rate would remain unchanged. During the time of transition there would only be small imbalances between CO_2 input and output. (Bear in mind that a reduction in atmospheric CO_2 by 50% to present-day levels over the last 20 million years, presuming constant ocean–atmosphere partition, would require a long-term imbalance of only about 10^{17} moles CO_2/year, or about 5% of the silicate weathering or volcanic degassing rates.) Such a regulated drawdown requires strong feedback, and that was provided by the climate–weathering–greenhouse connection.

Thus, according to this hypothesis, the progressive cooling of the Cenozoic climate was indeed the result of Himalayan uplift and its effect on atmospheric pCO_2, consistent with the conclusions of Raymo and others. However, global chemical erosion rates did not increase through the Cenozoic. Rather, global weatherability did. Erosion rates increased in the Himalayan region, but they decreased elsewhere as a result of effective teleconnection provided by the climate–weathering feedback.

Acknowledgments

This research was support by National Science Foundation grant EAR 92-200008. We thank R. Berner and W. Ruddiman for helpful comments.

REFERENCES

1. Raymo, M. E., Ruddiman, W. F., and Froelich, P. N. (1988). *Geology* **16**, p. 649.
2. Raymo, M. E. (1989). *Geology* **19**, p. 344.
3. Hodell, D. A., Mueller, P. A., McKenzie, J. A., and Mead, G. A. (1989). *Earth Planet. Sci. Lett.* **92**, p. 165.
4. Hodell, D. A., Mead, G. A., and Mueller, P. A. (1990). *Chem. Geol.* **80**, p. 1.
5. Capo, R. C., and DePaolo, D. J. (1990). *Science* **249**, p. 51.
6. Richter, F. M., Rowley, D. B., and DePaolo, D. J. (1992). *Earth Planet. Sci. Lett.* **109**, p. 11.
7. Brass, G. W. (1976). *Geochim. Cosmochim. Acta* **40**, p. 721.
8. Caldeira, K., Arthur, M. A., Berner, R. A., and Lasaga, A. C. (1993). *Nature* **361**, p. 123.
9. Volk, T. (1993). *Nature* **361**, p. 123.
10. Walker, J. C. G., Hays, P. B., and Kasting, J. F. (1981). *J. Geophys. Res.* **86**, p. 9776.
11. Berner, R. A., Lasaga, A. C., and Garrels, R. M. (1983). *Am. J. Sci.* **283**, p. 641.
12. François, L. M., and Walker, J. C. G. (1992). *Am. J. Sci.* **292**, p. 81.
13. Gibbs, M., and Kump, L. R. (1994). *Paleoceanography* **9**, p. 529.

14. Kominz, M. A. (1984). *Am. Assoc. Pet. Geol. Mem.* **36**, p. 109.
15. Larson, R. L. (1991). *Geology* **19**, p. 547.
16. Engebretson, D. C., Kelley, K. P., Cashman, H. J., and Richards, M. A. (1992). *GSA Today* **2**, p. 93.
17. Caldeira, K. (1992). *Nature* **357**, p. 578.
18. Kerrick, D. M., and Caldeira, K. (1993). *Chem. Geol.* **108**, p. 201.
19. Beck, R. A., Burbank, D. W., Sercombe, W. J., Olson, T. L., and Khan, A. M. (1995). *Geology* **23**, p. 387.
20. Caldeira, K. (1995). *Am. J. Sci.* **295**, p. 1077.
21. Palmer, M. R., and Elderfield, H. (1985). *Nature* **314**, p. 526.
22. Kump, L. R. (1989). *Am. J. Sci.* **289**, p. 390.
23. Berner, R. A., and Rye, D. (1992). *Am. J. Sci.* **292**, p. 136.
24. Berner, R. A. (1991). *Am. J. Sci.* **291**, p. 339.
25. Goddéris, Y. and François, L. M. (1995). *Chem. Geol.* **126**, p. 169.
26. Veizer, J. (1989). *Ann. Rev. Earth Planet. Sci.* **17**, p. 141.
27. Graham, D. W., Bender, M. L., Williams, D. F., and Keigwin, L. D. Jr. (1982). *Geochim. Cosmochim. Acta* **46**, p. 1281.
28. Chaudhuri, S., and Clauer, N. (1986). *Chem. Geol.* **59**, p. 293.
29. Holland, H. D. (1978). *The Chemistry of the Atmosphere and Oceans*, Wiley-Interscience, New York.
30. Blatt, H., and Jones, R. L. (1975). *Geol. Soc. Am. Bull.* **86**, p. 1085.
31. Meybeck, M. (1987). *Am. J. Sci.* **287**, p. 401.
32. Faure, G. (1986). *Principles of Isotope Geology*. John Wiley, New York.
33. Edmond, J. M., Measures, C., McDuff, R. E., Chan, L. H., Collier, R., Grant, B., Gordon, L. J., and Corliss, J. B. (1979). *Earth Planet. Sci. Lett.* **46**, p. 1.
34. Albarede, F., Michard, A., Monster, J. F., and Michard, G. (1981). *Earth Planet. Sci. Lett.* **55**, p. 229.
35. Berner, R. A. (1994). *Am. J. Sci.* **294**, p. 56.
36. Barron, E. J., Sloan, J. L. II, and Harrison, C. G. A. (1980). *Palaeogeogr. Palaeoclim. Palaeoecol.* **30**, p. 17.
37. Bluth, G. J. S. (1990). Effects of Paleogeology, Chemical Weathering, and Climate on the Global Geochemical Cycle of Carbon Dioxide. Ph.D. Dissertation, Pennsylvania State University, University Park, Pennsylvania.
38. Bluth, G. J. S., and Kump, L. R. (1991). *Am. J. Sci.* **291**, p. 284.
39. Bickle, M. J. (1994). *Nature* **367**, p. 699.
40. Scotese, C. R., Gahagan, L. M., and Larson, R. L. (1988). *Tectonophysics* **155**, p. 27.
41. Gaffin, S. R. (1987). *Am. J. Sci.* **287**, p. 596.
42. Arthur, M. A., Dean, W. E., and Claypool, G. E. (1985). *Nature* **315**, p. 216.
43. Dean, W. E., Arthur, M. A., and Claypool, G. E. (1986). *Mar. Geol.* **70**, p. 119.
44. Popp, B. N., Takigiku, T., Hayes, J. M., Louda, J. W., and Baker, E. W. (1989). *Am. J. Sci.* **289**, p. 436.
45. Rau, G. H., Takahashi, T., and Des Marais, D. J. (1898). *Nature* **341**, p. 516.
46. Rau, G. H., Froelich, P. N., Takahashi, T., and Des Marais, D. J. (1991). *Paleoceanography* **6**, p. 335.
47. Laws, E. A., Popp, B. N., Bidigare, R. R., Kennicutt, M. C., and Macko, S. A. (1995). *Geochim. Cosmochim. Acta* **59**, p. 1131.
48. Lasaga, A. C., Berner, R. A., and Garrels, R. M. (1985). In: *The Carbon Cycle and Atmospheric CO₂: Natural Variations Archean to Present*, (E. T. Sundquist and W. S. Broecker, eds.), pp. 397–411. American Geophysical Union, Washington, D.C.
49. Volk, T. (1987). *Am. J. Sci.* **287**, p. 763.
50. Edmond, J. M., Palmer, M. R., Measures, C. I., Grant, B., and Stallard, R. F. (1995). *Geochim. Cosmochim. Acta* **59**, p. 3301.
51. Davies, T. A., and Worsley, T. R. (1981). *Soc. Econ. Paleont. Mineral. Spec. Pub.* **32**, p. 169.
52. Delaney, M. L., and Boyle, E. A. (1988). *Paleoceanography* **3**, p. 137.
53. Opdyke, B. N., and Wilkinson, B. H. (1988). *Paleoceanography* **3**, p. 685.

54. Milliman, J. D. (1993). *Global Biogeochem. Cycles* **7**, p. 927.
55. Broecker, W. S., and Peng, T. S. (1982). *Tracers in the Sea.* Lamont-Doherty Geological Observatory, New York.
56. Berner, E. K., and Berner, R. A. (1987). *The Global Water Cycle: Geochemistry and Environment.* Prentice-Hall, Englewood Cliffs, NJ.
57. Morse, J. W., and Mackenzie, F. T. (1990). *Geochemistry of Sedimentary Carbonates.* Elsevier, Amsterdam.
58. Raymo, M. E., and Ruddiman, W. F. (1992). *Nature* **359**, p. 117.
59. Derry, L. A., and France-Lanord, C. (1996). *Paleoceanography* **11**, p. 267.
60. Delaney, M. L., and Filippelli, G. M. (1994). *Paleoceanography* **9**, p. 513.
61. Filippelli, G. M., and Delaney, M. L. (1994). *Paleoceanography* **9**, p. 643.
62. Palmer, M. R., and Edmond, J. (1992). *Geochim. Cosmochim. Acta* **56**, p. 2099.
63. Derry, L. A., and France-Lanord, C. (1996). *Earth Planet. Sci. Lett.* **142**, p. 59.
64. Barron, E. J. (1985). *Palaeogeogr. Palaeoclim. Palaeoecol.* **50**, p. 729.
65. Barron, E. J., and Washington, W. M. (1985). In: *The Carbon Cycle and Atmospheric CO₂: Natural Variations Archean to Present* (E. T. Sundquist and W. S. Broeker, eds.), pp. 546–553. American Geophysical Union, Washington, D.C.
66. Freeman, K. H., and Hayes, J. M. (1992). *Global Biogeochem. Cycles* **6**, p. 185.
67. Drever, J. I., and Zobrist, J. (1992). *Geochim. Cosmochim. Acta* **56**, p. 3209.
68. Armstrong, R. E. (1971) *Nature Phys. Sci.* **230**, p. 132.
69. Kump, L. R., and Alley, R. B. (1994). In: *Material Fluxes on the Surface of the Earth* (Board on Earth Sciences, ed.), pp. 46–60. National Academy Sciences, Washington, D.C.
70. Berner, R. A. (1992). *Geochim. Cosmochim. Acta* **56**, p. 3225.
71. Schwartzmann, D. W., and Volk, T. (1989). *Nature* **340**, p. 457.
72. Kump, L. R., and Volk, T. (1991). In: *Scientists on Gaia* (S. H. Schneider and P. J. Boston, eds.), pp. 191–199. MIT Press, Cambridge, Mass.
73. Stallard, R. F., and Edmond, J. M. (1983). *J. Geophys. Res.* **88**, p. 9671.

19

The Marine $^{87}Sr/^{86}Sr$ and $\delta^{18}O$ Records, Himalayan Alkalinity Fluxes, and Cenozoic Climate Models

Sean E. McCauley and Donald J. DePaolo

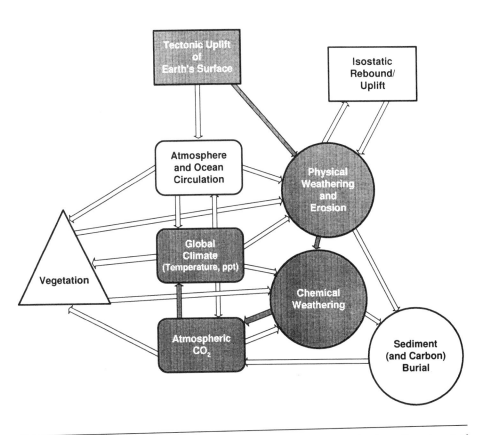

Sean E. McCauley and Donald J. DePaolo • Center for Isotope Geochemistry, Department of Geology and Geophysics, University of California at Berkeley, Berkeley, California 94720-4767.

Tectonic Uplift and Climate Change, edited by William F. Ruddiman. Plenum Press, New York, 1997.

1. INTRODUCTION

Oxygen isotope records of deep-sea paleotemperature indicate that the Cenozoic is a period of dramatic climatic change. Deep-ocean temperatures were highest 55 million years ago, and have declined by 12 to 15°C to the present.[1,2] The Earth's climate appears to have changed as dramatically during the past 55 million years as it has over any similar period in the last 500 million years.

Long-term climate change is believed to be controlled by subtle shifts in the chemical composition of the oceans,[3] which fix the CO_2 content of the atmosphere and thus ultimately determine the surface temperature of the Earth.[4-13] Shifts in ocean chemical composition are a result of chemical interactions between the hydrosphere and the rocks of the Earth's surface. These chemical interactions involve direct injection of volatiles to the atmosphere by degassing of magma and metamorphic rocks, chemical exchange associated with weathering of silicate rocks on the continents and the ocean floor, and chemical deposition and diagenesis of sediment in the oceans.

The effect on the oceans of these chemical interactions between the hydrosphere and the lithosphere can be expressed in terms of competing fluxes of dissolved inorganic C and cations to and from the ocean. The C and cation fluxes can be thought of (informally) as acidity and alkalinity fluxes, respectively. Adding acidity to the ocean has the effect of lowering its pH and raising its pCO_2; this in turn raises the pCO_2 of the atmosphere because the atmosphere tends to stay in equilibrium with surface ocean water. Adding alkalinity to the ocean raises its pH and decreases pCO_2 and causes the atmosphere to have a lower pCO_2. The ability to predict how atmospheric pCO_2 has changed in the geological past, and thus to undestand past climates of the Earth, depends on understanding how alkalinity and acidity fluxes to the oceans have changed in the geological past. As these fluxes cannot be measured directly, the objective has been instead to attempt to scale the fluxes to various observable quantities; among those used are sea level, seafloor spreading rates, sediment chemistry, and sedimentation rates.[4,5,7-9,14] To a large degree, the answer that one obtains for past atmospheric CO_2 contents is determined by how one scales the carbon and cation fluxes to observables.

The primary purpose of this paper is to investigate the carbon and cation fluxes to the oceans over the past 75 million years, and to evaluate: (1) the extent to which the climate changes over this time period correspond to them, and (2) what feedbacks or interrelationships between the fluxes are compatible with observations. The essential problem that is encountered once one has assumed this task is to answer the questions: (a) what measures are there of carbon and cation fluxes over this time period, and (b) what measures are there of climate over time period? We have chosen to focus on the past 75 million years because this is the only period for which there is a quantitative and continuous record of global climate change, in the form of the benthic foraminiferal $\delta^{18}O$ record, with relatively good time–stratigraphic control. We argue that the marine Sr isotope record, which is appropriately detailed for this time period, is a measure of cation fluxes to the oceans when suitably scaled. In addition, the last 150 million years is the only period for which there is a quantitative record of seafloor spreading rates. These are the records that represent the best hope for testing models of the relationships between tectonically driven geochemical processes and climate.

The idea of using measures of past tectonic activity to model past atmospheric CO_2 levels was first proposed and carried out by Berner et al.,[4] in a groundbreaking paper describing modeling that has since become known as BLAG. The models presented in the original paper were modified by Lasaga et al.,[5] and have since been further modified by Berner.[6,7,14] The essential thesis of the BLAG models is that the supply of C to the oceans largely controls climate, and that the supply of C scales with the global seafloor generation rate. As we show below, there is essentially only one independent variable in the BLAG models: the seafloor generation rate. The BLAG models are designed such that cation fluxes to the ocean are a function of the carbon supply; any increase in C supply is accompanied by an increase in cation fluxes (via continental weathering), stabilizing the system at a unique atmospheric CO_2 level that is appropriate for the C supply.

An alternative view is that the cation flux to the oceans is the driver of long-term climate. This idea is implicit in the work of Chamberlain[10] and has been recently championed by Raymo[11-13]; it is now commonly referred to as the Raymo hypothesis. The Raymo hypothesis is based largely on the observation that global climate, as reflected in the benthic foraminifera $\delta^{18}O$ curve, has cooled during a period when global cation fluxes appear to have been enhanced by weathering associated with rapid erosion of the Himalaya and Tibet. A problem with this model is that increased alkalinity fluxes to the oceans, to the extent that their effects are understood, should tend to increase ocean pH and decrease atmospheric CO_2 too rapidly, unless there is a way of concurrently absorbing excess alkalinity fluxes or increasing acidity fluxes. There are no obvious mechanisms for achieving this.

A useful model should be able to reproduce what is known about Cenozoic climate changes, carbon fluxes, and cation fluxes. As described below, the inferred values of paleoatmospheric pCO_2 are highly dependent on the model used to calculated them, and they are also strongly dependent on the accuracy of the "tectonic" records that are used as the model input. The BLAG model can reproduce some aspects of Cenozoic climate change, but the result depends primarily on which record of seafloor generation rate is used. The Raymo hypothesis can only work if there are essential features of the climate system that are not yet included in existing models. Perhaps the most fundamental issue concerns feedback mechanisms that act to keep the alkalinity and acidity fluxes to the oceans in near balance. The BLAG model incorporates a particular type of feedback which we believe is ad hoc, whereas the Raymo hypothesis works most straightforwardly if there is no feedback. Which idea is closer to the actual system behavior is critical. Climate change works in an entirely different way in the absence of strong feedbacks; the extent to which one view or the other can be accepted or rejected depends on the degree to which the system behavior is characterized adequately and the accuracy with which the fluxes to the ocean and the ocean's buffer capacity are known.

In Section 2 of this paper we review the geochemical and geological records for the Cenozoic that form the model inputs. In Section 3, we examine the construction of the BLAG models. In Sections 4 and 5, we use the marine Sr isotope record, evidence concerning the alkalinity flux associated with unroofing of the Himalaya, and a model of ocean chemistry, to develop a quantitative version of the Raymo hypothesis. In Sections 6 and 7, we discuss the degree to which the Raymo hypothesis can be considered compatible with observations and consider some aspects of the ocean system that may be important in resolving the broader issues concerning the controls on atmospheric pCO_2.

2. CENOZOIC RECORDS

2.1. Oxygen Isotopes

Oxygen isotope records of benthic foraminifera are the only continuous high-quality record of global climate that spans the Cenozoic (Fig. 1). Benthic foraminifera record bottom water temperatures and ice volume, both of which reflect high-latitude surface temperatures.[1] Typically the $\delta^{18}O$ data are corrected for the effect of changing continental ice storage in order to retrieve bottom water

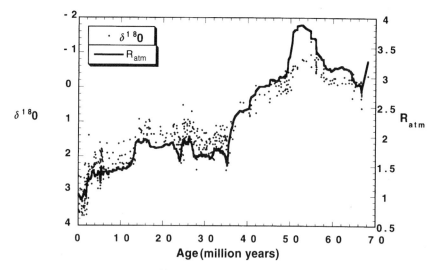

FIGURE 1.   Benthic foraminiferal $\delta^{18}O$ data, digitized from Miller et al.[1] for the last 65 million years. The $\delta^{18}O$ scale has been inverted so that warm periods (more negative values) plot upward. A 10-point running mean of the data has been transformed to pCO_2 (R_{atm}) using a function described in the text.

temperatures. Although there is considerable uncertainty about the magnitude and timing of permanent ice storage, it is generally believed that ice storage was minimal prior to about 35 million years ago, and that it increased substantially at about 35, 15, and 3 million years ago. Another significant source of uncertainty is in our inability to know whether the species-specific offsets from isotopic equilibrium with seawater ("vital effects") are constant through time.

For our purposes, the bottom water temperature may not necessarily be the only important parameter, since even with a minor adjustment of bottom water temperature, an increase in ice storage probably represents a considerable lowering of surface temperature; for example, this is almost certainly the case during Pleistocene glacial maxima as compared to interglacials. Thus we consider here the $\delta^{18}O$ record uncorrected for changes in ice volume as being a reasonable representation of global high-latitude temperature changes in the Cenozoic.

The $\delta^{18}O$ record shows a strong trend from warm high-latitude temperatures in the Paleocene to cold glacial conditions in the latter part of the Cenozoic.[1,2] The cooling of the climate in the last 55 million years is uneven: short-term oscillations are superimposed on the long-term trends. Steady cooling during the middle and late Eocene is followed by oscillating but otherwise steady values through the Oligocene and early Miocene.[1] There is a significant drop in temperature in the middle Miocene,[1] steadily decreasing temperature through the late Miocene and Pliocene and a more dramatic drop in the Pleistocene.

Unfortunately, little is known about the relationship between paleotemperature and paleo-pCO_2. We are aware of only one published estimate of paleo-pCO_2, based on organic $\delta^{13}C$ data,[15] but more recent studies show that this proxy is complex and resistant to simple interpretation.[16,17] Using a scaling argument of 5°C high-latitude temperature change per doubling of CO_2, and the fact that Eocene high-latitude temperatures were at least 10°C warmer than they are today, we estimate that the Eocene must have had at least four times the modern values pCO_2, and possibly much higher. We have transformed the $\delta^{18}O$ data in Fig. 1 to an estimate of the relative change in atmospheric CO_2 over time, using an exponential scaling function:

$$R_{\text{atm}} = \frac{pCO_2(t)}{pCO_2(0)} = 1.4^{[\delta^{18}O(0) - \delta^{18}O(t)]} \tag{1}$$

This function uses a ten-point running mean of the $\delta^{18}O$ data,[1] and is designed to achieve a four-fold increase in CO_2 at the Paleocene temperature maximum.

2.2. Strontium Isotopes

Over the past 10 years there have been a substantial number of papers published concerning the Sr isotopic evolution of the oceans during the Cenozoic.[18-24] These data, and our unpublished data for the Campanian–Paleocene period, now provide a relatively dense record extending from 75 million years ago to the present, with the exception of sparse coverage in the Eocene. There are minimal data for the time period spanning 24 to 18 million years ago, and the data that do exist for this time interval are not all in agreement, presumably because of difficulties with stratigraphic age assignments. A representative set of data is plotted in Fig. 2a, which shows the record as it now exists.

The interpretation of the Sr isotope curve in terms of weathering rates of continental rocks is controversial. Richter *et al.*[25] argue that there is good reason to believe that the increasing marine $^{87}Sr/^{86}Sr$ value beginning about 40 million years ago is a reflection of an increased dissolved Sr flux. If the dissolved Sr flux has increased during this period, it can plausibly be inferred that the fluxes of other cations (alkalinity) to the oceans may also have been increasing in the last 40 million years. Edmond[26] on the other hand, argues that the increasing marine $^{87}Sr/^{86}Sr$ value in that time is due entirely to a shift in the $^{87}Sr/^{86}Sr$ value of the rocks being weathered and that there has been no change in the dissolved Sr flux to the oceans. This issue is discussed in more detail below, but we believe it is reasonable to ask a slightly different question first: What evidence exists that the Sr isotope record has anything at all to do with climate variations?

Figure 2b illustrates one line of argument that suggests that the Sr curve is meaningful. Plotted on the figure is the derivative of the $^{87}Sr/^{86}Sr$ ratio versus age. When the derivative is relatively high, it should correspond to enhanced silicate weathering and therefore to a stronger tendency toward global cooling. Interestingly, the ages of all except one of the Cenozoic epoch and subepoch boundaries in the last 55 million years correspond closely to peaks in the $^{87}Sr/^{86}Sr$ derivative curve. One case that does not conform, the Paleocene–Eocene boundary 55 million years ago, occurs at the low point of the $^{87}Sr/^{86}Sr$ derivative and corresponds to the warmest period in the Cenozoic. Assuming that the boundaries represent major faunal changes, which are presumably a response to environmental changes, this diagram is suggestive that the Sr evolution curve is a proxy of climate change.

Other evidence that supports the relationship between the Sr record and erosion–weathering rates is illustrated in Fig. 3; it follows the arguments of Richter *et al.*[25,27] The steep portion of the Sr curve in the age range 20 to 15 million years, seen also as a peak on Fig. 3 centered at about 17 million years ago, corresponds very closely in time to a period of very rapid unroofing in the Himalaya and southern Tibet. Raymo *et al.*[11] also pointed out that the relatively steep part of the curve in the period from 5 million years ago to the present corresponds to a period of rapid erosion in the Himalaya.

2.3. Seafloor Generation Rates

The seafloor generation rate is thought to be a critical parameter for scaling the CO_2 outgassing rate both from the mantle and from subduction zone metamorphism. As we will show in the section on BLAG models, the seafloor generation rate chosen determines the pCO_2 response obtained from these

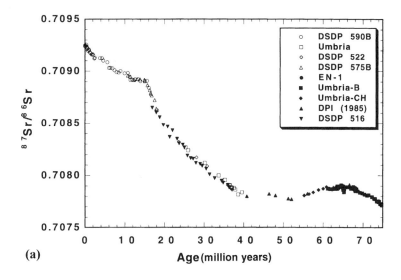

(a)

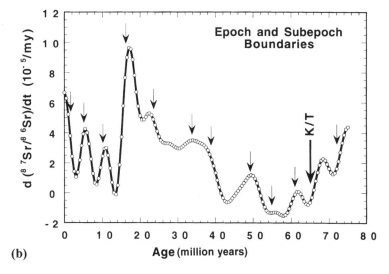

(b)

FIGURE 2. (a) Compilation of $^{87}Sr/^{86}Sr$ values from marine carbonates for the last 75 million years. (b) The derivative of the marine Sr isotope record reveals the times at which the seawater $^{87}Sr/^{86}Sr$ value was increasing most rapidly. Interestingly, most of the Cenozoic epoch and subepoch boundaries appear to correspond to maxima in the derivative; this may lend support to the idea that periods of rapid increase in seawater $^{87}Sr/^{86}Sr$ correspond to periods of increased chemical weathering, cooling, and biological change.

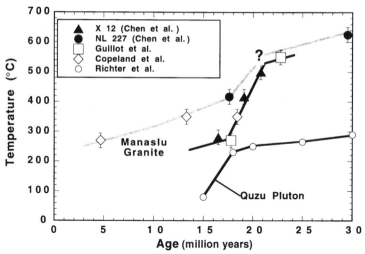

FIGURE 3.   Cooling histories of Himalayan K-feldspars, based on $^{40}Ar/^{39}Ar$ thermochronometry. These data suggest a period of rapid cooling and thus rapid uplift and erosion during the period from 20 to 15 million years ago. This period also shows the most rapid change in the marine $^{87}Sr/^{86}Sr$ record, indicating that the rate of silicate chemical weathering may have been very high owing to the rapid uplift in the Himalaya.

models. For this reason, it is important to ask whether any of the published generation rate records are capable of producing a pCO_2 history compatible with the $\delta^{18}O$ record.

Various methods have been used to measure past seafloor generation rates, each of which produces a significantly different estimate. The traditional approach uses the aged-dated pattern of magnetic reversals to calculate spreading rates for pieces of preserved ocean crust. Some authors present a world average spreading rate, in cm/year,[28] while others quote both spreading rates and ridge lengths, which can be used to compute a true seafloor generation rate in units of km^2/year (the sum of spreading rate times ridge length for each segment).[29,30] In Fig. 4a, we compare a number of published curves: The Southam and Hay[28] and Pitman[29] curves as interpreted by BLAG are presented; a more detailed analysis by Kominz[30] provides a data set of mean world spreading rates and total ridge lengths, which we have multiplied to generate maximum and minimum seafloor generation rate curves. Of these three papers, we prefer the Kominz curve for its clear presentation of the data, its level of detail, and its error analysis.

Further back in time, it is impossible to use the record of preserved ocean crust to estimate spreading rates. Another method uses an inversion of the eustatic sea-level curve to calculate ridge volumes and thus (presumably) seafloor generation rate throughout the Phanerozoic.[31] This estimate serves as the basis for the 1990/1991 version of the BLAG model.[6,7] We show the Gaffin curve in Fig. 4a to illustrate how little detail it contains for the period from 120 million years ago to the present. Given the number of assumptions required to extract seafloor generation rates from these data, this method is not desirable where other data are available.

Engebretson et al.[32] use the positions of subduction zones and the velocities of subducting plates to estimate the total amount of seafloor that has been subducted over the last 180 million years. This estimate uses newer plate reconstruction data and benefits by its use of subduction zone lengths, which are better preserved than ridge lengths. It accurately reproduces the sea-level curve of Haq et al.,[33] and appears to be the best estimate of the seafloor generation rate currently available. It was used as the proxy for CO_2 outgassing for the first 150 million years of the 1994 version of BLAG. The Engebretson record is shown in bold in Fig. 4a.

In Fig. 4b, we show a comparison between the Engebretson seafloor generation rate curve and the marine ^{87}Sr/^{86}Sr record (same data as Fig. 2a). There appears to be a strong inverse relationship between the generation rate and the Sr isotope curve, for the period from 75 to 40 million years ago. This suggests that the marine ^{87}Sr/^{86}Sr record is dominated by Sr from hydrothermal weathering of the ocean crust during this period, but not later in the Cenozoic. If our interpretation is correct, it also indicates that the Cenozoic seafloor generation rate curve is now well known.

3. DECONSTRUCTING BLAG

3.1. Structure

The models of Berner, Lasaga, and Garrels attempt to track the effects of volcanic CO_2 outgassing, continental weathering, hydrothermal alteration of the ocean crust, sedimentation, and sea-level change on the ocean chemical parameters that control atmospheric CO_2[4-7,14] in the BLAG model, the atmospheric pCO_2 is determined by the carbonate equilibria of the oceans, and it is assumed that the oceans are always saturated with calcite. The defining feature of the BLAG models is that continental weathering rates (i.e., the alkalinity flux to the oceans) are made a linear function of atmospheric CO_2 concentration. This feature makes the alkalinity flux adjust to the acidity flux on a timescale of about 10^5 years, so that the acidity flux determines the atmospheric CO_2 value. Furthermore, since the state of the system in the past is scaled to that of the modern system, the model result for atmospheric CO_2 in the past is entirely determined by the ratio of the acidity flux in the past to that today. The BLAG models scale the acidity flux to the seafloor generation rate, and, consequently, the dimensionless paleo-CO_2 concentration of the atmosphere (or R_{atm}) is largely determined by the dimensionless seafloor generation rate. And since the seafloor generation rate is effectively derived from the Phanerozoic "sea-level" history of Gaffin[31] there is little information in the BLAG models, which are primarily a recasting of the sea-level curve.

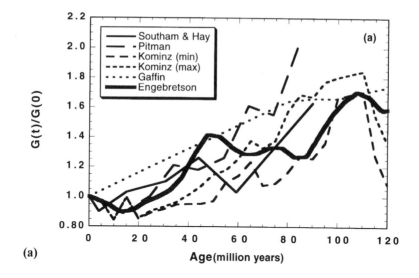

(a)

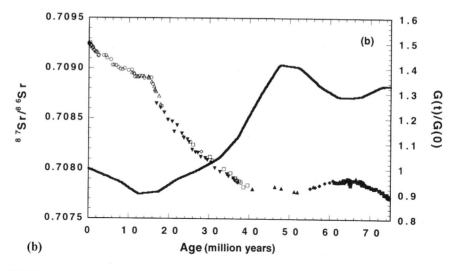

(b)

FIGURE 4. (a) A comparison of relative spreading rate and seafloor generation rate histories, estimated by various authors,[28-32] for the last 120 million years. Of these, the Engebretson curve is preferred for reasons discussed in the text. (b) A comparison of the Engebretson curve with the marine $^{87}Sr/^{86}Sr$ record. Between 40 and 75 million years ago, there are striking similarities between the seafloor generation rate and the $^{87}Sr/^{86}Sr$ record, suggesting that this period is dominated by volcanic Sr inputs.

3.2. Recasting the BLAG Models

Because the ocean is saturated with respect to calcite, dissolution of calcite on land, followed by subsequent delivery to the ocean, has no net effect on ocean chemistry:

$$CO_2 + H_2O \Leftrightarrow H_2CO_3 \text{ (Formation of carbonic acid)} \tag{2a}$$

$$H_2CO_3 + CaCO_3 \Rightarrow Ca^{2+} + 2HCO_3^- \text{ (Calcite dissolution on continents)} \tag{2b}$$

$$Ca^{2+} + 2HCO_3^- \Rightarrow CaCO_3 + CO_2 + H_2O \text{ (Calcite in precipitation in oceans)} \tag{2c}$$

All of the terms in these three equations cancel, and so we can ignore these components of the BLAG models without consequence.

The fluxes that matter are those of acidity and alkalinity between the ocean and silicate reservoirs. One component is the release of CO_2 to the ocean–atmosphere system from volcanism. BLAG assume that both midocean ridge and subduction zone CO_2 outgassing scale with the seafloor generation rate. A second type of acidity flux used in the BLAG models arises from hydrothermal alteration of ocean basalts. The exchange of seawater Mg^{2+} for basalt Ca^{2+} during high-temperature hydrothermal weathering, which is documented from sampling of black smokers,[34] causes $CaCO_3$ precipitation in the ocean. As shown in a modified form of Eq. (2c):

$$Ca^{2+}(aq) + HCO_3^-(aq) = CaCO_3(s) + H^+(aq) \tag{2d}$$

Each mole of Ca^{2+} that is added to the (already calcite-saturated) ocean causes 1 mole of $CaCO_3$ to be precipitated, and 1 mole of H^+ to be released. The net effect is to make the ocean more acidic; since the equilibrium pCO_2 of a calcite-saturated ocean increases with decreasing pH, the atmospheric pCO_2 increases. Hydrothermal weathering in the BLAG models is, like CO_2 outgassing, assumed to scale with the seafloor spreading rate.

Silicate weathering at low temperature involves the formation of clay minerals from primary silicate minerals. The net effect is that protons from the ocean are exchanged for cations from primary silicate minerals. Simplified weathering reactions can be written to show this exchange[35]:

$$CaAl_2Si_2O_8 + 2CO_2 + 3H_2O \Rightarrow Al_2Si_2O_5(OH)_4 + Ca^{2+} + 2HCO_3^- \tag{3a}$$

$$2NaAlSi_3O_8 + 2CO_2 + 6H_2O \Rightarrow Al_2Si_4O_{10}(OH)_2 + 2Na^+ + 2HCO_3^- + 2H_4SiO_4 \tag{3b}$$

In each of the reactions above, two equivalents of alkalinity are released. This has the effect of increasing pH, which in turn shifts carbonate speciation away from CO_2. Ultimately, the alkalinity that weathering adds to the ocean drives

$CaCO_3$ precipitation, through Eq. (2c). The effects of perturbing the ocean's carbonate system are subtle and must be examined using a chemical model that includes all of the relevant reactions.

In the original BLAG model,[4] weathering is described by a first-order rate expression, which is a function of a modern weathering rate estimate, the area of continents as a function of time, and a term that depends on the pCO_2 in the atmosphere. The pCO_2 dependence is a function that ensures that weathering rates increase when pCO_2 and temperature are high and decrease when pCO_2 and temperature are low. The weathering "feedback" is the primary stabilizing mechanism in the BLAG models. Continental area is the other parameter that modifies weathering in this model.

Because all of the relevant chemical fluxes are scaled to modern values, it is possible to reproduce the BLAG model results with an equation of the form

$$\frac{d(CO_2)_{atm}}{dt} = P(t) - A(t)(CO_2)_{atm}^n \tag{4}$$

where $P(t)$ is a CO_2 production term, which includes CO_2 outgassing and Mg/Ca exchange between ocean floor basalt and the ocean; $A(t)(CO_2)_{atm}^n$ represents the rate of continental silicate weathering [it has two subterms: $A(t)$ is an effective continental area term (modern = 1.0). In later versions of the BLAG model, this term is modified to include other factors, such as runoff, continental uplift, and changes in biology, which act to increase mineral surface area available to chemical attack, and $(CO_2)_{atm}^n$, which is the weathering feedback term.] The sensitivity of atmospheric pCO_2 levels is strongly dependent on the strength assigned to the pCO_2 feedback, which Berner et al.[4] show in their Fig. 7a, 7b. A value of $n = 0.21$ gives a good fit to the feedback function in the 1983 BLAG paper.[4]

Because the amount of CO_2 in the ocean–atmosphere system is small compared to the production and removal terms, for long timescales Eq. (3a) can be treated as being in steady state:

$$P(t) = A(t)(CO_2)_{atm}^n \tag{5}$$

This assumption requires the product of the area term and the CO_2 feedback term to always match the production term. It also means that atmospheric CO_2 is determined by the ratio $P(t)/A(t)$.

For the Cenozoic (and for most of the Phanerozoic as well), continental area depends primarily on sea level (in fact the latter is derived from the former), which in turn is a function of the seafloor generation rate. When seafloor generation is more rapid, sea level is higher and land area is smaller (see, e.g., Pitman[29] and Gaffin[31]). The result is an inverse relationship between seafloor generation rate and continental area. Thus we can write $A(t) = P(t)^{-r}$ for the 1983 and 1985 versions of BLAG.[4,5] The CO_2 production term is scaled to the seafloor generation rate, so Eq. (3b) can be rewritten as

$$R_{atm}(t) = G(t)^{(1+r)/n} \qquad (6)$$

where $R_{atm}(t)$ is the ratio of atmospheric CO_2 at time t to that of today (dimensionless CO_2 concentration) and $G(t)$ is the ratio of seafloor generation rate at time t to that of today (dimensionless seafloor generation rate). The dimensionless equation requires as input only a single parameter $-G(t)$. Using the value $r = 0.4$ we obtain

$$R_{atm}(t) = G(t)^{6.7} \qquad (7)$$

Our Figure 5a compares the CO_2 history given in the 1983 BLAG paper[4] and that based on Eq. (7) above. The difference between the two curves is small. Lasaga et al.[5] modified the 1983 BLAG model mainly by adding a term representing organic carbon burial. The result is to make the production term $P(t)$ vary approximately as $G(t)^{0.8}$, so that the new expression becomes:

$$R_{atm}(t) = G(t)^{5.7} \qquad (8)$$

Figure 5b shows a comparison between Eq. (8) and the Lasaga et al. result. The fit shows how the BLAG models depend, to first order, on the record of seafloor generation.

Berner[6,7] extends the BLAG model back through the entire Phanerozoic. This version of BLAG is simplified in that it tracks only the fluxes of C among three reservoirs: a carbonate reservoir, an organic reservoir, and an ocean–atmosphere reservoir. The model includes seven C fluxes, which are determined by a number of dimensionless parameters and rate constants, and a CO_2-dependent weathering feedback. As in previous BLAG models, the independent function is CO_2 released by volcanism and metamorphic degassing, with an additional term for organic weathering. These fluxes are tied primarily to the seafloor generation rate, which is from Gaffin.[31]

The flux of C associated with silicate weathering contains the same CO_2 feedback as in previous models, but instead of driving the variation in weathering solely with continental area there are four parameters that control the weathering function: continental area, river runoff, elevation, and a term for increasing weathering rates as a result of angiosperm evolution in the Mesozoic. Analysis of the continental area, elevation, and runoff parameters shows that they are all explicitly derived from geological records of paleosea level.[31,36−38]

Although there are additional variables considered in the model, such as a major peak in the δ^{13}C of seawater around 300 million years ago, an increase in the amount of calcite deposited in the deep ocean beginning around 150 million years ago, and the evolution of angiosperms around 130 million years ago, the Berner[7] model results can be reproduced by raising the seafloor spreading rate to an exponent of 3.4 from the present to 140 million years ago and an exponent of

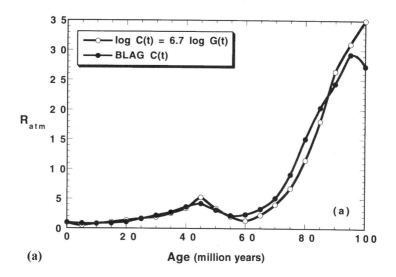

(a)

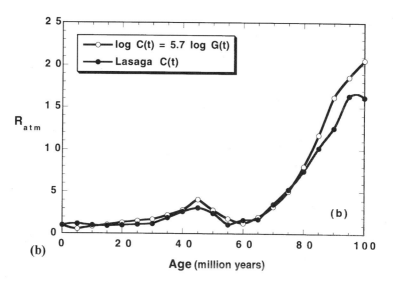

(b)

FIGURE 5.   (a) Comparison of the BLAG[4] model pCO$_2$ history with that obtained by raising the Southam and Hay spreading rate curve to an exponent of 6.7. (b) Comparison of the Lasaga[5] model pCO$_2$ history with that obtained by raising the Southam and Hay spreading rate curve to an exponent of 5.7.

6.4 from 150 to 570 million years ago. The change in exponential dependence in the last 140 million years is due to the introduction of the angiosperm term from 130 million years ago and a deep-ocean calcite deposition term after 150 million years ago, which change the dependence of weathering rate on atmospheric CO_2 concentration. Figure 6a compares our single-component model to the Berner R_{atm} curve.

In the 1994 version of BLAG (renamed GEOCARB II), an elevation term that depends on the marine ^{87}Sr/^{86}Sr record is added to the $A(t)$ term in the silicate weathering flux equation. Since the marine ^{87}Sr/^{86}Sr does not correlate well with inferred paleosea level, this term causes the deduced R_{atm} values to deviate slightly from those that would be predicted for a simple dependence on seafloor generation rate. Nonetheless, the GEOCARB II model results are not significantly different from those in Berner[7] and can be reproduced for the last 120 million years by raising the Engebretson seafloor generation rate[32] to an exponent of 2.2 (Fig. 6b). It is also noteworthy that the CO_2 feedback has been strengthened, which has the effect of reducing the size of the predicted R_{atm} variations.[14]

The BLAG models reduce to simple, single-variable models with the seafloor generation rate being the only independent variable. An analysis by Volk[39] came to a similar conclusion, although it did not recognize the connection between spreading rate and continental area. The exponents used here to scale R_{atm} to $G(t)$ are chosen to most closely represent the BLAG calculations, but in fact are poorly constrained, so that the absolute value of the R_{atm} variations is uncertain. The assumptions built into the model do not allow continental weathering to vary independently. Furthermore, because the weathering feedback is very efficient, R_{atm} varies only smoothly and over very long timescales. The BLAG models present an elegant solution for stabilizing the pCO_2 of the atmosphere; the primary question is whether the Earth works in the way they represent it.

3.3. BLAG versus Cenozoic Paleotemperatures

A necessary, but not a sufficient condition, of accepting any model construction is that it reproduce observed phenomena. Thus, a successful model of Cenozoic pCO_2 history should be consistent with the global temperature record as represented by the benthic foraminifera δ^{18}O curve. In particular, it should be able to correctly predict the major features of the Cenozoic O isotope record: the timing of the early Eocene temperature maximum, the cooling between 50 and 35 million years ago, and the relatively rapid changes 15 and 2 million years ago. In Fig. 7a, the original BLAG CO_2 curves[4] are juxtaposed with the benthic δ^{18}O record, recast as R_{atm} as shown in Fig. 1.[1] The model results show an overall trend toward cooler modern conditions, but predict a relatively cool episode in the early Eocene and a warm peak in the late Eocene. This produces a poor fit to the δ^{18}O record. The peak in CO_2 40 million years ago reflects the two large inflection points in the Southam and Hay spreading rate reconstruction used and seems to suggest that changes in spreading rate alone cannot explain the transition from

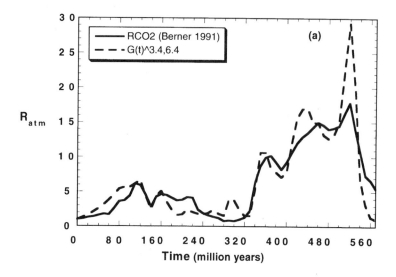

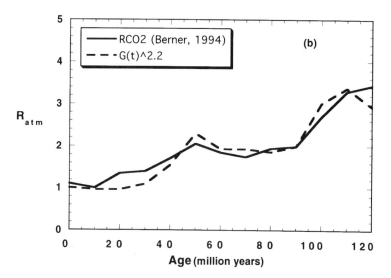

FIGURE 6.   (a) Comparison of the Berner[7] model pCO_2 history with that obtained by raising the Gaffin sea-level spreading rate inversion curve to an exponent of 3.4 prior to 150 million years ago, and an exponent of 6.4 after 150 million years. (b) Comparison of the Berner[14] model pCO_2 history for the last 120 million years with that obtained by raising the seafloor spreading rate curve used for this period to an exponent of 2.2. Note that the exponents used, and thus pCO_2 levels, have decreased in each successive model.

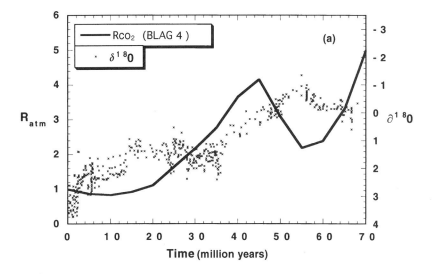

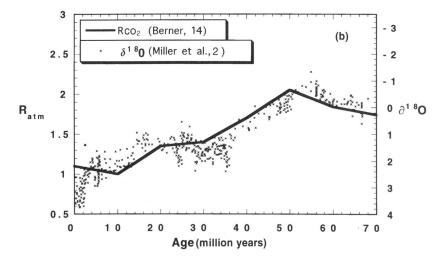

FIGURE 7. (a) Comparison of the BLAG[4] model pCO_2 and the benthic foraminiferal $\delta^{18}O$ record,[1] which we hypothesize should have a similar shape. (b) The 1994 model[14] shows improved agreement with the marine $\delta^{18}O$ record. This does not constitute a definitive test of the assumptions built into the model, however, because it contains sufficient adjustable parameters that it is unclear what drives the model.

warm to cold conditions during the Cenozoic. Similarly, the major climate shifts generally understood to have taken place 15 and 2 million years ago are nowhere evident in the 1983 BLAG curve.[4]

In the 1991 model,[7] the level of detail in the spreading rate used (and thus the resulting pCO_2) is generally insufficient to produce a pCO_2 history that resembles the $\delta^{18}O$ record. The 1994 model results[14] better match the $\delta^{18}O$ record (Fig. 7b).[1] The model's pCO_2 curve captures the Eocene temperature maximum, the major cooling trend from 50 million years ago, some of the Oligocene climate moderation and then the cooling during the Miocene. The most recent descent in the modern glacial climate is not accurately captured by the 1994 model.

The major improvement of the GEOCARB II model, particularly for the Paleogene period, is due to the use of the revised seafloor generation history of Engebretson et al.[32] There may also be some effect in the Neogene from the inclusion of the relief term in the weathering function, although this is less obvious. The GEOCARB II results, in contrast to those of BLAG,[4] suggest that seafloor generation can explain at least some of the major aspects of the Cenozoic climate record.

4. QUANTIFYING THE RAYMO HYPOTHESIS

The Raymo hypothesis states that CO_2 levels in the atmosphere respond to the amount of weathering, which in turn can be stimulated by continental orogeny.[11] In its early form, the Raymo hypothesis was invoked to explain only the latest Cenozoic cooling.[11] In a later paper, the more prolonged cooling since the early Eocene is attributed specifically to the Himalayan orogeny.[12] The Raymo hypothesis holds that accelerated erosion associated with mountain building processes can increase global weathering rates by increasing the supply of fresh mineral surfaces exposed to chemical attack. This idea gives less weight to temperature and pCO_2 as controls on the global silicate weathering rate, and so implicitly challenges the idea that climate is simply a function of the seafloor generation rate. A number of lines of evidence have been invoked to support the hypothesis, including the marine $\delta^{18}O$ and $^{87}Sr/^{86}Sr$ records.[11-13] An attempt to quantify the increase in global weathering rate associated with erosion of the Himalaya and Tibet was carried out by Richter et al.[26] The problem of rapid CO_2 drawdown that would result from an increase in silicate weathering rates is recognized, and various negative feedbacks have been suggested.[13,40] So far the details of the Raymo hypothesis have been left unquantified. In this section we reexamine the marine $^{87}Sr/^{86}Sr$ record and how it is related to the benthic $\delta^{18}O$ record, and try to better constrain the alkalinity fluxes generated by Himalayan erosion. This information is then used below to evaluate the Raymo model.

4.1. The Marine Strontium Isotope Record and Himalayan Alkalinity Fluxes

The most frequently invoked evidence for increased weathering rates comes from the marine Sr isotope ratio,[11-13,25] which began to increase dramatically

40 million years ago. The increase in ratio since the late Eocene cannot be due solely to changes in the hydrothermal flux of Sr and so must be the result of some shift in the ratio and/or size of the Sr flux coming from continents.[19,25,26,41-44] Himalayan uplift and erosion have been suggested as the cause of the change in the marine ^{87}Sr/^{86}Sr record, but the degree to which this change has been driven by an increased dissolved Sr flux as opposed to an increased ^{87}Sr/^{86}Sr ratio of riverine Sr has not been resolved.[12,25,26]

The equation that describes the Sr isotopic evolution of the oceans is[41]

$$N\left(\frac{dR_{SW}}{dt}\right) = \sum_n J_n(R_n - R_{SW}) \tag{9}$$

where N is the number of moles of Sr in the ocean, R_{SW} is the ^{87}Sr/^{86}Sr of seawater, J_n is the flux of Sr (moles/year) from source n, and R_n is the ^{87}Sr/^{86}Sr ratio of source n. Because the alkalinity contributed to the ocean by carbonate weathering has no effect on atmospheric CO_2, it is important to decouple the silicate and carbonate components of continental weathering in Eq. (9). The Sr fluxes we account for in our model are: (1) the ocean-floor hydrothermal flux, (2) the silicate weathering flux from continental cratons worldwide, (3) the silicate weathering flux from the Himalaya, and (4) the carbonate weathering flux. The ^{87}Sr/^{86}Sr values we choose are 0.703 for the ocean-floor hydrothermal flux[25,43] and 0.72 for average cratonic silicate rocks worldwide.[41] We choose a value of 0.742 for Sr from weathering of silicates from the Himalayan uplift, based on Bengal Fan sediment analyses.[45]

Choosing the appropriate Sr isotopic ratio for the carbonate weathering input and constraining its flux history is problematic, but not critical. The isotopic weighting factor for the carbonate Sr flux is small compared to those for the ocean-floor hydrothermal and silicate weathering Sr fluxes. We bracket the role of carbonate Sr by choosing two end-member cases: In the first, minimum impact case, we assume that most of the dissolved carbonate component comes from continental shelves and has a ^{87}Sr/^{86}Sr ratio that approximates that of the contemporary seawater; in this case the isotopic weighting factor is zero and the term associated with the carbonate-derived Sr influx is similarly zero. The maximum impact case uses a constant value of ^{87}Sr/^{86}Sr = 0.708, appropriate to carbonate sediment deposited during the early Cenozoic. To make the modern mass balance work we choose a constant flux of carbonate Sr that is 80% of the modern riverine Sr flux (3.3×10^{10} moles/year).[43] This, together with a 20% silicate Sr flux representing some combination of cratonic and Himalayan fluxes, yields an appropriate modern river ^{87}Sr/^{86}Sr ratio of 0.711.[43]

The expanded mass-balance equation we use is

$$N\frac{dR_{SW}}{dt} = J_{Hyd}(R_{Hyd} - R_{SW}) + J_{Crat}(R_{Crat} - R_{SW}) + J_{Him}(R_{Him} - R_{SW})$$

$$+ J_{Carb}(R_{Carb} - R_{SW}) \tag{10}$$

The hydrothermal Sr flux is scaled to seafloor generation:

$$J_{Hyd}(t) = J_{Hyd}(0)G(t) \tag{11}$$

where the $J_{Hyd}(0)$ is estimated to be 8×10^9 moles/year[25,43] and the $G(t)$ is the Engebretson seafloor generation curve.[32] We can solve for the size of one Sr flux at a time, if the others are specified. For the period from 75 to 40 million years ago, we solve for J_{Crat}. Because we are testing the hypothesis that Himalayan uplift has driven the increase in the marine Sr ratio, beginning 40 million years ago, we hold J_{Crat} constant and solve for J_{Him}. The Sr fluxes from silicate sources are shown in Fig. 8.

By decoupling the carbonate and silicate Sr fluxes, we obtain a value for J_{Him} that is roughly an order of magnitude smaller than that obtained by Richter.[25] Our value is an estimate of the silicate weathering flux, whereas the Richter value is a total dissolved Sr flux. The minimum ($R_{Carb} = 0.7092$) estimate of J_{Him} (4.8×10^8 moles/year) constitutes about 11% of the world silicate Sr flux, but only 1.5% of the world total Sr flux. The maximum ($R_{Carb} = 0.708$) estimate for J_{Him} (1.6×10^9 moles/year) constitutes roughly 32% of the silicate Sr flux and

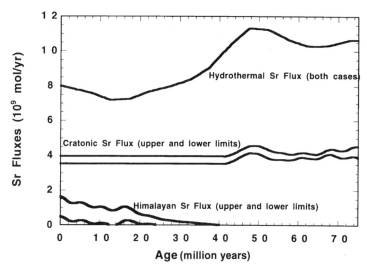

FIGURE 8. Sr fluxes used in, and derived from, two end-member cases of the mass balance described in the text. Both cases assume a constant carbonate Sr flux of 2.64×10^{10} moles/year. The difference between these two cases is the ratio assigned to the carbonate source: in the minimum Himalayan flux case, $R_{Carb} = R_{SW}$ at any given time; in the maximum Himalayan flux case $R_{Carb} = 0.708$ for the duration of the model. The hydrothermal Sr flux is scaled to the Engebretson *et al.*[32] seafloor generation curve in both cases. From 75 to 40 million years ago, we solve for the cratonic Sr flux (ratio = 0.72) necessary to account for the variation in the marine $^{87}Sr/^{86}Sr$ record; from 40 million years ago to the present, we hold the cratonic source constant, and solve for the flux of Himalayan Sr (ratio = 0.742) necessary to account for measured variation in the marine $^{87}Sr/^{86}Sr$ record. Choosing a carbonate $^{87}Sr/^{86}Sr$ value of 0.708 requires a significantly larger Himalayan Sr source.

4.8% of the world total Sr flux. These estimates compare well with the total measured Sr flux from the Ganges–Brahmaputra river system which is about 1.27×10^9 moles/year,[43] of which as estimated below, 40% (or 5.1×10^8 moles/year) comes from silicate sources.

4.2. Comparison of the Strontium Mass-Balance Results with the Oxygen Isotope Record

Our mass balance contains the Sr fluxes from MORB, background continental weathering, and the Himalaya. The continental Sr sources (J_{Crat} and J_{Him}) also contribute significant fluxes of the major cations Na$^+$, K$^+$, Mg^{2+}, and Ca^{2+}; these Sr sources can thus be thought of as "alkaline Sr sources." Conversely, the hydrothermal Sr source (J_{Hyd}) scales with seafloor spreading, which contributes CO_2 through direct outgassing and through Mg^{2+}–Ca^{2+} exchange; thus, this Sr source can be thought of as an "acid Sr source." Assuming that the scaling between these Sr fluxes and their respective acidity/alkalinity fluxes is roughly constant, the ratio of J_{Hyd} to $J_{Crat} + J_{Him}$ gives an index (Q_J) that may be a proxy for the acidity–alkalinity balance of the oceans:

$$Q_J = \frac{J_{Hyd}}{J_{Crat} + J_{Him}} \tag{12}$$

When the value of Q_J is high, we expect larger acid fluxes to drive the ocean and atmosphere toward higher pCO$_2$; low values of Q_J should drive the ocean and atmosphere toward lower pCO$_2$. This index is plotted against time and compared to the benthic δ^{18}O record in Fig. 9a. The spreading rate history used has no effect on the shape of the Q_J curve in the period from 75 to 40 million years ago, and only a minor effect from 40 million years ago to the present.

For a steady state BLAG-type model, Q_J is equivalent to the ratio $P(t)/A(t)$ and therefore should be directly related to R_{atm}:

$$R_{atm} = \left(\frac{Q_J(t)}{Q_J(0)}\right)^{-n} \tag{13}$$

The strength of the feedback determines the amplitude of the pCO$_2$ response, but the shape of the CO_2 curve is essentially the same as Q_J itself. Adopting the CO_2 feedback strength used in GEOCARB II ($n = 0.5$), we can transform Q_J into an estimated CO_2 history and compare it to that derived from the $-\delta^{18}$O record (Fig. 9b).

4.3. Modern Alkalinity Fluxes from Himalayan–Tibetan Plateau Rivers

To evaluate the impact of Himalayan uplift on climate, it is also important to directly assess the magnitude of the alkalinity fluxes from the Himalayan–Tibetan Plateau (HTP) region. Raymo and Ruddiman cite dissolved solids data

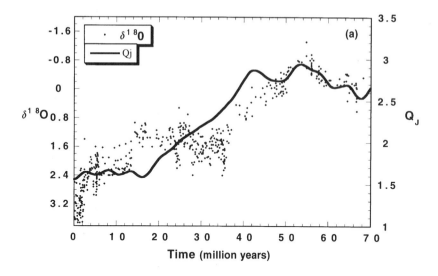

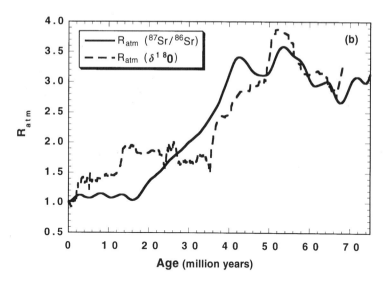

FIGURE 9. (a) A comparison Q_J, a measure of the relative acidity and alkalinity fluxes to the ocean, with the Cenozoic benthic O isotope record. (Miller *et al.*[2]). The model acidity–alkalinity flux seems to agree well with the $\delta^{18}O$ record during the Paleocene and Eocene. It then captures much of the temperature descent in the early Oligocene, but seems to diverge in the late Eocene and Miocene. While imperfect, Q_J demonstrates how a relatively straightforward interpretation of the $^{87}Sr/^{86}Sr$ record can capture much of the climatic variability recorded by the benthic $\delta^{18}O$ record. (b) Q_J converted to R_{atm}, using a BLAG-style pCO_2 feedback strength of $n = 0.5$. This CO_2 curve is compared to that derived from the $\delta^{18}O$ record [Eq. (1)]. Both are highly model-dependent, but yield a Eocene R_{atm} value of about four times modern.

for rivers draining the HTP region that show that these rivers supply 25% of the world dissolved solids to the oceans by mass, although the HTP region itself makes up only 4.2% of the Earth's land surface.[13,46,47] The Ganges, Brahmaputra, and Indus rivers, which drain the Himalaya, supply about 44% of the regional dissolved solids flux and 11% of the global flux.

Mass estimates of solute fluxes are insufficient to address the alkalinity budget problem, because they reflect dissolution of silicates, carbonates, and evaporites, along with aerosol inputs. It is difficult to reliably separate the silicate weathering from evaporite and carbonate dissolution in rivers,[48–50] but it is worth making an attempt because only the silicate component can cause changes in ocean chemistry and atmospheric pCO_2.

Solute concentration measurements on the Ganges–Brahmaputra river system by Sarin et al.[51] provide a data set that we can use to constrain the size of modern alkalinity fluxes. Table 1A shows data from the Ganges and Brahmaputra river systems, along with data from the Indus, Mekong, Chiang (Yangtze), and Huang He (Yellow) rivers.[52] To correct for aerosols and evaporite NaCl we subtract the Cl^- from Na^+ concentration of the river waters. Similarly, we subtract SO_4^{2-}, which most likely comes from evaporites (gypsum and anhydrite) and aerosol inputs, from Ca^{2+} ion concentrations (Table 1B).

To evaluate the Ca^{2+} and Mg^{2+} contributions from silicates, we use the ^{87}Sr/^{86}Sr ratio of Ganges–Brahmaputra river water (about 0.7213)[53] to estimate the fraction of Sr^{2+} in the rivers contributed by silicate weathering. The granites exposed in the High Himalayan Crystalline (HHC) series have extremely high ^{87}Sr/^{86}Sr ratios (between 0.74 and 0.76).[54,55] Bengal Fan sediments show an average ^{87}Sr/^{86}Sr ratio of 0.742, which must represent the average ratio of the silicate source rocks.[45] The appropriate mass balance is

$$(^{87}Sr/^{86}Sr)_{Sil} \times f_{Sr_{Sil}} + (^{87}Sr/^{86}Sr)_{Carb} \times (1 - f_{Sr_{Sil}}) = (^{87}Sr/^{86}Sr)_{River} \qquad (14a)$$

Solving for $f_{Sr_{Sil}}$

$$f_{Sr_{Sil}} = \frac{(^{87}Sr/^{86}Sr)_{River} - (^{87}Sr/^{86}Sr)_{Carb}}{(^{87}Sr/^{86}Sr)_{Sil} - (^{87}Sr/^{86}Sr)_{Carb}} \qquad (14b)$$

The ^{87}Sr/^{86}Sr value of the carbonate component must be between 0.7078 to 0.7092, and so we calculate that the silicate Sr flux is 37 to 40% of the total dissolved Sr in the Ganges–Brahmaputra river system. We choose an early Cenozoic value of 0.708 for the carbonate source in this mass balance to put an upper limit on the silicate component (Table 1B).

To solve for the fraction of Ca^{2+} from silicate weathering, we use typical Sr/Ca ratios and the above estimate of $f_{Sr_{Sil}}$

$$f_{Ca_{Sil}} = \frac{(f_{Sr_{Sil}}/(Sr/Ca)_{Sil}}{(f_{Sr_{Sil}}/(Sr/Ca)_{Sil}) + (f_{Sr_{Carb}}/(Sr/Ca)_{Carb})} \qquad (15)$$

TABLE 1.   Alkalinity Fluxes from Himalayan–Tibetan Plateau Rivers

A. HTP Fluxes (10^9 moles/year)[a]

River	K^+	Na^+	Ca^{2+}	Mg^{2+}	Cl^-	SO_4^{2-}
Ganges	27	172	249	113	55	35
Brahmaputra	29	56	214	96	19	64
Indus	12	93	157	55	48	65
Mekong	29	91	204	76	86	23
Chiang (Yangtze)	34	190	1190	286	123	198
Huang He (Yellow)	3	102	60	43	54	37
Totals:	134	704	2074	669	385	422

B. Sr Mass Balance[b]

River	$(^{87}Sr/^{86}Sr)_R$	$(^{87}Sr/^{86}Sr)_{Sl}$	$(^{87}Sr/^{86}Sr)_{Carb}$	$f_{Sr,Sil}$	$(Sr/Ca)_{Sil}$	$(Sr/Ca)_{Carb}$	$f_{Ca,Sil}$	$f_{Mg,Sil}$
Ganges/ Brahmaputra	0.7213	0.742	0.708	0.3912	0.0035	0.0011	0.1680	0.0840
Indus	0.7112	0.72	0.708	0.2667	0.0035	0.0011	0.1026	0.0513
Mekong	0.7102	0.72	0.708	0.1833	0.0035	0.0011	0.0659	0.0330
Chiang	0.7109	0.72	0.708	0.2417	0.0035	0.0011	0.0910	0.0455

C. Corrected Silicate Weathering Fluxes (10^9 moles/year)[c]

River	K^+	$Na^+ - Cl^-$	$(Ca^{2+} - SO_4^{2-})$ $\times f_{Ca,Sil}$	$Mg^{2+} \times f_{Mg,sil}$	Alkalinity
Ganges	27	117	36	9	235
Brahmaputra	29	37	25	8	133
Indus	12	45	9	3	82
Mekong	29	5	12	3	63
Chiang (Yangtze)	34	67	90	13	308
Huang He (Yellow)	3	48	2	2	59
Totals:	134	319	175	38	879
Global Silicate cation fluxes	1200	2800	2720	3020	15480
% Silicate weathering	11.2	11.4	6.4	1.3	5.7

[a]Modern solute fluxes from rivers that drain the Himalaya and Tibetan Plateau.[51,52]
[b]Riverine $^{87}Sr/^{86}Sr$ mass balance, used to estimate the fraction of Ca^{2+} and Mg^{2+} derived from silicate rocks in these rivers.
[c]River solute data, corrected for acid evaporite and aerosol inputs first, by subtracting Cl^- and SO_4^{2-} concentrations; next corrected for carbonate Ca^{2+} and Mg^{2+} using the correction factors $f_{Ca_{sil}}$ and $f_{Mg_{sil}}$ derived in (b).[4,59,60]

Marine carbonates have high total Sr concentrations (average 1000 ppm) but lower Sr/Ca ratios than granites (1.1×10^{-3}) (Palmer[42] and Maybeck[56]). Average granites have Sr/Ca ratios around 3.5×10^{-3} (Best[57]) but Himalayan leucogranites have Sr/Ca ratios as high as 10×10^{-3} (Vidal et al.[54]). Although the initial dissolution rates of various cations are quite different, as weathering proceeds toward steady state, all cations must be released according to their proportions in the primary minerals.[35] We make this assumption for Sr^{2+} and Ca^{2+}, and adopt an average granitic value of $(Sr/Ca)_{Sil}$ (3.5×10^{-3}) to put an upper limit on silicate Ca^{2+} fluxes in the Ganges–Brahmaputra river system. This yields a value for $f_{Ca_{Sil}}$ of 0.168 (Table 1B).

We can apply the same analysis to the Indus, Chiang, and Mekong rivers, using the measured river $^{87}Sr/^{86}Sr$ values and assuming an average cratonic $^{87}Sr/^{86}Sr$ of 0.72 (Brass[41]). In the case of the Indus, which has a riverine $^{87}Sr/^{86}Sr$ of 0.7112, we estimate $f_{Ca_{Sil}}$ to be 0.103. For the Mekong and Chiang, which have measured $^{87}Sr/^{86}Sr$ values of 0.710 and 0.710, we obtain estimates of $f_{Ca_{Sil}}$ of 0.066 and 0.091, respectively.[43] For the Huang He, we have no measured riverine $^{87}Sr/^{86}Sr$ value; we will adopt the same $f_{Ca_{Sil}}$ value as for the Chiang. Clearly, better knowledge of the silicate source rock $^{87}Sr/^{86}Sr$ values would help, but by choosing an average of 0.72, we hope to overestimate, rather than underestimate, the silicate Ca^{2+} flux (Table 1B).

Estimating the fraction of Mg^{2+} that can be attributed to silicate weathering is difficult. One approach is to recognize that Mg/Ca ratios in granites are about 0.5 (Best[57]), and simply guess that $f_{Mg_{Sil}} \approx 0.5 \times f_{Ca_{Sil}}$; this approach is shown in Table 1B. Although highly uncertain, it seems clear from the general approach that $f_{Mg_{Sil}}$ is less than 10% of the total dissolved Mg^{2+}.

We are reasonably confident of our corrections for the Ganges and Brahmaputra rivers, and somewhat less so for the Indus, but we may be overestimating the amount of silicate weathering represented by the Chinese rivers. The Tibetan Sedimentary Series (TSS) rocks that blanket the Tibetan Plateau are composed largely of carbonates, evaporites, and cation-depleted sediments that are less likely to contribute new alkalinity to the oceans.[58] Even after the corrections used here, our estimate of the Chiang's silicate alkalinity flux is larger than any other HTP river. It also has the highest fraction of alkalinity from Ca^{2+} and Mg^{2+}, indicating that it is still dominated by carbonate dissolution. The Mekong has an anomalously high flux of K^+ and the Huang He has a very high Na^+ flux; these may represent deposits of K and Na carbonate or other evaporites that we cannot correct for.

The total estimated flux for each cation is 11% or less of the world totals estimated by Berner (estimates from Morel and Hering[59] corrected for atmospheric cycling; based on Berner et al.[4,60]; data from Meybeck[61]). The total silicate alkalinity flux estimate for these six rivers is at most 8.8×10^{11} eq/year or less, corresponding to less than 6% of the world total. Of this, a maximum of 3.7×10^{11} eq/year (2.4% of the world total) comes from the Ganges–Brahmaputra river system; when the Indus is included, the estimated alkalinity flux from these three rivers is about 4.5×10^{11} eq/year (2.9% of the world total).

The total fluxes of silicate alkalinity from the HTP, at less than 6% of the world total, appear to be a small fraction of the estimated world total and out of proportion to the area of the HTP region by 50%, rather than the factor of five estimated by Raymo and Ruddiman.[13] The Ganges, Brahmaputra, and Indus rivers, which drain about 2% of total continental area, supply less than 3% of total silicate alkalinity. The data on global rates of silicate weathering are actually very sparse (most values quoted are traced to Meybeck[46,61]), so the fractional contribution of HTP rivers to the global flux has a substantial uncertainty that cannot be accurately assessed.

4.4. Estimating Paleoalkalinity Fluxes from Bengal Fan Sediments

Cation fluxes estimated from modern river dissolved loads are subject to seasonal and interannual variability and anthropogenic effects.[43] A longer-term view of the alkalinity flux from the Ganges–Brahmaputra river system can be obtained by comparing the chemistry of Bengal Fan sediments with that of the source rocks. ODP sites 717 and 718, located in the distal fan, sample sediments that span the last 20 million years. Isotopic evidence demonstrates that the clay and silt phases in Bengal Fan sediments derive overwhelmingly from granites and leucogranites of the HHC series. The mean $^{87}Sr/^{86}Sr$ ratio of fan sediments is 0.742, which is similar to HHC series granites and leuco-granites (0.74 to 0.76). The ε_{Nd} value of fan sediments, at -15.9, is similar to the values for the HHC, but significantly different from the TSS and Lesser Himalaya (LH) values.[45,55] Thus, significant contributions from the TSS and the LH are ruled out.

To estimate the dissolved cation flux associated with the sediments of the Bengal Fan, we assume that Al is conserved. We estimate the cation alkalinity deficit in the fan sediments by comparing the $(Na + K + 2Ca + 2Mg)/Al$ ratio (eq/mole Al) of the sediments to those of the likely source rocks. The total cation delivery to the ocean over the time period of deposition of the fan sediments is the average cation deficit per unit mass multiplied by the mass of the fan. The appropriate mass balance begins with these ideas:

$$eq_{Lost} = \left(\frac{eq}{mole\ Al} \right)_{Rocks} \times (moles\ Al)_{Total} - \left(\frac{eq}{mole\ Al} \right)_{Sediments} \times (moles\ Al)_{Total}$$

(16)

The sand fraction of the sediments appears to be largely unweathered, and so we assume they represent purely mechanical erosion. Chemical measurements of the sand fraction reveal that it has a cation eq/moles Al ratio of 1.01, which is somewhat higher than the Manaslu leucogranite (0.86 eq/mole Al),[54] but consistent with a primarily HHC source (Fig. 10a). The silt and clay fractions show significant cation loss, with an eq/mole Al ratio of 0.75 and a $\%Al_2O_3$ value of 17.3% (Fig. 10b). Using the sand fraction as the weighted average of source rock

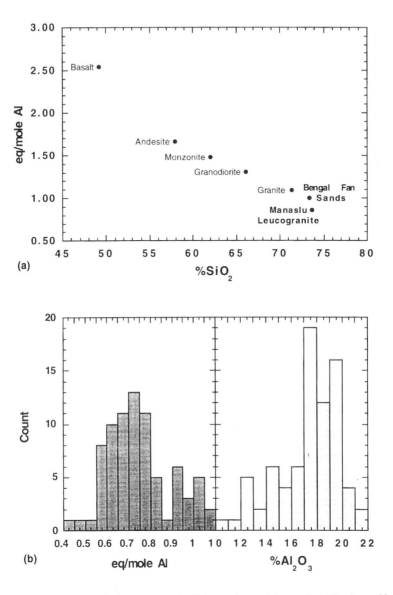

FIGURE 10.   (a) Cation equivalents per mole Al for major rock types, the Himalayan Manaslu leucogranite and Bengal Fan sands (average rock compositions from Best[57]). Cation contents of silicate rocks decreases with increasing %SiO$_2$. The Manaslu leucogranite's cation content indicates that High Himalayan Crystalline series have low initial cation contents in general. Sands plot very close, but somewhat above Manaslu leucogranite in cation eq/mole Al. Biasing toward quartz, and thus high SiO$_2$, in the sands has no effect on the mass balance used here. (b) Histograms of cation eq/mole Al and %Al$_2$O$_3$ in the silt/clay fraction of Bengal Fan sediments. The silt/clay fraction shows about 25% of cation equivalents are lost to solution, relative to the unweathered source rocks.

chemistry allows the mass balance above to be written as

$$eq_{Lost} = (Mass\ fan)\left[\left(\frac{eq}{mole\ Al}\right)_{Sand} - \left(\frac{eq}{mole\ Al}\right)_{Silt/Clay}\right]$$
$$\times \left(\frac{(1 - f_{Sand})(\%Al_2O_3)_{Silt/Clay}}{5098}\right) \qquad (17)$$

The mass of the fan has been estimated by seismic survey and density modeling to be 2.88×10^{22} g (Curray[62]). The fraction of sand in fan sediments is the most problematic variable in this mass balance. Our measurements on distal fan sediments at ODP sites 717 and 718 show roughly 7% sand, with substantial uncertainty. This estimate is almost certainly too low, owing to the distal nature of these sites. We will use this number as an upper estimate to the alkalinity fluxes from the Ganges–Brahmaputra river system. A best-guess of 50% sand for the fan as a whole is also made.

Using the values above, we calculate an upper limit to the total alkalinity loss from Himalayan weathering to be 2.4×10^{19} eq; our best-guess f_{sand} estimate changes this value to 1.3×10^{19} eq. Full error propagation analysis shows that these estimates are robust, and precise to about 50%. We are currently gathering more data on the Bengal Fan, and will present them in detail in a separate paper.

To convert this total alkalinity loss estimate to a flux, the major unknown is the age of initiation of Bengal Fan deposition. If we assume that the sediment has been deposited since 40 million years ago, the average flux would be 3.25×10^{11} eq/year. This estimates is lower than, but within the uncertainties of, the modern river flux we derived above. Assuming that the flux has increased over the past 40 million years and is a maximum at present, the two estimates are in remarkably close agreement.

5. THE IMPACT OF HIMALAYAN ALKALINITY FLUXES ON ATMOSPHERIC CO₂: AN OCEAN CHEMICAL MODEL

We have constructed a simple model of the ocean's alkalinity system to evaluate the effect of increased late Cenozoic alkalinity fluxes from the Himalaya. The model describes the carbonate equilibria in the ocean and assumes equilibrium with calcite sediments, as does the BLAG model. We start with the hydrogen ion mole balance (TOTH) for seawater (see Morel and Hering[59]), which can be written in the form

$$TOTH = [H_2CO_3^*] - [CO_3^{2-}] + [H^+] - [OH^-] \qquad (18)$$

or alternatively in the form:

$$TOTH = C_T - Na_T - K_T - 2Mg_T - 2Ca_T + Cl_T + 2SO_{4T} \qquad (19)$$

where C_T is the total dissolved carbon, and the other terms refer to total dissolved ion concentrations.

The species in Eq. (18) can be expressed in terms of hydrogen ion concentration, equilibrium constants, and total carbon concentration (C_T):

$$[OH^-] = K_W/[H^+] \tag{20a}$$

$$[H_2CO_3] = C_T \times \alpha_0 \tag{20b}$$

$$[CO_3^{2-}] = C_T \times \alpha_2 \tag{20c}$$

where K_W = The self-dissociation constant of water (corrected for seawater's ionic strength) and where

$$\alpha_0 = \left(1 + \frac{K_1}{[H^+]} + \frac{K_1 K_2}{[H^+]^2}\right)^{-1}$$

$$\alpha_2 = \left(\frac{[H^+]^2}{K_1 K_2} + \frac{[H^+]}{K_1} + 1\right)^{-1}$$

and where K_1 is the first dissociation constant of carbonic acid and K_2 is the second. The base cation and acid anion concentrations in Eq. (16) can be expressed as alkalinity:

$$\text{Alk} = Na_T + K_T + 2Mg_T + 2Ca_T - Cl_T - 2SO_{4T} \tag{21}$$

Substitution and rearrangement yields an expression that can be solved implicitly for $[H^+]$ by specifying concentrations of carbon, cations, and anions:

$$C_T(1 - \alpha_0 + \alpha_2) - \text{Alk} + \frac{K_W}{[H^+]} - [H^+] = 0 \tag{22}$$

To track the effect of an increased alkalinity flux to the oceans, we use a finite difference method, where the alkalinity and acidity fluxes are specified as a function of time. Since the equilibrium constants are pressure- and temperature-dependent, we must set P and T. For each time-step, apparent equilibrium constants for the carbonate system in seawater are calculated.[63,64] Addition of cations and C_T in the next time-step change the concentrations of cations and C_T in seawater, and the equation above is solved for $[H^+]$ with the new concentrations. All species concentrations are calculated and the new solution is tested for equilibrium with seawater. If the solution is supersaturated (at a depth of 2000 meters) a small amount of C_T and Ca_T is substracted to represent calcite precipitation. The H^+ ion concentration is determined and the model iterates until saturation is reached. If the system is undersaturated with respect to calcite,

Ca_T and C are added (corresponding to dissolution of calcite on the ocean floor) until saturation is reached.

The model reproduces modern ocean chemistry well. Atmospheric pCO_2 is accurate to 10% when surface ocean conditions are specified. Because the mathematics of subtracting cations and C_T from solution are the same as adding them to solution, the model can be run forward or backward in time. In the model results we discuss here, we take the modern ocean as a starting point, specify geochemical fluxes for the Cenozoic, and subtract these fluxes to see what state the ocean would have had in the past. All experiments focus on Himalayan "excess weathering" and assume that no other stabilizing processes exist.

The results of the modeling are shown in Figs. 11 and 12. Figure 11 shows the effects on ocean pH and pCO_2 of removing a given amount of alkalinity from modern seawater. The pH response is a titration curve; the pH changes rapidly as the equivalence point, corresponding to equal concentrations of the species HCO_3^- and CO_3^{2-}, is reached. The pCO_2 response is approximately quadratic with respect to the amount of alkalinity added or subtracted. The important conclusion to be drawn from Fig. 11 is that the reconstructed pCO_2 of the Eocene atmosphere, corresponding to the removal of 1.3×10^{19} eq of alkalinity from the oceans over the past 40–50 million years, is 40 times the modern value. This value is about 10 times larger than our estimate of Eocene pCO_2 based on the apparent temperature changes indicated by the $\delta^{18}O$ record.

Figure 12 shows the results of a calculation of the Cenozoic pCO_2 history, where the putative alkalinity flux from the Himalaya is scaled to the Q_J parameter

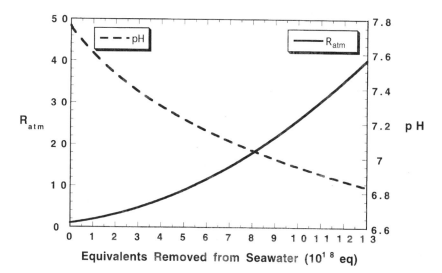

FIGURE 11. Time-independent response of ocean pCO_2 and pH to the addition of 1.3×10^{19} eq of alkalinity. This is equivalent to a flux of 3.25×10^{11} eq/year for 40 million years, as derived in the Bengal Fan mass balance. If no stabilizing or compensating mechanisms are invoked, the Paleocene ocean would have to start at fiftyfold modern CO_2 levels to accommodate this alkalinity flux.

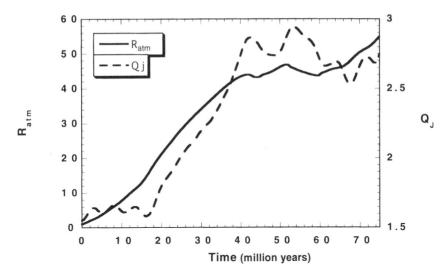

FIGURE 12.   Response of whole-ocean pCO_2 (solid line) to an alkalinity flux scaled to the Q_J parameter (dashed line), such that the total flux reaches 1.3×10^{19} equivalents 52 million years ago. CO_2 decreases monotonically as long as $Q_J < Q_J^*$, or the equilibrium point in the acidity–alkalinity balance. The monotonic character of the response does not fit our expectations for the climate system, but does indicate a very basic result: inflections in pCO_2 response only occur when strength of acidity and alkalinity fluxes reverse. This fact, together with climate records that indicate short- and long-term oscillations in pCO_2, could be interpreted as evidence for feedbacks that keep the fluxes relatively closely balanced but with enough freedom to get out of balance, and perhaps overshoot at times.

of Fig. 9, according to

$$F_{Alk}(t) = F_{Alk}(0)\frac{[Q_J(t)/Q_J^* - 1]}{[Q_J(0)/Q_J^* - 1]} \tag{23}$$

and such that

$$-\int_{t_{Him}}^{0} F_{Alk}(t)dt = 1.3 \times 10^{19}\,eq \tag{24}$$

where Q_J^* is the value of Q_J that corresponds to zero net alkalinity flux to the oceans (the alkalinity provided by continental weathering is exactly balanced by the acidity produced by volcanism and related processes) and t_{Him} is the age of the initiation of the excess Himalayan alkalinity flux. In the calculation shown we have chosen the value $Q_J^* = 2.55$ to correspond to zero net alkalinity flux. This value is an appropriate average for the Eocene, which is a time when the marine $^{87}Sr/^{86}Sr$ value was nearly constant for about 20 million years (Fig. 2a).

The calculated pCO_2 history has some features that correspond to the ocean paleotemperature record (Fig. 1) and some that do not. The calculated curve gives the highest value of pCO$_2$ about 52 million years ago and the lowest pCO$_2$ at the present, which is desired. The total change in the pCO$_2$ is large, as noted above. The curve shows pCO$_2$ dropping continuously from about 40 million years ago to the present. There is a more rapid decrease in calculated pCO$_2$ from 19 to 15 million years ago, which corresponds to the period of rapid marine ^{87}Sr/^{86}Sr increase. One feature of the δ^{18}O record that is not captured by the pCO$_2$ curve is the relatively large temperature shift between 55 and 40 million years ago. The model cannot reproduce this feature without rescaling the relationship between the alkalinity flux and Q_J for the period from 40 to 60 million years ago. The other feature of the δ^{18}O curve that is not captured by the model is the irregular decrease in temperature between 40 million years ago and the present. In particular there is no mechanism in the model to account for the apparent warming about 25 million years ago, for instance, since in the model the pCO$_2$ is continuously dropping from 40 million years ago to the present.

The calculated smooth shape of the pCO$_2$ curve does not match the more steplike behavior of the δ^{18}O record. One way to rationalize the relationship between the two would be if the ocean record responded in a nonlinear fashion to gradually dropping atmospheric pCO$_2$. At specific threshold values of pCO$_2$, the global surface temperature and/or the oceanic response to it as represented by the δ^{18}O record, may change drastically, whereas in other ranges of pCO$_2$ there might be little change. For example, the relatively large change in δ^{18}O about 35 million years ago followed a drop in calculated R_{atm} from about 45 about 50 million years ago to about 40 some 35 million years ago. The next major shift in δ^{18}O occurred about 15 million years ago after a drop of R_{atm} from 40 about 35 million years ago to about 12 some 15 million years ago. The last major δ^{18}O shift, about 2–3 million years ago, followed a drop in calculated R_{atm} from 12 around 15 million years ago to about 2 some 3 million years ago. In each case, the major response of the oceanic record may occur only after a certain value of R_{atm} is reached. As noted above and discussed further below, we think the absolute values of the calculated changes in R_{atm} are too large, but the calculated pattern of change of R_{atm} through the Cenozoic could be consistent with the δ^{18}O record.

6. DISCUSSION

From our modeling results, the most problematic aspect of the Raymo hypothesis appears to be that atmospheric pCO$_2$ in the Eocene is required to have been unrealistically high, in order to absorb the Himalayan alkalinity fluxes and arrive at the modern pCO$_2$ values. Exactly how high pCO$_2$ was in the Eocene is not known, and consequently we cannot evaluate fully how serious this problem is. In any case, there are several possible factors that are not included in our model that could come into play and result in a substantially lower estimate for Eocene pCO$_2$. There are three main issues. One is whether our estimates of the imbalance of the acidity and alkalinity fluxes to the ocean are correct. Another is whether there may be interocean processes that complicate the relationship between the chemical composition of the bulk ocean and that of the surface ocean, which is the part of the ocean that is likely to be in equilibrium with the atmosphere. The third issue concerns the buffer capacity of the ocean system and whether it is represented adequately in our model and in the BLAG–Berner models.

6.1. Cenozoic Alkalinity Fluxes

The most important conclusion to be drawn from our three Himalayan silicate weathering analyses is that earlier regional alkalinity fluxes have been overestimated. This means that the size of the alkalinity "imbalance" is reduced by roughly an order of magnitude. It is also possible that global alkalinity fluxes have been overestimated by a similar amount. Most estimates of global silicate weathering trace back to Meybeck,[46] who discusses the problem of correcting for silicate weathering but does not actually carry it out. We believe that the sedimentary fan analysis method developed here is preferable to river solute and bed-load estimates; we predict that if similar analyses are carried out on other major rivers, it will be shown that they also have much smaller silicate alkalinity fluxes than previously estimated. Because the sensitivity of the ocean's pCO_2 reflects the ratio of its total alkalinity to the global silicate alkalinity flux at any given time, reducing the global fluxes by an order of magnitude would increase pCO_2 stability by a similar amount.

Still, if the excess Himalayan silicate alkalinity flux (1.3×10^{19} eq) is not balanced somehow, the response of the ocean will remain too large. Another possibility that should be considered is that we have overestimated the net alkalinity flux "imbalance" from the Himalaya. The High Himalaya, which is being rapidly eroded, represents about 2% of the world's land area, but is a small fraction of southern Asia. It is possible that as a whole, the region is weathering less rapidly, relative to its area, than we have estimated here. It is also possible that global weathering rates have decreased, through a BLAG-style pCO_2 feedback, to compensate for excess Himalayan weathering. Both possibilities are difficult to assess.

Also, in our model we have included the alkalinity flux represented by dissolved, silicate-derived, Na^+ and K^+ from the Himalaya. These fluxes are not accounted for in the BLAG–Berner models as they are assumed to be always balanced or somehow neutralized. The sinks for these two cations are not known and constitute an important gap in our understanding of the alkalinity cycle. Na^+ from silicate weathering has a residence time in the ocean that is at least 10 times longer than K^+, indicating that sinks for Na^+ removal are inefficient. The Na^+ flux from Himalayan weathering represents about 25% of the total Himalayan alkalinity flux, and consequently has a major impact on calculated global pCO_2 in our model. If we were to include in our model only the alkalinity flux from riverine silicate-derived Ca^{2+}, it would reduce our calculated Eocene pCO_2 change by about a factor of five.

Variations in the size of the sedimentary organic C reservoir can also play a significant role in regulating atmospheric pCO_2 and have been used in the debate over controls of long-term climate.[6,7,14,40,65,66] Raymo[40] suggests that C burial rates during the Cenozoic have decreased to partially compensate for the increased alkalinity flux from the Himalaya. Derry and France-Lanord,[65] however, use a C mass balance that explicitly accounts for changing isotopic fractionations between the organic and inorganic components to argue that the late Cenozoic was characterized by a growth in the size of the organic C reservoir. This would imply that organic C burial rates increased in concert with the increasing silicate alkalinity fluxes. If Derry and France-Lanord are correct, then organic C burial rates cannot serve as an effective CO_2 "source" during this period. It has also been suggested there may have been an increased input of CO_2 to the atmosphere during the earliest stages of the India–Asian collision owing either to metamorphism of Tethyan carbonates[67] or to weathering of uplifted Tethyan sediments rich in organic matter.[68]

6.2. Surface/Deep Ocean Disequilibrium

Decoupling the deep ocean from the surface ocean pCO_2 is still another method for reducing the impact of increased weathering rates on atmospheric pCO_2. Our model is uniformitorial in the sense that it assumes that the present relationship between the surface ocean and the deep ocean applies for all time. If the whole ocean changes pCO_2 by a factor of 50, we have assumed that the surface ocean also changes pCO_2 by a factor of 50. However, there is a possibility that during warmer periods, because of decreased vertical mixing, the contrast between deep ocean and surface ocean pCO_2 could have been larger than it is today (see e.g., Knoll et al.[69]). If this were the case, the fiftyfold increase in bulk ocean pCO_2 predicted by our model could correspond to a smaller difference in atmospheric pCO_2. This feature could be added to the model, but would be purely speculative.

6.3 Heterogeneous Buffering Mechanisms

Another idea that has been considered, but abandoned in the past, is the concept of heterogeneous buffering between the aqueous phase and multiple solid phases in the ocean. Sillen[3] first noted that the apparent buffer capacity of seawater was small relative to the sizes of the acidity and alkalinity fluxes into the ocean, and thus that ocean pH should be highly sensitive to any imbalance. Sillen hypothesized that equilibrium reactions between seawater and clay sediments created a "pH-stat" in the ocean. The model predicted the reconstitution of a number of cation-rich solid phases (in addition to $CaCO_3$) in ocean sediments to provide sinks for Na, K, and Mg, which would also essentially fix the pH of the ocean at modern values. This "reverse weathering" hypothesis was popular until the 1970s, with various models predicting the nature of the equilibria between seawater and silicate phases in the oceans (see e.g., Sillen,[3,70] Mackenzie and Garrels,[71,72] Holland,[73] Hegelson and Mackenzie,[74] and Morel et al.[75]).

Eventually, the idea of heterogeneous buffering was abandoned because the predicted authigenic phases could not be identified in the sediments (or distinguished from detrital phases), nor could equilibrium be demonstrated between detrital phases and seawater. The 1979 discovery of black smoker vents in the deep sea,[34] and their effect on the Mg^{2+} budget of seawater, helped set the stage for the BLAG model construction, in which acidity and alkalinity fluxes are kinetically controlled processes (again tied primarily to the seafloor generation rate).

Recently, the idea of "reverse weathering" was revived in a paper that demonstrated significant amounts of K^+ being removed from seawater into a clay phase precipitated in Amazon Delta waters.[76] At the very least, this helps to identify a sink for K^+, which must exist based on its residence time in seawater. If equilibrium between clay phases of this type and seawater could be demonstrated, then we could begin to explore the resulting increase in ocean buffer capacity.

Following the methodology of Morel,[75] it is possible to derive the buffer capacity of an aquatic system in equilibrium with a number of solid phases. In this methodology, all of the chemical "species" (which actually exist) are expressed in terms of a minimum number of chemical "components" (which may or may not correspond to real species). Choosing the chemical components to be the solids with fixed activity, and the most abundant aquatic species (i.e., the "major species") simplifies calculation of the H^+ ion mole balance (TOTH). The "minor species" then are those lower-concentration species not required to be chemical components. (The reader is referred to Morel[75] for a detailed discussion of these ideas).

Buffer capacity is defined as the rate of change in TOTH with pH:

$$\beta_H = -\frac{\delta\text{TOTH}}{\delta\text{pH}} = -\left(\frac{\delta\text{pH}}{\delta\text{TOTH}}\right)^{-1} \tag{25}$$

where $\text{TOTH} = \Sigma_i \lambda_i[S_i]$, with $[S_i]$ the concentrations of the "minor species," and λ_i the coefficient of "minor species" S_i in TOTH. As the numerical value of TOTH increases, buffer capacity increases.

Morel's Minor Species theorem[75] states that if the chemical components are chosen as equilibrium solid phases and "major species," then the TOTH equation will be in terms of the "minor species" concentrations. It can be shown[59,75] that using these rules, an excellent approximation for the buffer capacity of the system is given by a very simple formula, related to the TOTH equation:

$$\beta_H = 2.3 \sum_i \lambda_i^2[S_i] \tag{26}$$

As cation-rich solid phases are added to the system, they become chemical components displacing the aqueous cations that were "major species." These cations cease to be "major species" and will instead become "minor species" included in the TOTH of the new system.[75] This causes the numerical value of the TOTH equation to increase, with the result that buffer capacity tends to increase dramatically with each additional solid phase that is in equilibrium with the solution. Conceptually, this means that the more solid phases that are capable of dissolving or precipitating in response to a pH change in solution, the better buffered the solution phase will be.

It can be shown that the TOTH equation used for a seawater–calcite system, where the $[Ca^{2+}]$ of the solution phase is controlled by equilibrium with calcite, is

$$TOTH = 2[H_2CO_3] + [HCO_3^-] + [H^+] - [OH^-] \qquad (27)$$

and the resulting buffer capacity is

$$\beta_H = 2.3\{4[H_2CO_3] + [HCO_3^-] + [H^+] + [OH^-]\} \approx 0.0042 \qquad (28)$$

If the clay phase identified by Michalopoulos and Aller[76] is in equilibrium with seawater, the appropriate TOTH and β_H would be:

$$TOTH = 0.937[K^+] + 2[H_2CO_3^*] + [HCO_3^-] - 2[H_2SiO_4^{2-}]$$
$$- [Al(OH)_4^-] + 3[Fe^{3+}] - [OH^-] + [H^+] \qquad (29)$$

$$\beta_H = 2.3\{0.878[K^+] + 4[H_2CO_3^*] + [HCO_3^-] + 4[H_2SiO_4^{2-}] + [Al(OH)_4^-]$$
$$+ 9[Fe^{3+}] + [OH^-] + [H^+]\} \approx 0.025 \qquad (30)$$

The resulting system would have a buffer capacity roughly six times larger than our simple model ocean, owing primarily to the inclusion of $[K^+]$ in the TOTH equation. This is in line with the roughly factor-of-ten increase in stability that is needed to accommodate the estimated Himalayan alkalinity fluxes we derived above.

While the idea that there are solid phases that approximate equilibrium with seawater and exert a significant influence on its pH stability is speculative and even somewhat heretical, it is exactly the type of stabilizing mechanism that would allow both weathering and CO_2 fluxes to vary somewhat independently, without wild swings in atmosphere pCO_2.

Overall, it seems probable that mechanisms could be found for reducing the change in pCO_2 from the Eocene to the present from the fiftyfold shown in Fig. 12 to something like a fivefold change. If these mechanisms can be shown to be plausible, then the Raymo hypothesis itself will have to be considered plausible as well.

7. OTHER CONSIDERATIONS

Our modeling suggests that it may be possible to explain some of the features of the Cenozoic paleotemperature record in terms of a long-term climate model where the alkalinity flux from the continents and the acidity flux from volcanism and related processes vary independently and can get out of balance by at least a few percent. The evidence in favor of a no-feedback or limited-feedback model is that there are good correlations between the marine $^{87}Sr/^{86}Sr$ record and the existing measures of long-term climate change. Our interpretation of the $^{87}Sr/^{86}Sr$ record calls for a CO_2 system that is dominated at times by volcanism and at other times by continental weathering. This suggests that either the ocean's pH buffering system has sufficient capacity to absorb these imbalances or that there are weaker feedbacks that may kick in after a few million years to bring the climate system to a new equilibrium. In either case, we believe a model of climate control that is less rigidly moderated by the CO_2 outgassing rate is required.

The idea that major climate shifts are driven by small, uncompensated imbalances in the alkalinity and acidity fluxes to the oceans makes predicting past pCO_2 levels extremely uncertain because it is difficult to determine the absolute values of any of the fluxes to a few percent, and imbalances at the few percent level can make all the difference. Although they cannot be measured directly, relatively small perturbations to the acidity or alkalinity fluxes should be observable in paleotemperature records. For example, the arrival of a mantle plume head at the Earth's surface can lead to a temporary basaltic eruption rate of ca. $1\,km^3/year$ over a period of one to several million years. This eruption rate is roughly 5% of the modern seafloor spreading rate, and if CO_2 degassing were proportional to eruption rate, the plume volcanism could cause a major global warming. A similar mechanism involving the reorganization of plate boundaries has also been mentioned in conjunction with the particularly warm period 55 million years ago.[77,78]

Clearly, stabilizing mechanisms are necessary to keep the system in relative balance; the nature and strength of these mechanisms are still the critical issues. Any model that attempts to predict past pCO_2 levels must choose among proposed mechanisms, and the results obtained will reflect those choices. Careful comparison of model results to data on past climates may help resolve which fluxes are driving the system and which processes are acting to stabilize pCO_2.

Another approach that would help would be to measure directly the chemical state of the oceans in the past. Such a data set would help answer the question of how much and on what timescale pCO_2 actually varies; the timing and pattern of chemical shifts in the ocean could indicate, by correlation to geological proxies such as the marine $^{87}Sr/^{86}Sr$ and seafloor generation records, which processes are driving the system at any given time. Methods have been proposed for measuring both paleo-pCO_2 and paleo-pH in the ocean,[16,79] but neither has yet yielded a detailed or reliable data set that can be used to constrain models. The importance of such data makes developing these methods a priority.

8. CONCLUSION

The problem of understanding the history and mechanisms of long-term climate is a truly difficult one as there is little to constrain it.

What is commonly "known" about the ocean–atmosphere system is that pCO_2, and thus climate, should be highly sensitive to imbalances in acidity and alkalinity fluxes to the ocean. We submit that even this sensitivity needs to be questioned: we do not believe that there is sufficient evidence to reject the hypothesis of heterogeneous buffering of the ocean, as outlined by Morel.[75]

If the system is highly sensitive to acidity–alkalinity flux imbalances, the size of the fluxes of acids and bases into and out of the ocean–atmosphere system are not known well enough to accurately "predict" paleo-pCO_2 levels, either today or in the past. It is unlikely that all of the relevant fluxes will ever be known with sufficient precision. We have attempted to develop an index of the acidity–alkalinity balance by reinterpreting the marine $^{87}Sr/^{86}Sr$ record in terms of a Himalayan Sr source. This index, Q_J, is roughly consistent with the marine $\delta^{18}O$ record, and thus cannot be used to reject the Raymo hypothesis.

Close examination of the actual alkalinity fluxes from the Himalayan region (both today and in the past) reveals two important points. The first is that chemical weathering in the Himalayan–Tibetan Plateau region represents roughly 5%, rather than 50%, of the world total silicate alkalinity fluxes (although this world total also needs closer examination). The second point is that a 5% imbalance appears to be more than enough to have driven the Cenozoic cooling.

Furthermore, the mechanisms that cause weathering and degassing rates to closely match each other are unclear and unproven, if they exist at all. We have outlined arguments as to why we believe the weathering feedback used in BLAG-style climate models is too strong and discussed alternative mechanisms.

Constructing elaborate models of paleo-pCO_2 is a questionable pursuit. The actual pCO_2 response of the system is unknown, and so the only checks on the results obtained are indirect. If sufficient adjustable parameters are introduced into a model (see, e.g., Berner[14]), it is possible to produce a paleo-pCO_2 curve that looks a lot like the $\delta^{18}O$ record. We have shown that an alternative formulation — with different assumptions and fewer adjustable parameters — does nearly as well in reproducing the $\delta^{18}O$ curve. Clearly, even if a model accurately predicted the behavior of the system, and could be checked against some proxy, the model would not uniquely determine the mechanisms operating. We warn against the logical fallacy of affirming model results as evidence that they are correctly constructed. Instead, we urge that models be used to develop better experiments, which might lead to a better understanding of both the history and mechanisms of long-term climate evolution.

Acknowledgments

This research was sponsored by the U.S. Department of Energy, Office of Energy Research, Environmental Sciences Division, Office of Health and Environmental Research, under appointment to the Graduate Fellowships for Global Change administered by Oak Ridge Institute for Science and Education. Thanks to C. F. Lanord and L. A. Derry for stimulating discussion of Bengal Fan isotopes and chemistry, and for use of preliminary data that showed a Bengal Fan mass balance was possible. Thanks also to A. J. Spivack and D. M. Bice for their helpful reviews of this manuscript.

REFERENCES

1. Miller, K. G., Fairbanks, R. G., and Mountain, G. S. (1987). *Paleoceanography* **2**, p. 1.
2. Zachos, J. C., Stott, L. D., and Lohmann, K. C. (1994). *Paleoceanography* **9**, p. 353.
3. Sillen, L. G. (1961). In: *Oceanography*, Vol 67. (M. Sears, ed.), p. 549. AAAS, Washington, D.C.
4. Berner, R. A., Lasaga, A. C., and Garrels, R. M. (1983). *Am. J. Sci.* **283**, p. 641.
5. Lasaga, A. C., Berner, R. A., and Garrels, R. M. (1985). In: *The Carbon Cycle and Atmospheric CO_2: Natural Variations Archean to Present*, Vol. 32 (E. T. Sundquist and W. S. Broecker, eds.), p. 397. American Geophysical Union, Washington, D.C.
6. Berner, R. A. (1990). *Science* **249**, p. 1382.
7. Berner, R. A. (1991). *Am. J. Sci.* **291**, p.339.
8. France-Lanord, C., Sheppard, S. M. F., and Le Fort, P. (1988). *Geochim. Cosmochim. Acta* **52**, p. 513.
9. Francois, L. M., and Walker, J. C. G. (1992). *Am. J. Sci.* **292**, p. 81.
10. Chamberlain, T. C. (1899). *J. Geol.* **7**, p. 545.
11. Raymo, M. E., Ruddiman, W. F., and Froelich, P. N. (1988). *Geology* **16**, p. 649.
12. Raymo, M. E. (1991). *Geology* **19**, p. 344.
13. Raymo, M. E., and Ruddiman, W. F. (1992). *Nature* **359**, p. 117.
14. Berner, R. A. (1994). *Am. J. Sci.* **294**, p. 56.
15. Arthur, M. A., Hinga, K. R., Pilson, E. Q., and Whitaker, E. (1991). *EOS, Trans. AGU* **72**, p. 166.
16. Hinga, K. R., Arthur, M. A., Pilson, M. E. Q., and Whitaker, D. (1994). *Global Biogeochem. Cycles* **8**, p. 91.
17. Laws, E. A., Popp, B. N., Bidigare, R. R., Kennicutt, M. C., and Macko, S. A. (1995). *Geochim. Cosmochim. Acta* **59**, p. 1131.
18. DePaolo, D. J., and Ingram, B. L. (1985). *Science* **227**, p. 938.
19. DePaolo, D. J. (1986). *Geology* **14**, p. 103.
20. Hess, J., Bender, M. L., and Schilling, J.-G. (1986). *Science* **231**, p. 979.
21. Hess, J., Bender, M. L., and Schilling, J.-G. (1991). *Earth Planet. Sci. Lett.* **103**, p. 133.
22. Richter, F. M., and DePaolo, D. J. (1987). *Earth Planet. Sci. Lett.* **83**, p. 27.
23. Richter, F. M., and DePaolo, D. J. (1988). *Earth Planet. Sci. Lett.* **90**, p. 382.
24. Farrell, J. W., Clemens, S. C., and Gromet, L. P. (1995). *Geology* **23**, p. 403.
25. Richter, F. M., Rowley, D. B., and DePaolo, D. J. (1992). *Earth Planet. Sci. Lett.* **109**, p. 11.
26. Edmond, J. M. (1992). *Science* **258**, p. 1594.
27. Richter, F. M., Lovera, O. M., Harrison, T. M., and Copeland, P. (1991). *Earth Planet. Sci. Lett.* **105**, p. 266.
28. Southam, J. R., and Hay, W. W. (1977). *J. Geophys. Res.* **82**, p. 3825.
29. Pitman, W. C., III (1978). *Geol. Soc. Am. Bull.* **89**, p. 1389.
30. Kominz, M. A. (1984). In: *Interregional Unconformities and Hydrocarbon Accumulation*, Vol. 36 (J. S. Schlee, ed.), p. 109. American Association of Petroleum Geologists, Tulsa.
31. Gaffin, S. (1987). *Am. J. Sci.* **287**, p. 596.
32. Engebretson, D. C., Kelley, K. P., Cashman, H. J., and Richards, M. A. (1992). *GSA Today* **2**, p. 93.
33. Haq, B. U., Hardenbol, J. and Vail, P. R. (1987). *Science* **235**, p. 1136.
34. Edmond, J. M., Measures, C., McDuff, R. E., Chan, L. H., Collier, R., and Grant, B. (1979). *Earth Planet. Sci. Lett.* **46**, p. 1.

35. Stumm, W. (1992). *Chemistry of the Solid-Water Interface: Processes at the Mineral–Water and Particle–Water Interface in Natural Systems.* John Wiley, New York.
36. Ronov, A. B. (1976). *Geochem. Int.* **13**, p. 172.
37. Barron, E. J., Sloan, J. L., and Harrison, C. G. A. (1980). *Palaeogr. Palaeoclim. Palaeoecol.* **30**, p. 17.
38. Tardy, Y., N'Kounkou, R., and Probst, J.-L. (1989). *Am. J. Sci.* **289**, p. 455.
39. Volk, T. (1987). *Am. J. Sci.* **287**, p. 763.
40. Raymo, M. E. (1994). *Paleoceanography* **9**, p. 399.
41. Brass, G. W. (1976). *Geochim. Cosmochim. Acta* **40**, p. 721.
42. Palmer, M. R., and Elderfield, H. (1985). *Nature* **314**, p. 326.
43. Palmer, M. R., and Edmond, J. M. (1989). *Earth Planet Sci. Lett.* **92**, p. 11.
44. Capo, R. C., and DePaolo, D. J. (1990). *Science* **249**, p. 51.
45. France-Lanord, C., Derry, L. A., and Michard, A. (1993). In: *Himalayan Tectonics,* Vol. 74 (P. J. Treloar and M. P. Searle, eds.), p. 603. The Geological Society, London.
46. Meybeck, M. (1976). *Hydrol. Sci. Bull.* **21**, p. 265.
47. Pinet, P., and Souriau, M. (1988). *Tectonics* **7**, p. 563.
48. Stallard, R. F., and Edmond, J. M. (1981). *J. Geophys. Res.* **86**, p. 9844.
49. Stallard, R. F., and Edmond, J. M. (1983). *J. Geophys. Res.* **88**, p. 9671.
50. Stallard, R. F., and Edmond, J. M. (1987). *J. Geophys. Res.* **92**, p. 8293.
51. Sarin, M. M., Krishnaswami, S., Dilli, K., Somayajulu, B. L. K., and Moore, W. S. (1989). *Geochim. Cosmochim. Acta* **53**, p. 997.
52. Hu, M., Stallard, R. F., and Edmond, J. M. (1982). *Nature* **298**, p. 550.
53. Krishnaswami, S., Trivedi, J. R., Sarin, M. M., Ramesh, R., and Sharma, K. K. (1992). *Earth Planet. Sci. Lett.* **109**, p. 243.
54. Vidal, P., Cocherie, A., and Le Fort, P. (1982). *Geochim. Cosmochim. Acta* **46**, p. 2279.
55. Deniel, C., Vidal, P., Fernandez, A., Le Fort, P., and Peucat, J.-J. (1987). *Contrib. Mineral Petrol.* **96**, p. 78.
56. Meybeck, M. (1988). In: *Physical and Chemical Weathering in Geochemical Cycles,* (A. Lerman and M. Meybeck, eds.), p. 247. Kluwer, Dordrecht.
57. Best, M. B. (1982). *Igneous and Metamorphic Petrology,* W. H. Freeman, New York.
58. Kidd, W. S. F., Yusheng, P., Chengfa, C., Coward, M. P., Dewey, J. F., Gansser, A., Molnar, P., Shackleton, R. M., and Yiyin, S. (1988). In: *The Geological Evolution of Tibet* (C. Chengfa, R. M. Shackleton, J. F. Dewey and Y. Jixang, eds.), p. 287. The Royal Society, London.
59. Morel, F. M. M., and Hering, J. G. (1993). *Principles and Applications of Aquatic Chemistry.* John Wiley, New York.
60. Berner, E. K., and Berner, R. A. (1987). *The Global Water Cycle.* Prentice-Hall, Englewood Cliffs, NJ.
61. Meybeck, M. (1979). *Rev. Geol. Dyn. Geogr. Phys.* **21**, p. 215.
62. Curray, J. R. (1994). *Earth Planet. Sci. Lett.* **125**, p. 371.
63. Roy, R. N., Roy, L. N., Vogel, K. M., Porter-Moore, C., Pearson, T., Good, C. E., Millero, F. J., and Campbell, D. M. (1993). *Mar. Chem.* **44**, p. 249.
64. Millero, F. J., and Sohn, M. L. (1992). *Chemical Oceanography,* CRC Press, Boca Raton, FL.
65. Derry, L. A., and France-Lanord, C. (1996). *Paleoceanography* **11**, p. 267.
66. Compton, J. S., and Mallinson, D. J. (1996). *Paleoceanography* **11**, p. 431.
67. Kerrick, D. M., and Caldeira, K. (1993). *Chem. Geol.* **108**, p. 201.
68. Beck, R. A., Burbank, D. W., Sercombe, W. J., Olson, T. L., and Khan, A. M. (1995). *Geology* **23**, p. 387.
69. Knoll, A. H., Bambach, R. K., Canfield, D. E., and Grotzinger, J. P. (1996). *Science* **273**, p. 452.
70. Sillen, L. G. (1963). *Sven. Kem. Tid.* **75**, p. 161.
71. Mackenzie, F. T., and Garrels, R. M. (1966). *Am. J. Sci.* **264**, p. 507.
72. Mackenzie, F. T., and Garrels, R. M. (1966). *J. Sed. Pet.* **36**, p. 1075.
73. Holland, H. D. (1965). *Proc. Natl. Acad. Sci.* **53**, p. 1173.
74. Helgeson, H. C., and Mackenzie, F. T. (1970). *Deep-Sea Res.* **17**, p. 877.
75. Morel, F. M. M., McDuff, R. E., and Morgan, J. J. (1976). *Mar. Chem.* **4**, p. 1.

76. Michalopoulos, P., and Aller, R. C. (1995). *Science* **270**, p. 614.
77. Owen, R. M., and Rea, D. K. (1985). *Science* **227**, p. 166.
78. Rea, D. K., Zochos, J. C., Owen, R. M., and Gingerich, P. D. (1990). *Palaeogr. Palaeoclim. Palaeoecol.* **79**, p. 117.
79. Spivack, A. J., You, C.-F., and Smith, H. J. (1993). *Nature* **363**, p. 149.

V

Synthesis

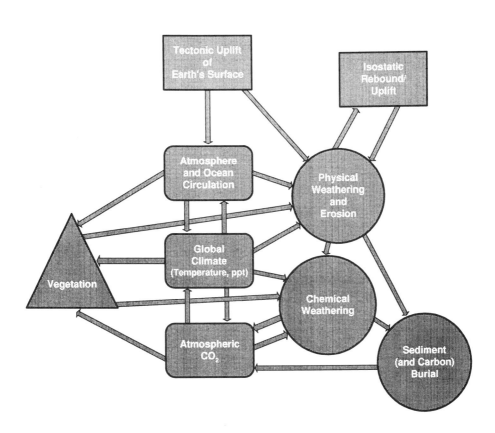

20

The Uplift–Climate Connection: A Synthesis

William F. Ruddiman, Maureen E. Raymo, Warren L. Prell, and John E. Kutzbach

1. INTRODUCTION

Viewed in their entirety, the chapters in this volume confirm earlier hypotheses[1–5] that tectonic uplift has played a prominent role in climatic change during the Cenozoic. This chapter synthesizes evidence published in this volume, supplemented by other publications in recent years, to update the status of the uplift–climate hypotheses. It also incorporates criticisms of the hypotheses and integrates relevant information from meterological, climatological, geomorphic, and sedimentological sources.

The synthesis is divided into a series of sections that develop the case for the uplift hypothesis in stepwise fashion: Section 2 reviews evidence of Cenozoic uplift and the unusual height of modern orography. Section 3 summarizes the effects of Cenozoic uplift on large-scale atmospheric circulation and climate. Section 4 covers the effects of Cenozoic uplift and altered atmospheric circulation on physical and chemical weathering. Section 5 reviews the argument for increased chemical weathering in uplifted regions as the main cause of the gradual Cenozoic CO_2 decrease and discusses the feedback effects resulting from CO_2 change. Section 6 summarizes the combined orographic and CO_2 effects of uplift on global climate during the Cenozoic, reviews other sources of Cenozoic climatic forcing, and briefly discusses unresolved issues.

William F. Ruddiman • Department of Environmental Sciences, University of Virginia, Charlottesville, Virginia 22903.   *Maureen E. Raymo* • Department of Earth, Atmospheric, and Planetary Sciences, Massachusetts Institute of Technology, Cambridge, Massachusetts 02139.   *Warren L. Prell* • Department of Geological Sciences, Brown University, Providence, Rhode Island 02912.   *John E. Kutzbach* • Center for Climatic Research, University of Wisconsin-Madison, Madison, Wisconsin 53706.

Tectonic Uplift and Climate Change, edited by William F. Ruddiman. Plenum Press, New York, 1997.

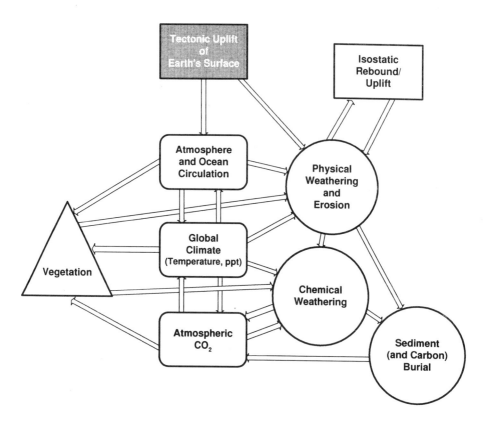

2. CENOZOIC UPLIFT

This section focuses on the starting point for uplift theory: the extent of Cenozoic uplift of the Earth's surface, and the related issue of whether modern elevations are unusual compared to longer geologic history.

2.1. Cenozoic Uplift Created Most Modern High-Elevation Terrain

Unambiguous evidence of uplift is provided by marine sediments previously deposited at or just below sea level and now found at high elevations. As noted in this volume (Chapters 2 and 3) and elsewhere, Cretaceous marine carbonates are found atop extensive areas of the Asian Tibetan Plateau, the South American Altiplano and Eastern Cordillera, the high plateau and mountain terrain of western North America, and the Alpine–Himalayan mountain belt. These marine sediments indicate unambiguously that high plateaus and mountains that exist in many regions today have been created (uplifted) since the middle to late Cretaceous.

Other geologic evidence supports this conclusion, principally shortening of the crust by thrust faulting under compressional tectonic conditions. If isostatic compensation prevails, crustal shortening and stacking of faulted slivers of crust will produce simultaneous uplift of mountains and plateaus along with formation of a thick, low-density crustal root. Major post-Cretaceous crustal shortening has occurred since the Eocene in Tibet (Chapter 2), since the Paleocene in the Altiplano and Eastern Cordilla (Chapter 3), from the Mesozoic through the Eocene in the American West[6] and since the Paleocene in the Alps.[7]

Deep-crustal or upper-mantle processes may also play a role in tectonic uplift and subsidence. These include heating of the upper-mantle lithosphere by delamination,[8] addition of low-density basaltic melts to the lower lithosphere, and changes in the angle of subduction of lithosphere beneath continents. These deep-seated processes are difficult to evaluate conclusively because they generally involve rocks that are unavailable for direct examination.

Another potential indicator of uplift of the Earth's surface is rapid denudation of high terrain. This can be evaluated by examining sediment masses deposited in adjacent basins and by applying geochemical techniques that reconstruct rates of removal of erosional overburden. Although climate change can also cause erosion and complicate interpretation of uplift histories (Section 5.3), denudation histories provide supporting evidence for Cenozoic uplift (Chapters 2 and 3).

Even though marine sediments show that much of the high topography existing today has a youthful (i.e., Cenozoic) origin, this is not a sufficient basis for claiming that the modern configuration is unique. The problem is that the highest topography during *any* interval of geologic time is inevitably recent in origin. This results from the action of erosion in continually wearing down high topography and reducing older inactive orogenic belts to lower elevations. Erosion works exponentially faster in youthful, high-elevation orogenic belts.[9,10] Consequently, in order to demonstrate that uplift is a major factor in explaining the major climatic cooling of the Cenozoic, it is necessary to demonstrate not just that most modern orography has been created by youthful (Cenozoic) uplift, but that it is unusually massive compared to earlier geologic eras when global climate was warmer. Both elevation and areal extent (i.e., the total "mass") of uplifted terrain are important parts of the uplift–climate hypotheses. Therefore, the remainder of this chapter will emphasize elevation, but will also note changes in areal extent where appropriate.

2.2. Modern Orography Is Unusually High Compared to Geologic History

The dominant topographic feature on the continents today is the Tibetan Plateau at a mean elevation of 5 km over an area of more than 4 million km^2 (Fig. 1). Tibet formed from the collision of India and Asia, beginning in some regions around 55 million years ago and under way along the entire front by 40 million years ago. Although the exact timing of subsequent surface uplift of Tibet is uncertain, a major regional-scale increase in elevation is thought to have been under way by 20 million years ago[11–15] (see Chapter 2 and 12).

The Tibetan Plateau is an atypically high and massive feature compared to long spans of geologic history. Over the 570 million years of Phanerozoic time,

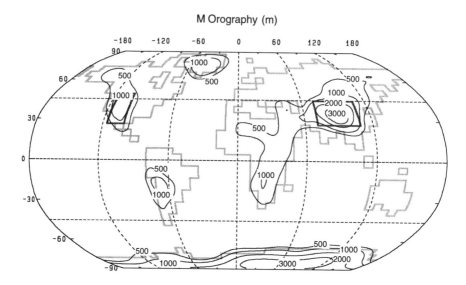

FIGURE 1.   Smoothed orography of continents used in typical GCM experiments. Major features include: Tibetan Plateau, East and South African plateaus, Western Cordillera of North and South America, and the Antarctic and Greenland ice sheets. Owing to spectral truncation and smoothing, contours (shown in meters) extend over some ocean gridpoints.

the only interval of possibly comparably elevated terrain is between 320 to 240 million years ago, during the continental collisions that assembled the Pangaean supercontinent. These collisions produced mountain belts and plateaus in Europe[16] and eastern North America.[17] The presence of the Tibetan Plateau alone thus suggests that modern orography is unusually high and massive compared to "typical" geologic history.

The marine $^{87}Sr/^{86}Sr$ ratio provides another means of inferring the past existence of large plateaus (Fig. 2). Positive excursions in this ratio are attributed to large-scale continental collisions. These collisions produce both a distinctive kind of metamorphism and an increase in weathering of the continents, and these factors drive the $^{87}Sr/^{86}Sr$ ratio higher[18,19] (see Chapters 2, 12, 14, and 19). The modern Sr isotope value exceeds any in the last 500 million years, suggesting that unusually elevated orography today is producing unusually strong weathering and metamorphism. The interval of continental collisions during Pangaean assembly is the only other interval with nearly comparable $^{87}Sr/^{86}Sr$ ratios in the last 500 million years.

After the Tibetan Plateau, the next largest emergent feature on Earth today is the Antarctic ice sheet (Fig. 1). Although both Antarctic and Greenland ice result from climatic cooling and are not a direct part of the rock uplift story, these ice plateaus contribute to the unusually high mean elevations of the planet today. Both came into existence in the Cenozoic. An ice sheet developed in east Antarctica by 36 million years ago.[20] Prior to this, there had been no ice sheets

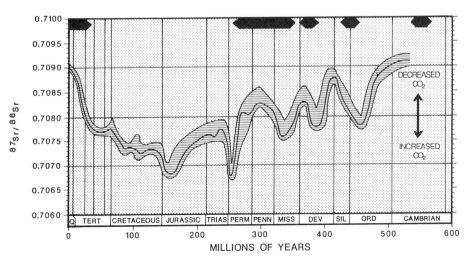

FIGURE 2.   Comparison of long-term $^{87}Sr/^{86}Sr$ seawater ratio (solid line in middle of region with horizontal ruling) with intervals of known glaciation (black bars at top) during the last 525 million years (from Raymo[18]). High $^{87}Sr/^{86}Sr$ values generally correlate with intervals of glaciation.

on Earth for over 200 million years, since the assembly of Pangaea (Fig. 2). There has been an ice sheet on Greenland for only a few million years.[21–22]

For the remaining high-elevation regions on Earth (Fig. 1), the evidence is more equivocal as to whether modern orography is uniquely massive. High terrain in these regions has been formed by tectonic processes acting at rates closer to a steady state, compared to the highly sporadic collisions between large continents. This implies that high-elevation terrain could have been created continuously in these (or other) regions at earlier times, but subsequently (and continuously) reduced by erosion and by deep-seated thermal processes. Although the balance between uplift and erosion makes it very difficult to reconstruct orography over most of geologic time, such reconstructions are more feasible during the middle Cretaceous, about 100 million years ago. A comparison of middle Cretaceous versus modern elevations is especially pertinent to the central theme of this book because climate during the middle Cretaceous was in a warm "greenhouse" state as opposed to the cold "icehouse" world of today. It would thus further strengthen the case for the uplift–climate hypothesis if regions outside of Tibet had orography that was lower or less extensive in area during the middle Cretaceous compared to today.

The modern Andean mountains and Altiplano–Puna Plateau result from subduction of ocean crust under South America. Because subduction has been going on for over 200 million years, it can be argued that Cordilleran-style mountains have long existed at near-constant elevations because of a balance between orogenic uplift denudation. However, evidence presented in Chapter 3 indicates that both the high Altiplano–Puna Plateau and the Eastern Cordillera were created by post-Paleocene crustal shortening.[23,24] This represents a major eastward expansion of the Cordilleran chain compared to the middle Cretaceous, when the only high orography was in the Western Cordillera. Whatever the elevation and lateral continuity of the Western Cordillera during the middle Cretaceous, the conclusion seems unavoidable that eastward expansion of orogeny in the Cenozoic greatly increased the orographic mass of the Andes.

Subduction has also occurred along the western margin of North America for a long time. In the middle Cretaceous, a mountainous fold-thrust belt existed in areas of the far West (California, Nevada, and Idaho) that remain mountainous today. In contrast, an interior seaway covered broad regions of Colorado, Wyoming, Utah, and Montana that now form the top of a topographic bulge extending from the High Plains (at 1 km) westward beneath the Rockies to the Colorado Plateau (at 2 km). As in South America, a large eastward expansion of high terrain has thus occurred since the middle Cretaceous. Many geologists infer that surface uplift of this former Cretaceous seaway occurred mainly during crustal shortening produced by the Laramide orogeny 75–45 million years ago,[6] and paleobotanical (leaf-margin) data support this view.[25] Some parts of this region, however, such as the Colorado Plateau and High Plains, show little compressive deformation of the upper crust. This has led to two views: (1) uplift occurred during the Laramide orogeny but shallow subduction caused a more diffuse style of tectonic buckling and uplift[6]; (2) uplift occurred later in the Cenozoic owing to deep-seated processes.[8,26] In either case, all agree that this broad region of the American West is much higher today than it was in the middle Cretaceous. This substantial increase in elevation and mass has to be balanced against likely post-Cretaceous losses of elevation in the western fold-thrust belt, especially the Basin and Range. Although unproven, the total orographic mass of western North America today may well exceed that in the middle Cretaceous.

The extensive East African and South African plateaus lie at mean elevations above 1 km (Fig. 1). Cretaceous marine deposits present around the northern and eastern margins of Africa show that there has been at least 100–200 m of Cenozoic uplift over a broad region.[27] Evidence of crustal shortening is absent because no subduction occurs beneath Africa. Past elevation changes are inferred mainly from radiometrically dated volcanic construction of the plateaus, from dating rift-valley faulting assumed to be synchronous with uplift and from dating paleoplanation surfaces. The strongest case for youthful uplift is in the Kenyan and Ethiopian portions of east Africa, where large volcanic edifices have been built in the last 30–20 million years[28,29] (see Chapter 4). In contrast, the South African Plateau is widely ascribed to post-Pangaean rifting over the late Jurassic Karroo hotspot plume,[30] and its mean elevation is thought to have been reduced by later cooling and erosional retreat of marginal escarpments. Changes in overall paleoelevation in Africa since the mid-Cretaceous reflect a balance between increased elevation of young plateaus in east Africa and elevational losses of the older South African Plateau. It is unclear which of these changes in mass has been larger.

No other modern mountain belt has a mass comparable to those already discussed. The long Alpine belt west of the Himalaya formed during closure of the Tethys seaway, with major crustal shortening and uplift of the Alps from the early to middle Cenozoic.[7,32] But this orogenic belt is discontinuous, with few regions exceeding 1 km in elevation, and thus it has little impact on large-scale atmospheric circulation (Fig. 1).

In summary, the Tibetan Plateau is an unusually massive orographic feature, and its existence alone makes modern orography unusual compared to a large portion of geologic time. It is also true that the area (and presumably the mass) of high plateau and mountain terrain in the Andean region of South America is much greater now than it was in the middle Cretaceous. In western North America, the areal extent of high terrain has also expanded far to the east since the middle Cretaceous, although this appears to have been offset by a lowering of elevations in parts of the Far West. From a global perspective, the net Cenozoic increases in high terrain in active orogens (mainly Tibet and the Andes) far exceed the likely elevational losses in older eroding orogens and volcanic plateaus. Modern global orography is thus not only youthful in origin, but also unusually high compared to longer-term geologic history. This means that uplift is a viable source of forcing for the Cenozoic climate changes discussed below.

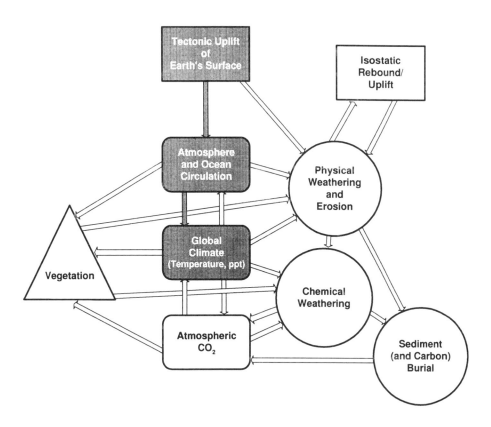

3. EFFECTS OF CENOZOIC UPLIFT ON CIRCULATION AND CLIMATE

This section summarizes results from climate modeling studies and modern climatological observations about the effects of uplift on Earth's circulation and climate. Evidence of regional uplift from Section 2 is then used to infer the potential effects of Cenozoic uplift on circulation and climate. Geologic evidence of these effects is discussed in Sections 4–6.

3.1. Uplift Has Cooled Midlatitude Plateau Regions

Uplift elevates the Earth's surface to altitudes at which temperatures are markedly cooler[33] (Fig. 3a). The atmosphere cools at the free-air lapse rate of 6.5°C per km of altitude, although this rate varies near ground level owing to seasonal and local effects. Geoscientists have long hypothesized that uplift could cool Earth's climate in this way,[34,35] but the mountains cited in these earlier hypotheses (Alps, Apennines) were very small in extent and unlikely to have had more than local climatic effects. Birchfield and Weertman[1] extended this concept to the Tibetan Plateau, noting that cooling atop such a massive feature, augmented by the resulting increase in winter snowcover, could have a larger-scale impact on Northern Hemisphere circulation.

General Circulation Model (GCM) sensitivity experiments have explored the potential cooling effects of high plateaus in southern Asia and the American West (Fig. 4). Experiments with and without plateau orography show that temperatures differ by more than 16°C over Tibet and more than 8°C across the American West. The zonal mean temperature difference for land at these latitudes (30°–40°N) exceeds 10°C (Fig. 4), and the mean hemispheric change over land is 8°C.

Because the Tibetan Plateau has been created in the last 50 million years (Section 2), these GCM "mountain/no-mountain" simulations (Fig. 4) can be used to approximate the cooling impact (> 16°C) of Cenozoic uplift in southern Asia. Tibetan uplift should also markedly affect the zonal mean cooling at 30°–40°N (Fig. 4). In North America, latest Cretaceous and Cenozoic uplift of the western interior from the Great Plains to the Colorado Plateau also has the potential to have produced significant regional cooling, but the regional-scale climatic impact of this uplift may have been offset by a decrease in elevations in regions in the far West that were uplifted during earlier Mesozoic orogenic activity but subsequently collapsed (Section 2). In South America, substantial net regional cooling is likely to have occurred during the middle and late Cenozoic owing to broad-scale uplift of the Altiplano and Eastern Cordillera (Section 2), regardless of the history of the older Western Cordillera.

3.2. Uplift Has Increased Meandering of the Standing Waves

High terrain contributes to the presence of quasi-stationary ("standing") meanders in the upper-westerly circulation by adding to the flow perturbations caused by land–sea temperature contrasts.[36] In the Northern Hemisphere,

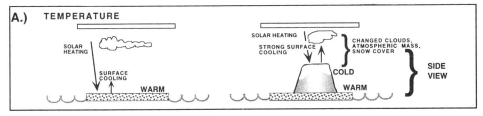

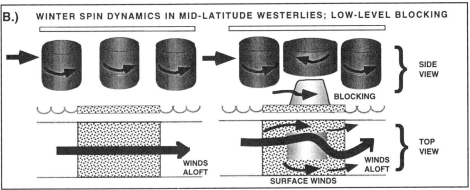

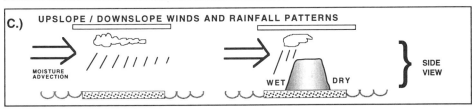

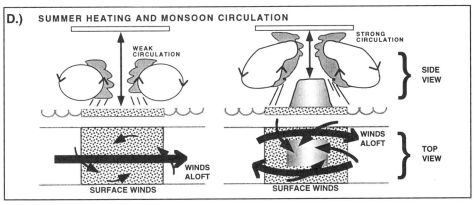

FIGURE 3. Large-scale effects of plateau and mountain uplift on climate: (a) cooling of high terrain owing to effect of lapse rate; (b) winter deflection of midlatitude surface westerlies and jet stream by orographic blocking; (c) orographic precipitation resulting from interception of zonal winds by rising terrain; and (d) monsoonal circulation and precipitation caused by heating of rising plateaus (after Kutzbach et al.[33]).

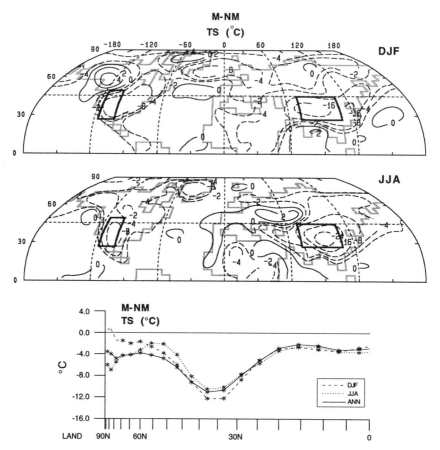

FIGURE 4. A GCM experiment ("mountain minus no-mountain" differences) showing tempera-ture changes caused by the presence of orography: (top) winter temperature changes; (middle) summer temperature changes; (bottom) zonal-mean temperature changes. Asterisks show changes significant at the 95% level (after Chapter 9).

mountains and plateaus form obstacles that block lower-level westerly flow and divert it northward and upward, with a return flow to the south downstream of high terrain (Fig 3b). The resulting flow and pressure contrasts in the vicinity of the high terrain propagate the diversion of the westerlies upward to jet stream levels. Orography enhancement of meandering is greatest in winter when dynamic effects linked to the strong zonal mean flow are largest. Early GCM experiments by Manabe and Terpstra[37] confirmed that orography enhances meanders in the standing waves, and recent estimates[38] suggest that orography accounts for two-thirds of their amplitude.

Ruddiman and colleagues[2,3,39] first noted that this concept may be appli-cable to recent geologic history, and that pervasive late Cenozoic uplift may have enhanced the amplitude of standing-wave meanders, including southward meanders in east-central North America, which would be favorable to glaciation (Fig. 5). Evidence of late Cenozoic uplift of Tibet summarized in Section 2 suggests that the amplitude of standing waves in the Northern Hemisphere may have increased, and GCM studies[3] suggest that uplift in Asia may have induced southward meanders far downstream over eastern North America. Within North America, however, the initial impact of high-mountain topography on the standing waves probably dates back to earlier Mesozoic orogenic activity.

SLP (mb)

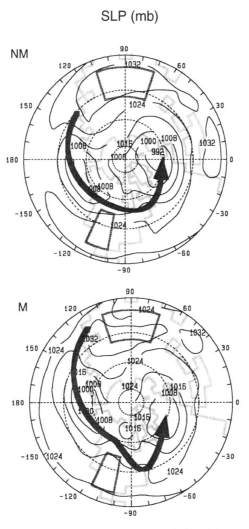

FIGURE 5. A GCM experiment showing increased meandering of the storm-track axis owing to the presence of plateaus in Asia and North America. Mean storm-track position (arrows) is defined by maximum variability in daily surface pressure associated with traveling low-pressure centers: (top) no orography; (bottom) full orography (from Ruddiman and Kutzbach[2]).

3.3. Uplift Has Increased Orographic Precipitation

High terrain generates orographic precipitation in two ways. First, as zonal winds meet orographic obstacles too wide to pass around, the air masses are lifted and cooled, causing condensation of water vapor and precipitation (Fig. 3c). Downwind from the high terrain, warming of subsiding air masses increases retention of water vapor, inhibits precipitation, and creates arid rainshadows. These effects switch "polarity" with the direction of the zonal winds. The Andes of South America have sufficient latitudinal extent that rainfall maxima occur east of the Northern Andes in the latitudes of the easterly trade winds and west of the Southern Andes in westerly wind latitudes[40,41] (Fig. 6).

A different kind of orographically controlled precipitation occurs in monsoons: in this case, high plateaus create their own circulation pattern by drawing in moisture-laden winds (Fig. 3d). The best example is the powerful south Asian monsoon.[42] Sensible heating of the elevated (> 5km) Tibetan plateau surface initiates upward motion of heated air, lowering surface pressures locally. Air is then drawn in toward the low-pressure center. Because this flow comes from the tropical and subtropical Indian Ocean, it carries abundant moisture. The monsoonal air masses are drawn up against the high terrain, cool, and release monsoonal rains. The accompanying release of latent heat enhances rising motion, which further lowers the surface pressure and increases the inflow of moist winds.

Monsoonal rains are important in other regions as well. Climatological analyses indicate that the Bolivian Altiplano at a mean elevation of 3600 m generates a summer monsoon circulation by similar mechanisms.[43-45] Monsoonal rainfall is very heavy on the northeastern margin of the Antiplano, a region

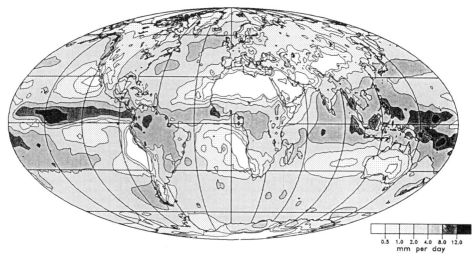

FIGURE 6. Global mean annual precipitation, in mm/day (from Legates[40] and Legates and Wilmot[41]).

that generates significant runoff for the Amazon River. Weaker summer monsoon circulations exist on other continents,[45] but none has an areal extent comparable to Asia. They also lack sufficient high topography to focus precipitation (in the case of northwest Africa) or proximal warm oceans to provide abundant moisture (in the case of western North America).

In all cases, topographically generated precipitation is greater on the lower and middle (windward) slopes than at higher elevations. This is due to progressive loss of water vapor with altitude. In addition, the high plateau of north-central Tibet lies in the rainshadow of the High Himalaya to the south, and the Bolivian Altiplano in the rainshadow of the Cordillera to the east.

Cenozoic uplift (Section 2) must have greatly enhanced precipitation rates in two regions: (1) in southern Asia, where GCM results show that uplift of Tibet has created the powerful summer monsoon precipitation[2,3,33,46,47] (Fig. 7); and (2) in South America, where uplift of the Bolivian Altiplano in the Central Andes has probably created the monsoonal precipitation regime, and where uplift of the Eastern Cordillera in the Northern Andes has enhanced orographic interception of precipitation from zonal easterly winds. Southern Asia and northern South America are two global-scale centers of precipitation over land mapped by Legates and colleagues[40,41] (Fig. 6). These areas also dominate global runoff from the continents compiled by UNESCO[48] (Fig. 8). Both regions tap into the abundant moisture available along the seasonally shifting Inter-Tropical Convergence Zone (ITCZ) and provide a strong orographic focus for summer precipitation.

3.4. Uplift Has Increased Seasonal, Orbital, and Flood-Event Precipitation

A monsoon is a seasonal reversal of low-level wind circulation often associated with high-elevation plateaus. In the Northern Hemisphere, these winds spin counterclockwise around the low-pressure cells centered over the plateaus in summer and clockwise around the high-pressure cells centered over (or north of) the plateaus in winter. This change in direction reverses low-level wind trajectories seasonally by up to 180°. For example, summer winds east of Tibet are dominated by warm moist air masses from the Indian Ocean to the south, whereas winter winds in the same region are dominated by cold dry air masses from the Asian interior to the north (Fig. 9). Cenozoic uplift of Tibet and the Bolivian Altiplano has enhanced monsoonal circulations and thus increased the seasonality of precipitation in these areas.

Uplift has also amplified the sensitivity of precipitation and runoff responses to orbital-scale insolation forcing. Modeling results for the Tibetan region (Chapter 8) suggest that the range of runoff variation in response to orbital insolation forcing is a factor of two larger in the full-orography than in the quarter-orography simulations, primarily due to precipitation and runoff increases during insolation maxima. This tectonic amplification during the Cenozoic should have created a highly pulsed hydrologic cycle in southern Asia, with rainfall maxima mainly during precession-dominated insolation maxima every 21,000

M-NM

Precipitation (mm/day)

(a)
JJA
(mm/day)

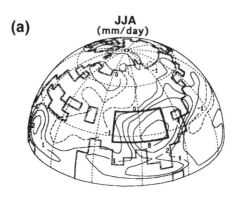

(b)
DJF
(mm/day)

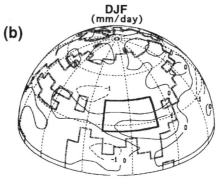

(c)
ANN
(% change)

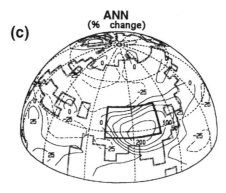

FIGURE 7. A GCM experiment showing changes in mean annual precipitation (in mm/day) owing to the presence of high plateau and mountain topography in Asia: (a) June–August; (b) December–February; (c) mean annual (shown as % change). Differences shown are those between "M" simulations (full orography) and "NM" simulations (no orography) (after Kutzbach *et al.*[33]).

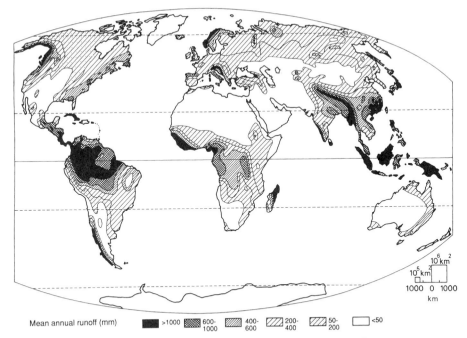

Mean annual runoff (mm)   ■ >1000   ▦ 600-1000   ▨ 400-600   ▨ 200-400   ▨ 50-200   ☐ <50

FIGURE 8.   Mean annual runoff in mm, from UNESCO.[48]

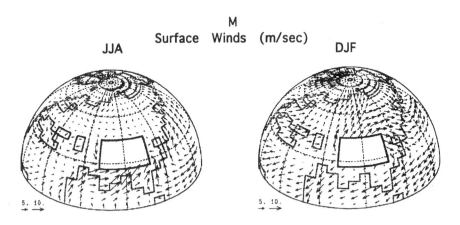

FIGURE 9.   A GCM experiment showing seasonally reversing surface winds in Asia: (left) June–August; (right) December–February (after Kutzbach et al.[33]).

years. The same trend shoud also have affected the uplifted Altiplano in South America during the Cenozoic.

The global distribution of flood-prone regions compiled by Hayden[49] (Fig. 10) is highly correlated with maps of mean annual precipitation (Fig. 6). As noted above, GCM experiments indicate that Cenozoic uplift in southern Asia and South America has increased mean precipitation levels in those regions and that more frequent high-precipitation events should occur in southern Asia, implying more frequent flooding. To the extent that increasing precipitation in these wet areas is seasonally and orbitally focused, as it is in the south Asian monsoon, the likelihood of increases in major flood events during the Cenozoic is even larger.

3.5. *Uplift Has Caused Drying of Midlatitude Continental Interiors*

Uplift has contributed to large-scale regional drying of continental interiors at midlatitudes[2,3,33,46,47,50] (see Chapters 5–7). In summer, focusing of monsoonal precipitation on the southeastern (Himalayan) margin of the Tibetan Plateau is accompanied by drying north and west of Tibet (Fig. 7). This drying in part reflects rainshadow effects behind the rising Himalaya and Tibetan Plateau. It is also related to subsidence of dry air in partial compensation for strong rising motion in the monsoonal regions. Because of counterclockwise flow around the Tibetan low-pressure center, hot dry air is advected from the Asian interior southwestward over the Caspian Sea, Arabia, and northwestern Africa. Uplift of Tibet should have produced summer drying across this large area mainly in the last 20 million years.

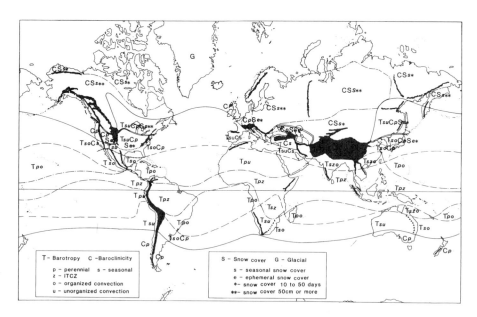

FIGURE 10.   Regions of frequent heavy flooding (in black), from Hayden[49]

GCM sensitivity experiments suggest that large midlatitude areas between 30°–60°N in Asia and North America are drier in winter if there is mountain and plateau orography[2,3,43,49] (Fig. 7b,c). This drying reflects several factors discussed in Chapter 5: orographic anchoring of stationary winter westerly waves in positions that steer cyclonic storms away from continental interiors and toward east-coast margins of continents; prevailing subsidence just downstream of the high topography; and rainshadow effects behind rising topography. Evidence discussed in Section 2 suggests that Cenozoic uplift in Asia should have caused a winter drying of midlatitudes, but that in western North America the winter drying probably began earlier than the Cenozoic, owing to initial uplift of high Cordilleran orography.

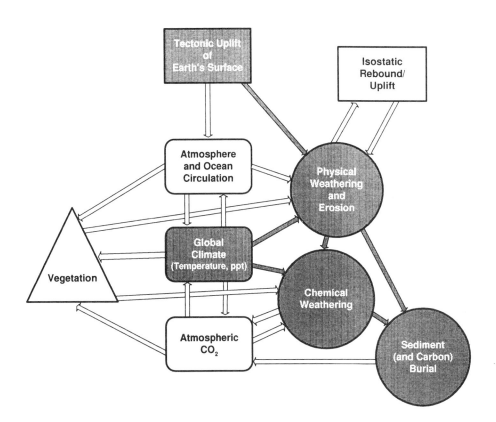

4. EFFECTS OF UPLIFT ON PHYSICAL AND CHEMICAL WEATHERING

Tectonic uplift of the Earth's surface (Section 2), and the changes it causes in circulation and climate (Section 3), combine to enhance physical and chemical weathering rates in orogenic belts.

4.1. Cenozoic Uplift Has Increased Physical Weathering in Orogenic Belts

Regional denudation can be estimated by examining the particulate load carried by rivers[51–53] (Fig. 11). Mean rates of basin denudation (Fig. 12) are highly correlated to mean or maximum basin elevation,[10,53,54] but elevation alone is not the critical factor. Instead, denudation is maximized in regions of greatest relief[9,53–57] (Fig. 13). This distinction is best demonstrated for the Tibetan and Altiplano Plateaus, where the erosion that occurs on the flat plateau tops is modest in scale,[58] whereas denudation is extremely rapid on steep slopes over a range of elevations.[59] Both elevation and relief tend to be proxy measures of youthful tectonism.[53]

Several tectonic factors favor strong physical erosion of recently uplifted orogenic margins (Chapter 10): compressional faulting exposes brecciated sedimentary rocks at the surface, and earthquakes enhance mass-wasting processes that dislodge debris from uplift-steepened slopes. Climatic factors are also important: heavy precipitation is focused on plateau and mountain margins (Section 3), vegetation cover is scarce at high altitudes, and glacial–periglacial weathering processes are important.[60] All of these "climatic" processes are a function of high elevation and relief (and thus of recent uplift). Other climatic factors may also enhance erosion, such as increased seasonal or orbital focusing of precipitation[61–64] and increased frequency of large floods.[65,66] The influence of these latter factors has not yet been fully quantified and deserves further study.

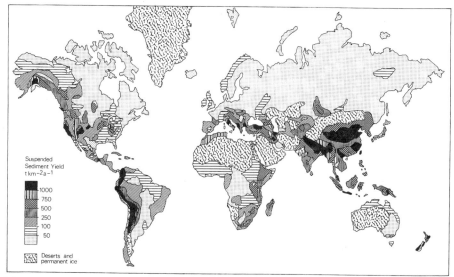

Suspended
Sediment Yield
t km^{-2}a^{-1}

1000
750
500
250
100
50

Deserts and
permanent ice

FIGURE 11.   Present-day suspended sediment yield of rivers in tons per km^2 per year (from Summerfield,[56] after Walling and Webb[52]).

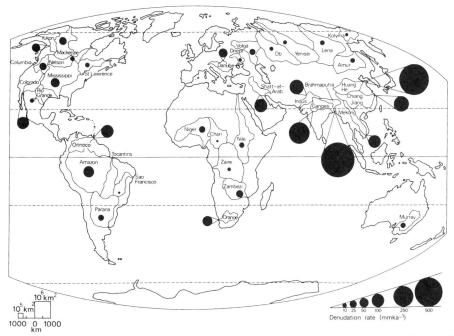

FIGURE 12. Present-day rates of denudation for major drainage basins in mm per thousand years (from Summerfield and Hulton[57]).

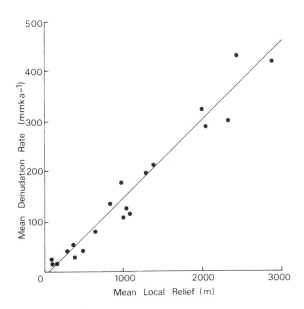

FIGURE 13. Correlation between mean annual rates of denudation and local relief (from Ahnert[9]).

Climate is not, however, entirely dependent on elevation or relief. For plateau margins of similar elevation and relief, erosion is much stronger on the rainy windward side facing major sources of moisture (the southern margin of Tibet) than on the dry leeward northern margin. This helps to account for the more modest correlation of denudation rates to runoff and precipitation,[57] as does the heavy ITCZ-related precipitation that occurs in some low-elevation tropical regions. Recent papers have begun to investigate the role of climatically distinct erosional histories in altering the style of tectonics on different orogenic margins.[67,68]

In summary, the primary control on physical weathering is ultimately tectonic: high-elevation plateaus or mountains maximize erosion on their windward margins. Climatic factors largely (but not entirely) linked to tectonic uplift are less important, but still significant.

4.2. Cenozoic Uplift Has Increased Global Physical Weathering

Accumulations of terrigenous sediment in the oceans suggest but do not prove that global-scale denudation rates increased during the Cenozoic. The largest masses of young terrigenous sediment in the oceans are in the Indian Ocean Bengal and Indus fans, where deposition began about 40 million years ago,[69] increased around 20 million years ago,[13] and increased again some 10 million years ago.[70,71] On the Amazon Cone, terrigenous sedimentation has greatly increased during the last 10 million years,[72] an interval for which fission-track evidence indicates very large increases in erosional unroofing of the northeast margin of the Bolivian Antiplano, an important source area for Amazon River sediments.[59,72] Late Miocene diversion by uplift of East Andean drainage from the Carribbean to the Atlantic[24,73] is implicated in the advent of Amazon Cone sedimentation.

Compilations of mean accumulation rates of terrigenous sediment from Deep-Sea Drilling sites[75] suggest a global-scale increase in late Cenozoic terrigenous fluxes (Fig. 14). Although the long-term increase must in part reflect differential preservation of younger sediments, this cannot explain the abrupt rise in clastic deposition in the last 15 million years. Other sediment-flux compilations also show greater late Cenozoic accumulation of terrigenous sediment,[76] although any such compilation is inevitably incomplete because of sediments lost into trenches on subducting margins.

In the end, the best argument for increased global mean erosion in the Cenozoic is based on two fundamental observations. First, physical weathering is strongest on the margins of young orogens (Section 4.1), as shown by global patterns of suspended sediment yields (Fig. 11) and denudation rates (Fig. 12). As a result, modern denudation rates on each continent are highest in orogens created or enlarged during the Cenozoic: the Himalaya, Andes, Alps, and American Cordillera. Second, Cenozoic tectonics have brought modern-day orography to elevations and masses unusual compared to the long span of geologic history, particularly in southern Asia and the Andes–Altiplano (Section

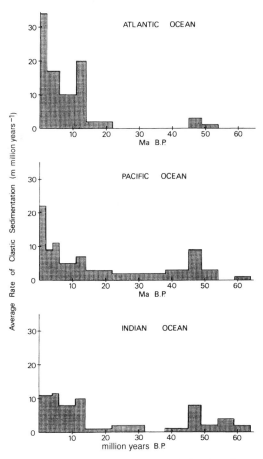

FIGURE 14.   Mean deposition rates (mm per million years) of clastic terrigenous sediment in sites drilled by DSDP in the deep ocean (from Davies et al.[75]).

2.2). Unusually high and massive orography must produce unusually rapid global-scale denudation.

4.3. Cenozoic Uplift Has Increased Chemical Weathering in Orogenic Belts

Almost a century ago, Chamberlin[77] suggested that tectonic uplift in orogenic belts causes increased chemical weathering. Long overlooked by the geoscience community, this concept was independently derived by Raymo[4,5,78] and its basic validity is now widely accepted (see Chapters 10–19). Although dissolved river loads tend to be dominated by the product of limestone dissolution (and to a lesser extent evaporites), the order-of-magnitude slower hydrolysis of silicates[79] that extracts CO_2 from the atmosphere (and thus effects climate) is the main focus of this section.

Key evidence of strong chemical weathering in orogenic regions comes from South America, where the Amazon River emerges from the Andes already carrying 70–80% of the dissolved chemical load that it discharges to the Atlantic Ocean.[80,81] Relatively little chemical weathering occurs in the Amazon lowlands, which are covered by thick residual soils that inhibit chemical weathering of fresh bedrock below.

This observation also holds up in global-scale comparisons. For rivers draining large basins, dissolved loads are most strongly correlated with basin relief and runoff, with the two factors together accounting for over 60% of the variance.[57] Studies showing weaker correlations between chemical weathering and relief used less diagnostic measures such as the mean elevation of basins or of entire continents[82–85] (Section 4.1).

The heavy precipitation typical on the margins of high terrain (Section 3.3) enhances chemical weathering because water is fundamental to both dissolution and hydrolysis reactions. The generalized correlation between total river solute loads (including bicarbonates) and precipitation or runoff[10,57,86] in the basins they drain (Fig. 15) is thus in part a measure of high precipitation on steep slopes (Fig. 6).

A more critical reason for strong chemical weathering on plateau and mountain margins is the increased surface area of fresh unweathered rock exposed

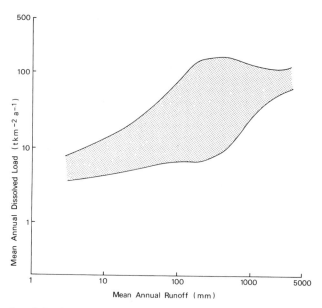

FIGURE 15.   Correlation between mean-annual dissolved load of rivers and runoff (from Summerfield[56], after Walling and Webb[86]). Dissolved load includes products of both carbonate–evaporite dissolution and silicate hydrolysis.

by rapid physical weathering[57] (Section 4.1). The importance of exposing new surface area is demonstrated by laboratory experiments over geologically short (multiyear) intervals[87] and by field evidence relevant to geologic timescales. Blum (Chapter 11) examined granitic debris in glacial moraines of varying ages and showed that chemical weathering rates are very rapid immediately after initial exposure, but then decrease exponentially through time, and are very low 100,000 years after initial exposure. This suggest that mass wasting and other slope-dependent processes that produce fresh debris at recurrence intervals of less than 100,000 years can exponentially enhance rates of chemical weathering. Bluth and Kump[88] found that chemical weathering of basalts is more rapid in Iceland than Hawaii, despite much colder temperatures, because thick leached soils in Hawaii retard chemical weathering compared to the much greater degree of exposure of basalts in Iceland. These two field studies are each normalized to a single type of silicate rock and avoid complications caused by rapid carbonate dissolution. They also bracket the full compositional range of silicate rocks found in orogenic belts.

It is not entirely clear where chemical weathering of orogenic belts is concentrated. Mass wasting and physical weathering initially expose and fragment fresh rock mainly at high and middle elevations where slopes are steepest. Precipitation is generally heaviest on lower and middle slopes. At lower elevations, other factors such as warmer temperatures and retention of CO_2 as carbonic acid in soils are favorable to chemical weathering. Overbank flooding, sediment reworking by shifts in river channels, and changes in sea level can expose and reexpose unweathered sediments on lowland floodplains and deltas. The locus of maximum chemical weathering may vary with factors specific to individual orogens.

Cenozoic uplift has strongly enhanced chemical weathering in at least two regions. In southern Asia, uplift of Tibet and the Himalaya during the last 40 million years is widely viewed as having caused a regional increase in chemical weathering (see Chapters 12, 15, 16, 18, and 19). The steep rise in the oceanic $^{87}Sr/^{86}Sr$ ratio over the last 40 million years (Fig. 2) is mainly attributed to Himalayan chemical weathering and river input: although 50% or more of the rise in the mean ocean $^{87}Sr/^{86}Sr$ value may reflect delivery of more radiogenic debris (higher $^{87}Sr/^{86}Sr$ ratio) from the Himalaya, the rest is ascribed to increased chemical weathering fluxes from the Himalaya and other active orogens.[15,89] McCauley and DePaolo (Chapter 19) attempt to distinguish Asian Sr fluxes owing to carbonate dissolution from those due to silicate hydrolysis and infer a substantial increase in the latter during the last 50 million years. In addition, GCM experiments based on Tibetan uplift show increased monsoonal runoff from Asia at levels compatible with Sr-based estimates[14] (see Chapters 7 and 8).

The second region where a major increase in chemical weathering has probably occurred is the Eastern Andes and Antiplano, where Cenozoic uplift[23] (Chapter 3) has been accompanied by deep erosion of the Eastern Cordilleran margins[59] and by delivery of a huge volume of sediment to the Atlantic via the Amazon River system.[24,72]

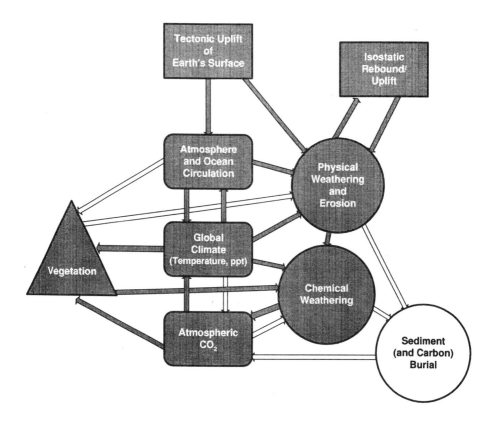

5. EFFECTS OF UPLIFT-DRIVEN CHANGES IN CO₂ AND CLIMATE

This section reviews evidence that increased chemical weathering driven by uplift explains the hypothesized CO_2 decrease of the last 30–40 million years, reviews modeling evidence of the climatic impact of such a decrease, and summarizes positive and negative climatic feedbacks that could be triggered by these CO_2 changes.

5.1. Uplift-Driven Chemical Weathering Has Lowered CO₂ Levels

The strongest indication that atmospheric CO_2 levels have decreased during the Cenozoic is the progressive expansion of glacial ice at both poles. Evidence for glacial ice on east Antarctica dates back to 36 million years ago,[20] and the Greenland ice sheet formed in the last several million years.[21,22] Mountain glaciers are first recorded in South America between 7 and 4.5 million years ago[90] and in Alaska about 5 million years ago.[91] Cyclic development of midlatitude continental-scale ice sheets in the Northern Hemisphere began 2.7 million years ago, with larger maxima after 0.9 million years.[92,93] Vegetation has progressively shifted toward cold-tolerant forms in the Arctic (see Chapter 9), while disappearing entirely on Antarctica.[94] Bipolar development of glacial ice and cold-adapted polar vegetation requires a global-scale explanation best provided by a CO_2 decrease. In addition, terrestrial $\delta^{13}C$ evidence of a shift from C_3 to C_4 vegetation that occurred simultaneously on four continents around 7 million years ago suggests that atmospheric CO_2 dropped below 500 ppm, the threshold below which the C_4 photosynthetic pathway becomes more efficient[95] (see Chapter 13). Other long-term CO_2 proxies such as $\delta^{13}C$ ratios of marine organic C have also been proposed as evidence for decreasing CO_2, but diagenesis or vital effects may invalidate their use.[96] The remainder of this chapter assumes that a decrease in atmospheric CO_2 is the primary control on the bipolar Cenozoic cooling, even though final proof of this is not yet at hand.

Over tectonic timescales, two factors are regarded as the primary controls on atmospheric CO_2: (1) input from volcanoes and hydrothermal sources; and (2) removal by chemical weathering of silicate rock (Fig. 16). These two processes determine the mean CO_2 content of the ocean, which interacts with and controls the smaller reservoir of CO_2 in the atmosphere.

The rate of global volcanism is the primary source of forcing for the BLAG class of models.[97,98] It is proposed that the mean global rate of seafloor spreading controls rates of CO_2 outgassing at ocean ridge crests and along zones of

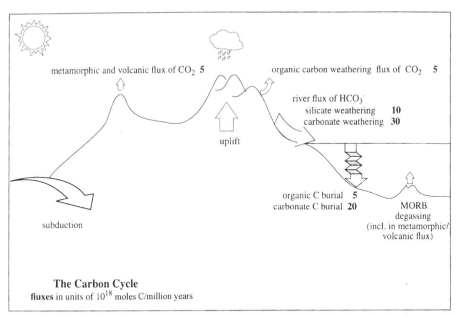

metamorphic and volcanic flux of CO_2 5

organic carbon weathering flux of CO_2 5

river flux of HCO_3^-
silicate weathering   **10**
carbonate weathering **30**

uplift

organic C burial **5**
carbonate C burial **20**

subduction

MORB
degassing
(incl. in metamorphic/
volcanic flux)

The Carbon Cycle
fluxes in units of 10^{18} moles C/million years

FIGURE 16.   Schematic representation (from Chapter 18) of processes hypothesized to control CO_2 levels in the atmosphere over long tectonic (million-year or longer) timescales: (1) rates of seafloor spreading (controlling rates of CO_2 input from subduction-zone volcanoes and along the midocean ridge system; (2) rates of chemical weathering of silicate rock on land (controlling CO_2 removal in recently uplifted mountain belts and plateaus); and (3) rates of C burial in the ocean.

subduction and collision. Because global mean spreading rates declined rapidly until 30 million years ago,[99] the BLAG model simulates a decrease in CO_2 input up to that point that could have contributed to observed early Cenozoic cooling.[97,98] During the last 30 million years, however, when the bulk of Cenozoic cooling occurred, the mean spreading rate shows no net change (Fig. 17). As a result, the BLAG "tectonic-input" mechanism cannot account for a decrease in atmospheric CO_2 levels or explain global cooling over the last 30 million years. If the estimated history of midplate volcanism[100] is added to the model input, the situation worsens, with no net CO_2 change apparent all the way back to 40 million year ago (Fig. 17; also see Chapter 19). In short, changing inputs of CO_2 linked to spreading rates and plume volcanism cannot explain the climatic cooling of the last 40 to 30 million years, the interval for which tectonic and climatic data are the best constrained in all of geologic history. Another explanation is required.

The other widely acknowledged factor affecting atmospheric CO_2 is the rate of removal of CO_2 by chemical weathering of silicate rocks. Raymo[4,5,78] proposed that increased chemical weathering in actively uplifting orogenic belts during the Cenozoic would lower atmospheric CO_2 and cool global climate. This hypothesis emphasized uplift and weathering in the Himalayan–Tibetan region of Asia and the Andes–Altiplano region of South America.

Evidence in this volume offers some support for this hypothesis. McCauley and DePaolo (Chapter 19) estimate that increases in weathering-driven alkalinity fluxes from the Himalaya to the oceans over the last 50 million years are more than sufficient to cause a major Cenozoic decrease in atmospheric CO_2. Edmond and Huh (Chapter 14) note that modern rates of fixation of atmospheric CO_2 in many Andean basins are larger than in those draining the Himalaya, although they make no attempt to quantify Cenozoic changes in Andean weathering fluxes.

It is extremely difficult to quantify even the present-day chemical weathering budgets within orogenic regions (and elsewhere). Data from any orogenic regions are too sparse, unsystematic, anthropogenically disturbed, and uneven in type and quality to constrain modern chemical dissolved loads adequately (see Chapter 14). Further, the few larger rivers that are well studied do not dominate the global budgets.[85] Although chemical weathering is generally estimated to be one-sixth as strong as physical weathering on a global basis,[101–102] this mean value does not discriminate climatically significant silicate hydrolysis from carbonate dissolution. Nor can large-scale averages be applied to individual basins within orogenic belts; the variability of lithologies between basins, as well as basin-specific changes in tectonic setting and climate, require a basin-by-basin analysis.

More difficult still is the challenge of reconstructing changes in chemical fluxes during the Cenozoic. Regional uplift histories and their accompanying climatic effects are understood only in basic outline (see Chapters 2 and 3). Raymo and colleagues[4,5,78] proposed that globally significant increases in chemical weathering are recorded in marine sediments by changes in geochemical indices such as the mean ocean $^{87}Sr/^{86}Sr$ value and the carbonate compensation depth (CCD). As noted previously, the mean ocean $^{87}Sr/^{86}Sr$ value is influenced

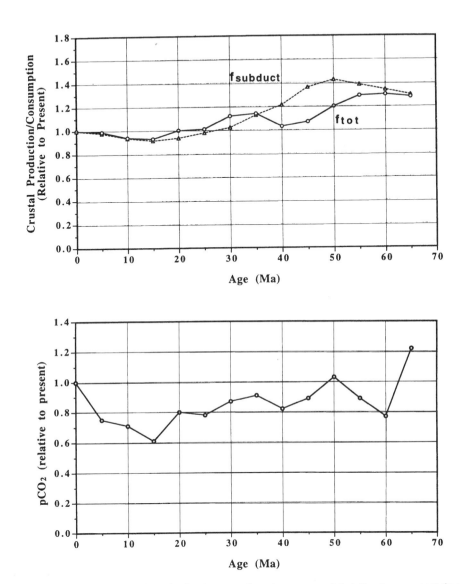

FIGURE 17. (Top) Rates of production/consumption of ocean crust (relative to a present-day value of 1) during the last 70 million years (from Chapter 18). Rate based on global mean seafloor spreading rate shown by $f_{subduct}$. Rate shown as f_{total} also includes estimated midplate plume volcanism. Both trends show no net change since 25–30 million years ago and small increases in the last 15 million years. Bottom: Simulated atmospheric CO_2 over the last 70 million years from a geochemical model driven mainly by rates of crustal consumption shown above (from Chapter 18; also see Chapter 19).

not only by changes in chemical weathering during the late Cenozoic, but also by time-varying changes in regional lithology (such as formation and weathering of radiogenic leucogranites). Hence, it cannot be used as a simple index of global (or even regional) weathering rates. The same is probably true of the $^{187}Os/^{186}Os$ index (see Chapter 17). Deepening of the CCD, although plausibly affected by increased river alkalinity fluxes to the oceans, may also have other (or additional) explanations (see Chapter 18). Despite these complications, widespread support is evident in this volume (see Chapters 14, 15, 17, 18, and 19) for the Raymo–Chamberlin hypothesis that increased chemical weathering in orogenic belts is the most likely explanation for the inferred CO_2 decrease over the last 30–40 million years. In tacit confirmation of this, the development of BLAG-type models in recent years has incorporated forcing from tectonic uplift (using the $^{87}Sr/^{86}Sr$ record as a proxy), as well as longer-term forcing from solar luminosity and the evolution of vegetation.[103,104]

Other proposed causes of long-term changes in CO_2 may also play a role during the Cenozoic. High rates of CO_2 release from the Tethys region owing to metamorphism of C-rich ocean sediments might contribute to higher CO_2 levels during the early Cenozoic[105] but cannot explain the continuing decrease over the last 30 million years. Mass-balance analysis of the Sr isotope record of silicate clastic sediment fractions suggest that metamorphic decarbonation reactions in orogenic belts could be a larger source of CO_2 to the atmosphere than the volcanic degassing input used in BLAG models.[106] Some C model studies suggest that increased burial of organic C in the late Cenozoic has drawn significant amounts of CO_2 out of the atmosphere.[107] Although this trend has been questioned owing to insufficient constraints on data used as inputs to these models (see Chapter 16), this possibility is well worth pursuing. Another possible mechanism involves low-temperature alteration of ocean basalts driven by ocean pH changes[108]; however, weathering is a very weak function of pH within the range normal for the deep ocean.[109]

5.2. Uplift-Driven CO_2 Cooling and Drying

Arrhenius[110] first assessed the planetary-scale "greenhouse" impact of CO_2 on global temperature. He noted that CO_2 impacts on temperature are enhanced at low and middle latitudes by related changes in water vapor and at high latitudes by albedo-temperature feedbacks associated with snow and ice. In recent decades, many climate-modeling studies have examined possible future "greenhouse" changes in climate owing to CO_2 increases from fossil-fuel burning.[111] GCM studies indicate that CO_2-driven changes in sea-ice extent greatly amplify winter temperature changes at high latitudes.[112] CO_2-driven climatic feedbacks related to slow-responding ice sheets are not yet simulated in GCM models, and the full effect of clouds remains uncertain.[113]

Arrhenius attributed the long-term Cenozoic cooling to a slow decline in atmospheric CO_2, here confirmed to be most plausibly explained by increased chemical weathering in uplifted orogenic areas of southern Asia and South

America. Model experiments that simulate future "greenhouse" climate changes owing to fossil-fuel burning can be reversed in sense to isolate "reverse greenhouse" changes owing to CO_2-driven Cenozoic cooling. Experiments described in Chapter 9 suggest that a late Cenozoic decrease from twice-modern to modern CO_2 levels would cool climate in all seasons (Fig. 18), with the largest cooling at high latitudes in winter being due to advancing sea ice, a much smaller Arctic cooling in summer, and a substantial cooling in all seasons at lower and middle latitudes. These results are similar to experiments from other GCMS,[112] although more recent models tend to show less cooling.

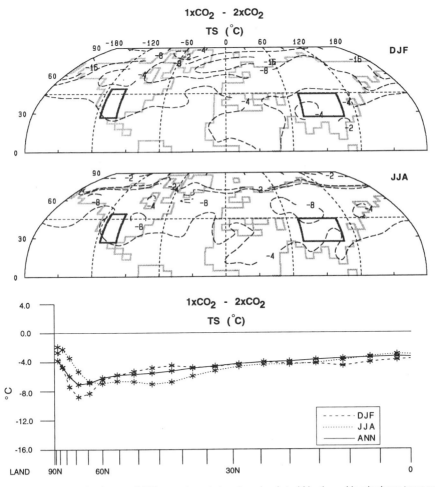

FIGURE 18. Results from a GCM experiment showing simulated Northern Hemisphere temperature changes in °C owing to CO_2 decrease from twice-modern to modern values: (top) changes in winter; (middle) changes in summer; (bottom) zonal mean seasonal and annual changes (after Chapter 9). Asterisks show changes significant at the 95% level.

The "reverse-greenhouse" climatic changes associated with falling CO_2 values during the Cenozoic should also include a decrease in the amount of water vapor in the atmosphere.[110,111] GCM experiments described in Chapter 9 simulate diminished precipitation and general drying over the continents, especially poleward of 40°N in winter and 55°N in summer (Fig. 19). The decrease in precipitation at low latitudes is not significant relative to the large variability of the model control case, and GCM "greenhouse" experiments differ

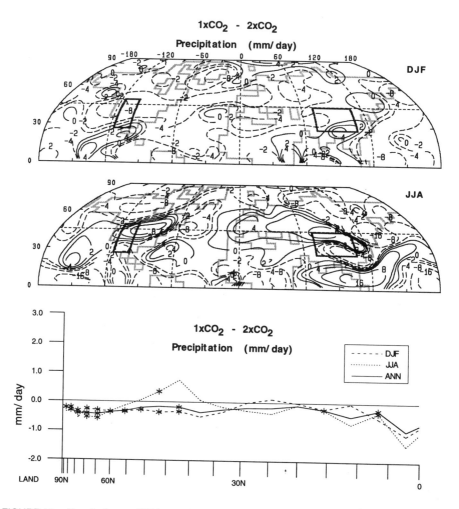

FIGURE 19.   Results from a GCM experiment showing simulated Northern Hemisphere precipitation changes in mm/day owing to CO_2 decrease from twice-modern to modern values: (top) changes in winter; (middle) changes in summer; (bottom) zonal mean seasonal and annual changes (after Chapter 9). Asterisks show changes significant at the 95% level.

greatly in their simulations of regional precipitation changes.[112] These observations suggest the need for caution in accepting these precipitation simulations as geologically meaningful. All GCM simulations, however, indicate significant winter drying north of 40°N, suggesting that this winter-season trend may be applicable to geologic history. The winter drying trend is additive to that produced in continental interiors by the orographic effects of uplift (see Chapters 5–7).

5.3. Climatic Feedbacks Driven by Decreasing CO_2

Late Cenozoic changes in climate caused by uplift-driven decreases in CO_2 have the potential to provide both positive and negative feedback to climate changes caused by the direct orographic effects of uplift.

Molnar and England[114,115] noted that some presently elevated regions that have been heavily denuded in the late Cenozoic were probably tectonically inactive when denudation began. They proposed a climatically driven "rejuvenation" of older, tectonically inactive mountain belts, a process that can provide positive feedback to the initial uplift–weathering concept. The feedback sequence is this: true uplift of the Earth's surface in orogenically active regions causes climatic change (such as CO_2 cooling), which leads to erosional reactivation of older mountain belts. Increased erosion in older mountain chains then enhances CO_2 drawdown, causing additional cooling, glaciation, and erosion. Isostatic rebound of uneroded mountain peaks to still higher elevations[114–117] produces more weathering, and so on.

The clearest mechanism proposed by Molnar and England for increasing denudation in older mountains is lowering of the snowline onto mountains owing to long-term CO_2-induced cooling (Fig. 18). This initiates or enhances mountain glaciation, with intense denudation and fragmentation of rocks by physical weathering (Section 4.1). This in turn causes increased exposure of rock to chemical weathering and thus enhanced CO_2 drawdown. Blum (Chapter 11) found that fluctuations of mountain glaciers at periods of 100,000 years or less will greatly enhance chemical weathering.

Molnar and England also proposed that there was an increase in late Cenozoic physical erosion due to greater moisture advection by stronger winds and increased storms caused by steeper latitudinal temperature gradients. This explanation is open to question because the water vapor content of the atmosphere would decrease with lowered CO_2. GCM experiments with lower CO_2 levels (Fig. 19) simulate reduced, rather than increased, precipitation at middle and high latitudes where temperature gradients are largest.[112] Also, field data suggest that modern denudation rates are not strongly dependent on runoff variability (a proxy for "storminess").[57]

A second set of positive feedbacks may be associated with the latest Cenozoic appearance of continental ice sheets in the Northern Hemisphere. Blum (Chapter 11) shows that increased exposure of freshly eroded glacial debris in high latitudes owing to large orbital-scale oscillations in ice-sheet limits may have a large impact on the marine $^{87}Sr/^{86}Sr$ record and even have some effect on global chemical weathering. These ice-sheet fluctuations may expose additional fresh rock debris in periglacial regions marginal to the ice-sheet limits and in mountain belts where snow and ice lines drop even lower. Also, as shown by COHMAP,[118] orbitally driven fluctuations in Northern Hemisphere ice sheets may cause orographic diversions of the jet stream and periodically enhance precipitation in some previously dry regions such as the American Southwest.

All of these positive feedbacks associated with CO_2-induced cooling during the last half of the Cenozoic would further enhance rock exposure at high latitudes and high altitudes and in both orogenically active and inactive mountain regions. Altough this increase in amount of weatherable material will promote greater chemical weathering in these regions, and thus potentially cause even greater global cooling, negative weathering feedbacks associated with the same climatic changes may simultaneously be working in the opposite direction.

Walker and colleagues[119] proposed that the climate system is buffered against extreme climatic fluctuations by a negative feedback that is also linked to silicate weathering. This negative weathering feedback comes about through changes in "greenhouse" climatic parameters: temperature, precipitation, and vegetation. If, for example, climate begins to cool (for any reason), the decreased temperature and associated decreases in precipitation and vegetation cover slow the rate of silicate weathering and thus the rate of removal of CO_2 from the atmosphere. The climate ends up cooling, but not nearly as much as it would have without the negative feedback. The slow Cenozoic decline of CO_2 thus may also have strengthened mechanisms that tended to oppose uplift-induced cooling.

The temperature–weathering link has long been used as the primary source of negative feedback in BLAG-style models,[97,98,103,104,120] for which tectonically driven changes in CO_2 outgassing are assumed to be the main driver of climate change. For the late Cenozoic, where uplift-induced weathering in orogenic belts may instead be the best explanation for climatic cooling, weathering can still be invoked as a major negative feedback if it is shown that the overall rate of chemical weathering decreases significantly outside of the orogenic regions.

Evidence for negative temperature–weathering feedback is derived in part from laboratory experiments.[85,87] These indicate that weathering rates of silicate minerals increase by a factor of 2 to 4 per 10°C increase in temperature. It is, however, difficult to decide how to apply such results in nature, because of complications such as the choice of which silicate minerals to emphasize.[121] Laboratory results[85] also indicate some dependence of weathering rates on pH, which is linked both to precipitation and to carbonic acid provided by vegetation.[122]

Some field studies report little correlation between dissolved river loads and temperature,[57,123] but these do not fully exclude carbonate dissolution. Bluth and

Kump[88] found greater weathering of Icelandic basalts in a cold/dry climate than of Hawaiian basalts in a warm/wet climate, opposite in sense to the temperature–weathering feedback, but here again the degree of rock exposure may be the controlling variable. Edmond and Huh (Chapter 14) compared river dissolved loads from low-lying shields in the warm tropics to those in Siberia and found comparable weathering rates despite very different climates, but Berner and Berner (Chapter 15) claim that small amounts of carbonate in the Asian basins invalidate conclusions about the effect of temerature on silicate weathering. Other studies that have normalized for lithology find that temperature and weathering are positively correlated.[79,102,124]

Although these field studies remain open to varying interpretations, results from laboratory studies indicate that there must be some level of negative weathering feedback. The fact that other factors (rock exposure, precipitation) dominate in a given basin does not exclude the action of an underlying negative weathering feedback. How strong a factor this feedback has been at a global scale during the Cenozoic remains an open question. Sundquist[125] suggested that it could have been a major factor in moderating an uplift-driven climatic cooling. Papers in this volume variously support this view (Chapters 15 and 18), oppose it (Chapter 14), and remain unconvinced (Chapter 19).

The GCM-simulated amount of cooling of Northern Hemisphere land masses based on an estimated drop in CO_2 values from twice-modern to modern values is $4°–8°C$ (Fig. 18). Given laboratory-based evidence that silicate weathering rates change by a factor of 2–4 for a $10°C$ change in temperature,[84,87] this implies something in the range of a 50 to 300% decrease in rates of chemical weathering owing to falling temperatures over the last 20 million years. Because this cooling is global in scope, such a reduction would be effective everywhere on the globe, enhanced or reduced locally by smaller-scale changes in precipitation (Fig. 19) and vegetation cover (see Chapter 9). Temperature-dependent reductions in weathering rates would also presumably be at work as an underlying factor even in actively uplifting orogenic regions, although swamped by much larger increases in chemical weathering that are due to greater rock exposure and precipitation.

In summary, chemical weathering of orogenic regions of Asia and South America increased during the Cenozoic owing to greater rock exposure and precipitation in the uplifting regions. In addition, glacial and periglacial processes active in high-altitude and high-latitude regions promoted positive feedback via increased physical weathering, greater exposure of fresh rock, and thus stronger chemical weathering. At the same time, the CO_2-driven reduction of global temperatures caused an opposing tendency toward decreased chemical weathering. It is thus possible that localized increases in chemical weathering in orogenic and glacial–periglacial regions are approximately balanced by larger-scale decreases in weathering elsewhere[125] (see Chapter 18), resulting in little net change in total global chemical weathering.

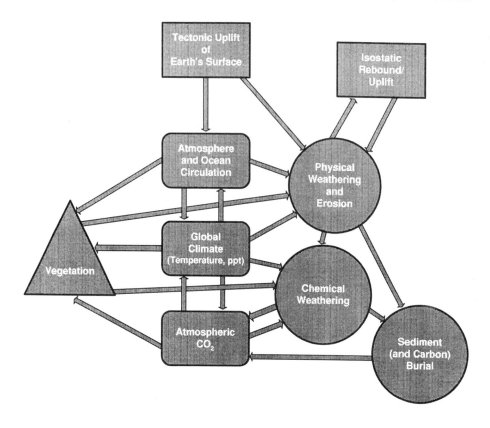

6. HYPOTHESES OF CENOZOIC CLIMATE CHANGE

This section summarizes the current status of the uplift–climate hypotheses, briefly reviews other proposed sources of climate forcing, and notes several ongoing issues that require additional research.

6.1. Summary of Updated Uplift–Climate Hypothesis

As discussed previously, uplift of the south Asian Tibetan Plateau and South American Altiplano and Eastern Cordillera have created features with a total orographic mass that is unusual for most of geologic history (Section 2). This uplift has had large-scale impacts on atmospheric and oceanic circulation summarized in Section 3: cooling and drying atop broad plateau surfaces at low-middle latitudes; increased meandering of standing waves in the tropospheric winds; focusing of heavy monsoonal and other orographic precipitation on the margins of uplifted terrain; and drying of continental interiors and northern midlatitudes. The steepening of tectonically active plateau and mountain slopes and focusing of precipitation on their margins increases mass wasting and physical weathering, which in turn enhances chemical weathering (Section 4).

Increased weathering of silicate minerals in uplifted regions draws CO_2 from the atmosphere and extends the climatic effects of uplift to a global scale. This uplift–weathering control of CO_2[4,5,77,78] is at the center of the uplift–climate hypothesis. Decreasing CO_2 during the Cenozoic has driven numerous other climatic responses, most prominently a major poleward-intensified cooling that created mountain glaciers and continental ice sheets. These physical and chemical changes were accompanied by large-scale vegetation changes, including circum-Arctic southward expansion of tundra and boreal forest, and large-scale replacement of forests and woodlands by grasslands at lower-middle latitudes (see Chapter 9).

Several of these changes are summarized by GCM simulations of the combined impact of plausible changes in orography and CO_2 during the last 20 million years shown in Figs. 20 and 21. These include: the widespread cooling

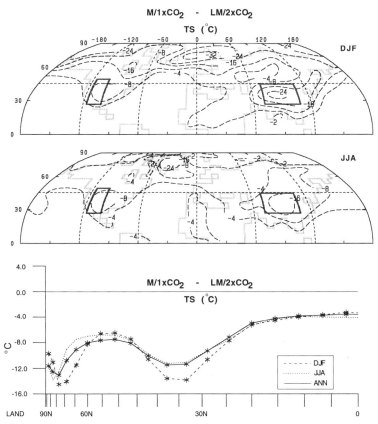

FIGURE 20.   Results from a GCM experiment showing simulated Northern Hemisphere temperature changes in °C owing to plausible changes over the last 20 million years in atmospheric CO_2 (from twice-modern to modern values) and in orography (selective increases in Asia and elsewhere): (top) changes in winter; (middle) changes in summer; (bottom) zonal mean seasonal and annual changes (after Chapter 9). Asterisks show changes significant at the 95% level.

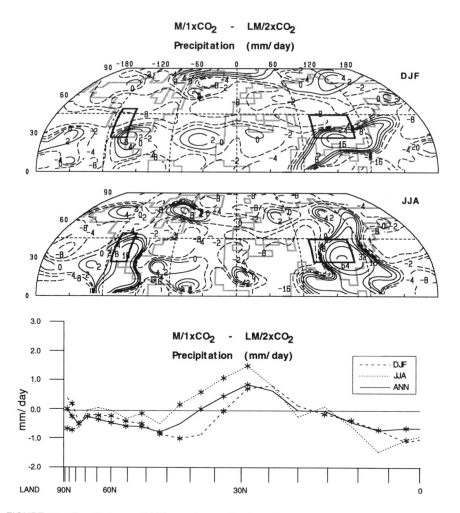

FIGURE 21. Results from a GCM experiment showing simulated Northern Hemisphere changes ir. precipitation in mm/day owing to plausible changes over the last 20 million years in atmospheric CO_2 (from twice-modern to modern values) and in orography (selective increases in Asia and elsewhere): (top) changes in winter; (middle) changes in summer; (bottom) zonal mean seasonal and annual changes (after Chapter 9). Asterisks show changes significant at the 95% level.

owing to reduced CO_2, the amplified cooling atop plateaus owing to local uplift, the enhanced precipitation on plateau margins owing to uplift-induced monsoon circulations, and the general reduction in precipitation away from plateaus owing to both falling CO_2 and orography. This list of climatic responses to uplift encompasses many first-order paleoenvironmental changes that have occurred on Earth during the Cenozoic. This indicates that uplift may be the major source of climate forcing during this interval, as suggested in early versions of the uplift–climate hypotheses.[1-4]

Several aspects of the initial hypotheses have not survived intact. The earlier versions[2,3] invoked accelerating uplift over the last 5 to 10 million years, but this may largely be an artifact of climatic effects that overprint the erosional and paleobotanical evidence used to infer uplift.[114,115] Hsu[126] used paleobotanical evidence to argue that most Tibetan uplift occurred in the last 3 million years, a view often expressed in the Chinese literature.[127] Fort[128] summarized arguments that refute the suggestion of extremely recent large-scale uplift.

Earlier uplift-hypothesis estimates[4] of increased erosional fluxes over the last 5 million years have also been found to be too large[15] (see Chapter 19). Subsequent versions of the initial hypotheses[5,46] moved toward the view advanced here: that progressive uplift in several regions over most of the Cenozoic provides an explanation of the gradual climatic deterioration that finally culminated in large late Pliocene and Pleistocene glaciations.

Earlier papers[2,3] also cited geomorphic and paleobotanical evidence for late Cenozoic uplift of western North America. This assumption has been brought into question on a geomorphic basis by Molnar and England[114,115] and by paleoelevation reconstructions that use paleobotanical techniques based on leaf shapes rather than vegetation composition.[25,129] This has led to increasing prominence (but not full acceptance) of the view that large-scale uplift of the Cretaceous foreland basin of western North America dates to the Laramide orogeny 75 to 45 million years ago,[6] not to the late Cenozoic, and that the fold-thrust belt farther west has been high-standing since at least the Cretaceous, with likely late Cenozoic subsidence (extensional collapse) of regions such as the Basin and Range. It is thus questionable whether late Cenozoic uplift in North America could have played any significant role in affecting large-scale climate,[2,3] although late Cenozoic uplift in Tibet certainly did.

Subsequent studies have also increased the potential scope of the uplift–weathering hypothesis. For example, regional or global climatic changes (mainly cooling) caused by uplift may have activated an important feedback loop involving intensified erosion and isostatic uplift of mountains that had for some time been orogenically inactive[114,115] (Section 5); as a result, increased physical and chemical erosion in these regions may have further reduced CO_2. Also, decreasing CO_2 levels during the latest Miocene appear to have significantly affected vegetation changes through physiological (fertilization) mechanisms (see Chapters 9, 13) that are independent of previously inferred climatic effects of uplift.[2,46]

A valid criticism of the uplift–weathering theory is the lack of negative feedback sufficient to keep the delicately balanced atmospheric CO_2 reservoir from emptying out and freezing the entire Earth (Section 5). Sundquist[125] proposed that "reverse greenhouse" climate changes may reduce chemical weathering and CO_2 consumption in areas outside of orogenic belts, thus moderating the CO_2 and temperature decreases originally caused by uplift. Support for this idea emerges in Chapters 15 and 19, but not in Chapter 14. Other possible negative feedbacks (reviewed in Chapters 12, 16, 18 and 19) include changes in the organic C subcycle or in "reverse silicate weathering" in the ocean.

Geologic evidence also supports the potential viability of the uplift–climate hypothesis over the last several hundred million years. Two previous intervals of large-scale continental collision and widespread plateau–mountain uplift (700–600 and 320–240 million years ago) were also times of widespread glaciation.[16,18]

6.2. Other Sources of Cenozoic Climate Forcing

Other factors have certainly been important in driving global or regional climate changes farther back in time, including: variations in rates of CO_2 outgassing caused by changes in mean spreading rate[97,98] and/or variations in the C content of subducted material,[105] motion of continents with respect to latitudinal temperature gradients,[130] and the slow increase in solar luminosity over all of Earth's history.[103,104,131]

Large-scale climate forcing mechanisms specific to the middle and late Cenozoic have also been proposed. One hypothesis is a cooling of climate by volcanically ejected sulfate aerosols,[132] but there are difficulties with this concept. Because the effects of individual explosions diminish by half every year and are largely confined to one hemisphere, it would take volcanic explosions on a nearly continuous (annual) basis in both hemispheres to impact global climate significantly over millions of years. More critically, the more likely long-term impact of such explosions would be to add CO_2 to the atmosphere and warm (not cool) the Earth, as in the BLAG hypothesis (Section 5.1).

Some scientists question whether CO_2 changes are actually the main cause of Cenozoic climatic cooling. GCM simulations of warm Cretaceous and Eocene intervals overheat the tropics and overchill both the poles and continental interiors compared to proxy temperature data.[133,134] To explain these discrepancies, Barron and colleagues[133,134] invoked increased ocean heat transport poleward to cool the tropics, warm the poles, and bring model-generated climates in line with observations. Although there are questions about the accuracy of the proxy data[135,136] and about the mechanisms by which the ocean–atmosphere system could achieve larger poleward heat fluxes despite lowered temperature gradients,[137,138] this hypothesis remains under consideration.

Following this same line of argument, it has been suggested[139,140] that the latest Cenozoic cooling of the Arctic region requires a decrease in poleward heat transport in the Atlantic. This hypothesis was invoked because oceanic CO_2 proxies ($\delta^{13}C$ values in marine organic C) suggested too little CO_2 change over the latest Cenozoic to cool the Arctic. This conclusion can be questioned on several grounds. First, recent evidence of strong vital effects in these $\delta^{13}C$ values makes their use as CO_2 proxies problematic.[96] Second, the idea that changes in poleward heat transport might be due to orographically induced changes in wind forcing of the surface ocean[138,141] is brought into question by the ocean model simulations in Chapter 6 indicating that northward heat transport in the Atlantic owing to higher Asian (but not North American) orography may have increased (not decreased) during the latest Cenozoic. Finally, modeling experiments described in Chapter 9 suggest that the Cenozoic cooling over the last 20 million years that would have been caused by a plausible drop in atmospheric CO_2 levels appears to be sufficient to explain the observed changes in temperature-sensitive circum-Arctic vegetation since the early-middle Miocene, with no obvious need for altered ocean heat transport.

Another possible tectonic explanation for changes in poleward heat transport involves ocean "gateways," by opening or closing of isthmuses and shoaling or deepening of sills. Based on evidence of mid-Cenozoic opening of unimpeded circum-Antarctic ocean circulation,[142] Kennett[143] inferred that this tectonic change would cause a regional climatic cooling and formation of an ice sheet on Antarctica. However, results from combined experiments with an ocean model and an energy balance model[144] do not confirm a significant effect of circum-Antarctic circulation on Antarctic climate, although they do indicate a large impact on bottom water production. As for the Northern Hemisphere, ocean modeling results indicate that closing of the Panamanian isthmus between 10 and 3 million years ago[145] would have caused a regional warming of the high-latitude North Atlantic,[146] but such a warming is opposite in sense both to the flow required by the poleward heat-transport hypothesis and to the long-term cooling actually observed. However, the accompanying rerouting of salt northward may have increased formation of northern-component deep water in the late Cenozoic.[146] Finally, episodically enhanced subsidence of the Norwegian–Greenland Sill during the last 20 million years may have altered rates of formation of northern-source deep water and possibly contributed to short-term regional temperature changes, but episodic sill deepening is not thought to have played a role in the longer-term climatic cooling.[147]

Viewed as a group, hypotheses involving changes in prominent oceanic "gateways" do not explain ongoing cooling during the last half of the Cenozoic. They are discontinuous events (with discrete beginnings and ends) occurring in widely scattered locations, rather than an ongoing source of global climate forcing. They are likely, however, to have had important effects on the formation of deep and bottom water, and they may have played a role in regional climatic perturbations to longer-term trends.

The Cenozoic plate motions that are responsible for uplift have also had other regional-scale climatic effects. Aside from the northward motion of India into Asia, the largest change in latitudinal terms over the last 40 million years is the northward movement of Australia. Although motion away from the poles is generally thought to be opposite in sense to that needed to aid Cenozoic cooling, Australia probably became warmer and drier owing to this movement into the subtropics. Other continental motions have been mainly latitude-parallel (further opening of the Atlantic and closing of the Pacific), with uncertain climatic impacts.

6.3. Ongoing Issues

Although observational data and model simulations verify in a general way that uplift has driven many observed climatic changes over the last 40 million years, many uncertainties remain, and a few of these are noted here. The uplift history of Tibet remains uncertain even at the broadest (10-million-year) time resolution (see Chapter 2) because most exploration has concentrated on the Himalaya and southernmost Tibet. Most of central and northern Tibet has never been geologically mapped, and such studies are needed. Debate also continues over how much uplift occurred in western North America either before and during late Cretaceous and early Cenozoic crustal shortening[6] or later in the Cenozoic as a result of deep-seated processes.[26] Further development of techniques such as temperature-sensitive leaf-margin indices[25,129,147] may resolve this issue. Accurate uplift histories are critical both as a boundary condition for climate model simulations and for comparison against climatic trends. Better knowledge is also needed of the controls, rates and histories of physical and chemical weathering, particularly in regions of Cenozoic uplift. More comprehensive drilling of Indian Ocean fans, the Amazon Cone, and other terrigenous sediment piles would help achieve this. Development (or validation) of reliable proxies of the long-term history of CO_2 change is another obvious first-order need.

Numerous improvements in climate models are needed. Foremost among these is better simulation of the sensitivity of global climate to CO_2, with cloud amount, height, and distribution still the main uncertainties. Low gridbox resolution currently permits simulations of broad plateau-scale climatic effects, but not of finer mountain-scale effects, except in "nested" regional models. Precipitation is inadequately simulated, as is the frictional "drag" of mountains on winds. Improvements in treatment of interactive soil moisture, vegetation, runoff, snowcover, and sea ice are needed (and are rapidly emerging). Ocean circulation models are still far cruder instruments than atmospheric models, and for this reason the effects of uplift on the ocean were little emphasized in this summary.

Despite the general agreement over uplift effects on climate during the broader span of the middle and late Cenozoic, disagreements have emerged at finer resolution. One currently contentious issue is the degree of match or mismatch during the last 20 million years among indices of Himalayan weathering tied to Tibetan uplift (the $^{87}Sr/^{86}Sr$ ratio and rates of sediment burial on the Bengal and Indus fans and in the Siwaliks) and the $\delta^{18}O$ record as an index of global climate change (see Chapters 8, 12, and 19).

Another recent issue is whether or not uplift has continued to force global climatic cooling during the last 5–10 million years, ironically the very interval most emphasized in the original uplift hypotheses.[2–4] Pervasive extensional faulting in southern Tibet during the last several million years is viewed as evidence of gravitational "collapse" and even possibly some lowering of elevations,[149,150] although only a single dated fault actually constrained the age of initiation of extension. If true, this view could mean there has been some reduction of climatic forcing attributable to Asian orography in the last several million years. Evidence of late Cenozoic extension atop the Altiplano[23] could indicate a similar trend in the other region of major Cenozoic uplift.

One problem with this argument is that recent evidence suggests that the timing of initiation of extensional faulting on Tibet is as early as 14 million years ago.[151] By inference, this would push the age at which the plateau reached its maximum height and then began to collapse back toward the middle of the total interval of India–Asia convergence. This timing would also disagree with an array of evidence of major uplift subsequent to this date.

In addition, plate convergence continues today both in the continent–continent collision in southern Asia and the ocean–continent convergence of South America. This convergence requires some kind of tectonic accommodation (most likely crustal shortening, thickening, and uplift) somewhere in these orogens. In Tibet, this may involve "extrusion" of crust eastward into China causing higher elevations there[152,153] or crustal thickening (and uplift) along the largely unexplored northern margins of the plateau.[154] In the Andes, shortening during the latest Cenozoic appears to be achieved mainly by crustal thickening (and uplift) along the eastern margins of the Altiplano and Andes[23] (see Chapter 3).

This pattern of progressive outward expansion of plateaus by crustal shortening is different from that modeled in most GCM studies, which use pistonlike *upward* movement of plateaus of constant lateral dimensions. Outward growth should increase the overall temperature–albedo effect of the plateaus and (in the case of Tibet) will also increase the orographic interception of zonal winds. As GCMs achieve higher gridbox resolution, they will become useful for experiments testing whether or not outward plateau growth enhances monsoonal circulations in the same sense that upward growth does.

Another possibility is that the pervasive Cenozoic cooling of the last few million years has occurred in large part because of its own positive feedbacks. Increased physical weathering and rock exposure in glaciated mountain regions and in areas periodically eroded by continental ice sheets[155] (see Chapter 11) may be enhancing chemical weathering in those regions and could conceivably be a significant contributor to the global cooling of the last few million years.

REFERENCES

1. Birchfield, G. E., and Weertman, J. (1983). *Science* **219**, p. 284.
2. Ruddiman, W. F., and Kutzbach, J. E. (1989). *J. Geophys. Res.* **94**, p. 18409.
3. Kutzbach, J. E., Guetter, P. J., Ruddiman, W. F., and Prell, W. L. (1989). *J. Geophys. Res.* **94**, p. 18393.
4. Raymo, M. E., Ruddiman, W. F., and Froelich, P. N. (1988). *Geology* **16**, p. 649.
5. Raymo, M. E., and Ruddiman, W. F. (1992). *Nature* **359**, p. 117.
6. Coney, P. J., and Harms, T. A. (1984). *Geology* **12**, p. 550.
7. Pfiffner, D. A., Frei, W., Valasek, P., Sfauble, M., Levato, L., Dubois, L., Schmid, S. M., and Smithson, S. P. (1990). *Tectonics* **9**, p. 1327.
8. Bird, P. (1979). *J. Geophys. Res.* **84**, p. 7561.
9. Ahnert, F. (1970). *Am. J. Sci.* **268**, p. 243.
10. Pinet, P., and Souriau, M. (1988). *Tectonics* **7**, p. 563.
11. Zietler, P. K. (1985). *Tectonics* **4**, p. 127.

12. Johnson, N. M., Stix, J., Tauxe, L., Cerveny, P. F., and Tahirkheli, R. A. K. (1985). *J. Geol.* **93**, p. 27.
13. Cochran, J. (1990). *Ocean Drilling Project Sci. Results* **116**, p. 397.
14. Harrison, T. M., Copeland, P., Kidd, W., and An, Y. (1992). *Science* **255**, p. 1663.
15. Richter, F. M., Rowley, D. B., and DePaolo, D. J. (1992). *Earth Planet. Sci. Lett.* **109**, p. 11.
16. Menard, G., and Molnar, P. (1988). *Nature* **334**, p. 235.
17. Dewey, J. F., and Burke, K. C. A. (1973). *J. Geol.* **81**, p. 683.
18. Raymo, M. E. (1991). *Geology* **19**, p. 344.
19. Edmond, J. M. (1992). *Science* **258**, p. 1594.
20. Zachos, J. C., Breza, J. R., and Wise, S. W. (1992). *Geology* **20**, p. 569.
21. Jansen, E., and Sjohom, J. (1991). *Nature* **349**, p. 600.
22. Funder, S., Abrahamsen, N., Bennike, O., and Feyling-Hansen, R. W. (1984). *Geology* **13**, p. 542.
23. Isacks, B. L. (1988). *J. Geophys. Res.* **93**, p. 3211.
24. Hoorn, C., Guerrero, J., Sarmiento, G. A., and Lorente, M. A. (1995). *Geology* **23**, p. 237.
25. Chase, C. G., Gregory, K. M., Parrish, J. T., and Decelles, P. G. In: *Tectonic Boundary Conditions for Climate Reconstructions* (T. J. Crowley and K. Burke, eds.), Oxford University Press (in press).
26. Morgan, P., and Swanberg, C. A. (1985). *J. Geodyn.* **3**, p. 39.
27. Bond, G. C. (1979). *Tectonophysics* **61**, p. 285.
28. Saggerson, E. P., and Baker, B. H. (1965). *Quart. J. Roy. Soc. Lond.* **121**, p. 51.
29. Williams, L. A. J. (1978). In: *Petrology and Geochemistry of Continental Rifts*, Proceedings of the NATO Advanced Study Institute on Paleorift Systems, No. 36 (H. J. Neumann and I. B. Ramberg, eds.). Reidel, Dordrecht.
30. White, R. S., and McKenzie, D. (1989). *J. Geophys. Res.* **94**, p. 7685.
31. Westaway, R. (1993). *Earth Planet. Sci. Lett.* **119**, p. 331.
32. Trumpy, R. (1980). In: *Geology of Switzerland* (R. Trumpy, ed.), 140 pp. Wepf, Basel.
33. Kutzbach, J. E., Press, W. L., and Ruddiman, W. F. (1993). *J. Geol.* **101**, p. 177.
34. Dana, J. D. (1856). *Am. J. Sci.* **22**, p. 305.
35. Ramsay, W. (1924). *Geol. Mag.* **61**, p. 152.
36. Bolin, B. (1950). *Tellus* **2**, p. 184.
37. Manabe, S., and Terpstra, T. B. (1974). *J. Atmos. Sci.* **31**, p. 3.
38. Nigam, S., Held, I. M., and Lyons, S. W. (1988). *J. Atmos. Sci.* **45**, p. 1433.
39. Ruddiman, W. F., Raymo, M. E., and McIntyre, A. (1986). *Earth Planet. Sci. Lett.* **80**, p. 117.
40. Legates, D. R. (1987). *Publ. Climatol.* **40**, 84 pp.
41. Legates, D. R., and Willmot, C. J. (1990). *Int. J. Climatol.* **10**, p. 111.
42. Hahn, D. G., and Manabe, S. (1975). *J. Atmos. Sci.* **32**, p. 1515.
43. Gutman, G. J., and Schwerdtfeger, W. (1965). *Meteorol. Rundsch.* **18**, p. 69.
44. Rao, G. V., and Erdogan, S. (1989). *Boundary Layer Meteorol.* **46**, p. 13.
45. Meehl, G. (1992). *Ann. Rev. Earth Planet. Sci.* **20**, p. 85.
46. Ruddiman, W. F., and Kutzbach, J. E. (1990). *Sci. Amer.* **264**, p. 66.
47. Prell, W. L., and Kutzbach, J. E. (1992). *Nature* **360**, p. 647.
48. UNESCO (1978). *Atlas of the World Water Balance*, UNESCO, Paris.
49. Hayden, B. C. (1988). In: *Flood Geomorphology* (V. R. Baker, R. C. Kochel, and P. C. Patton, eds.), pp. 13–26. John Wiley, New York.
50. Broccoli, A. J., and Manabe, S. (1992). *J. Clim.* **5**, p. 1181.
51. Milliman, J. D., and Meade, R. H. (1983). *J. Geol.* **91**, p. 1.
52. Walling, D. E., and Webb, B. W. (1983). In: *Background to Paleohydrology* (K. J. Gregory, ed.), p. 69–100. John Wiley, New York.
53. Milliman, J. D., and Syvitski, J. P. M. (1992). *J. Geol.* **100**, p. 525.
54. Schumm, S. A. (1963). U.S. Geological Survey Professional Paper 454-H, p. 1–13.
55. Carson, M. A., and Kirby, M. K. (1972). *Hillslope Form and Processes.* Cambridge University Press, New York.
56. Summerfield, M. A. (1991). *Global Geomorphology: An Introduction to the Study of Landforms.* Longman, Essex.

57. Summerfield, M. A., and Hulton, N. J. (1994). *J. Geophys. Res.* **99**, p. 13871.
58. Fielding, E., Isacks, B., Barazangi, M., and Duncan, C. (1994). *Geology* **22**, p. 163.
59. Masek, J. G., Isacks, B. L., Gubbels, T. L., and Fielding, E. J. (1994). *J. Geophys. Res.* **99**, p. 13941.
60. Saunders, I., and Young, A. (1983). *Earth Surf. Proc. Landforms* **8**, p. 473.
61. Langbein, W. B., and Schumm, S. A. (1958). *EOS, Trans. AGU* **39**, p. 1076.
62. Corbel, J. (1959). *Zeitschr. Geomorph.* **3**, p. 1.
63. Wilson, L. (1973). *Am. J. Sci.* **273**, p. 335.
64. Ohmori, H. (1983). *Bull. Dep. Geog. Univ. Tokyo* **15**, p. 77.
65. Baker, V. R. (1988). In: *Flood Geomorphology* (V. R. Baker, R. C. Kochel, and P. C. Patton, eds.), pp. 81–95. John Wiley, New York.
66. Komar, P. D. (1988). In: *Flood Geomorphology* (V. R. Baker, R. C. Kochel, and P. C. Patton, eds.), pp. 97–111. John Wiley, New York.
67. Hoffman, P. F., and Grotzinger, J. P. (1993). *Geology* **21**, p. 195.
68. Willet, S., Beaumont, C., and Fullsack, P. (1993). *Geology* **21**, p. 371.
69. Curray, J. R. (1994). *Earth Planet. Sci. Lett.* **125**, p. 371.
70. Rea, D. K. (1992). In: *Synthesis of Results from Scientific Drilling in the Indian Ocean* (R. A. Duncan, D. K. Rea, R. B. Kidd, U von Rad, and J. K. Weissel, eds.), pp. 387–402. American Geophysical Union Monograph 70.
71. Amano, K., and Taira, A. (1992). *Geology* **20**, p. 391.
72. Leg 154 Shipboard Party. (1995). *EOS Trans. AGU.* **76**, p. 41.
73. Benjamin, M. T., Johnson, N. M., and Naeser, C. W. (1987). *Geology* **15**, p. 680.
74. Harrington, H. J. (1962). *Am. Assoc. Petr. Geol. Bull.* **46**, p. 1773.
75. Davies, T.A., Hay, W. W., Southam, J. R., and Worsley, T. R. (1977). *Science* **197**, p. 53.
76. Ronov, A. B. (1990). In: *The Earth's Sedimentary Shell: Quantitative Patterns of Its Structure, Compositions and Evolution* (A. A. Yaroshevskii, ed.), 80 pp. Nauka, Moscow.
77. Chamberlin, T. C. (1899). *J. Geol.* **7**, p. 545.
78. Raymo, M. E. (1991). *Geology* **19**, p. 344.
79. Meybeck, M. N. (1987). *Am. J. Sci.* **287**, p. 401.
80. Gibbs, R. J. (1967). *Geo. Soc. Am. Bull.* **78**, p. 1203.
81. Stallard, R. F., and Edmond, J. M. (1983). *J. Geophys. Res.* **88**, p. 9671.
82. Meybeck, M. N. (1976). *Hydrol. Sci. Bull.* **21**, p. 265.
83. Holland, H. D. (1978). *The Chemistry of the Atmosphere and Oceans.* John Wiley, New York.
84. Berner, E. K., and Berner, R. A. (1987). *The Global Water Cycle.* Prentice-Hall, Englewood Cliffs, NJ.
85. Lasaga, A. C., Soler, J. M., Ganor, J., Burch, T. E., and Nagy, K. (1994). *Geochim. Cosmochim. Acta* **58**, p. 2361.
86. Walling, D. E., and Webb, B. W. (1986). In: *Solute Processes* (S. T. Trudgill, ed.), pp. 251–327. John Wiley, New York.
87. Lasaga, A. C. (1984). *J. Geophys. Res.* **89**, p. 4009.
88. Bluth, G. J. S., and Kump, L. R. (1994). *Geochim. Cosmochim. Acta* **58**, p. 2341.
89. Harris, N. (1995). *Geology* **23**, p. 795.
90. Mercer, J. H., and Suter, J. (1981). *Palaeogeogr. Palaeoclim. Palaeoecol.* **38**, p. 185.
91. Armentrout, J. M. (1983). In: *Glacial Marine Sedimentation* (B. F. Molnia, ed.), pp. 629–666. Plenum, New York.
92. Shackleton, N. J., Backman, J., Zimmerman, H., Kent, D. V., Hall, M. A., Roberts, D. G., Schnitker, D., Baldauf, J. G., Desprairies, A., Homrighausen, R., Huddlestun, P., Keene, J. B., Kaltenback, A. J., Krumsiek, K. A. O., Morton, A. C., Murray, J. W., and Westberg-Smith, J. (1984). *Nature* **307**, p. 620.
93. Shackleton, N. J., and Opdyke, N. D. (1973). *Quat. Res.* **3**, p. 39.
94. Wolfe, J. A. (1987). *Geophys. Monogr. AGU* **32**, p. 357.
95. Cerling, T. E., Wang, Y., and Quade, J. (1993). *Nature* **361**, p. 344.
96. Laws, E. A., Popp, B. N., Bidigare, R. B., Kennicutt, M. C., and Macko, S. A. (1995). *Geochim. Cosmochim. Acta* **59**, p. 1131.
97. Berner, R. A., Lasaga, A. C., and Garrels, R. M. (1983). *Am. J. Sci.* **283**, p. 641.

98. Lasaga, A. C., Berner, R. A., and Garrels, R. M. (1985). *Geophys. Monogr. AGU* **32**, p. 397.
99. Engebretson, D. C., Kelley, K. P., Cashman, H. J., and Richards, M. A. (1992). *Geol. Soc. Am. Today* **2**, p. 93.
100. Larson, R. L. (1991). *Geology* **19**, p. 547.
101. Stallard, R. F. (1988). In: *Physical and Chemical Weathering in Geochemical Cycles*, (A. Lerman and M. Meybeck, eds.), pp. 225–246. Kluwer, Dordrecht.
102. Meybeck, M. (1988). In: *Physical and Chemical Weathering in Geochemical Cycles*, (A. Lerman and M. Meybeck, eds.). pp. 247–252. Kluwer, Dordrecht.
103. Berner, R. A. (1993). *Science* **261**, p. 68.
104. Berner, R. A. (1994). *Am. J. Sci.* **294**, p. 56.
105. Kerrick, D. M., and Caldeira, K. (1994). *Geol. Soc. Am. Today* **4**, p. 57.
106. Bickle, M. J. (1995). *Nature* **367**, p. 699.
107. Derry, L. A., and France-Lanord, C. (1996). *Paleoceanography* **11**, p. 267.
108. Francois, L. M., and Walker, J. C. G. (1992). *Am. J. Sci.* **292**, p. 81.
109. Caldeira, K. (1995). *Am. J. Sci.* **295**, p. 1077.
110. Arrhenius, S. (1986). *Philosoph. Mag.* **41**, p. 237.
111. Mitchell, J. F. B. (1987). *Rev. Geophys.* **27**, p. 115.
112. Schlesinger, M. E., and Mitchell, J. F. B. (1987). *Rev. Geophys.* **25**, p. 760.
113. Ramanathan, V. (1988). *Science* **240**, p. 293.
114. Molnar, P., and England, P. (1990). *Nature* **346**, p. 29.
115. England, P., and Molnar, P. (1990). *Geology* **18**, p. 1173.
116. Wager, L. R. (1933). *Nature* **132**, p. 28.
117. Holmes, A. (1965). *Principles of Physical Geology*, Nelson, London.
118. COHMAP Project Members (1988). *Science* **241**, p. 1043.
119. Walker, J. C. G., Hays, P. B., and Kastings, J. F. (1981). *J. Geophys. Res.* **86**, p. 9776.
120. Berner, R. A. (1991). *Am. J. Sci.* **291**, p. 339.
121. Brady, P. V. (1991). *J. Geophys. Res.* **96**, p. 18101.
122. Volk, T. (1987). *Am. J. Sci.* **287**, p. 763.
123. Peters, N. E. (1984). U.S. Geological Survey Water-Supply Paper 2228.
124. White, A. F., and Blum, A. E. (1995). *Geochim. Cosmochim. Acta* **59**, p. 1729.
125. Sundquist, E. T. (1991). *Quat. Sci. Rev.* **10**, p. 283.
126. Hsu, J. (1976). *Paleobotanist* **25**, p. 131.
127. Li, J. J., Wen, S., Zhang, Q., Wang, F., Zheng, B. X., and Li, B. (1979). *Sci. Sin.* **22**, p. 1314.
128. Fort, M. (1996). *Palaeogeogr. Palaeoclim. Palaeoecol.* **120**, p. 123.
129. Wolfe, J. A. (1993). *Bull. U.S. Geol. Surv.* **2040**.
130. Barron, E. J. (1981). *Geol. Rundsch.* **70**, p. 737.
131. Sagan, C., and Mullen, G. (1972). *Science* **177**, p. 52.
132. Kennett, J. P., and Thunell, R. C. (1975). *Science* **187**, p. 497.
133. Barron, E. J., Thompson, S. L., and Schneider, S. H. (1981). *Science* **212**, p. 501.
134. Barron, E. J. (1987). *Paleoceanography* **2**, p. 729.
135. Zachos, J. C., Stott, L. D., and Lohmann, K. C. (1994). *Paleoceanography* **9**, p. 353.
136. Wilson, P. A., and Opdyke, B. N. (1996). *Geology* **24**, p. 555.
137. Sloan, L. C., Walker, J. C. G., and Moore, T. C., Jr. (1995). *Paleoceanography* **10**, p. 347.
138. Crowley, T. C. (1996). *Mar. Micropal.* **27**, p. 3.
139. Dowsett, H. J., Cronin, T. M., Poore, R. Z., Thompson, R. S., Whatley, R. C., and Wood, A. M. (1992). *Science* **258**, p. 1133.
140. Raymo, M. E., Grant, B., Horowitz, M., and Rau, G. H. (1996). *Mar. Micropal.* **27**, p. 313.
141. Rind, D., and Chandler, M. (1991). *J. Geophys. Res.* **96**, p. 7437.
142. Barker, P. F., and Burrell, J. (1977). *Mar. Geol.* **25**, p. 15.
143. Kennett, J. P. (1977). *J. Geophys. Res.* **82**, p. 3843.
144. Mikolajewicz, U., Maeir-Reimer, E., Crowley, T. J., and Kim, K.-Y. (1993). *Paleoceanography* **8**, p. 409.
145. Keigwin, L. D. (1982). *Science* **217**, p. 350.
146. Maier-Reimer, E., Mikolajewicz, U., and Crowley, T. (1990). *Paleoceanography* **5**, p. 349.

147. Wright, J. D., and Miller, K. G. (1996). *Paleoceanography* **11**, p. 157.
148. Forest, C. E., Molnar, P., and Emanuel, K. A. (1995). *Nature* **374**, p. 347.
149. England, P., and Houseman, G. (1989). *J. Geophys. Res.* **94**, p. 17561.
150. Molnar, P., England, P., and Martinod, J. (1993). *Rev. Geophys.* **31**, p. 357.
151. Coleman, M., and Hodges, K. (1995). *Nature* **374**, p. 49.
152. Molnar, P., and Tapponnier, P. (1975). *Science* **189**, p. 419.
153. Lave, J., Avouac, J. P., Lacassin, R., Tapponnier, P., and Montagner, J. P. (1996). *Earth Planet. Sci. Lett.* **140**, p. 83.
154. Molnar, P., Burchfield, B. C., K'uangyi, L., and Ziyun, Z. (1987). *Geology* **15**, p. 249.
155. Hodell, D. A., Mead, G. A., and Mueller, P. A. (1990). *Chem. Geol.* **80**, p. 1.

Index